Lecture Notes in Statistics

Volume 223

Series Editors
Peter Diggle, Department of Mathematics, Lancaster University, Lancaster, UK
Ursula Gather, Dortmund, Germany
Scott Zeger, Baltimore, MD, USA

Lecture Notes in Statistics (LNS) includes research work on topics that are more specialized than volumes in Springer Series in Statistics (SSS). The series editors are currently Peter Bickel, Peter Diggle, Stephen Fienberg, Ursula Gather, and Scott Zeger. Ingram Olkin was an editor of the series for many years.

More information about this series at http://www.springer.com/series/694

Carlos A. Coelho • Barry C. Arnold

Finite Form Representations for Meijer G and Fox H Functions

Applied to Multivariate Likelihood Ratio Tests Using Mathematica®, MAXIMA and R

Springer

Carlos A. Coelho
Department of Mathematics
Universidade NOVA de Lisboa
Caparica, Portugal

Barry C. Arnold
Department of Statistics
University of California
Riverside
CA, USA

ISSN 0930-0325 ISSN 2197-7186 (electronic)
Lecture Notes in Statistics
ISBN 978-3-030-28789-4 ISBN 978-3-030-28790-0 (eBook)
https://doi.org/10.1007/978-3-030-28790-0

Mathematics Subject Classification (2010): 62E15, 62E10, 62H05, 62H10, 62H15, 65D20, 44A20, 33E20

This Springer imprint is published by the registered company Springer Nature Switzerland AG.
The registered company address is: Gewerbestrasse 11, 6330 Cham, Switzerland

Preface

Meijer G and Fox H functions are extraordinary tools to represent distributions of random variables and statistics, in particular when these statistics have the same distribution as that of a sum or a product of given independent random variables. However, although at present there are available a number of softwares that can compute the Meijer G function or have implementations for its computation, the evaluation of these functions can still remain extremely slow and memory consuming. This is especially true when the number of elementary random variables involved in those sums or products grows to just moderately large numbers. The aim of this book is to provide tools to identify particular situations in which these computations can be done in a much faster way, by using closed finite form representations. Many of these instances correspond to distributions of likelihood ratio statistics of interest in multivariate analysis. The availability and benefits of these finite closed form representations for probability density and cumulative distribution functions are generally overlooked in the available multivariate analysis literature. As such this book is intended to make a useful contribution to the practical accurate implementation and use of such tests and to promote their more common usage by providing an easy and extremely fast way to obtain exact quantiles and p-values for the associated statistics.

The idea is not to downplay those softwares which do a wonderful job providing functions or modules that enable the general computation of the Meijer G and Fox H functions but rather to show that there are general ways to recognize situations in which it is possible to obtain closed finite form representations for those functions. These efficient and effective alternatives may (and should) indeed be included as integrated parts of future implementations of the Meijer and Fox functions.

One may ask: "why the interest in likelihood ratio statistics?" Likelihood ratio testing is very appealing from an intuitive point of view, since when we use this procedure to test for a given value of a given distribution parameter, we compare the likelihood, or as Wilks calls it in his 1962 *Mathematical Statistics* book (Wilks, S.S.: Mathematical Statistics, 2nd edn. J. Wiley & Sons Inc., New York (1962), Sect. 13.3), the plausibility, of a given value of this parameter versus another possible value, given that we have a given sample from the distribution in hand. Once we take our probability distribution or mass function as our model for the data and we look at it as a function of the parameters in that distribution, we make it our likelihood function. Thus, as Wilks explains, if by looking through the entire parameter space we are not able to find a value for that parameter that gives the likelihood function a sufficiently larger value than the values that this function takes when we consider the value or values of this parameter under the null hypothesis, then we will be inclined to not reject this null hypothesis, and vice versa. Moreover, as Muirhead (Muirhead, R.J.: Aspects of Multivariate Statistical Theory, 2nd edn. J. Wiley & Sons Inc., New York (2005), Sect. 6.1) brings to our attention "In many hypothesis testing problems in Multivariate Analysis there is no uniformly most powerful or uniformly most powerful unbiased test," but one interesting and useful feature is the invariance of such testing procedures under a given group of transformations. It happens that most likelihood ratio testing procedures used in multivariate analysis are indeed invariant under linear transformations. This invariance property of the likelihood ratio tests, making some of them invariant to scale changes, others to location changes, and yet others to scale and location changes, is a very attractive feature, for example, in situations where one may be willing to compare similar studies, involving similar variables which may have been measured using different units of measurement. Invariance is indeed a quite interesting and desirable property for a test statistic since, for example, scale invariance means that changes in the units in which variables are measured will not affect the computed value of the statistic, as long as all variables undergo similar conversions in units and these conversions involve only multiplication by given factors. Location invariance is also an important and desirable property since it means that the computed value of the statistic is not affected by shifts in the measuring scale. If a test enjoys these two types of invariance, location and scale, this means that the computed value of the statistic will remain unchanged, as for example in cases where the variables involved are temperatures, in which case the computed value of the test statistic will be the same whether we use temperatures measured in degrees Fahrenheit or Celsius. In multivariate analysis, likelihood ratio tests are quite often not the most powerful tests, but as Muirhead (2005, Sect. 6.1) states "In some interesting situations it turns out that within this class [of invariant tests] there exists a test which is uniformly most powerful, and such a test is called, of course, a uniformly most powerful invariant test. Often such a test, if it exists, is the same as the likelihood ratio test." Two other important attributes of these tests, which add to their importance in practice, are that (1) although they are usually designed for underlying multivariate normal distributions, for many of the testing procedures, their distributions under the corresponding null hypotheses remain the same for underlying elliptically contoured distributions, and (2) these tests have optimal asymptotic properties for large samples, under the same conditions for which asymptotically normally distributed maximum likelihood estimators exist, as noted by Wilks (1962, Sect. 13.3).

Rather than supply proofs for all of the many likelihood ratio tests appearing in the book, we have decided to include only a few representative examples of such proofs. Most of these are placed in Appendices, since some readers may want to skip them, at least on a first reading. But we include them for readers who may use them to acquire a more complete understanding of the tests and procedures being addressed.

In case someone still questions the focus on multivariate analysis, we would only point out that indeed the world and life are themselves intrinsically multivariate. Considering just a single variable is, most of the time, an effort towards an artificial simplification thought to be methodologically necessary rather than a real objective. Moreover, all univariate models and tests may be obtained as particular cases of their multivariate counterparts, as, for example, is the case with the common univariate Regression (simple or multiple) and Analysis of Variance models, which appear as particular cases of one of the models considered in Chap. 5.

The present book is intended to be useful both for more theoretically inclined and for more application-oriented statisticians. It is hoped that it will help furthering their understanding of the testing procedures addressed and some of their peculiarities, and that it will enable an easy and accurate practical implementation of the tests. It is also intended to motivate the unveiling of other possible cases for which the Meijer G and Fox H functions may have closed finite form representations and to foster the development of sharp approximations for other cases of these functions.

Caparica, Portugal Carlos A. Coelho
Riverside, CA, USA Barry C. Arnold
July 2019

Acknowledgments

This work was supported by Fundação para a Ciência e a Tecnologia (Portuguese Foundation for Science and Technology) through projects UID/MAT/00297/2013 and UID/MAT/00297/2019 (Centro de Matemática e Aplicações—CMA-FCT/UNL).

Contents

Acronyms and Notation

List of Abbreviations and Acronyms Used

bdiag	Block-diagonal (used to denote a block-diagonal matrix)
c.d.f.	Cumulative distribution function
c.f.	Characteristic function
EGIG	Exponentiated Generalized Integer Gamma (distribution)
GIG	Generalized Integer Gamma (distribution)
l.r.	Likelihood ratio
l.r.t.	Likelihood ratio test
m.l.e.	Maximum likelihood estimator
p.d.f.	Probability density function
r.v.	Random variable

List of Some Notation Used

A'	The transpose of the matrix A (the prime denotes "transpose")
$\|A\|$	Determinant of the (square) matrix A
$A \otimes B$	The Kronecker product of the matrices A and B
$Beta(a, b)$	Representation for a beta distribution with parameters a and b
$\mathbb{C}$	The set of complex numbers
$\mathbb{C}^p$	The p-dimensional complex space (if $\underline{\mu}$ is a $p\times 1$ complex vector, we may write $\underline{\mu} \in \mathbb{C}^p$)
$\stackrel{d}{\equiv}$	Means "equivalent in distribution to"
E_{nm}	A matrix of 1's with dimensions $n\times m$, with E_{n1} representing a column vector of 1's with dimension $n\times 1$, and E_{1p} representing a row vector of 1's with dimension $1\times p$ $\big($note that $E_{nm}E_{mp} = mE_{np}\big)$
$\Gamma(r, \lambda)$	Representation for a gamma distribution with shape parameter r and rate parameter λ (see Appendix 2.A for the p.d.f. of this distribution)
I_p	The identity matrix of order p

$Im(\cdot)$	The imaginary part of the argument
$j = 1{:}p$	Is used as a short notation for $j = 1, \ldots, p$, namely in the expressions for the Meijer G and Fox H functions and also in the expressions for the p.d.f. and c.d.f. of the GIG and EGIG distributions
$Logbeta(a, b)$	Representation for the distribution of a random variable X, for which e^{-X} has a $beta(a, b)$ distribution
$N_p(\underline{\mu}, \Sigma)$	Real p-variate normal distribution with mean vector $\underline{\mu}$ and covariance matrix Σ
$CN_p(\underline{\mu}, \Sigma)$	Complex p-variate normal distribution with mean vector $\underline{\mu}$ and variance-covariance matrix Σ
$\mathbb{N}$	The set of positive integers
$\mathbb{N}_0$	The set of nonnegative integers
$'$ (prime)	The prime denotes the transpose of a vector or a matrix—e.g., A' denotes the transpose of the matrix A
$\mathbb{R}$	The set of real numbers
$\mathbb{R}^+$	The set of positive real numbers
$\mathbb{R}^p$	The p-dimensional real space (if $\underline{\mu}$ is a $p\times 1$ real vector, we may write $\underline{\mu} \in \mathbb{R}^p$)
$Re(\cdot)$	The real part of the argument
$tr(A)$	Trace of the (square) matrix A, which is the sum of its diagonal elements
$\underline{X}$	The underscore indicates that it represents a vector, which, given the capital letter used, would be a random vector
$\underline{X}'$	Represents the transpose of the vector $\underline{X}$
$X^{\#}$	Transpose of complex conjugate of X, where X is a matrix
$\underline{X}^{\#}$	Transpose of complex conjugate of $\underline{X}$, where $\underline{X}$ is a vector
$vec(X)$	Represents the vectorization of the matrix X, that is, the vector obtained by stacking or piling up the successive columns of the matrix X
$W_p(n, \Sigma)$	Real Wishart distribution with n degrees of freedom and parameter matrix Σ, for a matrix of dimensions $p\times p$
$CW_p(n, \Sigma)$	Complex Wishart distribution with n degrees of freedom and parameter matrix Σ, for a matrix of dimensions $p\times p$
$\sim$	"distributed as"
$\widetilde{\{\cdot\}}$	Represents the "contraction" of the argument set to the corresponding set of unique values, e.g., $\widetilde{\{1, 2, 3, 2, 5, 1\}} \equiv \{1, 2, 3, 5\}$
$\overset{\approx}{\{\cdot\}}$	Represents the "contraction" of the argument set, made in accordance with the "contraction" of a $\widetilde{\{\cdot\}}$ paired set, by adding the values on the argument set for "contracted" values in $\widetilde{\{\cdot\}}$, e.g., $\overset{\approx}{\{2, 4, 6, 1, 3, 5\}}$, in accordance with the paired set $\widetilde{\{1, 2, 3, 2, 5, 1\}}$, gives $\{7, 5, 6, 3\}$

Chapter 1
Setting the Scene

Abstract Although the Fox H and Meijer G functions yield very handy representations for the probability density functions and cumulative distribution functions of several distributions and distributions of products of independent random variables, they are not computationally efficient and most of the time not even utilizable in practice because of serious difficulties found in their efficient computational implementation, even when using the most up-to-date software. This way, looking for alternative representations, namely finite form ones, is a most useful and desirable goal.

Keywords Finite form · Fox H function · GIG (Generalized Integer Gamma) distribution · Meijer G function

The Fox H and Meijer G functions may yield attractive representations for the p.d.f.'s (probability density functions) and c.d.f.'s (cumulative distribution functions) of several distributions and distributions of products of independent r.v.'s (random variables). However, they are usually not computationally efficient and most of the time they cannot be used in practice because of difficulties found in their efficient computational implementation, even when using the most up-to-date software. Consequently, certain finite form representations of these special functions are preferable since they are more easily implemented and provide better or comparable accuracy with greatly, often enormously, reduced computation time.

The selling point of these representations is that they are *finite form* representations as opposed to the more commonly encountered representations in the form of infinite series or unsolved integral equations, which in practice will have to be truncated in the first case or numerically approximated in the second case.

In Multivariate Analysis many statistics of interest admit representations in terms of products of independent Beta r.v.'s. The corresponding distribution functions can be naturally represented in terms of Meijer G or Fox H functions, and it is precisely in some of these situations that suitable finite form representations can be directly developed and profitably used.

C. A. Coelho, B. C. Arnold, *Finite Form Representations for Meijer G and Fox H Functions*, Lecture Notes in Statistics 223,
https://doi.org/10.1007/978-3-030-28790-0_1

By using two extended multiplication formulas for the Gamma function we will be able to show the equivalence of the distribution of certain multiple products of independent Beta r.v.'s with rational second parameters, subject to certain restrictions, to the distribution of products of positive powers of independent Beta r.v.'s with integer valued second parameters. Then we establish the equivalence of the distribution of such products to that of the exponential of a GIG (Generalized Integer Gamma) distributed r.v. (Coelho, 1998, 1999). From this result and the definitions of the Meijer *G* and Fox *H* functions as inverse Mellin transforms we will then be able to establish the equivalences between some instances of these two functions and the expressions for the p.d.f.'s and the c.d.f.'s of exponentials of GIG distributed r.v.'s. The GIG distribution is the distribution of the sum of independent Gamma r.v.'s all with integer valued shape parameters, and both its p.d.f. and its c.d.f., as well as the p.d.f. and the c.d.f. of the exponential of a GIG distributed r.v. have very manageable finite form representations.

The establishment of these results is useful in helping to recognize a number of situations, not previously identified, and as such, not able to be sorted out by the available software, in which the Meijer *G* and Fox *H* functions have alternative finite and consequently much more manageable representations, through the p.d.f. and c.d.f. of a GIG or EGIG (Exponentiated GIG) distribution (Arnold et al., 2013). As such, the application of these results leads to huge gains in computation time and, frequently, improved precision.

More important is the fact that the application of these results yields, in a number of situations, finite closed form expressions for both the p.d.f. and the c.d.f. of a number of well-known, and some not so well-known, l.r.t. (likelihood ratio test) statistics used in Multivariate Analysis, allowing in this way for accurate and very fast evaluations of these distributions and simple and fast computation of quantiles and p-values. Specific examples may be found in Chap. 5. In that chapter, where more than 40 different likelihood ratio tests applicable in Multivariate Analysis are addressed, for each statistic addressed, its p.d.f. and c.d.f. are given, for the general case, in terms of Meijer *G* or Fox *H* functions, according to the case which makes it simpler to use one or the other of these functions. Then, the p.d.f. and c.d.f of each of these statistics are also expressed in closed finite form through the EGIG p.d.f. and c.d.f. for the particular cases that may be addressed through this distribution, based on the results in Theorems 3.1–3.3 in Chap. 3 and Corollaries 4.1–4.5 in Chap. 4. These corollaries also yield alternative representations of these p.d.f.'s and c.d.f.'s in terms of the Meijer *G* and Fox *H* functions for the particular cases for which the EGIG representation is available, but for the sake of keeping the length of the book within sensible bounds, these representations are not reported, it being possible to easily obtain them from the statement of the parameters for the EGIG distribution of each test statistic. We may remark here that in general these particular cases cover most situations of applicability of the statistic or even all possible cases, as it happens with all statistics used with the complex multivariate Normal distribution that are addressed in this book.

Although the distributions of these l.r.t. statistics are derived under normality assumptions, the results in Sects. 4.5, 5.7, 7.9, and 13.8 and mainly in Sects. 8.11,

9.11, 10.11, and 11.8 of Anderson (2003), as well as the results in Chmielewski (1980), Jensen and Good (1981), Kariya (1981), Anderson et al. (1986), Anderson and Fang (1990), and Fang and Zhang (1990) allow us, in most cases, to extend such results to elliptically contoured distributions.

For each test in Chap. 5, modules programmed in Mathematica®, MAXIMA, and R are provided to compute the p.d.f., c.d.f., quantiles, and p-values, and although for each test a few quantile tables are provided in the book's supplementary material web site (https://sites.google.com/site/meijerfoxfiniteforms/), the authors strongly advise the use of the modules to compute quantiles and p-values, which, given the present ubiquitous availability of personal computers, laptops, and notebooks, provides a much easier way to obtain such quantiles and p-values for any combination of parameter values, with the desired precision. A note is due here: while all random sample matrices in the text have their rows defined by the variables and the columns defined by the sample units, i.e., they are of the form $p \times n$, where p is the number of variables and n the sample size, all sample matrices used in the computation modules are of the form $n \times p$, with their columns defined by the variables and their rows defined by the sample units. This option was taken because while it seemed more convenient in terms of exposition and notation to have in the text the sample matrices of the type $p \times n$, on the other hand, the majority of all data tables, files, and matrices are commonly given in the $n \times p$ form. More precise notes on these modules and their use are available along the several subsections in Chap. 5, for each test addressed there, while in Chap. 6 the whole packages in Mathematica®, MAXIMA, and R are described in more detail. The full content of the three packages is available on the book's supplementary material web site.

Two brief notes that are in order here are that whenever we generally refer to "the m.l.e. (maximum likelihood estimator) of Σ" or "the m.l.e. of μ" or the m.l.e. of any other parameter, we are indeed referring to the m.l.e. of that parameter under the alternative hypothesis and that whenever we refer to "the distribution of Λ" or "the distribution of the l.r.t. (likelihood ratio test) statistic," we indeed refer to the distribution of Λ or of the l.r.t. statistic under H_0, that is, under the null hypothesis.

References

Anderson, T.W.: An Introduction to Multivariate Statistical Analysis, 3rd edn. J. Wiley, New York (2003)

Anderson, T.W., Fang, K.-T.: Inference in multivariate elliptically contoured distributions based on maximum likelihood. In: Fang, K.-T., Anderson, T.W. (eds.) Statistical inference in elliptically contoured and related distributions, pp. 201–216. Allerton Press, Inc., New York (1990)

Anderson, T.W., Fang, K.-T., Hsu, H.: Maximum-likelihood estimates and likelihood-ratio criteria for multivariate elliptically contoured distributions. Can. J. Stat. **14**, 55–59 (1986)

Arnold, B.C., Coelho, C.A., Marques, F.J.: The distribution of the product of powers of independent uniform random variables – a simple but useful tool to address and better understand the structure of some distributions. J. Multivar. Anal. **113**, 19–36 (2013)

Chmielewski, M.A.: Invariant scale matrix hypothesis tests under elliptically symmetry. J. Multivar. Anal. **10**, 343–350 (1980)

Coelho, C.A.: The generalized integer Gamma distribution – a basis for distributions in multivariate statistics. J. Multivar. Anal. **64**, 86–102 (1998)
Coelho, C.A.: Addendum to the paper "The generalized integer Gamma distribution – a basis for distributions in multivariate statistics". J. Multivar. Anal. **69**, 281–285 (1999)
Fang, K.T., Zhang, Y.-T.: Generalized Multivariate Analysis. Springer, New York (1990)
Jensen, D.R., Good, I.J.: Invariant distributions associated with matrix laws under structural symmetry. J. R. Stat. Soc. Ser B **43**, 327–332 (1981)
Kariya, T.: Robustness of multivariate tests. Ann. Stat. **9**, 1267–1275 (1981)

Chapter 2
The Meijer *G* and Fox *H* Functions

Abstract In this chapter besides briefly introducing the Meijer *G* and Fox *H* functions through their usual definitions, the authors open the way to the establishment of several cases where there are finite representations for these functions which have not been previously identified and that as such are not recognized by any of the available software. These cases are related to the distribution of three extended products of independent Beta random variables, which are treated in Chap. 3.

Keywords Beta distribution · Characteristic functions · Inverse Mellin transform · Mellin transform · Mellin-Barnes contour integral · Products of Beta r.v.'s

2.1 The Meijer *G* and Fox *H* Functions as Inverse Mellin Transforms

Although Meijer (1936a,b) first defined the Meijer *G* function as a series, related to the generalized hypergeometric function (see Luke (1975, Sect. 5.3, expr. (5))), a more encompassing definition, based on a Mellin-Barnes contour integral, was given later by the same author (Meijer, 1941, 1946). We will use this latter definition.

Let $\mathbb{N}$ denote the set of positive integers and $\mathbb{N}_0$ denote the set of nonnegative integers. For $m, n, p, q \in \mathbb{N}_0$, with $m \le q$ and $n \le p$, and for $a_k - b_j \notin \mathbb{N}$ $(k = 1, \ldots, n;\; j = 1, \ldots, m)$ and $z \neq 0$, the Meijer *G* function with arguments $m, n, p, q \in \mathbb{N}_0$, and any real or complex $a_1, \ldots, a_p$ and $b_1, \ldots, b_q$, is defined, for $s \in \mathbb{C}$, as

$$G^{m,n}_{p,q}\left(\begin{array}{c}\{a_j\}_{j=1:p}\\ \\ \{b_j\}_{j=1:q}\end{array}\middle|\, z\right) = \frac{1}{2\pi\mathrm{i}}\oint_L \frac{\left\{\prod_{j=1}^{m}\Gamma(b_j+s)\right\}\left\{\prod_{j=1}^{n}\Gamma(1-a_j-s)\right\}}{\left\{\prod_{j=m+1}^{q}\Gamma(1-b_j-s)\right\}\left\{\prod_{j=n+1}^{p}\Gamma(a_j+s)\right\}}\, z^{-s}\, ds\,, \tag{2.1}$$

where any empty product is taken as evaluating to 1.

C. A. Coelho, B. C. Arnold, *Finite Form Representations for Meijer G and Fox H Functions*, Lecture Notes in Statistics 223,
https://doi.org/10.1007/978-3-030-28790-0_2

Actually, in (2.1), instead of using the original form in Meijer (1946, I, expr. (2)), which is also used by Erdélyi et al. (1953, Sect. 5.3), Luke (1969, Sect. 5.2; 1975, Sect 5.3) and Askey and Daalhaus (2010), we prefer to use the more convenient form in Mathai and Saxena (1973, Sect. 1.1), Mathai (1993, Sect. 2.1), Mathai and Haubold (2008, Sect. 1.8), and Prudnikov et al. (1990, Sect. 8.2) which may be obtained from the former one just by replacing s by $-s$, or rather, by making the change of variable $s \to s^* = -s$.

In (2.1) we need to have the poles of $\Gamma(b_j + s)$ $(j = 1, \ldots, m)$, which are the points $s = -b_j - \nu$ $(\nu = 0, 1, \ldots)$, separated from the poles of $\Gamma(1 - a_k - s)$ $(k = 1, \ldots, n)$, which are the points $s = 1-a_k+\nu^*$ $(\nu^* = 0, 1, \ldots)$. This condition requires $a_k - b_j \notin \mathbb{N}$ for $k = 1, \ldots, n$; $j = 1, \ldots, m$ and allows for the existence of paths of integration which are contours that separate the poles of $\Gamma(b_j + s)$ from those of $\Gamma(1 - a_k - s)$.

There are three different possible paths of integration L commonly considered. These are listed for example in Sect. 5.3 of Erdélyi et al. (1953), Sect. 1.1. of Mathai and Saxena (1973), Sect. 5.2 of Luke (1969), Sect. 5.3 of Luke (1975), Sect. 8.2.1 of Prudnikov et al. (1990), or Sect. 2.1 of Mathai (1993). The particular path used in this book is either (i) a loop starting and ending at $+\infty$ and encircling all poles $\Gamma(b_j + s)$, $j = 1, \ldots, m$, once in the negative direction, but none of the poles of $\Gamma(1 - a_k - s)$, $k = 1, \ldots, n$, which makes the integral to converge for $|z| < 1$, if $p = q \geq 1$, or for all z if $q \geq 1$ and $q > p$, or (ii) a path going from $c - \mathrm{i}\infty$ to $c + \mathrm{i}\infty$ so that all the poles of $\Gamma(1 - a_k - s)$ $(k = 1, \ldots, m)$ lie to the right of the path and all poles of $\Gamma(b_j + s)$ $(j = 1, \ldots, m)$ lie to the left of the path, so that if $|\arg z| = 0$ the integral converges absolutely for $p = q$ and $Re(\mu) < -1$, for $\mu = \sum_{j=1}^{q} b_j - \sum_{j=1}^{p} a_j$, and where this usual requirement about μ is indeed spurious since it is actually not necessary to be verified in order that the integral in (2.1) converges, which is assured if the path or contour of integration is adequately chosen. For a sketch of possible adequate paths or contours of integration see Roach (1997), where the definition of the Meijer G function used is that in Meijer (1946, I, expr. (2)).

The definition in (2.1) is indeed a Mellin-Barnes contour integral (Pincherle, 1888; Mellin, 1891, 1900, 1902a,b, 1904; Barnes, 1907, 1908; Mellin, 1910; Erdélyi et al., 1953, Sect. 1.19) and it allows us to view the Meijer G function as an inverse Mellin transform.

In fact, taking the Mellin transform (Mellin, 1887, 1899, 1900, 1902b, 1904, 1910; Mathai, 1993, expr. (1.8.1)) of the nonnegative r.v. (random variable) X as $E(X^{s-1})$, for any $s \in \mathbb{C}$, the integrand in (2.1), excluding the power of z, is, for $a_j, b_j > 0$, $m = q = p$, and $n = 0$, the Mellin transform of a product of p independent $Beta(a_j, b_j)$ random variables, since if we take

$$Z = \prod_{j=1}^{p} Y_j \,, \qquad \text{where} \qquad Y_j \sim Beta(a_j^*, b_j^*) \text{ are } p \text{ independent r.v.'s}, \tag{2.2}$$

from the expression of the h-th moment of a Beta distributed r.v. we have

$$E(Z^{s-1}) = \prod_{j=1}^{p} E(Y_j^{s-1}) = \prod_{j=1}^{p} \frac{\Gamma(a_j^* + b_j^*)}{\Gamma(a_j^*)} \frac{\Gamma(a_j^* + s - 1)}{\Gamma(a_j^* + b_j^* + s - 1)}.$$

Then, from the inversion formula for the Mellin transform (Mellin, 1891, expr. (35); Cahen, 1894; Mellin, 1900, 1902b, expr. (1); Mellin, 1902a, expr. (19); Mellin, 1904, expr. (1); Mellin, 1910, expr. after (21); Mathai and Rathie, 1971, expr. (4); Shah and Rathie, 1974, expr. (2.2); Mathai, 1993, expr. (1.8.2)) we have the p.d.f. (probability density function) of Z given by

$$f_Z(z) = \frac{1}{2\pi \mathrm{i}} \oint_L E\left(Z^{s-1}\right) z^{-s}\, ds\,, \tag{2.3}$$

where L is usually taken as path (ii) on the previous page, but where it may actually be any one of the paths usually used for the definition of the Meijer G function (see Mellin (1910, expr. after (21)); Mathai and Rathie (1971, expr. (4)); Shah and Rathie (1974, expr. (2.2))), so that from (2.3) and the definition of the Meijer G function in (2.1), we immediately obtain the p.d.f. of Z in (2.2) as

$$\begin{aligned} f_Z(z) &= \frac{1}{2\pi \mathrm{i}} \oint_L \prod_{j=1}^{p} \frac{\Gamma(a_j^* + b_j^*)}{\Gamma(a_j^*)} \frac{\Gamma(a_j^* + s - 1)}{\Gamma(a_j^* + b_j^* + s - 1)} z^{-s}\, ds \\ &= \left\{ \prod_{j=1}^{p} \frac{\Gamma(a_j^* + b_j^*)}{\Gamma(a_j^*)} \right\} \frac{1}{2\pi \mathrm{i}} \oint_L \prod_{j=1}^{p} \frac{\Gamma(a_j^* + s - 1)}{\Gamma(a_j^* + b_j^* + s - 1)} z^{-s}\, ds \\ &= \left\{ \prod_{j=1}^{p} \frac{\Gamma(a_j^* + b_j^*)}{\Gamma(a_j^*)} \right\} G^{p,0}_{p,p}\left(\begin{array}{c} \{a_j^* + b_j^* - 1\}_{j=1:p} \\ \{a_j^* - 1\}_{j=1:p} \end{array} \middle|\, z \right), \end{aligned} \tag{2.4}$$

where $z \in]0, 1[$ denotes the running value of the r.v. Z in (2.2), and which is the result in Section 4.4.2 of Springer (1979) and also the same as expression (5.2.4) in Mathai and Saxena (1973) if we take into account that straight from the definition of the Meijer G function in (2.1) we have

$$z^c\, G^{m,n}_{p,q}\left(\begin{array}{c} \{a_j\}_{j=1:p} \\ \{b_j\}_{j=1:q} \end{array} \middle|\, z \right) = G^{m,n}_{p,q}\left(\begin{array}{c} \{a_j + c\}_{j=1:p} \\ \{b_j + c\}_{j=1:q} \end{array} \middle|\, z \right) \quad \text{for } c \in \mathbb{C}, \tag{2.5}$$

which is expression (15) in Prudnikov et al. (1990, Subsect. 8.2.2) or expression (1.2.7) in Mathai and Saxena (1973).

Then, by simple integration and the use of (2.5) above, it is easy to see that the corresponding c.d.f. will be given by

$$
\begin{aligned}
F_Z(z) &= \left\{\prod_{j=1}^{p} \frac{\Gamma(a_j^*+b_j^*)}{\Gamma(a_j^*)}\right\} \int_0^z G_{p,p}^{p,0}\left(\begin{array}{c} \{a_j^*+b_j^*-1\}_{j=1:p} \\ \{a_j^*-1\}_{j=1:p} \end{array} \middle| \, u\right) du \\
&= \left\{\prod_{j=1}^{p} \frac{\Gamma(a_j^*+b_j^*)}{\Gamma(a_j^*)}\right\} \frac{1}{2\pi \mathrm{i}} \oint_L \prod_{j=1}^{p} \frac{\Gamma(a_j^*-1+s)}{\Gamma(a_j^*+b_j^*-1+s)} \int_0^z u^{-s}\, du\, ds \\
&= \left\{\prod_{j=1}^{p} \frac{\Gamma(a_j^*+b_j^*)}{\Gamma(a_j^*)}\right\} \frac{1}{2\pi \mathrm{i}} \oint_L \prod_{j=1}^{p} \frac{\Gamma(a_j^*-1+s)}{\Gamma(a_j^*+b_j^*-1+s)} \frac{z^{-s+1}}{1-s}\, ds \\
&= \left\{\prod_{j=1}^{p} \frac{\Gamma(a_j^*+b_j^*)}{\Gamma(a_j^*)}\right\} \frac{1}{2\pi \mathrm{i}} \oint_L \prod_{j=1}^{p} \frac{\Gamma(a_j^*-1+s)}{\Gamma(a_j^*+b_j^*-1+s)} \frac{\Gamma(1-s)}{\Gamma(2-s)} z^{-s+1}\, ds \\
&= \left\{\prod_{j=1}^{p} \frac{\Gamma(a_j^*+b_j^*)}{\Gamma(a_j^*)}\right\} z\, G_{p+1,p+1}^{p,1}\left(\begin{array}{c} \{0, \{a_j^*+b_j^*-1\}_{j=1:p}\} \\ \{\{a_j^*-1\}_{j=1:p}, -1\} \end{array} \middle| \, z\right) \\
&= \left\{\prod_{j=1}^{p} \frac{\Gamma(a_j^*+b_j^*)}{\Gamma(a_j^*)}\right\} G_{p+1,p+1}^{p,1}\left(\begin{array}{c} \{1, \{a_j^*+b_j^*\}_{j=1:p}\} \\ \{\{a_j^*\}_{j=1:p}, 0\} \end{array} \middle| \, z\right).
\end{aligned}
\tag{2.6}
$$

However, in order to address the distribution of the r.v.

$$
Z^* = \prod_{j=1}^{p} Y_j^{c_j}, \tag{2.7}
$$

where Y_j are the same r.v.'s as in (2.2), and in the general case the a_j^*'s, b_j^*'s, and c_j's are some positive reals, we need to resort to the use of the Fox H function (Fox, 1961).

The H function, for $m, n, p, q \in \mathbb{N}_0$, $s \in \mathbb{C}$, $z \neq 0$, positive reals α_k and β_j and real or complex a_k and b_j, such that $a_k - (b_j + \nu)\alpha_k/\beta_j \notin \mathbb{N}$ for $k = 1, \ldots, p$; $j = 1, \ldots, q$, is defined as (Fox, 1961; Mathai and Saxena, 1978, Sect. 1.1; Prudnikov et al., 1990, Sect. 8.3; Mathai, 1993, Sect. 3.11; Mathai and Haubold, 2008, Sect. 1.9) (see also Springer (1979, Sect. 6.2) and Braaksma (1963) for an equivalent definition)

$$
H_{p,q}^{m,n}\left(\begin{array}{c} \{(a_j, \alpha_j)\}_{j=1:p} \\ \{(b_j, \beta_j)\}_{j=1:q} \end{array} \middle| \, z\right) = \\
\frac{1}{2\pi \mathrm{i}} \oint_L \frac{\left\{\prod_{j=1}^{m} \Gamma(b_j+\beta_j s)\right\}\left\{\prod_{j=1}^{n} \Gamma(1-a_j-\alpha_j s)\right\}}{\left\{\prod_{j=m+1}^{q} \Gamma(1-b_j-\beta_j s)\right\}\left\{\prod_{j=n+1}^{p} \Gamma(a_j+\alpha_j s)\right\}} z^s\, ds\,,
\tag{2.8}
$$

where the paths of integration are similar to those used for the Meijer G function and where the condition $a_k - (b_j + \nu)\alpha_k/\beta_j \notin \mathbb{N}$ $(k = 1, \ldots, p;\ j = 1, \ldots, q)$ assures that no pole of $\Gamma(b_j + \beta_j s)$ coincides with any pole of $\Gamma(1 - a_j - \alpha_j s)$.

The Fox H function is defined, among other situations, when $\sum_{j=1}^{q} \beta_j - \sum_{j=1}^{p} \alpha_j = 0$ for $|z| < \beta^{-1}$, where

$$\beta = \prod_{j=1}^{p} \alpha_j^{\alpha_j} \prod_{j=1}^{q} \beta_j^{-\beta_j}.$$

See Braaksma (1963) and Mathai (1972) and also the references mentioned above for further details on the conditions of existence of the Fox H function.

It is then easy to see that the H function reduces to the G function if all α_j and β_j equal 1. Also, when all $\alpha_j = \beta_j = c$, it is easy to see that

$$H^{m,n}_{p,q}\left(\begin{matrix} \{(a_j, c)\}_{j=1:p} \\ \{(b_j, c)\}_{j=1:q} \end{matrix} \middle| \, z \right) = \frac{1}{c}\, G^{m,n}_{p,q}\left(\begin{matrix} \{a_j\}_{j=1:p} \\ \{b_j\}_{j=1:q} \end{matrix} \middle| \, z^{1/c} \right).$$

Moreover, as Mathai and Haubold (2008, Sec. 1.9) remark, when all α_j and β_j are rationals it is always possible to reduce the Fox H function to a Meijer G function. Prudnikov et al. (1990, Subsect. 8.3, expr. 22) provide an explicit expression for the conversion of a Fox H function with all rational α_j and β_j to a Meijer G function.

Using a similar procedure to the one used in (2.4) with the Meijer G function, we may easily see that the p.d.f. of Z^* in (2.7), for general positive real a_j^*, b_j^*, and c_j, as stated in Mathai et al. (2010, Sect. 4.2.1), is given by

$$f_{Z^*}(z) = \left\{ \prod_{j=1}^{p} \frac{\Gamma(a_j^* + b_j^*)}{\Gamma(a_j^*)} \right\} H^{p,0}_{p,p}\left(\begin{matrix} \{(a_j^* + b_j^* - c_j, c_j)\}_{j=1:p} \\ \{(a_j^* - c_j, c_j)\}_{j=1:p} \end{matrix} \middle| \, z \right), \tag{2.9}$$

while using a similar procedure to the one used in (2.6), and the equivalent of (2.5) for the Fox H function, which, for $c \in \mathbb{C}$, is (see for example Mathai and Saxena (1978, expr. (1.2.4)), Prudnikov et al. (1990, Subsect. 8.3.2, expr. 8), or Mathai (1993, expr. (3.11.5)))

$$z^c\, H^{m,n}_{p,q}\left(\begin{matrix} \{(a_j, \alpha_j)\}_{j=1:p} \\ \{(b_j, \beta_j)\}_{j=1:q} \end{matrix} \middle| \, z \right) = H^{m,n}_{p,q}\left(\begin{matrix} \{(a_j + c\alpha_j, \alpha_j)\}_{j=1:p} \\ \{(b_j + c\beta_j, \beta_j)\}_{j=1:q} \end{matrix} \middle| \, z \right),$$

it is not too hard to show that the c.d.f. of Z^* will be given by

$$F_{Z^*}(z) = \left\{ \prod_{j=1}^{p} \frac{\Gamma(a_j^* + b_j^*)}{\Gamma(a_j^*)} \right\} H^{p,1}_{p+1,p+1}\left(\begin{matrix} \left\{(1, 1), \{(a_j^* + b_j^*, c_j)\}_{j=1:p}\right\} \\ \left\{\{(a_j^*, c_j)\}_{j=1:p}, (0, 1)\right\} \end{matrix} \middle| \, z \right). \tag{2.10}$$

There are a number of situations where the Meijer $G^{p,0}_{p,p}$ and $G^{p,1}_{p+1,p+1}$ as well as the Fox $H^{p,0}_{p,p}$ and $H^{p,1}_{p+1,p+1}$ functions may be given finite form representations through the p.d.f. and c.d.f. of a GIG (Generalized Integer Gamma) or an EGIG (Exponentiated Generalized Integer Gamma) distribution (Coelho, 1998; Arnold et al., 2013), whose shape parameters count the multiplicities of the poles of the integrand functions of the corresponding Meijer G and Fox H functions. The most simple case among these is whenever the b_j^* in (2.2) are positive integers. This is the situation addressed in the next subsection, opening the way to the more general and elaborate situations addressed in Chaps. 3 and 4 where the b_j^* are taken to be rational. These results will then be used in Chap. 5 where it is shown how the exact distributions of a large number of l.r.t. statistics of interest in Multivariate Analysis have p.d.f.'s and c.d.f.'s which may feature closed finite form representations from which it is simple to obtain exact p-values and quantiles.

2.2 Straightforward Instances of Meijer *G* and Fox *H* Functions with Alternative Finite Representations

Springer (1979) states in Theorem 4.4.1 that there is a finite form representation for the p.d.f. in (2.4) when all a_j^*'s and b_j^*'s are integers. However, this is actually true if just the b_j^*'s are positive integers, since in this case we may write the c.f. (characteristic function) of $W = -\log Z$, for Z in (2.2), as

$$\begin{aligned}\Phi_W(t) &= E\left(e^{\mathrm{i}tW}\right) = E\left(e^{-\mathrm{i}t\log Z}\right) = E\left(Z^{\mathrm{i}t}\right) = E\left(\prod_{j=1}^{p} Y_j^{\mathrm{i}t}\right) = \prod_{j=1}^{p} E\left(Y_j^{\mathrm{i}t}\right)\\ &= \prod_{j=1}^{p}\frac{\Gamma(a_j^*+b_j^*)}{\Gamma(a_j^*)}\,\frac{\Gamma(a_j^*-\mathrm{i}t)}{\Gamma(a_j^*+b_j^*-\mathrm{i}t)} = \prod_{j=1}^{p}\prod_{\ell=0}^{b_j^*-1}(a_j^*+\ell)(a_j^*+\ell-\mathrm{i}t)^{-1}\end{aligned}$$

which is the c.f. of a GIG distribution which, in the case in which all parameters $a_j^*+\ell$ ($\ell=0,\ldots,b_j^*-1$; $j=1,\ldots,p$) are different, will be a GIG distribution of depth $\sum_{j=1}^{p} b_j^*$, with rate parameters $a_j^*+\ell$ and shape parameters all equal to 1. In case some of the parameters $a_j^*+\ell$ are equal it will be a GIG distribution of depth $g<\sum_{j=1}^{p} b_j^*$, with rate parameters

$$\{d_h\}_{h=1:g} = \widetilde{\left\{\widetilde{\left\{a_j^*+\ell\right\}}_{\substack{j=1:p\\ \ell=0:b_j^*-1}}\right\}}, \tag{2.11}$$

where $\widetilde{\{\,\cdot\,\}}$ denotes the "contraction" of the set, that is, the set of the $g \leq \sum_{j=1}^{p} b_j^*$ different rate parameters $a_j^* + \ell$ $(\ell = 0, \ldots, b_j^* - 1;\ j = 1, \ldots, p)$ and shape parameters

$$\{r_h\}_{h=1:g} = \widetilde{\widetilde{\left\{ \{1\}_{\substack{j=1:p \\ \ell=0:b_j^*-1}} \right\}}} \tag{2.12}$$

where $\widetilde{\widetilde{\{\,\cdot\,\}}}$ denotes the "contraction" of the argument set, corresponding to the "contraction" of the set on the right-hand side of (2.11), that is, the set of the shape parameters r_h corresponding to the rate parameters d_h, with r_h being the number of times the value d_h occurs on the right-hand side of (2.11), which is the multiplicity of the poles of the integrand function of the corresponding Meijer G function.

As such, using the definition of the G function in (2.1) and the notation in Appendix for the p.d.f. and c.d.f. of the GIG and EGIG distributions, we may write

$$G^{p,0}_{p,p}\left(\begin{array}{c} \{a_j^* + b_j^* - 1\}_{j=1:p} \\ \{a_j^* - 1\}_{j=1:p} \end{array} \middle|\, z \right) =$$
$$\left\{ \prod_{j=1}^{p} \frac{\Gamma(a_j^*)}{\Gamma(a_j^* + b_j^*)} \right\} f^{EGIG}\left(z \,\middle|\, \widetilde{\widetilde{\left\{ \{1\}_{\substack{j=1:p \\ \ell=0:b_j^*-1}} \right\}}} ; \widetilde{\left\{ \{a_j^* + \ell\}_{\substack{j=1:p \\ \ell=0:b_j^*-1}} \right\}} ; g \leq \sum_{j=1}^{p} b_j^* \right), \tag{2.13}$$

and

$$G^{p,1}_{p+1,p+1}\left(\begin{array}{c} 1, \{a_j^* + b_j^*\}_{j=1:p} \\ \{a_j^*\}_{j=1:p}, 0 \end{array} \middle|\, z \right) =$$
$$\left\{ \prod_{j=1}^{p} \frac{\Gamma(a_j^*)}{\Gamma(a_j^* + b_j^*)} \right\} F^{EGIG}\left(z \,\middle|\, \widetilde{\widetilde{\left\{ \{1\}_{\substack{j=1:p \\ \ell=0:b_j^*-1}} \right\}}} ; \widetilde{\left\{ \{a_j^* + \ell\}_{\substack{j=1:p \\ \ell=0:b_j^*-1}} \right\}} ; g \leq \sum_{j=1}^{p} b_j^* \right). \tag{2.14}$$

But, in addition to these cases, it is also true that there are finite form representations for the Meijer G function, through the expressions for the p.d.f. and c.d.f. of the EGIG distribution, when the a_j^*'s and b_j^*'s are rationals which satisfy certain relations. These finite form representations are particularly useful for the representation of the distributions of quite complicated products of independent Beta r.v.'s, which represent the exact distribution of many l.r.t. statistics used in Multivariate Analysis. It is cases such as these that will be addressed in the next chapters.

As mentioned in the first chapter, it will be by using two extended multiplication formulas for the Gamma function that we will be able to show the equivalence of the distribution of two multiple products of independent Beta r.v.'s with rational second parameters, which follow some amenable rules, to the distribution of products of positive powers of independent Beta r.v.'s with integer second parameters and showing then that the p.d.f. and c.d.f. of the distribution of these products have highly manageable finite form representations through the p.d.f. and c.d.f. of the GIG or EGIG distributions.

This is so since, in a similar manner to what happens with the Meijer G function, it is not too hard to show that when in (2.7) all b_j^* are positive integers, there is a finite form representation for the Fox H function through the p.d.f. or the c.d.f. of the GIG or EGIG distribution, since for $W^* = -\log Z^*$ and integer valued b_j^*'s we may write

$$
\begin{aligned}
\Phi_{W^*}(t) &= E\left(e^{\mathrm{i}tW^*}\right) = E\left(e^{-\mathrm{i}t\log Z^*}\right) = E\left(Z^{*-\mathrm{i}t}\right) \\
&= E\left(\prod_{j=1}^{p} Y_j^{-\mathrm{i}tc_j}\right) = \prod_{j=1}^{p} E\left(Y_j^{-\mathrm{i}tc_j}\right) \\
&= \prod_{j=1}^{p} \frac{\Gamma(a_j^*+b_j^*)}{\Gamma(a_j^*)}\,\frac{\Gamma(a_j^*-\mathrm{i}tc_j)}{\Gamma(a_j^*+b_j^*-\mathrm{i}tc_j)} = \prod_{j=1}^{p}\prod_{\ell=0}^{b_j^*-1}(a_j^*+\ell)(a_j^*+\ell-\mathrm{i}tc_j)^{-1} \\
&= \prod_{j=1}^{p}\prod_{\ell=0}^{b_j^*-1}\left(\frac{a_j^*+\ell}{c_j}\right)\left(\frac{a_j^*+\ell}{c_j}-\mathrm{i}t\right)^{-1}
\end{aligned}
$$

which is the c.f. of a GIG distribution which in case all $(a_j^*+\ell)/c_j$ $(\ell = 0,\dots,b_j^*-1;\ j=1,\dots,p)$ are different will be a GIG distribution of depth $\sum_{j=1}^{p} b_j^*$, with rate parameters $(a_j^*+\ell)/c_j$ and shape parameters all equal to 1, and which in case some of the parameters $(a_j^*+\ell)/c_j$ are equal will be a GIG distribution of depth $g < \sum_{j=1}^{p} b_j^*$, with rate parameters

$$
\{d_h\}_{h=1:g} = \widetilde{\left\{\left\{\frac{a_j^*+\ell}{c_j}\right\}_{\substack{j=1:p\\ \ell=0:b_j^*-1}}\right\}}, \tag{2.15}
$$

where, as in (2.11), $\widetilde{\{\cdot\}}$ denotes the "contraction" of the argument set, that is, the set of the $g \le \sum_{j=1}^{p} b_j^*$ different rate parameters $(a_j^*+\ell)/c_j$ $(\ell = 0,\dots,b_j^*-1;\ j = 1,\dots,p)$, and with shape parameters

$$
\{r_h\}_{h=1:g} = \overset{\approx}{\left\{\{1\}_{\substack{j=1:p\\ \ell=0:b_j^*-1}}\right\}}
$$

where $\widetilde{\widetilde{\{\,\cdot\,\}}}$ denotes, as in (2.12), the "contraction" of the argument set, corresponding to the "contraction" of the set on the right-hand side of (2.15), that is, $\{r_h\}_{h=1,\dots,g}$ is the set of shape parameters corresponding to the rate parameters d_h, with r_h, being the number of times the value d_h occurs on the right-hand side of (2.15).

As such, using the definition of the H function in (2.8) and the notation in Appendix for the p.d.f. and c.d.f. of the GIG and EGIG distributions, we may write

$$H^{p,0}_{p,p}\left(\begin{array}{c}\{(a_j^*+b_j^*-c_j,c_j)\}_{j=1:p}\\ \{(a_j^*-c_j,c_j)\}_{j=1:p}\end{array}\middle|\, z\right)=$$
$$\left\{\prod_{j=1}^{p}\frac{\Gamma(a_j^*)}{\Gamma(a_j^*+b_j^*)}\right\} f^{EGIG}\left(z\,\middle|\,\widetilde{\widetilde{\left\{\{1\}_{\substack{j=1:p\\ \ell=0:b_j^*-1}}\right\}}};\widetilde{\widetilde{\left\{\left\{\frac{a_j^*+\ell}{c_j}\right\}_{\substack{j=1:p\\ \ell=0:b_j^*-1}}\right\}}};g\leq\sum_{j=1}^{p}b_j^*\right), \tag{2.16}$$

and

$$H^{p,1}_{p+1,p+1}\left(\begin{array}{c}\left\{(1,1),\{(a_j^*+b_j^*,c_j)\}_{j=1:p}\right\}\\ \left\{\{(a_j^*,c_j)\}_{j=1:p},(0,1)\right\}\end{array}\middle|\, z\right)=$$
$$\left\{\prod_{j=1}^{p}\frac{\Gamma(a_j^*)}{\Gamma(a_j^*+b_j^*)}\right\} F^{EGIG}\left(z\,\middle|\,\widetilde{\widetilde{\left\{\{1\}_{\substack{j=1:p\\ \ell=0:b_j^*-1}}\right\}}};\widetilde{\widetilde{\left\{\left\{\frac{a_j^*+\ell}{c_j}\right\}_{\substack{j=1:p\\ \ell=0:b_j^*-1}}\right\}}};g\leq\sum_{j=1}^{p}b_j^*\right). \tag{2.17}$$

There are many examples in the literature where authors used Meijer G or Fox H functions to represent the distribution of products of r.v.'s (see for example Springer and Thompson (1979), Podolski (1972), Mathai (1972), Carter and Springer (1977), Salo et al. (2006)) or the distribution of l.r.t. statistics used in Multivariate Analysis (see for example Consul (1969), Pillai et al. (1969), Pillai and Jouris (1971), Pillai and Nagarsenker (1971), Mathai (1973), Nagarsenker and Das (1975), Gupta (1976), Davis (1979), Gupta and Rathie (1983), Nagar et al. (1985), Cardeño and Nagar (2001)). However, the Meijer G and Fox H functions are indeed only alternative representations for Mellin-Barnes contour integrals, which still have to be solved numerically and although they provide a very general framework and an extraordinarily handy notation for such distributions, they are not very helpful in obtaining representations for those distributions which are adequate for practical purposes, namely when we want to carry out tests involving such statistics, where we need to compute quantiles and/or p-values in an easy and fast way. That is so, given the fact that even nowadays the most up-to-date symbolic softwares still have problems computing instances of the Meijer G function related to the distributions mentioned above, which are only computable for small values of the parameters, as it will be shown in Chap. 4, while the Fox H function is usually not even implemented in such softwares, or when it is, that is through its relation with the Meijer G function.

Although more than 45 years have passed over the words in Mathai (1973) where the author says that "it is evident that a G-function is only a contour integral representation and a statement to the effect that the density of a particular test criterion is a G-function does not carry any meaning beyond giving the moment expression for the test criterion. The problem of getting the distribution is not solved unless the G-function is put into computable forms or in terms of elementary Special Functions" and that "As remarked earlier, (2.4) is only an integral representation and the distribution is obtained only if the G-function in (2.4) is put into tractable forms," these words continue to state an absolute truth in our days.

This is the reason why the main aim of this book is to identify situations in which instances of the Meijer G and Fox H functions may have finite form representations which may then allow for an easy and fast computation of quantiles and p-values for the corresponding distributions. We will focus on distributions of products of independent Beta r.v.'s, given the importance that these distributions have because of their relation with the distributions of l.r.t. statistics used in Multivariate Analysis.

In the next chapter we will establish the results necessary to identify a number of other situations of great practical interest where both the Meijer G and Fox H functions have finite form representations, and then in Chap. 4 we will establish the finite form of those representations through the representation of the p.d.f. or c.d.f. of GIG or EGIG distributions. In Chap. 5 we will illustrate how the distributions of many l.r.t. statistics with interest in Multivariate Analysis may be setup under the framework designed in these two previous chapters.

Appendix: Notation and Expressions for the Probability Density and Cumulative Distribution Functions of the Gamma, GIG, and EGIG Distributions

This appendix is used to establish the notation used concerning the p.d.f. and c.d.f. of the GIG (Generalized Integer Gamma) and the EGIG (Exponentiated Generalized Integer Gamma) distributions.

We say that the r.v. (random variable) X has a Gamma distribution with shape parameter r (> 0) and rate parameter λ (> 0), and we will denote this fact by $X \sim \Gamma(r, \lambda)$, if the p.d.f. of X is

$$f_X(x) = \frac{\lambda^r}{\Gamma(r)}\, e^{-\lambda x}\, x^{r-1} \quad (x > 0)\,.$$

Let $X_j \sim \Gamma(r_j, \lambda_j)$ $(j = 1, \ldots, p)$ be a set of p independent r.v.'s and consider the r.v.

$$W = \sum_{j=1}^{p} X_j \,.$$

In case all the $r_j \in \mathbb{N}$, the distribution of W is what we call a GIG distribution (Coelho, 1998, 1999). If all the λ_j are different, W has a GIG distribution of depth p, with shape parameters r_j and rate parameters λ_j, with p.d.f.

$$f_W(w) = f^{\,GIG}\Big(w \,\big|\, \{r_j\}_{j=1:p}; \{\lambda_j\}_{j=1:p}; p\Big) = K \sum_{j=1}^{p} P_j(w)\, e^{-\lambda_j w}\,, \quad (w > 0) \tag{2.18}$$

and c.d.f.

$$F_W(w) = F^{GIG}\Big(w \,\big|\, \{r_j\}_{j=1:p}; \{\lambda_j\}_{j=1:p}; p\Big) = 1 - K \sum_{j=1}^{p} P_j^*(w)\, e^{-\lambda_j w}\,, \quad (w > 0) \tag{2.19}$$

where

$$K = \prod_{j=1}^{p} \lambda_j^{r_j}\,, \qquad P_j(w) = \sum_{k=1}^{r_j} c_{j,k}\, w^{k-1}$$

and

$$P_j^*(w) = \sum_{k=1}^{r_j} c_{j,k}(k-1)! \sum_{i=0}^{k-1} \frac{w^i}{i!\,\lambda_j^{k-i}}\,,$$

with

$$c_{j,r_j} = \frac{1}{(r_j - 1)!} \prod_{\substack{i=1 \\ i \neq j}}^{p} (\lambda_i - \lambda_j)^{-r_i}\,, \quad j = 1, \ldots, p\,, \tag{2.20}$$

and

$$c_{j,r_j-k} = \frac{1}{k} \sum_{i=1}^{k} \frac{(r_j - k + i - 1)!}{(r_j - k - 1)!} R(i, j, p) c_{j,r_j-(k-i)}\,, \quad (k = 1, \ldots, r_j - 1;\; j = 1, \ldots, p) \tag{2.21}$$

where

$$R(i,j,p)=\sum_{\substack{k=1\\k\neq j}}^{p} r_k(\lambda_j-\lambda_k)^{-i} \quad (i=1,\dots,r_j-1)\,. \tag{2.22}$$

In case some of the λ_j assume the same value as other λ_j's, the distribution of W still is a GIG distribution, but in this case with a reduced depth. In this more general case, let $\{\lambda_\ell;\ell=1,\dots,g(\leq p)\}$ be the set of different λ_j's and let $\{r_\ell;\ell=1,\dots,g(\leq p)\}$ be the set of the corresponding shape parameters, with r_ℓ being the sum of all r_j $(j\in\{1,\dots,p\})$ which correspond to the λ_j assuming the value λ_ℓ. In this case W will have a GIG distribution of depth g, with shape parameters r_ℓ and rate parameters λ_ℓ $(\ell=1,\dots,g)$.

Let us consider the r.v. $Z=e^{-W}$. Then the r.v. Z has what Arnold et al. (2013) call an Exponentiated Generalized Integer Gamma (EGIG) distribution of depth p, with p.d.f.

$$\begin{aligned} f_Z(z) &= f^{EGIG}\Big(z\,\Big|\,\{r_j\}_{j=1:p};\{\lambda_j\}_{j=1:p};p\Big)\\ &= f^{GIG}\Big(-\log z\,\Big|\,\{r_j\}_{j=1:p};\{\lambda_j\}_{j=1:p};p\Big)\frac{1}{z} \qquad (0<z<1) \end{aligned} \tag{2.23}$$

and c.d.f.

$$\begin{aligned} F_Z(z) &= F^{EGIG}\Big(z\,\Big|\,\{r_j\}_{j=1:p};\{\lambda_j\}_{j=1:p};p\Big)\\ &= 1-F^{GIG}\Big(-\log z\,\Big|\,\{r_j\}_{j=1:p};\{\lambda_j\}_{j=1:p};p\Big) \qquad (0<z<1)\,. \end{aligned} \tag{2.24}$$

References

Arnold, B.C., Coelho, C.A., Marques, F.J.: The distribution of the product of powers of independent uniform random variables – a simple but useful tool to address and better understand the structure of some distributions. J. Multivar. Anal. **113**, 19–36 (2013)

Askey, R.A., Daalhaus, A.B.O.: Generalized Hypergeometric Functions and Meijer *G*-function. In: Olver, W.J., Lozier, D.W., Boisvert, R.F., Clark, C.W. (eds.) NIST Handbook of Mathematical Functions, pp. 403–418. National Institute of Standards and Technology, U.S. Dept. of Commerce and Cambridge University Press, New York (2010)

Barnes, E.W.: The asymptotic expansion of integral functions defined by generalised hypergeometric series. Proc. Lond. Math. Soc. **s2-5** 59–116 (1907)

Barnes, E.W.: A new development of the theory of hypergeometric functions. Proc. Lond. Math. Soc. **s2-6** 141–177 (1908)

Braaksma, B.L.J.: Asymptotic expansions and analytic continuations for a class of Barnes-integrals. Compos. Math. **15**, 239–341 (1963)

Cahen, E.: Sur la fonction $\zeta(x)$ de Riemann et sur des fonctions analogues. Ann. Sci. École Normale Supérieure **11** 75–164 (1894)

Cardeño, L., Nagar, D.K.: Testing block sphericity of a covariance matrix. Divulgaciones Matemáticas **9**, 25–34 (2001)

Carter, B.D., Springer, M.D.: The distribution of products, quotients and powers of independent H-function variates. SIAM J. Appl. Math. **33**, 542–558 (1977)

Coelho, C.A.: The generalized integer Gamma distribution – a basis for distributions in multivariate statistics. J. Multivar. Anal. **64**, 86–102 (1998)

Coelho, C.A.: Addendum to the paper "The generalized integer Gamma distribution – a basis for distributions in multivariate statistics". J. Multivar. Anal. **69**, 281–285 (1999)

Consul, P.C.: The exact distributions of likelihood criteria for different hypotheses. In: Krishnaiah, P. R. (ed.) Multivariate Analysis II, pp. 171–181. Academic Press, New York (1969)

Davis, A.W.: On the differential equation for Meijer's $G_{p,p}^{p,0}$ function, and further tables of Wilks's likelihood ratio criterion. Biometrika **66**, 519–531 (1979)

Erdélyi, A., Magnus, W., Oberhettinger, F., Tricomi, F.G.: Higher Transcendental Functions (The Bateman Manuscript Project), vol. I. McGraw-Hill, New York (1953)

Fox, C.: The G and H functions as symmetrical kernels. Trans. Amer. Math. Soc. **98**, 395–429 (1961)

Gupta, A.K.: Nonnull distribution of Wilks' statistic for MANOVA in the complex case. Commun. Statist. Simulation Comput. **5**, 177–188 (1976)

Gupta, A.K., Rathie, P.N.: Nonnull distributions of Wilks' Lambda in the complex case. Statistics **43**, 443–450 (1983)

Luke, Y.L.: The Special Functions and Their Approximations, vol. I. Academic Press, New York (1969)

Luke, Y.L.: Mathematical Functions and Their Approximations. Academic Press, New York (1975)

Mathai, A.M.: The exact non-central distribution of the generalized variance. Ann. Inst. Statist. Math. **24**, 53–65 (1972)

Mathai, A.M.: A few remarks about some recent articles on the exact distributions of multivariate test criteria: I. Ann. Inst. Statist. Math. **25**, 557–566 (1973)

Mathai, A.M.: A Handbook of Generalized Special Functions for Statistical and Physical Sciences. Oxford University Press, New York (1993)

Mathai, A.M., Haubold, H.J.: Special Functions for Applied Scientists. Springer, New York (2008)

Mathai, A.M., Rathie, P.N.: The Exact Distribution of Wilks' Criterion. Ann. Math. Statist. **42**, 1010–1019 (1971)

Mathai, A.M., Saxena, R.K.: Generalized Hypergeometric Functions with Applications in Statistics and Physical Sciences. Lecture Notes in Mathematics, vol. 348. Springer, New York (1973)

Mathai, A.M., Saxena, R.K.: The H-function with Applications in Statistics and Other Disciplines. Wiley, New York (1978)

Mathai, A. M., Saxena, R. K., Haubold, H. J.: The H-Function – Theory and Applications. Springer, New York (2010)

Meijer, C.S.: Über Whittakersche bzw. Besselsche Funktionen und deren Produkte. Nieuw Archief voor Wiskunde **18**(2), 10–29 (1936a)

Meijer, C.S.: Neue Integraldarstellungen aus der Theorie der Whittakerschen und Hankelschen Funktionen. Math. Annalen **112**(2), 469–489 (1936b)

Meijer, C.S.: Multiplikationstheoreme für die Funktion $G_{p,q}^{m,n}(z)$. Proc. Koninklijk Nederlandse Akademie van Weteenschappen **44**, 1062–1070 (1941)

Meijer, C.S.: On the G-function I–VIII. Proc. Koninklijk Nederlandse Akademie van Weteenschappen **49**, 227–237, 344–356, 457–469, 632–641, 765–772, 936–943, 1063–1072, 1165–1175 (1946)

Mellin, H.: Über einen zusammenhang zwischen gewissen linearen differential- und differenzengleichungen. Acta Math. **9**, 137–166 (1887)

Mellin, H.: Zur theorie der linearen Differenzengleichungen erster Ordnung. Acta Math. **15**, 317–384 (1891)

Mellin, H.: Über eine Verallgemeinerung der Riemannscher Funktion $\zeta(s)$. Acta Soc. Sci. Fenn. **24**, 1–50 (1899)

Mellin, H.: Eine Formel für den Logarithmus transcendenter Functionen von endlichem Geschlecht. Acta Soc. Sci. Fenn. **29** 1–49 (1900)

Mellin, H.: Über den Zusammenhang Zwischen den Linearen Differential- und Differenzengleichungen. Acta Math. **25**, 139–164 (1902)

Mellin, H.: Eine Formel für den Logarithmus transcendenter Functionen von endlichem Geschlecht. Acta Math. **25**, 165–183 (1902)

Mellin, H.: Die Dirichlettschen Reihen, die zahlentheoretischen Funktionen die unendlichen Produkte von endlichem Geschlecht. Acta Math. **28** 37–64 (1904)

Mellin, H.: Abriß einer einheitlichen Theorie der Gamma- und hypergeometrischen Funktionen. Math. Ann. **68**, 305–337 (1910)

Nagar, D.K., Jain S.K., Gupta A.K.: Distribution of LRC for testing sphericity of a complex multivariate Gaussian model. Int. J. Math. Math. Sci. **8**, 555–562 (1985)

Nagarsenker, B.N., Das, M.M.: Exact distribution of sphericity criterion in the complex case and its percentage points. Commun. Stat. **4**, 363–374 (1975)

Pillai, K.C.S., Jouris, G.M.: Some distribution problems in the multivariate complex Gaussian case. Amm. Math. Statist. **42**, 517–525 (1971)

Pillai, K.C.S., Nagarsenker, B.N.: On the distribution of the sphericity test criterion in classical and complex Normal populations having unknown covariance matrix. Ann. Math. Statist. **42**, 764–767 (1971)

Pillai, K.C.S., Al-Ani, S., Jouris, G.M.: On the distributions of the ratios of the roots of a covariance matrix and Wilks' criterion for test of three hypotheses. Ann. Math. Statist. **40**, 2033–2040 (1969)

Pincherle, S.: Sulle funzioni ipergeometriche generalizzate. Atti R. Accad. Lincei, Rend. Cl. Sci. Fis. Mat. Natur. **4** 694–700, 792–799 (1888)

Podolski, H.: The distribution of a product of n independent random variables with generalized Gamma distribution. Demonstratio Math. **IV**, 119–123 (1972)

Prudnikov, A.P., Brychkov, Y.A., Marichev, O.J.: Integrals and Series, vol. 3: More Special Functions. Gordon and Breach, Newark (1990)

Roach, K.: Meijer G function representations. In: Küchlin W. (eds) Proceedings of the 1997 International Symposium on Symbolic and Algebraic Computation, pp. 205–211. ACM Press, New York (1997). ISBN: 0-89791-875-4

Salo, J., El-Sallabi, H.M., Vainikainen, P.: The distribution of the product of independent Rayleigh random variables. IEEE Trans. Antennas Propag. **54**, 639–643 (2006)

Shah, M.C., Rathie, P.N.: Exact distribution of product of generalized F-variates. Can. J. Stat. **2**, 13–24 (1974)

Springer, M.D.: The Algebra of Random Variables. Wiley, New York (1979)

Springer, M.D., Thompson, W.E.: The distribution of products of Beta, Gamma and Gaussian random variables. SIAM J. Appl. Math. **18**, 721–737 (1970)

Chapter 3
Multiple Products of Independent Beta Random Variables with Finite Form Representations for Their Distributions

Abstract In this chapter the authors consider three multiple products of independent Beta random variables which are shown to have equivalent representations as the exponential of sums of independent integer Gamma r.v.'s, and as such have finite form representations for their distributions.

Keywords Characteristic function · EGIG distribution · Extended multiplication formula · Gamma distribution · GIG distribution · Product of Beta r.v.'s

3.1 A First Multiple Product and Its Particular Case of Interest

In this section we will show how by working on the c.f. (characteristic function) of the negative logarithm of multiple products of independent Beta r.v.'s, whose second parameters are rational and follow some rules, we are able to establish that in these cases the distributions of these products have closed finite forms for their p.d.f.'s and c.d.f.'s, with quite simple and manageable expressions, actually in the form of p.d.f.'s and c.d.f.'s of the exponential of GIG distributed r.v.'s.

Theorem 3.1 *For positive integers* n_ν, $k_{\nu\ell}$, *and* $m_{\nu\ell}$ ($\nu=1,\dots,m^*$; $\ell=1,\dots,n_\nu$), *and for real* $a_\nu > \sum_{\ell=1}^{n_\nu} k_{\nu\ell}\big/\min_\ell\{k_{\nu\ell}\}$ ($\nu = 1,\dots,m^*$), *let*

$$Z = \prod_{\nu=1}^{m^*}\prod_{\ell=1}^{n_\nu}\prod_{j=1}^{k_{\nu\ell}} Y_{\nu\ell j} \tag{3.1}$$

where

$$Y_{\nu\ell j} \sim Beta\left(a_\nu - \frac{j+\sum_{r=1}^{\ell-1}k_{\nu r}}{k_{\nu\ell}},\ \frac{m_{\nu\ell}}{k_{\nu\ell}}\right),\quad \nu=1,\dots,m^*;\ \ell=1,\dots,n_\nu;\ j=1,\dots,k_{\nu\ell},$$

C. A. Coelho, B. C. Arnold, *Finite Form Representations for Meijer G and Fox H Functions*, Lecture Notes in Statistics 223,
https://doi.org/10.1007/978-3-030-28790-0_3

are independent r.v.'s. Then, for $p_\nu = \sum_{\ell=1}^{n_\nu} k_{\nu\ell}$*,*

$$Z \stackrel{\mathrm{d}}{\equiv} \prod_{\nu=1}^{m^*}\prod_{\ell=1}^{n_\nu}\prod_{j=1+\sum_{r=1}^{\ell-1}k_{\nu r}}^{\sum_{r=1}^{\ell}k_{\nu r}} Y^*_{\nu\ell j} \stackrel{\mathrm{d}}{\equiv} \prod_{\nu=1}^{m^*}\prod_{j=1}^{p_\nu} Y^{**}_{\nu j} \stackrel{\mathrm{d}}{\equiv} \prod_{\nu=1}^{m^*}\prod_{\ell=1}^{n_\nu}\left(Y^{***}_{\nu\ell}\right)^{k_{\nu\ell}} \tag{3.2}$$
$$\stackrel{\mathrm{d}}{\equiv} \prod_{\nu=1}^{m^*}\prod_{\ell=1}^{n_\nu}\prod_{j=0}^{m_{\nu\ell}-1} e^{-W_{\nu\ell j}}\,,$$

or, for $W = -\log Z$*,*

$$W \stackrel{\mathrm{d}}{\equiv} \sum_{\nu=1}^{m^*}\sum_{\ell=1}^{n_\nu}\sum_{j=0}^{m_{\nu\ell}-1} W_{\nu\ell j}\,, \tag{3.3}$$

where "$\stackrel{\mathrm{d}}{\equiv}$*" stands for "is equivalent in distribution to," or "is stochastically equivalent to,"*

$$Y^*_{\nu\ell j} \sim Beta\left(a_\nu - \frac{j}{k_{\nu\ell}}, \frac{m_{\nu\ell}}{k_{\nu\ell}}\right), \qquad Y^{**}_{\nu j} \sim Beta\left(a_\nu - \frac{j}{k_{\nu j^*_{\nu j}}}, \frac{m_{\nu j^*_{\nu j}}}{k_{\nu j^*_{\nu j}}}\right),$$

$$Y^{***}_{\nu\ell} \sim Beta\left(a_\nu k_{\nu\ell} - \sum_{r=1}^{\ell} k_{\nu r},\, m_{\nu\ell}\right) \quad \text{and} \quad W_{\nu\ell j} \sim Exp\left(a_\nu + \frac{j - \sum_{r=1}^{\ell} k_{\nu r}}{k_{\nu\ell}}\right),$$

with

$$j^*_{\nu j} = 1 + \#\, of\ elements\ in\ \{j - \underline{k}_\nu\} > 0\,,$$

for $\nu = 1,\dots,m^*$ *and* $j = 1,\dots,p_\nu$*, where*

$$\underline{k}_\nu = \left\{k^*_{\nu h}\right\}_{h=1:n_\nu}, \quad \text{with} \quad k^*_{\nu h} = \sum_{r=1}^{h} k_{\nu r}\,,$$

and where the Beta and Exponential r.v.'s involved in any product in (3.2) or in the triple summation in (3.3) are all independent.

Proof The first equivalence in (3.2) is obtained by reversing the indexation in j and the second one by grouping the $Y_{\nu\ell j}$ r.v.'s in ℓ and j and re-indexing.

The establishment of the third and fourth equivalences in (3.2) needs the use of an extended version of the usual multiplication formula for the Gamma function which, for any $z, c \in \mathbb{C}$ and $n \in \mathbb{N}$, may be written as

$$\prod_{j=1}^{n} \Gamma\left(z+\frac{c-j}{n}\right) = \Gamma(nz+c-n)\,(2\pi)^{\frac{n-1}{2}}\, n^{\frac{1}{2}-nz-c+n}\,. \tag{3.4}$$

This extended multiplication formula may be very easily derived from the more common multiplication formula for the Gamma function, which uses $n \in \mathbb{N}$ in place of $c \in \mathbb{C}$ and which, for any $z \in \mathbb{C}$ and $n \in \mathbb{N}$, may be written as

$$\prod_{j=1}^{n} \Gamma\left(z+\frac{j-1}{n}\right) = \prod_{j=1}^{n} \Gamma\left(z+\frac{n-j}{n}\right) = \Gamma(nz)\,(2\pi)^{\frac{n-1}{2}}\, n^{\frac{1}{2}-nz}\,. \tag{3.5}$$

Indeed, (3.4) may be immediately derived from (3.5) since we may write

$$\prod_{j=1}^{n} \Gamma\left(z+\frac{c-j}{n}\right) = \prod_{j=1}^{n} \Gamma\left(z+\frac{c-n}{n}+\frac{n-j}{n}\right).$$

Then, from the first expression for the distribution of Z in the statement of the Theorem, using (3.4), and

$$\frac{\Gamma(a+n)}{\Gamma(a)} = \prod_{\ell=0}^{n-1}(a+\ell) \qquad \text{(for every integer } n \text{ and any complex } a\text{)}\,, \tag{3.6}$$

the c.f. of $W = -\log Z$ may be written as

$$\Phi_W(t) = \prod_{\nu=1}^{m^*}\prod_{\ell=1}^{n_\nu}\prod_{j=1}^{k_{\nu\ell}} \frac{\Gamma\left(a_\nu - \frac{j+\sum_{r=1}^{\ell-1}k_{\nu r}}{k_{\nu\ell}} + \frac{m_{\nu\ell}}{k_{\nu\ell}}\right)}{\Gamma\left(a_\nu - \frac{j+\sum_{r=1}^{\ell-1}k_{\nu r}}{k_{\nu\ell}}\right)} \frac{\Gamma\left(a_\nu - \frac{j+\sum_{r=1}^{\ell-1}k_{\nu r}}{k_{\nu\ell}} - \mathrm{i}t\right)}{\Gamma\left(a_\nu - \frac{j+\sum_{r=1}^{\ell-1}k_{\nu r}}{k_{\nu\ell}} + \frac{m_{\nu\ell}}{k_{\nu\ell}} - \mathrm{i}t\right)}$$

$$= \prod_{\nu=1}^{m^*}\prod_{\ell=1}^{n_\nu}\prod_{j=1}^{k_{\nu\ell}} \frac{\Gamma\left(a_\nu + \frac{m_{\nu\ell}-\sum_{r=1}^{\ell}k_{\nu r}}{k_{\nu\ell}} + \frac{k_{\nu\ell}-j}{k_{\nu\ell}}\right)}{\Gamma\left(a_\nu - \frac{\sum_{r=1}^{\ell}k_{\nu r}}{k_{\nu\ell}} + \frac{k_{\nu\ell}-j}{k_{\nu\ell}}\right)}$$

$$\times \frac{\Gamma\left(a_\nu - \frac{\sum_{r=1}^{\ell}k_{\nu r}}{k_{\nu\ell}} + \frac{k_{\nu\ell}-j}{k_{\nu\ell}} - \mathrm{i}t\right)}{\Gamma\left(a_\nu + \frac{m_{\nu\ell}-\sum_{r=1}^{\ell}k_{\nu r}}{k_{\nu\ell}} + \frac{k_{\nu\ell}-j}{k_{\nu\ell}} - \mathrm{i}t\right)}$$

$$
=\prod_{\nu=1}^{m^*}\prod_{\ell=1}^{n_\nu}\frac{\Gamma\left(a_\nu k_{\nu\ell}+m_{\nu\ell}-\sum_{r=1}^{\ell}k_{\nu r}\right)}{\Gamma\left(a_\nu k_{\nu\ell}-\sum_{r=1}^{\ell}k_{\nu r}\right)}
\times\frac{\Gamma\left(a_\nu k_{\nu\ell}-\sum_{r=1}^{\ell}k_{\nu r}-k_{\nu\ell}\mathrm{i}t\right)}{\Gamma\left(a_\nu k_{\nu\ell}+m_{\nu\ell}-\sum_{r=1}^{\ell}k_{\nu r}-k_{\nu\ell}\mathrm{i}t\right)} \tag{3.7}
$$

$$
=\prod_{\nu=1}^{m^*}\prod_{\ell=1}^{n_\nu}\prod_{j=0}^{m_{\nu\ell}-1}\left(a_\nu k_{\nu\ell}-\sum_{r=1}^{\ell}k_{\nu r}+j\right)\left(a_\nu k_{\nu\ell}-\sum_{r=1}^{\ell}k_{\nu r}+j-k_{\nu\ell}\mathrm{i}t\right)^{-1}
$$

$$
=\prod_{\nu=1}^{m^*}\prod_{\ell=1}^{n_\nu}\prod_{j=0}^{m_{\nu\ell}-1}\left(a_\nu+\frac{j-\sum_{r=1}^{\ell}k_{\nu r}}{k_{\nu\ell}}\right)\left(a_\nu+\frac{j-\sum_{r=1}^{\ell}k_{\nu r}}{k_{\nu\ell}}-\mathrm{i}t\right)^{-1}, \tag{3.8}
$$

where (3.7) is the c.f. of the sum of the negative logarithm of the Beta r.v.'s $Y^{***}_{\nu\ell}$ in the third product in (3.2), while (3.8) is the c.f. of W in (3.3), or of the negative logarithm of the product of the r.v.'s $e^{-W_{\nu\ell j}}$ in the fourth product in (3.2).

The last equivalence in (3.2) shows that the exact distribution of the original product of independent Beta r.v.'s is what Arnold et al. (2013) call an EGIG (Exponentiated Generalized Integer Gamma) distribution, of depth at most $\sum_{\nu=1}^{m^*}\sum_{\ell=1}^{n_\nu}m_{\nu\ell}$. □

The particular case for $k_{\nu\ell}=k_\nu$ and $m_{\nu\ell}=m_\nu$, for all $\ell=1,\dots,n_\nu$ yields the result in the following theorem.

Theorem 3.2 *If in Theorem 3.1 we have* $k_{\nu\ell}=k_\nu$ *and* $m_{\nu\ell}=m_\nu$, *for all* $\ell=1,\dots,n_\nu$, *we will have, for* $a_\nu>n_\nu$,

$$
Z=\prod_{\nu=1}^{m^*}\prod_{\ell=1}^{n_\nu}\prod_{j=1}^{k_\nu}Y_{\nu\ell j}\,, \tag{3.9}
$$

where

$$
Y_{\nu\ell j}\sim Beta\left(a_\nu+1-\ell-\frac{j}{k_\nu},\,\frac{m_\nu}{k_\nu}\right),\qquad \nu=1,\dots,m^*;\,\ell=1,\dots,n_\nu;\,j=1,\dots,k_\nu,
$$

are independent r.v.'s. Then, for $p_\nu = n_\nu k_\nu$,

$$Z \stackrel{d}{\equiv} \prod_{\nu=1}^{m^*}\prod_{\ell=1}^{n_\nu}\prod_{j=1+k_\nu(\ell-1)}^{k_\nu\ell} Y_{\nu j}^* \stackrel{d}{\equiv} \prod_{\nu=1}^{m^*}\prod_{j=1}^{p_\nu} Y_{\nu j}^* \stackrel{d}{\equiv} \prod_{\nu=1}^{m^*}\prod_{\ell=1}^{n_\nu}\left(Y_{\nu\ell}^{***}\right)^{k_\nu}$$
$$\stackrel{d}{\equiv} \prod_{\nu=1}^{m^*}\prod_{j=1}^{m_\nu+k_\nu(n_\nu-1)} e^{-W_{\nu j}}\,, \qquad (3.10)$$

or, for $W = -\log Z$,

$$W \stackrel{d}{\equiv} \sum_{\nu=1}^{m^*}\sum_{j=1}^{m_\nu+k_\nu(n_\nu-1)} W_{\nu j}\,, \qquad (3.11)$$

where, for $\nu = 1, \dots, m^*$; $j = 1, \dots, n_\nu\, k_\nu$, *and* $\ell = 1, \dots, n_\nu$,

$$Y_{\nu j}^* \sim Beta\left(a_\nu - \frac{j}{k_\nu}, \frac{m_\nu}{k_\nu}\right), \qquad Y_{\nu\ell}^{***} \sim Beta\big((a_\nu - \ell)k_\nu,\, m_\nu\big)$$

and, for $\nu = 1, \dots, m^*$; $j = 1, \dots, m_\nu + k_\nu(n_\nu - 1)$,

$$W_{\nu j} \sim \Gamma\left(r_{\nu j},\, a_\nu - n_\nu + \frac{j-1}{k_\nu}\right), \qquad (3.12)$$

with

$$r_{\nu j} = \begin{cases} h_{\nu j} & j = 1, \dots, k_\nu \\ h_{\nu j} + r_{\nu, j-k_\nu} & j = k_\nu + 1, \dots, m_\nu + k_\nu(n_\nu - 1)\,, \end{cases} \qquad (3.13)$$

for $\nu = 1, \dots, m^*$, *where, for* $j = 1, \dots, m_\nu + k_\nu(n_\nu - 1)$,

$$h_{\nu j} = (\#\ of\ elements\ in\ \{p_\nu, m_\nu\} \geq j) - 1\,, \quad \nu = 1, \dots, m^*\,, \qquad (3.14)$$

which shows that in this particular case the exact distribution of Z is an EGIG distribution of depth at most $\sum_{\nu=1}^{m^*} m_\nu + k_\nu(n_\nu - 1)$, *with shape parameters* $r_{\nu j}$ *and rate parameters* $a_\nu - n_\nu + (j-1)/k_\nu$ ($\nu = 1, \dots, m^*$; $j = 1, \dots, m_\nu + k_\nu(n_\nu - 1)$).

All r.v.'s in any product in (3.10) or in the double summation in (3.11) are independent.

Proof Because the result in this theorem is a particular case of the one in Theorem 3.1, the main part of the proof follows similar lines. Then the expression for the shape parameters $r_{\nu j}$ is obtained either by identifying the different c.f.'s of Exponential distributions that appear in the c.f. of W and counting how many of each of them occur, in a manner in all similar to the one used by Coelho (1998, 2006) and by Arnold et al. (2013), or, equivalently, by identifying the poles of the integrand function of the Meijer G function, or of the c.f. of W, and counting their multiplicities, as in Wald and Brookner (1941). □

3.2 A Second Multiple Product

A second type of multiple product of independent Beta r.v.'s with closed finite form representation for their p.d.f. and c.d.f. is presented in Theorem 3.3.

Theorem 3.3 *For positive integers k_ν, n_ν, nonnegative integers s_ν and real $a_\nu > n_\nu k_\nu$ ($\nu = 1, \ldots, m^*$), let*

$$Z = \prod_{\nu=1}^{m^*} \prod_{\ell=1}^{n_\nu} \prod_{j=1}^{k_\nu} Y_{\nu\ell j} \tag{3.15}$$

where, for $\nu = 1, \ldots, m^*$; $\ell = 1, \ldots, n_\nu$; $j = 1, \ldots, k_\nu$,

$$Y_{\nu\ell j} \sim Beta\left(a_\nu - \frac{(\ell-1)k_\nu + j}{n_\nu}, \frac{j + (\ell-1)k_\nu + \ell + s_\nu - 1}{n_\nu}\right),$$

are independent r.v.'s.

Then,

$$\begin{aligned} Z &\stackrel{d}{\equiv} \prod_{\nu=1}^{m^*} \prod_{\ell=1}^{n_\nu} \prod_{j=1+k_\nu(\ell-1)}^{k_\nu \ell} Y^*_{\nu\ell j} \stackrel{d}{\equiv} \prod_{\nu=1}^{m^*} \prod_{j=1}^{n_\nu k_\nu} Y^{**}_{\nu j} \stackrel{d}{\equiv} \prod_{\nu=1}^{m^*} \prod_{j=1}^{k_\nu} \left(Y^{***}_{\nu j}\right)^{n_\nu} \\ &\stackrel{d}{\equiv} \prod_{\nu=1}^{m^*} \prod_{\ell=1}^{n_\nu k_\nu + s_\nu} e^{-W_{\nu\ell}}, \end{aligned} \tag{3.16}$$

or, for $W = -\log Z$,

$$W \stackrel{d}{\equiv} \sum_{\nu=1}^{m^*} \sum_{\ell=1}^{n_\nu k_\nu + s_\nu} W_{\nu\ell}\,, \tag{3.17}$$

where

$$Y^*_{\nu\ell j} \sim Beta\left(a_\nu - \frac{j}{n_\nu}, \frac{j + \ell + s_\nu - 1}{n_\nu}\right),$$

$$Y^{**}_{\nu j} \sim Beta\left(a_\nu - \frac{j}{n_\nu}, \frac{j + s_\nu - 1 + \text{mod}^*(j, n_\nu)}{n_\nu}\right),$$

and

$$Y^{***}_{\nu j} \sim Beta\big(a_\nu n_\nu - n_\nu j,\, n_\nu j + s_\nu\big), \qquad W_{\nu\ell} \sim \Gamma\left(r_{\nu\ell},\, a_\nu + \frac{s_\nu - \ell}{n_\nu}\right),$$

where, for $\nu = 1, \dots, m^*$,

$$r_{\nu\ell} = \begin{cases} k_\nu & \ell = 1, \dots, s_\nu \\ k_\nu + 1 + \left\lfloor \frac{s_\nu - \ell}{n_\nu} \right\rfloor & \ell = s_\nu + 1, \dots, n_\nu k_\nu + s_\nu\,. \end{cases} \tag{3.18}$$

and, for $j = 1, \dots, n_\nu k_\nu$,

$$\operatorname{mod}^*(j, n_\nu) = \begin{cases} \operatorname{mod}(j, n_\nu) & \text{if } \operatorname{mod}(j, n_\nu) \neq 0 \\ n_\nu & \text{if } \operatorname{mod}(j, n_\nu) = 0\,. \end{cases} \tag{3.19}$$

The Beta random variables involved in any double or triple product in (3.16) are all independent, as well as the random variables $W_{\nu\ell}$ *in (3.16) and (3.17).*

The exact distribution of Z is thus, in this case, an EGIG distribution of depth at most $\sum_{\nu=1}^{m^*} n_\nu k_\nu + s_\nu$, *with rate parameters* $a_\nu - \frac{s_\nu - j}{n_\nu}$ *and shape parameters* $r_{\nu j}$ $(j = 1, \dots, n_\nu k_\nu + s_\nu;\ \nu = 1, \dots, m^*)$.

Proof The first distributional equivalence in (3.16) is obtained by reversing the indexing in j, but while the second equivalence is obtained only by re-indexing, the third one needs the use of an extended product expression for the Gamma function. This extended or double product expression for the Gamma function is

$$\begin{aligned}\prod_{\ell=1}^{n}\prod_{j=1}^{k} \Gamma\left(a - \frac{\ell-1}{n}k - \frac{j}{n}\right) &= \prod_{j=1}^{k}\left\{\Gamma(an - nj)\,(2\pi)^{\frac{n-1}{2}}\, n^{\frac{1}{2}-(an-nj)}\right\} \\ &= \left\{\prod_{j=1}^{k}\Gamma(an - nj)\right\}(2\pi)^{\frac{n-1}{2}k}\, n^{\frac{k}{2}-kan+n\frac{k(k+1)}{2}}\,,\end{aligned} \tag{3.20}$$

which for $k = 1$ yields the usual product expression for the Gamma function and which when used on the expression for the c.f. of $W = -\log Z$ obtained from (3.15) yields

$$\Phi_W(t) = \prod_{\nu=1}^{m^*}\prod_{\ell=1}^{n_\nu}\prod_{j=1}^{k_\nu} \frac{\Gamma\left(a_\nu + \frac{\ell-1}{n_\nu} + \frac{s_\nu}{n_\nu}\right)}{\Gamma\left(a_\nu - \frac{(\ell-1)k_\nu}{n_\nu} - \frac{j}{n_\nu}\right)}\, \frac{\Gamma\left(a_\nu - \frac{(\ell-1)k_\nu}{n_\nu} - \frac{j}{n_\nu} - \mathrm{i}t\right)}{\Gamma\left(a_\nu + \frac{\ell-1}{n_\nu} + \frac{s_\nu}{n_\nu} - \mathrm{i}t\right)} \tag{3.21}$$

$$= \prod_{\nu=1}^{m^*}\prod_{j=1}^{k_\nu} \frac{\Gamma(a_\nu n_\nu + s_\nu)}{\Gamma(a_\nu n_\nu - n_\nu j)}\, \frac{\Gamma(a_\nu n_\nu - n_\nu j - n_\nu \mathrm{i}t)}{\Gamma(a_\nu n_\nu + s_\nu - n_\nu \mathrm{i}t)} \tag{3.22}$$

$$=\prod_{\nu=1}^{m^*}\prod_{j=1}^{k_\nu}\prod_{\ell=0}^{s_\nu+j\,n_\nu-1}(a_\nu n_\nu - j\,n_\nu+\ell)(a_\nu n_\nu - j\,n_\nu+\ell-n_\nu \mathrm{i}t)^{-1} \qquad (3.23)$$

$$=\prod_{\nu=1}^{m^*}\prod_{j=1}^{k_\nu}\prod_{\ell=0}^{s_\nu+jn_\nu-1}\left(a_\nu - j+\frac{\ell}{n_\nu}\right)\left(a_\nu - j+\frac{\ell}{n_\nu}-\mathrm{i}t\right)^{-1} \qquad (3.24)$$

$$=\prod_{\nu=1}^{m^*}\prod_{j=1}^{k_\nu}\prod_{\ell=1}^{s_\nu+jn_\nu}\left(a_\nu + \frac{s_\nu-\ell}{n_\nu}\right)\left(a_\nu + \frac{s_\nu-\ell}{n_\nu}-\mathrm{i}t\right)^{-1} \qquad (3.25)$$

$$=\prod_{\nu=1}^{m^*}\prod_{\ell=1}^{s_\nu+k_\nu n_\nu}\left(a_\nu + \frac{s_\nu-\ell}{n_\nu}\right)^{r_{\nu\ell}}\left(a_\nu + \frac{s_\nu-\ell}{n_\nu}-\mathrm{i}t\right)^{-r_{\nu\ell}}. \qquad (3.26)$$

From (3.21) to (3.22) we have used (3.20) and from (3.22) to (3.23) we use (3.6). Then, from (3.23) to (3.24) we only divide each term by n_ν, while from (3.24) to (3.25) we only reverse the indexing in ℓ. Then, finally from (3.25) to (3.26) we only have to either identify the different Exponential c.f.'s that appear in the c.f. of W and then count how many of each of them occur, in a similar manner to the one used by Coelho (1998, 2006) and by Arnold et al. (2013), or, equivalently, to identify the poles of the integrand function of the Meijer G function, or of the c.f. of W, and count their multiplicity, as in Wald and Brookner (1941). We may note that while the c.f. in (3.21) is the c.f. of the negative logarithm of the product in (3.15), the c.f. in (3.22) is the c.f. of the negative logarithm of the third product in (3.16) and the c.f. in (3.26) is the c.f. of the sum in (3.17). This shows that the distribution of W is a GIG distribution of depth at most $\sum_{\nu=1}^{m^*} k_\nu n_\nu + s_\nu$, with rate parameters $a_\nu - \frac{s_\nu - j}{n_\nu}$ and shape parameters $r_{\nu\ell}$ ($\ell = 1,\dots,k_\nu n_\nu + s_\nu$; $\nu = 1,\dots,m^*$), which shows that the exact distribution of Z is thus, in this case, an EGIG distribution with the same depth and the same rate and shape parameters. □

The particular case of Theorem 3.2 for $m^* = 1$ was studied by Coelho (2006), while the particular case for $k_\nu = 2$ was studied by Coelho (1998) and Arnold et al. (2013). The particular case of $m^* = 1$ in Theorem 3.3 was also treated in Arnold et al. (2013).

3.3 Applications of the Results in Theorems 3.1–3.3

In the following chapter the results in Theorems 3.1–3.3 are used to obtain the equivalence between several instances of the Meijer G and Fox H functions and their finite form representations obtained through the use of the p.d.f. and c.d.f. of EGIG distributions.

References

Arnold, B.C., Coelho, C.A., Marques, F.J.: The distribution of the product of powers of independent uniform random variables – a simple but useful tool to address and better understand the structure of some distributions. J. Multivar. Anal. **113**, 19–36 (2013)

Coelho, C.A.: The generalized integer Gamma distribution – a basis for distributions in multivariate statistics. J. Multivar. Anal. **64**, 86–102 (1998)

Coelho, C.A.: The exact and near-exact distributions of the product of independent Beta random variables whose second parameter is rational. J. Combin. Inform. System Sci. **31**, 21–44 (2006)

Wald, A., Brookner, R.J.: On the distribution of Wilks' statistic for testing the independence of several groups of variates. Ann. Math. Stat. **12**, 137–152 (1941)

Chapter 4
Finite Form Representations for Extended Instances of Meijer G and Fox H Functions

Abstract In this chapter the authors use the results in the three theorems in the previous chapter to obtain the p.d.f.'s and c.d.f.'s of all the products of independent Beta r.v.'s in those theorems in terms of the Meijer G and the Fox H functions as well as their finite forms based on the EGIG (Exponentiated Generalized Integer Gamma) p.d.f. and c.d.f., this way obtaining the equivalences among several instances of the Meijer G and Fox H functions and their finite representations through the EGIG p.d.f. and c.d.f..

Keywords Distribution of products of Beta r.v.'s · EGIG distribution · Gamma random variables

In the corollaries in this chapter we use the notation in (2.1) and (2.8) for the Meijer G and Fox H functions and the notation in (2.13) or (2.16) and (2.14) or (2.17) for the p.d.f. and c.d.f. of the EGIG distribution.

4.1 Two Corollaries Based on Theorems 3.1 and 3.2 in the Previous Chapter

From the distributional results in Theorems 3.1 and 3.2 in the previous chapter, two useful Corollaries can be obtained. They give the expressions for the p.d.f. and c.d.f. of the products of independent Beta r.v.'s in those two theorems in terms of the Meijer G and Fox H functions (Meijer, 1936a,b, 1941, 1946; Erdélyi et al., 1953; Fox, 1961), as well as in terms of the p.d.f. and c.d.f. of an EGIG distribution, thus establishing the equivalence of these representations. Given the finite form of these latter representations we are able in this way to obtain finite

C. A. Coelho, B. C. Arnold, *Finite Form Representations for Meijer G and Fox H Functions*, Lecture Notes in Statistics 223,
https://doi.org/10.1007/978-3-030-28790-0_4

form representations for the cases of the Meijer G and Fox H functions that appear in these corollaries.

Corollary 4.1 *For the general case in Theorem 3.1 we may write, for $0 < z \le 1$, the p.d.f. of Z as,*

$$\left\{\prod_{\nu=1}^{m^*}\prod_{\ell=1}^{n_\nu}\prod_{j=1}^{k_{\nu\ell}}\frac{\Gamma\left(a_\nu-\frac{j-m_{\nu\ell}+\sum_{r=1}^{\ell-1}k_{\nu r}}{k_{\nu\ell}}\right)}{\Gamma\left(a_\nu-\frac{j+\sum_{r=1}^{\ell-1}k_{\nu r}}{k_{\nu\ell}}\right)}\right\}$$

$$\times\, G^{p^*,0}_{p^*,p^*}\left(\begin{array}{c|c}\left\{a_\nu-\frac{j-m_{\nu\ell}+\sum_{r=1}^{\ell-1}k_{\nu r}}{k_{\nu\ell}}-1\right\}_{\substack{\nu=1:m^*\\ \ell=1:n_\nu\\ j=1:k_{\nu\ell}}} & \\ & z\\ \left\{a_\nu-\frac{j+\sum_{r=1}^{\ell-1}k_{\nu r}}{k_{\nu\ell}}-1\right\}_{\substack{\nu=1:m^*\\ \ell=1:n_\nu\\ j=1:k_{\nu\ell}}} & \end{array}\right)$$

$$=\left\{\prod_{\nu=1}^{m^*}\prod_{\ell=1}^{n_\nu}\frac{\Gamma\left(a_\nu k_{\nu\ell}+m_{\nu\ell}-\sum_{r=1}^{\ell}k_{\nu r}\right)}{\Gamma\left(a_\nu k_{\nu\ell}-\sum_{r=1}^{\ell}k_{\nu r}\right)}\right\}$$

$$\times\, H^{p^{**},0}_{p^{**},p^{**}}\left(\begin{array}{c|c}\left\{\left((a_\nu-1)k_{\nu\ell}+m_{\nu\ell}-\sum_{r=1}^{\ell}k_{\nu r},\,k_{\nu\ell}\right)\right\}_{\substack{\nu=1:m^*\\ \ell=1:n_\nu}} & \\ & z\\ \left\{\left((a_\nu-1)k_{\nu\ell}-\sum_{r=1}^{\ell}k_{\nu r},\,k_{\nu\ell}\right)\right\}_{\substack{\nu=1:m^*\\ \ell=1:n_\nu}} & \end{array}\right)$$

$$=f^{EGIG}\left(z\,\middle|\,\widetilde{\widetilde{\left\{\{1\}_{\substack{\nu=1:m^*\\ \ell=1:n_\nu\\ j=0:m_{\nu\ell}-1}}\right\}}}\,;\,\widetilde{\left\{\left\{a_\nu+\frac{j-\sum_{r=1}^{\ell}k_{\nu r}}{k_{\nu\ell}}\right\}_{\substack{\nu=1:m^*\\ \ell=1:n_\nu\\ j=0:m_{\nu\ell}-1}}\right\}}\,;\,g\le\sum_{\nu=1}^{m^*}\sum_{\ell=1}^{n_\nu}m_{\nu\ell}\right),$$

where we use the notation in (2.11) and (2.12) and where we also use

$$p^*=\sum_{\nu=1}^{m^*}\sum_{\ell=1}^{n_\nu}k_{\nu\ell}\,,\qquad\text{and}\qquad p^{**}=\sum_{\nu=1}^{m^*}n_\nu\,,$$

and we may write the c.d.f. of Z as

$$\left\{\prod_{\nu=1}^{m^*}\prod_{\ell=1}^{n_\nu}\prod_{j=1}^{k_{\nu\ell}}\frac{\Gamma\left(a_\nu-\frac{j-m_{\nu\ell}+\sum_{r=1}^{\ell-1}k_{\nu r}}{k_{\nu\ell}}\right)}{\Gamma\left(a_\nu-\frac{j+\sum_{r=1}^{\ell-1}k_{\nu r}}{k_{\nu\ell}}\right)}\right\}$$

$$\times\, G^{p^*,1}_{p^*+1,p^*+1}\left(\begin{array}{c|c} 1,\left\{a_\nu-\frac{j-m_{\nu\ell}+\sum_{r=1}^{\ell-1}k_{\nu r}}{k_{\nu\ell}}\right\}_{\substack{\nu=1:m^*\\ \ell=1:n_\nu\\ j=1:k_{\nu\ell}}} & \\ & z \\ \left\{a_\nu-\frac{j+\sum_{r=1}^{\ell-1}k_{\nu r}}{k_{\nu\ell}}\right\}_{\substack{\nu=1:m^*\\ \ell=1:n_\nu\\ j=1:k_{\nu\ell}}},\ 0 & \end{array}\right)$$

$$=\left\{\prod_{\nu=1}^{m^*}\prod_{\ell=1}^{n_\nu}\frac{\Gamma\left(a_\nu k_{\nu\ell}+m_{\nu\ell}-\sum_{r=1}^{\ell}k_{\nu r}\right)}{\Gamma\left(a_\nu k_{\nu\ell}-\sum_{r=1}^{\ell}k_{\nu r}\right)}\right\}$$

$$\times\, H^{p^{**},1}_{p^{**}+1,p^{**}+1}\left(\begin{array}{c|c}\left\{(1,1),\left\{(a_\nu k_{\nu\ell}+m_{\nu\ell}-\sum_{r=1}^{\ell}k_{\nu r},k_{\nu\ell})\right\}_{\substack{\nu=1:m^*\\ \ell=1:n_\nu}}\right\} & \\ & z \\ \left\{\left\{(a_\nu k_{\nu\ell}-\sum_{r=1}^{\ell}k_{\nu r},k_{\nu\ell})\right\}_{\substack{\nu=1:m^*\\ \ell=1:n_\nu}},(0,1)\right\} & \end{array}\right)$$

$$=F^{EGIG}\left(z\,\middle|\,\approx\!\left[\{1\}_{\substack{\nu=1:m^*\\ \ell=1:n_\nu\\ j=0:m_{\nu\ell}-1}}\right]^{\approx};\ \sim\!\left[\left\{a_\nu+\frac{j-\sum_{r=1}^{\ell}k_{\nu r}}{k_{\nu\ell}}\right\}_{\substack{\nu=1:m^*\\ \ell=1:n_\nu\\ j=0:m_{\nu\ell}-1}}\right]^{\sim};\ g\le\sum_{\nu=1}^{m^*}\sum_{\ell=1}^{n_\nu}m_{\nu\ell}\right).$$

Proof Since this is a corollary of Theorem 3.1 in the previous chapter, there is indeed no need for a formal proof. In order to obtain the expressions for the p.d.f. and c.d.f. of Z in terms of the Meijer G function one only has to consider expressions (2.4), (2.6), (2.13), and (2.14) in Chap. 2 and the distribution of the r.v. Z in (3.1) and that of the r.v.'s $Y_{\nu\ell j}$ in that same expression, while to obtain the p.d.f. and c.d.f. of Z in terms of the Fox H function one only has to consider expressions (2.9), (2.10), (2.16), and (2.17) in Chap. 2 and the distribution of the r.v. Z in (3.2) and that of the r.v.'s $Y^{***}_{\nu\ell}$ in that same expression. Then, in order to obtain the expressions for the p.d.f. and the c.d.f. of Z in terms of the EGIG p.d.f. and c.d.f., one has to consider the distribution of Z as given by (3.2) in terms of the r.v.'s $W_{\nu\ell j}$ together with the distribution of these r.v.'s, and the expressions for the p.d.f. and c.d.f. of the EGIG distribution in (2.23) and (2.24) in Appendix of Chap. 2 and the fact that in Theorem 3.1 the last expression for the distribution of the r.v. Z shows that this distribution is the same as that of the exponential of a sum of independent Exponential r.v.'s with the rate parameters given in the body of Theorem 3.1. Since some of these parameters may eventually bear equal values, this yields for Z the EGIG distribution with p.d.f. and c.d.f. given in the corollary.

□

Corollary 4.2 *For the particular case in Theorem 3.2 we may write the p.d.f. of Z as*

$$\left\{\prod_{\nu=1}^{m^*}\prod_{\ell=1}^{n_\nu}\prod_{j=1}^{k_\nu}\frac{\Gamma\left(a_\nu+1-\ell+\frac{m_\nu-j}{k_\nu}\right)}{\Gamma\left(a_\nu+1-\ell-\frac{j}{k_\nu}\right)}\right\}$$

$$\times\, G^{p^*,0}_{p^*,p^*}\left(\begin{array}{c}\left\{a_\nu+1-\ell+\frac{m_\nu-j}{k_\nu}-1\right\}_{\substack{\nu=1:m^*\\ \ell=1:n_\nu\\ j=1:k_\nu}}\\ \left\{a_\nu+1-\ell-\frac{j}{k_\nu}-1\right\}_{\substack{\nu=1:m^*\\ \ell=1:n_\nu\\ j=1:k_\nu}}\end{array}\middle|\, z\right)$$

$$=\left\{\prod_{\nu=1}^{m^*}\prod_{\ell=1}^{n_\nu}\frac{\Gamma\big((a_\nu-\ell)k_\nu+m_\nu\big)}{\Gamma\big((a_\nu-\ell)k_\nu\big)}\right\}$$

$$\times\, H^{p^{**},0}_{p^{**},p^{**}}\left(\begin{array}{c}\big\{((a_\nu-\ell-1)k_\nu+m_\nu,\,k_\nu)\big\}_{\substack{\nu=1:m^*\\ \ell=1:n_\nu}}\\ \big\{((a_\nu-\ell-1)k_\nu,\,k_\nu)\big\}_{\substack{\nu=1:m^*\\ \ell=1:n_\nu}}\end{array}\middle|\, z\right)$$

$$=f^{\,EGIG}\left(z\,\middle|\,\widetilde{\widetilde{\left\{\{r_{\nu j}\}_{\substack{\nu=1:m^*\\ j=1:m_\nu+k_\nu(n_\nu-1)}}\right\}}};\ \widetilde{\left\{\left\{a_\nu-n_\nu+\frac{j-1}{k_\nu}\right\}_{\substack{\nu=1:m^*\\ j=1:m_\nu+k_\nu(n_\nu-1)}}\right\}};\ g\leq\sum_{\nu=1}^{m^*}m_\nu+k_\nu(n_\nu-1)\right),$$

and the c.d.f. as

$$\left\{\prod_{\nu=1}^{m^*}\prod_{\ell=1}^{n_\nu}\prod_{j=1}^{k_\nu}\frac{\Gamma\left(a_\nu+1-\ell+\frac{m_\nu-j}{k_\nu}\right)}{\Gamma\left(a_\nu+1-\ell-\frac{j}{k_\nu}\right)}\right\}$$

$$\times\, G^{p^*,1}_{p^*+1,p^*+1}\left(\begin{array}{c}\left\{1,\left\{a_\nu+1-\ell+\frac{m_\nu-j}{k_\nu}\right\}_{\substack{\nu=1:m^*\\ \ell=1:n_\nu\\ j=1:k_\nu}}\right\}\\ \left\{\left\{a_\nu+1-\ell-\frac{j}{k_\nu}\right\}_{\substack{\nu=1:m^*\\ \ell=1:n_\nu\\ j=1:k_\nu}},0\right\}\end{array}\middle|\, z\right)$$

$$
= \left\{ \prod_{\nu=1}^{m^*} \prod_{\ell=1}^{n_\nu} \frac{\Gamma\left((a_\nu-\ell)k_\nu+m_\nu\right)}{\Gamma\left((a_\nu-\ell)k_\nu\right)} \right\}
$$

$$
\times H^{p^{**},1}_{p^{**}+1,p^{**}+1}\left(\left. \begin{Bmatrix} (1,1), \{((a_\nu-\ell)k_\nu+m_\nu, k_\nu)\}_{\substack{\nu=1:m^* \\ \ell=1:n_\nu}} \\ \{((a_\nu-\ell)k_\nu, k_\nu)\}_{\substack{\nu=1:m^* \\ \ell=1:n_\nu}}, (0,1) \end{Bmatrix} \right| z \right)
$$

$$
= F^{EGIG}\left(z \,\middle|\, \overset{\approx}{\left\{ \{r_{\nu j}\}_{\substack{\nu=1:m^* \\ j=1:m_\nu+k_\nu(n_\nu-1)}} \right\}}; \widetilde{\left\{ \left\{a_\nu - n_\nu + \frac{j-1}{k_\nu}\right\}_{\substack{\nu=1:m^* \\ j=1:m_\nu+k_\nu(n_\nu-1)}} \right\}}; \; g \le \sum_{\nu=1}^{m^*} m_\nu + k_\nu(n_\nu-1) \right),
$$

for

$$
p^* = \sum_{\nu=1}^{m^*} n_\nu k_\nu\,, \quad \text{and} \quad p^{**} = \sum_{\nu=1}^{m^*} n_\nu\,,
$$

and where $0 < z \le 1$ *represents the running value of the r.v.* Z*,* $r_{\nu j}$ *is given by (3.13)–(3.14), and*

$$
\overset{\approx}{\left\{ \{r_{\nu j}\}_{\substack{\nu=1:m^* \\ j=1:m_\nu+k_\nu(n_\nu-1)}} \right\}}
$$

is the ordered set which h-th element ($h = 1, \ldots, g \le \sum_{\nu=1}^{m^*} m_\nu + k_\nu(n_\nu-1)$) *is the sum of all* $r_{\nu j}$ *associated with rate parameters* $a_\nu - n_\nu + \frac{j-1}{k_\nu}$ *with the same value as the h-th element of the ordered set*

$$
\widetilde{\left\{ \left\{a_\nu - n_\nu + \frac{j-1}{k_\nu}\right\}_{\substack{\nu=1:m^* \\ j=1:m_\nu+k_\nu(n_\nu-1)}} \right\}}.
$$

Proof The proof of this corollary follows similar lines to those of the proof of Corollary 4.1, replacing the occurrences of Theorem 3.1 by Theorem 3.2, and those of (3.1) and (3.2) respectively by (3.9) and (3.10) and the references to the r.v.'s $Y^{***}_{\nu\ell}$ and $W_{\nu\ell j}$ in (3.2) respectively by references to r.v.'s $Y^{***}_{\nu\ell}$ and $W_{\nu j}$ in (3.10). □

Through the use of (3.4), the product of Gamma functions that precedes the p.d.f. and c.d.f. representations in Corollary 4.1 in terms of Meijer G functions may be alternatively written as

$$\prod_{\nu=1}^{m^*}\prod_{\ell=1}^{n_\nu} k_{\nu\ell}^{-m_{\nu\ell}}\,\frac{\Gamma\left(a_\nu k_{\nu\ell}+m_{\nu\ell}-\sum_{r=1}^{\ell}k_{\nu r}\right)}{\Gamma\left(a_\nu k_{\nu\ell}-\sum_{r=1}^{\ell}k_{\nu r}\right)},$$

while the product of Gamma functions that precedes the p.d.f. and c.d.f. representations in terms of Meijer G functions in Corollary 4.2 may be alternatively written as

$$\prod_{\nu=1}^{m^*}\prod_{\ell=1}^{n_\nu} k_\nu^{-m_\nu}\,\frac{\Gamma\left((a_\nu-\ell)k_\nu+m_\nu\right)}{\Gamma\left((a_n u-\ell)k_\nu\right)},$$

this way giving to these products a more similar representation to the one used in the expressions involving the Fox H function.

Also, the Meijer G functions associated with the p.d.f. and the c.d.f. of Z in the statement of Corollary 4.1 are derived from the product of independent Beta r.v.'s in (3.1), but we may derive equivalent expressions involving Meijer G functions by considering the second product of independent Beta r.v.'s in (3.2), as

$$\left\{\prod_{\nu=1}^{m^*}\prod_{j=1}^{p_\nu}\frac{\Gamma\left(a_\nu-\frac{j-m_{\nu j^*_{\nu j}}}{k_{\nu j^*_{\nu j}}}\right)}{\Gamma\left(a_\nu-\frac{j}{k_{\nu j^*_{\nu j}}}\right)}\right\} G^{p^*,0}_{p^*,p^*}\left(\begin{array}{c}\left\{a_\nu-\frac{j-m_{\nu j^*_{\nu j}}}{k_{\nu j^*_{\nu j}}}-1\right\}_{\substack{\nu=1:m^*\\ j=1:p_\nu}}\\ \left\{a_\nu-\frac{j}{k_{\nu j^*_{\nu j}}}-1\right\}_{\substack{\nu=1:m^*\\ j=1:p_\nu}}\end{array}\middle|\, z\right)$$

for the p.d.f. of Z, and

$$\left\{\prod_{\nu=1}^{m^*}\prod_{j=1}^{p_\nu}\frac{\Gamma\left(a_\nu-\frac{j-m_{\nu j^*_{\nu j}}}{k_{\nu j^*_{\nu j}}}\right)}{\Gamma\left(a_\nu-\frac{j}{k_{\nu j^*_{\nu j}}}\right)}\right\} G^{p^*,1}_{p^*+1,p^*+1}\left(\begin{array}{c}\left\{1,\left\{a_\nu-\frac{j-m_{\nu j^*_{\nu j}}}{k_{\nu j^*_{\nu j}}}\right\}_{\substack{\nu=1:m^*\\ j=1:p_\nu}}\right\}\\ \left\{\left\{a_\nu-\frac{j}{k_{\nu j^*_{\nu j}}}\right\}_{\substack{\nu=1:m^*\\ j=1:p_\nu}},0\right\}\end{array}\middle|\, z\right)$$

for the c.d.f. of Z, where $p_\nu=\sum_{\ell=1}^{n_\nu}k_{\nu\ell}$ ($\nu=1,\dots,m^*$) and $p^*=\sum_{\nu=1}^{m^*}p_\nu$ are defined as in Theorem 3.1 and Corollary 4.1.

Concerning the statement of the p.d.f. and c.d.f. of Z in Corollary 4.2, by considering the second product in (3.10), we may obtain equivalent representations for the p.d.f. and c.d.f. in terms of Meijer G functions as

$$\left\{\prod_{\nu=1}^{m^*}\prod_{j=1}^{p_\nu}\frac{\Gamma\left(a_\nu-\frac{j-m_\nu}{k_\nu}\right)}{\Gamma\left(a_\nu-\frac{j}{k_\nu}\right)}\right\} G^{p^*,0}_{p^*,p^*}\left(\begin{array}{c}\left\{a_\nu-\frac{j-m_\nu}{k_\nu}-1\right\}_{\substack{\nu=1:m^*\\ j=1:p_\nu}}\\ \left\{a_\nu-\frac{j}{k_\nu}-1\right\}_{\substack{\nu=1:m^*\\ j=1:p_\nu}}\end{array}\middle|\, z\right)$$

for the p.d.f., and

$$\left\{\prod_{\nu=1}^{m^*}\prod_{j=1}^{p_\nu}\frac{\Gamma\left(a_\nu-\frac{j-m_\nu}{k_\nu}\right)}{\Gamma\left(a_\nu-\frac{j}{k_\nu}\right)}\right\} G^{p^*,1}_{p^*+1,p^*+1}\left(\begin{array}{c}\left\{1,\left\{a_\nu-\frac{j-m_\nu}{k_\nu}\right\}_{\substack{\nu=1:m^*\\ j=1:p_\nu}}\right\}\\ \left\{\left\{a_\nu-\frac{j}{k_\nu}\right\}_{\substack{\nu=1:m^*\\ j=1:p_\nu}},0\right\}\end{array}\middle|\, z\right)$$

for the c.d.f., where $p_\nu = n_\nu k_\nu$ and $p^* = \sum_{\nu=1}^{m^*} p_\nu$.

The computability of these alternative expressions for the Meijer G functions is similar to that of the previous versions, with very similar computing times to those exhibited by the versions in the body of Corollaries 4.1 and 4.2.

4.2 A Third Corollary Based on Theorem 3.3 in the Previous Chapter

Similar to what happens in Sect. 4.1, in this section we obtain a third corollary from the distributional results in Theorem 3.3 in the previous chapter. This corollary gives us expressions for the p.d.f. and c.d.f. of the products of independent Beta r.v.'s in that theorem in terms of the Meijer G and Fox H functions, as well as in terms of the p.d.f. and c.d.f. of the EGIG distribution. By establishing the equivalence of these representations, and given the finite form of the latter representations we are able in this way to obtain finite form representations for the cases of the Meijer G and Fox H functions for which Theorem 3.3 applies and which will prove to be extremely efficient and thus useful in computational terms.

Corollary 4.3 *For the case in Theorem 3.3 we may write the p.d.f. of Z, for* $0 < z \leq 1$, *as*

$$\left\{ \prod_{\nu=1}^{m^*} \prod_{\ell=1}^{n_\nu} \prod_{j=1}^{k_\nu} \frac{\Gamma\left(a_\nu + \frac{\ell+s_\nu-1}{n_\nu}\right)}{\Gamma\left(a_\nu - \frac{(\ell-1)k_\nu+j}{n_\nu}\right)} \right\} G^{p^*,0}_{p^*,p^*}\left(\begin{array}{c} \left\{a_\nu + \frac{\ell+s_\nu-1}{n_\nu} - 1\right\}_{\substack{\nu=1:m^* \\ \ell=1:n_\nu \\ j=1:k_\nu}} \\ \left\{a_\nu - \frac{(\ell-1)k_\nu+j}{n_\nu} - 1\right\}_{\substack{\nu=1:m^* \\ \ell=1:n_\nu \\ j=1:k_\nu}} \end{array} \middle| \, z \right)$$

$$= \left\{ \prod_{\nu=1}^{m^*} \prod_{\ell=1}^{n_\nu} \frac{\Gamma\left(a_\nu n_\nu + s_\nu\right)}{\Gamma\left(a_\nu n_\nu - n_\nu j\right)} \right\} H^{p^{**},0}_{p^{**},p^{**}}\left(\begin{array}{c} \left\{((a_\nu - 1)n_\nu + s_\nu, n_\nu)\right\}_{\substack{\nu=1:m^* \\ j=1:k_\nu}} \\ \left\{((a_\nu - 1)n_\nu - j\, n_\nu, n_\nu)\right\}_{\substack{\nu=1:m^* \\ j=1:k_\nu}} \end{array} \middle| \, z \right)$$

$$= f^{EGIG}\left(z \,\middle|\, \widetilde{\widetilde{\left\{ \{r_{\nu\ell}\}_{\substack{\nu=1:m^* \\ \ell=1:n_\nu k_\nu+s_\nu}} \right\}}} ; \widetilde{\left\{ \left\{a_\nu + \frac{s_\nu - \ell}{n_\nu}\right\}_{\substack{\nu=1:m^* \\ \ell=1:n_\nu k_\nu+s_\nu}} \right\}} ; \; g \leq \sum_{\nu=1}^{m^*} n_\nu k_\nu + s_\nu \right),$$

and the c.d.f. as

$$\left\{ \prod_{\nu=1}^{m^*} \prod_{\ell=1}^{n_\nu} \prod_{j=1}^{k_\nu} \frac{\Gamma\left(a_\nu + \frac{\ell+s_\nu-1}{n_\nu}\right)}{\Gamma\left(a_\nu - \frac{(\ell-1)k_\nu+j}{n_\nu}\right)} \right\} G^{p^*,1}_{p^*+1,p^*+1}\left(\begin{array}{c} \left\{1, \left\{a_\nu + \frac{\ell+s_\nu-1}{n_\nu}\right\}_{\substack{\nu=1:m^* \\ \ell=1:n_\nu \\ j=1:k_\nu}} \right\} \\ \left\{ \left\{a_\nu - \frac{(\ell-1)k_\nu+j}{n_\nu}\right\}_{\substack{\nu=1:m^* \\ \ell=1:n_\nu \\ j=1:k_\nu}}, 0 \right\} \end{array} \middle| \, z \right)$$

$$= \left\{ \prod_{\nu=1}^{m^*} \prod_{\ell=1}^{n_\nu} \frac{\Gamma\left(a_\nu n_\nu + s_\nu\right)}{\Gamma\left(a_\nu n_\nu - n_\nu j\right)} \right\}$$

$$\times H^{p^{**},1}_{p^{**}+1,p^{**}+1}\left(\begin{array}{c} \left\{(1,1), \left\{(a_\nu n_\nu + s_\nu, n_\nu)\right\}_{\substack{\nu=1,\dots,m^* \\ j=1,\dots,k_\nu}} \right\} \\ \left\{ \left\{(a_\nu n_\nu - j\, n_\nu, n_\nu)\right\}_{\substack{\nu=1,\dots,m^* \\ j=1,\dots,k_\nu}}, (0,1) \right\} \end{array} \middle| \, z \right)$$

$$= F^{EGIG}\left(z \,\middle|\, \overset{\approx}{\left\{ \{r_{\nu\ell}\}_{\substack{\nu=1:m^* \\ \ell=1:n_\nu k_\nu+s_\nu}} \right\}} ; \widetilde{\left\{ \left\{ a_\nu + \frac{s_\nu-\ell}{n_\nu} \right\}_{\substack{\nu=1:m^* \\ \ell=1:n_\nu k_\nu+s_\nu}} \right\}} ; \; g \le \sum_{\nu=1}^{m^*} n_\nu k_\nu + s_\nu \right),$$

for

$$p^* = \sum_{\nu=1}^{m^*} n_\nu\, k_\nu\,, \qquad \text{and} \qquad p^{**} = \sum_{\nu=1}^{m^*} k_\nu\,,$$

$r_{\nu\ell}$ given by (3.18) and where

$$\widetilde{\left\{ \left\{ a_\nu + \frac{s_\nu-\ell}{n_\nu} \right\}_{\substack{\nu=1:m^* \\ \ell=1:n_\nu k_\nu+s_\nu}} \right\}} \quad \text{and} \quad \overset{\approx}{\left\{ \{r_{\nu\ell}\}_{\substack{\nu=1:m^* \\ \ell=1:m_\nu+k_\nu(n_\nu-1)}} \right\}}$$

have similar definitions to the corresponding vectors in Corollary 4.2.

Proof It is enough to follow similar lines to those in the proof of Corollary 4.1, replacing the occurrences of Theorem 3.1 by Theorem 3.3, and those of (3.1) and (3.2) respectively by (3.15) and (3.16) and the references to the r.v.'s $Y^{***}_{\nu\ell}$ and $W_{\nu\ell j}$ in (3.2) respectively by references to r.v.'s $Y^{***}_{\nu j}$ and $W_{\nu\ell}$ in (3.16). □

Now, through the use of (3.20), also in the case of this corollary, the product of Gamma functions that precedes the p.d.f. and c.d.f. representations in terms of Meijer G functions may be alternatively written as

$$\prod_{\nu=1}^{m^*}\prod_{\ell=1}^{n_\nu} k_\nu^{-m_\nu} \frac{\Gamma\big((a_\nu-\ell)k_\nu+m_\nu\big)}{\Gamma\big((a_\nu-\ell)k_\nu\big)}\,.$$

And also the Meijer G functions associated with the p.d.f. and the c.d.f. of Z in the statement of Corollary 4.3, which are derived from the product of independent Beta r.v.'s in (3.15), find equivalent expressions involving Meijer G functions derived from the second product of independent Beta r.v.'s in (3.16), as

$$\left\{ \prod_{\nu=1}^{m^*}\prod_{j=1}^{p_\nu} \frac{\Gamma\left(a_\nu - \frac{s_\nu-1+\text{mod}^*(j,n_\nu)}{n_\nu}\right)}{\Gamma\left(a_\nu - \frac{j}{n_\nu}\right)} \right\} \times G^{p^*,0}_{p^*,p^*}\left(\begin{array}{c} \left\{ a_\nu - \frac{s_\nu-1+\text{mod}^*(i,n_\nu)}{n_\nu} - 1 \right\}_{\substack{\nu=1,\dots,m^* \\ j=1,\dots,p_\nu}} \\ \left\{ a_\nu - \frac{j}{n_\nu} - 1 \right\}_{\substack{\nu=1,\dots,m^* \\ j=1,\dots,p_\nu}} \end{array} \middle|\, z \right)$$

for the p.d.f., and

$$\left\{\prod_{\nu=1}^{m^*}\prod_{j=1}^{p_\nu}\frac{\Gamma\left(a_\nu-\frac{s_\nu-1+\text{mod}^*(j,n_\nu)}{n_\nu}\right)}{\Gamma\left(a_\nu-\frac{j}{n_\nu}\right)}\right\}$$

$$\times\, G^{p^*,1}_{p^*+1,p^*+1}\left(\begin{array}{c}\left\{1,\left\{a_\nu-\frac{s_\nu-1+\text{mod}^*(i,n_\nu)}{n_\nu}\right\}_{\substack{\nu=1,\dots,m^*\\ j=1,\dots,p_\nu}}\right\}\\ \left\{\left\{a_\nu-\frac{j}{n_\nu}\right\}_{\substack{\nu=1,\dots,m^*\\ j=1,\dots,p_\nu}},0\right\}\end{array}\middle|\, z\right)$$

for the c.d.f. of Z, where the function mod$^*(\cdot,\cdot)$ is defined in (3.19).

4.3 Further Corollaries Based on Combinations of Theorems 3.1–3.3

We may combine the results in Theorems 3.1 and 3.3, or rather, the results in Corollaries 4.1 and 4.3 in the previous sections to identify even more elaborate situations for which there is a useful equivalence between the representations which use the Meijer G or Fox H functions and the representations using the EGIG p.d.f. or c.d.f.. Although right now these cases may seem a bit artificial, we will see in the next chapter that they are in fact very useful in handling the exact distribution of several l.r.t. statistics used in Multivariate Analysis.

Corollary 4.4 *For a r.v. Z whose distribution is that of the product of two independent products of independent Beta r.v.'s, one of the type of that in Theorem 3.1 and the other one of the type of that in Theorem 3.3, using for all the parameters of the r.v.'s in Theorem 3.3 a starred notation, and m^{**} for m^*, the p.d.f. is given by*

$$\left\{\prod_{\nu=1}^{m^*}\prod_{\ell=1}^{n_\nu}\prod_{j=1}^{k_{\nu\ell}}\frac{\Gamma\left(a_\nu-\frac{j-m_{\nu\ell}+\sum_{r=1}^{\ell-1}k_{\nu r}}{k_{\nu\ell}}\right)}{\Gamma\left(a_\nu-\frac{j+\sum_{r=1}^{\ell-1}k_{\nu r}}{k_{\nu\ell}}\right)}\right\}\left\{\prod_{\nu=1}^{m^{**}}\prod_{\ell=1}^{n^*_\nu}\prod_{j=1}^{k^*_\nu}\frac{\Gamma\left(a^*_\nu+\frac{\ell+s^*_\nu-1}{n^*_\nu}\right)}{\Gamma\left(a^*_\nu-\frac{(\ell-1)k^*_\nu+j}{n^*_\nu}\right)}\right\}$$

$$\times\, G^{p^*,0}_{p^*,p^*}\left(\begin{array}{c}\left\{\left\{a_\nu-\frac{j-m_{\nu\ell}+\sum_{r=1}^{\ell-1}k_{\nu r}}{k_{\nu\ell}}-1\right\}_{\substack{\nu=1:m^*\\ \ell=1:n_\nu\\ j=1:k_{\nu\ell}}},\left\{a^*_\nu+\frac{\ell+s^*_\nu-1}{n^*_\nu}-1\right\}_{\substack{\nu=1:m^{**}\\ \ell=1:n^*_\nu\\ j=1:k^*_\nu}}\right\}\\ \left\{\left\{a_\nu-\frac{j+\sum_{r=1}^{\ell-1}k_{\nu r}}{k_{\nu\ell}}-1\right\}_{\substack{\nu=1:m^*\\ \ell=1:n_\nu\\ j=1:k_{\nu\ell}}},\left\{a^*_\nu-\frac{(\ell-1)k^*_\nu+j}{n^*_\nu}-1\right\}_{\substack{\nu=1:m^{**}\\ \ell=1:n^*_\nu\\ j=1:k^*_\nu}}\right\}\end{array}\middle|\, z\right)$$

$$= \left\{ \prod_{\nu=1}^{m^*} \prod_{\ell=1}^{n_\nu} \frac{\Gamma\left(a_\nu k_{\nu\ell} + m_{\nu\ell} - \sum_{r=1}^{\ell} k_{\nu r}\right)}{\Gamma\left(a_\nu k_{\nu\ell} - \sum_{r=1}^{\ell} k_{\nu r}\right)} \right\} \left\{ \prod_{\nu=1}^{m^{**}} \prod_{\ell=1}^{n_\nu^*} \frac{\Gamma\left(a_\nu^* n_\nu^* + s_\nu^*\right)}{\Gamma\left(a_\nu^* n_\nu^* - n_\nu^* j\right)} \right\}$$

$$\times H_{p^{**},p^{**}}^{p^{**},0}\left(\begin{Bmatrix} \left\{\left((a_\nu-1)k_{\nu\ell}+m_{\nu\ell}-\sum_{r=1}^{\ell} k_{\nu r}, k_{\nu\ell}\right)\right\}_{\substack{\nu=1:m^*\\ \ell=1:n_\nu}}, \left\{\left((a_\nu^*-1)n_\nu^*+s_\nu^*, n_\nu^*\right)\right\}_{\substack{\nu=1:m^{**}\\ j=1:k_\nu^*}} \\ \left\{\left((a_\nu-1)k_{\nu\ell}-\sum_{r=1}^{\ell} k_{\nu r}, k_{\nu\ell}\right)\right\}_{\substack{\nu=1:m^*\\ \ell=1:n_\nu}}, \left\{\left((a_\nu^*-1)n_\nu^*-jn_\nu^*, n_\nu^*\right)\right\}_{\substack{\nu=1:m^{**}\\ j=1:k_\nu^*}} \end{Bmatrix} \middle| z \right)$$

$$= f^{EGIG}\left(z \,\middle|\, \widetilde{\widetilde{\left\{ \{1\}_{\substack{\nu=1:m^*\\ \ell=1:n_\nu\\ i=0:m_{\nu\ell}-1}}, \{r_{\nu j}^*\}_{\substack{\nu=1:m^{**}\\ j=1:n_\nu^* k_\nu^*+s_\nu^*}} \right\}}}; \right.$$

$$\widetilde{\left\{ \left\{a_\nu + \frac{i-\sum_{r=1}^{\ell} k_{\nu r}}{k_{\nu\ell}}\right\}_{\substack{\nu=1:m^*\\ \ell=1:n_\nu\\ i=0:m_{\nu\ell}-1}}, \left\{a_\nu^* + \frac{s_\nu^*-j}{n_\nu^*}\right\}_{\substack{\nu=1:m^{**}\\ j=1:n_\nu^* k_\nu^*+s_\nu^*}} \right\}};$$

$$\left. g \le \sum_{\nu=1}^{m^*} \sum_{\ell=1}^{n_\nu} m_{\nu\ell} + \sum_{\nu=1}^{m^{**}} n_\nu^* k_\nu^* + s_\nu^* \right),$$

and the c.d.f. by

$$\left\{ \prod_{\nu=1}^{m^*} \prod_{\ell=1}^{n_\nu} \prod_{j=1}^{k_{\nu\ell}} \frac{\Gamma\left(a_\nu - \frac{j-m_{\nu\ell}+\sum_{r=1}^{\ell-1} k_{\nu r}}{k_{\nu\ell}}\right)}{\Gamma\left(a_\nu - \frac{j+\sum_{r=1}^{\ell-1} k_{\nu r}}{k_{\nu\ell}}\right)} \right\} \left\{ \prod_{\nu=1}^{m^{**}} \prod_{\ell=1}^{n_\nu^*} \prod_{j=1}^{k_\nu^*} \frac{\Gamma\left(a_\nu^* + \frac{\ell+s_\nu^*-1}{n_\nu^*}\right)}{\Gamma\left(a_\nu^* - \frac{(\ell-1)k_\nu^*+j}{n_\nu^*}\right)} \right\}$$

$$\times G_{p^*+1,p^*+1}^{p^*,1}\left(\begin{Bmatrix} 1, \left\{a_\nu - \frac{j-m_{\nu\ell}+\sum_{r=1}^{\ell-1} k_{\nu r}}{k_{\nu\ell}}\right\}_{\substack{\nu=1:m^*\\ \ell=1:n_\nu\\ j=1:k_{\nu\ell}}}, \left\{a_\nu^* + \frac{\ell+s_\nu^*-1}{n_\nu^*}\right\}_{\substack{\nu=1:m^{**}\\ \ell=1:n_\nu^*\\ j=1:k_\nu^*}} \\ \left\{a_\nu - \frac{j+\sum_{r=1}^{\ell-1} k_{\nu r}}{k_{\nu\ell}}\right\}_{\substack{\nu=1:m^*\\ \ell=1:n_\nu\\ j=1:k_{\nu\ell}}}, \left\{a_\nu^* - \frac{(\ell-1)k_\nu^*+j}{n_\nu^*}\right\}_{\substack{\nu=1:m^{**}\\ \ell=1:n_\nu^*\\ j=1:k_\nu^*}}, 0 \end{Bmatrix} \middle| z \right)$$

$$= \left\{ \prod_{\nu=1}^{m^*} \prod_{\ell=1}^{n_\nu} \frac{\Gamma\left(a_\nu k_{\nu\ell} + m_{\nu\ell} - \sum_{r=1}^{\ell} k_{\nu r}\right)}{\Gamma\left(a_\nu k_{\nu\ell} - \sum_{r=1}^{\ell} k_{\nu r}\right)} \right\} \left\{ \prod_{\nu=1}^{m^{**}} \prod_{\ell=1}^{n_\nu^*} \frac{\Gamma\left(a_\nu^* n_\nu^* + s_\nu^*\right)}{\Gamma\left(a_\nu^* n_\nu^* - n_\nu^* j\right)} \right\}$$

$$\times H^{p^{**},1}_{p^{**}+1,p^{**}+1}\left(\begin{array}{l} \left\{ (1,1), \left\{ \left(a_\nu k_{\nu\ell} + m_{\nu\ell} - \sum_{r=1}^{\ell} k_{\nu r}, k_{\nu\ell}\right) \right\}_{\substack{\nu=1:m^* \\ \ell=1:n_\nu}}, \left\{ \left(a_\nu^* n_\nu^* + s_\nu^*, n_\nu^*\right) \right\}_{\substack{\nu=1:m^{**} \\ j=1:k_\nu^*}} \right\} \\ \left\{ \left\{ \left(a_\nu k_{\nu\ell} - \sum_{r=1}^{\ell} k_{\nu r}, k_{\nu\ell}\right) \right\}_{\substack{\nu=1:m^* \\ \ell=1:n_\nu}}, \left\{ \left(a_\nu^* n_\nu^* - n_\nu^* j, n_\nu^*\right) \right\}_{\substack{\nu=1:m^{**} \\ j=1:k_\nu^*}}, (0,1) \right\} \end{array} \middle| \, z \right)$$

$$= F^{EGIG}\left(z \,\middle|\, \widetilde{\widetilde{\left\{ \{1\}_{\substack{\nu=1:m^* \\ \ell=1:n_\nu \\ i=0:m_{\nu\ell}-1}}, \{r_{\nu j}\}_{\substack{\nu=1:m^{**} \\ j=1:n_\nu^* k_\nu^* + s_\nu^*}} \right\}}}; \right.$$

$$\widetilde{\left\{ \left\{ a_\nu + \frac{i - \sum_{r=1}^{\ell} k_{\nu r}}{k_{\nu\ell}} \right\}_{\substack{\nu=1:m^* \\ \ell=1:n_\nu \\ i=0:m_{\nu\ell}-1}}, \left\{ a_\nu^* + \frac{s_\nu^* - j}{n_\nu^*} \right\}_{\substack{\nu=1:m^{**} \\ j=1:n_\nu^* k_\nu^* + s_\nu^*}} \right\}};$$

$$\left. g \leq \sum_{\nu=1}^{m^*} \sum_{\ell=1}^{n_\nu} m_{\nu\ell} + \sum_{\nu=1}^{m^{**}} n_\nu^* k_\nu^* + s_\nu^* \right),$$

for

$$p^* = \sum_{\nu=1}^{m^*} \sum_{\ell=1}^{n_\nu} k_{\nu\ell} + \sum_{\nu=1}^{m^{**}} n_\nu^* k_\nu^* \quad \text{and} \quad p^{**} = \sum_{\nu=1}^{m^*} n_\nu + \sum_{\nu=1}^{m^{**}} k_\nu^* .$$

Proof Instead of a formal proof we provide an argument which shows how to build the expressions of the p.d.f. and c.d.f. of Z in this case, from the expressions of the p.d.f.'s and c.d.f.'s in Corollaries 4.1 and 4.3. Given the definitions of the Meijer G and the Fox H functions all we have to do in order to obtain the expressions for the p.d.f. and the c.d.f. of Z based on these two functions, for this case, is to combine into a single product the two products of Gamma functions that appear in the expressions for the p.d.f. or the c.d.f. of Z in Corollary 4.1 and in Corollary 4.3 and then compose together into a single set of parameters the sets of parameters that appear in the numerators of the integrand functions in the definition of the Meijer G or the Fox H functions in those two corollaries, doing the same with the sets of parameters that appear in the denominator of those functions. In what concerns the representation of the p.d.f. and c.d.f. of Z based on the p.d.f. and c.d.f.

of the EGIG distribution, we obtain the set of rate parameters as the union of the sets of rate parameters for the representations based on the EGIG distribution in Corollary 4.1 and in Corollary 4.3 and then the set of shape parameters by doing the corresponding "contraction" on the set resulting from the union of the corresponding shape parameters, as indicated in Sect. 2.1. □

In a similar manner we may also combine the results in Theorems 3.2 and 3.3. For these results, which will be applied to the distribution of some l.r.t. statistics in Sect. 5.3 of Chap. 5, we will only state the expressions of the p.d.f. and c.d.f. in terms of the EGIG p.d.f. and c.d.f. avoiding the rather lengthy expressions in terms of the Meijer G and Fox H functions, which, given their computational inefficiency, will indeed never be used in any computations.

Corollary 4.5 *For a r.v. Z whose distribution is that of the product of two independent products of independent Beta r.v.'s, one of the type of that in Theorem 3.2 and the other one of the type of that in Theorem 3.3, using for all the parameters of the r.v.'s in Theorem 3.3 a starred notation, and m^{**} for m^*, the expression for its p.d.f., based on the EGIG p.d.f., is given by*

$$f^{EGIG}\left(z \,\middle|\, \widetilde{\widetilde{\left\{ \{r_{\nu j}\}_{\substack{\nu=1:m^*\\ j=1:m_\nu+k_\nu(n_\nu-1)}}, \{r^*_{\nu j}\}_{\substack{\nu=1:m^{**}\\ j=1:n^*_\nu k^*_\nu+s^*_\nu}} \right\}}};\right.$$

$$\widetilde{\left\{ \left\{ a_\nu + \frac{i-\sum_{r=1}^{\ell} k_{\nu r}}{k_{\nu\ell}} \right\}_{\substack{\nu=1:m^*\\ \ell=1:n_\nu\\ i=0:m_{\nu\ell}-1}}, \left\{ a^*_\nu + \frac{s^*_\nu - j}{n^*_\nu} \right\}_{\substack{\nu=1:m^{**}\\ j=1:n^*_\nu k^*_\nu+s^*_\nu}} \right\}};$$

$$\left. g \le \sum_{\nu=1}^{m^*} m_\nu + k_\nu(n_\nu-1) + \sum_{\nu=1}^{m^{**}} n^*_\nu k^*_\nu + s^*_\nu \right),$$

and the expression for its c.d.f., based on the EGIG c.d.f., is

$$F^{EGIG}\left(z \,\middle|\, \widetilde{\widetilde{\left\{ \{r_{\nu j}\}_{\substack{\nu=1:m^*\\ j=1:m_\nu+k_\nu(n_\nu-1)}}, \{r^*_{\nu j}\}_{\substack{\nu=1:m^{**}\\ j=1:n^*_\nu k^*_\nu+s^*_\nu}} \right\}}};\right.$$

$$\widetilde{\left\{ \left\{ a_\nu + \frac{i-\sum_{r=1}^{\ell} k_{\nu r}}{k_{\nu\ell}} \right\}_{\substack{\nu=1:m^*\\ \ell=1:n_\nu\\ i=0:m_{\nu\ell}-1}}, \left\{ a^*_\nu + \frac{s^*_\nu - j}{n^*_\nu} \right\}_{\substack{\nu=1:m^{**}\\ j=1:n^*_\nu k^*_\nu+s^*_\nu}} \right\}};$$

$$\left. g \le \sum_{\nu=1}^{m^*} m_\nu + k_\nu(n_\nu-1) + \sum_{\nu=1}^{m^{**}} n^*_\nu k^*_\nu + s^*_\nu \right).$$

Proof The proof of the result in this corollary would be obtained in a manner similar to that used for the proof of Corollary 4.4. □

We may note that, in a similar way, we may also combine results from Theorem 3.1 with results in Theorem 3.2 or even two different products but both of the type of those in any of Theorem 3.1, 3.2, or 3.3, or yet results in all three Theorems 3.1–3.3, and for all these possible combinations obtain both the representations through the Meijer G or Fox H functions and through the EGIG p.d.f. and c.d.f.. These latter formulations display all the computational advantages addressed in more detail in the next section. In particular, the GIG/EGIG formulation being the only one that generally can be used for the computation of quantiles.

4.4 Huge Gains in Computation Times

In order to illustrate the huge gains in computation time that are possible to obtain when using the p.d.f. and c.d.f. formulation based on the EGIG p.d.f. and c.d.f. instead of the Meijer G function formulation, several "scenarios" or different combinations of parameters were considered for r.v.'s in Theorems 3.1–3.3.

All computations were done with Mathematica®, version 9.0, release 9.0.0.0, as the only software running. The computer used for the calculations, unless otherwise indicated, was a laptop with an Intel® Core™ 2 Duo P8700 CPU running at 2.53GHz, with 4GB of RAM and running the Windows 7 Home Premium 64-bit operating system. Versions 10.0, release 10.0.0.0, and 11.0, release 11.0.1.0, of Mathematica® were also sparingly used, giving very similar computing times.

All computing times are reported in seconds, rounded to the third decimal place. While the computing times for the EGIG version of the p.d.f.'s and c.d.f.'s, given their very small values, are the average of one hundred runs, the computing times for the Mathematica® Meijer G function module, given their much larger values, are the average of only three runs. For "scenario" I.8, given its huge running time, when using the Meijer G function implementation of Mathematica®, only one run was made for both the p.d.f. and the c.d.f. under this implementation.

The computations were done with the Mathematica® modules in Appendix. In this Appendix these modules are listed, together with a description of their arguments and simple examples of the commands used for their implementation.

For r.v.'s Z in Theorem 3.1 we consider twelve different combinations of parameters or "scenarios," whose p.d.f.'s and c.d.f.'s are given by Corollary 4.1. These "scenarios," or rather, the sets of parameters used for these "scenarios," are listed in Table 4.1, together with the values z where the p.d.f.'s and c.d.f.'s are computed.

Table 4.1 Parameters for the scenarios used to obtain computation times for distributions in the first part of Corollary 4.1 for r.v.'s in Theorem 3.1

Scen.	m^*	n_ν and a_ν	$k_{\nu\ell}$ and $m_{\nu\ell}$	z
I.1	2	$n_\nu = \{3, 5\}$	$k_{\nu\ell} = \{\{3, 1, 4\}, \{2, 5, 3, 2, 2\}\}$	1×10^{-1}
		$a_\nu = \{16.3, 18.6\}$	$m_{\nu\ell} = \{\{2, 2, 3\}, \{1, 5, 2, 4, 3\}$	
I.2	2	$n_\nu = \{3, 5\}$	$k_{\nu\ell} = \{\{6, 6, 6\}, \{3, 4, 5, 6, 7\}\}$	4×10^{-1}
		$a_\nu = \{45.3, 25.6\}$	$m_{\nu\ell} = \{\{2, 2, 3\}, \{2, 2, 3, 3, 4\}\}$	
I.3	3	$n_\nu = \{3, 4, 2\}$	$k_{\nu\ell} = \{\{2, 2, 2\}, \{2, 2, 2, 2\}, \{2, 2\}\}$	1×10^{-6}
		$a_\nu = \{4, 8, 10\}$	$m_{\nu\ell} = \{\{17, 17, 17\}, \{9, 9, 9, 9\}, \{5, 5\}\}$	
I.4	3	$n_\nu = \{3, 5, 4\}$	$k_{\nu\ell} = \{\{3, 1, 4\}, \{2, 5, 3, 2, 2\}, \{4, 1, 2, 5\}\}$	2×10^{-1}
		$a_\nu = \{16.3, 18.6, 15.3\}$	$m_{\nu\ell} = \{\{2, 2, 3\}, \{1, 5, 2, 4, 3\}, \{1, 3, 4, 5\}\}$	
I.5	3	$n_\nu = \{3, 2, 3\}$	$k_{\nu\ell} = \{\{3, 4, 6\}, \{6, 7\}, \{5, 2, 3\}\}$	4×10^{-1}
		$a_\nu = \{45.3, 12.3, 22.3\}$	$m_{\nu\ell} = \{\{4, 2, 3\}, \{4, 5\}, \{2, 3, 4\}\}$	
I.6	3	$n_\nu = \{3, 2, 3\}$	$k_{\nu\ell} = \{\{5, 7, 6\}, \{6, 7\}, \{6, 5, 6\}\}$	3×10^{-1}
		$a_\nu = \{45.3, 12.3, 22.3\}$	$m_{\nu\ell} = \{\{4, 2, 3\}, \{9, 5\}, \{2, 3, 4\}\}$	
I.7	3	$n_\nu = \{3, 2, 3\}$	$k_{\nu\ell} = \{\{9, 7, 6\}, \{6, 7\}, \{6, 5, 9\}\}$	3×10^{-1}
		$a_\nu = \{45.3, 12.3, 22.3\}$	$m_{\nu\ell} = \{\{4, 2, 3\}, \{9, 5\}, \{2, 3, 4\}\}$	
I.8	3	$n_\nu = \{3, 2, 3\}$	$k_{\nu\ell} = \{\{7, 8, 9\}, \{6, 7\}, \{6, 5, 9\}\}$	8×10^{-2}
		$a_\nu = \{45.3, 12.3, 22.3\}$	$m_{\nu\ell} = \{\{4, 2, 3\}, \{9, 5\}, \{2, 3, 4\}\}$	
I.9	3	$n_\nu = \{3, 2, 3\}$	$k_{\nu\ell} = \{\{9, 7, 9\}, \{9, 7\}, \{6, 8, 9\}\}$	3×10^{-1}
		$a_\nu = \{45.3, 12.3, 22.3\}$	$m_{\nu\ell} = \{\{4, 2, 3\}, \{9, 5\}, \{2, 3, 4\}\}$	
I.10	3	$n_\nu = \{3, 2, 3\}$	$k_{\nu\ell} = \{\{9, 7, 11\}, \{12, 7\}, \{6, 10, 9\}\}$	3×10^{-1}
		$a_\nu = \{45.3, 12.3, 22.3\}$	$m_{\nu\ell} = \{\{4, 2, 3\}, \{9, 5\}, \{2, 3, 4\}\}$	
I.11	4	$n_\nu = \{3, 4, 3, 5\}$	$k_{\nu\ell} = \{\{3, 4, 6\}, \{2, 3, 3, 5\}, \{1, 2, 3\}, \{2, 3, 4, 5, 2\}\}$	2×10^{-1}
		$a_\nu = \{45.3, 12.3, 22.3, 67.3\}$	$m_{\nu\ell} = \{\{4, 2, 3\}, \{4, 5, 2, 3\}, \{2, 3, 4\}, \{3, 4, 7, 2, 3\}\}$	
I.12	5	$n_\nu = \{3, 4, 3, 5, 4\}$	$k_{\nu\ell} = \{\{3, 4, 6\}, \{2, 3, 3, 5\}, \{1, 2, 3\}, \{2, 3, 4, 5, 2\}, \{1, 2, 3, 4\}\}$	2×10^{-1}
		$a_\nu = \{45.3, 12.3, 22.3, 67.3, 34.5\}$	$m_{\nu\ell} = \{\{4, 2, 3\}, \{4, 5, 2, 3\}, \{2, 3, 4\}, \{3, 4, 7, 2, 3\}, \{4, 3, 2, 1\}\}$	

In Tables 4.2 and 4.3 we may analyze respectively the computing times for the p.d.f. and the c.d.f. for the scenarios in Table 4.1, when using the EGIG formulation and the Meijer G function formulation and in Table 4.4, the computing times for the quantiles, for the EGIG formulation. In these tables, as well as in Tables 4.5, 4.6, 4.7, 4.8, 4.9, 4.10, 4.11, 4.12, 4.13, 4.14, 4.15, 4.16, 4.17, 4.18, and 4.19 the following abbreviations are used.

Abbreviation	Meaning
Scen.	Scenario
p.d.	Precision digits (the number of precision digits used in the computation)
c. time	Computing time
s.v.	Starting value
eps	Epsilon (the upper-bound for the difference between the two last values found for the quantiles)

Table 4.2 Computing times (in seconds) for p.d.f. formulations based on the EGIG and on the Meijer G function implementations in Mathematica®, for the scenarios in Table 4.1, relative to the results in the first part of Corollary 4.1—computations done at the running values z in Table 4.1

		EGIG		Meijer G		
Scen.	Result	p.d.	c. time (a)	p.d.	c. time (b)	(b)/(a)
I.1	$8.7650963635\times10^{-1}$	21	0.005	22	795.075	159,015.0
I.2	$3.4123807297\times10^{+0}$	27	0.005	28	2853.945	570,789.0
I.3	$4.2225479403\times10^{+4}$	36	0.036	36	(>10,800.000)[a]	(>300,000.0)
I.4	$3.2209740218\times10^{-1}$	43	0.009	44	4770.979	530,108.8
I.5	$6.1644345125\times10^{-1}$	25	0.007	26	4275.301	610,757.3
I.6	$6.0455742873\times10^{-1}$	30	0.011	30	7420.312	674,573.8
I.7	$6.1500941303\times10^{-1}$	30	0.011	30	8061.160	732,832.7
I.8	$2.6227213190\times10^{+0}$	23	0.011	23	>43,200.000[b]	>3,927,272.7
I.9	$5.2827798743\times10^{-1}$	30	0.011	31	2583.704	234,882.2
I.10	$4.6041565757\times10^{-1}$	31	0.012	31	(>18,000.000)[c]	(>1,500,000.0)
I.11	$1.6165834565\times10^{+0}$	27	0.018	28	4977.633	276,535.2
I.12	$2.2636145605\times10^{-1}$	28	0.024	29	10,706.115	446,088.1

[a]After 3 h of computing the memory fills up and the computer stalls
[b]No result was yet obtained after more than 12 h of computing
[c]After 5 h of computing generates a huge memory dump, with a message of "MemoryAllocation-Failure"

As may be seen from Tables 4.2 and 4.3, while the computing times for the EGIG formulations show extremely small values, shorter than 5 hundredths of a second for both the p.d.f. and the c.d.f. for all the scenarios, displaying moreover a very small variation, the computing times for the Meijer G function based formulations, which use the native Mathematica® Meijer G function module, besides showing quite large variations, most commonly require more than an hour, with ratios of computing times between the Meijer G implementation and the EGIG formulation attaining in many cases a few hundreds of thousands fold, and in one case even almost four million fold. Version 11.2.0.0 of Mathematica® on a Intel® Core™ i5-7200U CPU running at 3.08GHz, with 8GB of RAM and running the Windows 10 Home 64-bit operating system, was sporadically used and it gave ratios of computing times even far more favorable to the EGIG implementations.

The "assisted" computation of quantiles, that is, the one where we provide a quite good starting value, is easily carried out with a simple implementation of a Newton

Table 4.3 Computing times (in seconds) for c.d.f. formulations based on the EGIG and on the Meijer G function implementations in Mathematica®, for the scenarios in Table 4.1, relative to the results in the first part of Corollary 4.1—computations done at the running values z in Table 4.1

		EGIG		Meijer G		
Scen.	Result	p.d.	c. time (a)	p.d.	c. time (b)	(b)/(a)
I.1	$1.4972280165\times10^{-2}$	20	0.005	21	1164.048	232,809.6
I.2	$1.8517340428\times10^{-1}$	26	0.006	27	3818.249	636,374.8
I.3	$9.6797117892\times10^{-1}$	33	0.041	33	(>10,800.000)[a]	(>262,414.6)
I.4	$9.9262423376\times10^{-1}$	41	0.010	42	6064.196	606,419.6
I.5	$9.7703836564\times10^{-1}$	24	0.008	25	4572.295	571,536.9
I.6	$9.8041750353\times10^{-1}$	28	0.011	29	7957.814	723,437.6
I.7	$9.8003464469\times10^{-1}$	28	0.011	29	8797.505	799,773.2
I.8	$4.5392048183\times10^{-2}$	22	0.011	22	>43,200.000[b]	>3,927,272.7
I.9	$9.8314923065\times10^{-1}$	29	0.011	29	2748.238	249 839.8
I.10	$9.8553729157\times10^{-1}$	29	0.012	29	(>18,000.000)[c]	(>1,500,000.0)
I.11	$9.5456690583\times10^{-1}$	25	0.019	26	5415.746	285,039.3
I.12	$9.9583679193\times10^{-1}$	27	0.025	28	11,203.259	448,130.4

[a] After 3 h of computing the memory fills up and the computer stalls
[b] No result was yet obtained after more than 12 h of computing
[c] After 5 h of computing generates a huge memory dump, with a message of "MemoryAllocationFailure"

method which uses the formulation of the p.d.f. as the derivative of the c.d.f. to implement the root finding algorithm. This implementation displays, for these cases, of finding quantiles for these distributions, a far more stable and efficient behavior than the `FindRoot` function of Mathematica®. Given the very large computing times obtained with the Mathematica® implementation of the Meijer G function, and since these quantile computations will involve a few computations of the c.d.f. and the p.d.f., it is unthinkable in practice to implement them with the Mathematica® Meijer G function module, since they would then take, even in the simplest cases, an inadmissible amount of time just to obtain a single quantile and it would not be possible to implement them at all in the cases for which no value of the c.d.f. was possible to be obtained with the Meijer formulation. On the other hand, when using the modules with the EGIG p.d.f. and c.d.f. implementations these quantile computations do not take more than a few hundredths of a second even for the more complicated scenarios in Table 4.1, as may be seen from the computing times in Table 4.4.

For each scenario the same starting value was used for both the 0.05 and the 0.01 quantiles. This starting value may be easily obtained after a couple of computations of the c.d.f., which, when using the EGIG implementation, will not take more than a couple of seconds.

Alternatively, a fully automated version of the module for the computation of quantiles may be used. This module does not require any starting value nor the specification of the precision to be used in the computations, since it will internally

Table 4.4 Computing times (in seconds) for quantiles computed using the formulation based on the EGIG implementation in Mathematica®, for the scenarios in Table 4.1, relative to the results in the first part of Corollary 4.1

			$\alpha = 0.05$			$\alpha = 0.01$		
Scen.	s.v.	eps	Quantile	p.d.	c. time	Quantile	p.d.	c. time
I.1	0.10	10^{-12}	$1.2455330812\times10^{-01}$	23	0.012	$9.3450609527\times10^{-02}$	23	0.012
I.2	0.30	10^{-11}	$3.4264483255\times10^{-01}$	30	0.014	$2.9330283232\times10^{-01}$	30	0.012
I.3	4×10^{-10}	10^{-21}	$5.7403952099\times10^{-10}$	16	0.069	$8.0859370796\times10^{-11}$	16	0.074
I.4	0.03	10^{-12}	$3.6985682021\times10^{-02}$	45	0.022	$2.5674244617\times10^{-02}$	42	0.023
I.5	0.10	10^{-12}	$1.3070742556\times10^{-01}$	20	0.017	$9.7756356011\times10^{-02}$	21	0.016
I.6	0.07	10^{-12}	$8.1163004573\times10^{-02}$	26	0.021	$5.8784992207\times10^{-02}$	26	0.021
I.7	0.07	10^{-12}	$8.1512934901\times10^{-02}$	26	0.021	$5.9062257498\times10^{-02}$	27	0.022
I.8	0.07	10^{-12}	$8.1702227526\times10^{-02}$	26	0.022	$5.9203289503\times10^{-02}$	26	0.022
I.9	0.07	10^{-12}	$7.7777262424\times10^{-02}$	27	0.021	$5.5913669078\times10^{-02}$	27	0.022
I.10	0.07	10^{-12}	$7.4552709095\times10^{-02}$	27	0.022	$5.3207660257\times10^{-02}$	27	0.023
I.11	0.05	10^{-12}	$5.6094322599\times10^{-02}$	24	0.036	$3.9950308294\times10^{-02}$	24	0.038
I.12	0.03	10^{-12}	$4.0994387468\times10^{-02}$	26	0.046	$2.9043009634\times10^{-02}$	25	0.044

Table 4.5 Computing times (in seconds) for quantiles computed using the formulation based on the EGIG implementation in Mathematica®, for the scenarios in Table 4.1, for the fully automated module

	Quantile			Quantile	
Scen.	0.05	0.01	Scen.	0.05	0.01
I.1	0.039	0.026	I.7	0.071	0.061
I.2	0.048	0.042	I.8	0.064	0.063
I.3	0.307	0.316	I.9	0.078	0.067
I.4	0.054	0.053	I.10	0.067	0.070
I.5	0.053	0.048	I.11	0.104	0.100
I.6	0.063	0.057	I.12	0.110	0.113

determine both. The results obtained with this module are exactly the same as the ones in Table 4.4 and the computing times obtained for scenarios I.1–I.12 are listed in Table 4.5.

In Table 4.6 are listed the computing times for the p.d.f., c.d.f. and quantiles for scenarios I.1–I.12 in Table 4.1 obtained when using a small laptop with a N280 CPU running at 1.66 GHz with 1GB of RAM and running the Windows 7 Starter operating system (with service pack 1). All these computing times are for the EGIG implementation of the distribution in Corollary 4.1 and they are listed in order to show how even using such a less powerful laptop we get extremely low computing times for the p.d.f. and c.d.f. in all cases as well as very low computing times for all quantiles, even when using the fully automated module.

We may note that the computation of quantiles when using the less powerful N280 CPU based laptop may require a slightly higher number of precision digits for some of the scenarios.

For r.v's whose distribution is described in Theorem 3.2 and whose p.d.f. and c.d.f. are given by Corollary 4.2, thirteen different scenarios were chosen. These are listed in Table 4.7.

Table 4.6 Computing times (in seconds) for the p.d.f., c.d.f., and quantiles computed with the formulation based on the EGIG implementation in Mathematica®, for the scenarios in Table 4.1, using an ASUS Eee PC, with 1GB of RAM, running on an Intel Atom CPU N280 at 1.66GHz with Windows 7 starter (service pack 1)

			Quantiles					
			Assisted				Fully automated	
Scen.	p.d.f.	c.d.f.	0.05	p.d.	0.01	p.d.	0.05	0.01
I.1	0.022	0.024	0.059	25	0.053	25	0.186	0.124
I.2	0.025	0.027	0.060	30	0.055	30	0.223	0.202
I.3	0.160	0.182	0.298	17	0.320	17	1.551	1.593
I.4	0.039	0.043	0.098	45	0.095	42	0.265	0.258
I.5	0.032	0.034	0.076	24	0.067	24	0.258	0.232
I.6	0.046	0.048	0.092	26	0.093	26	0.302	0.282
I.7	0.046	0.048	0.094	26	0.094	27	0.350	0.307
I.8	0.046	0.048	0.097	26	0.097	26	0.321	0.316
I.9	0.047	0.049	0.093	27	0.100	27	0.381	0.329
I.10	0.050	0.052	0.096	27	0.102	27	0.345	0.355
I.11	0.078	0.081	0.157	26	0.165	25	0.481	0.481
I.12	0.106	0.109	0.202	26	0.192	25	0.508	0.631

Table 4.7 Parameters for the scenarios used to obtain computation times for distributions in the second part of Corollary 4.1 for r.v.'s in Theorem 3.2

Scen.	m^*	n_ν and a_ν	k_ν and m_ν	z
II.1	1	$n_\nu = \{7\},\ a_\nu = \{71\}$	$k_\nu = \{2\},\ m_\nu = \{10\}$	0.313
II.2	1	$n_\nu = \{7\},\ a_\nu = \{78\}$	$k_\nu = \{2\},\ m_\nu = \{11\}$	0.303
II.3	1	$n_\nu = \{7\},\ a_\nu = \{85\}$	$k_\nu = \{2\},\ m_\nu = \{12\}$	0.313
II.4	1	$n_\nu = \{7\},\ a_\nu = \{106\}$	$k_\nu = \{2\},\ m_\nu = \{15\}$	0.313
II.5	1	$n_\nu = \{7\},\ a_\nu = \{176\}$	$k_\nu = \{2\},\ m_\nu = \{25\}$	0.330
II.6	1	$n_\nu = \{12\},\ a_\nu = \{181\}$	$k_\nu = \{2\},\ m_\nu = \{15\}$	0.330
II.7	2	$n_\nu = \{4, 2\},\ a_\nu = \{45.3, 12.3\}$	$k_\nu = \{7, 5\}\ m_\nu = \{5, 5\}$	0.500
II.8	2	$n_\nu = \{4, 5\},\ a_\nu = \{45.3, 12.3\}$	$k_\nu = \{7, 5\}\ m_\nu = \{5, 5\}$	0.100
II.9	2	$n_\nu = \{4, 5\},\ a_\nu = \{45.3, 12.3\}$	$k_\nu = \{7, 5\}\ m_\nu = \{7, 5\}$	0.100
II.10	3	$n_\nu = \{4, 2, 3\}$	$k_\nu = \{7, 5, 4\}\ m_\nu = \{5, 5, 2\}$	0.200
		$a_\nu = \{45.3, 12.3, 25.6\}$		
II.11	3	$n_\nu = \{5, 4, 3\}$	$k_\nu = \{7, 5, 4\}\ m_\nu = \{5, 5, 2\}$	0.200
		$a_\nu = \{45.3, 12.3, 25.6\}$		
II.12	3	$n_\nu = \{5, 4, 3\}$	$k_\nu = \{7, 5, 8\}\ m_\nu = \{5, 5, 2\}$	0.100
		$a_\nu = \{45.3, 12.3, 25.6\}$		
II.13	4	$n_\nu = \{4, 2, 3, 2\}$	$k_\nu = \{7, 5, 4, 3\}\ m_\nu = \{5, 5, 2, 2\}$	0.200
		$a_\nu = \{45.3, 12.3, 25.6, 32.1\}$		

The Mathematica® implementation of the Meijer G function is, for these cases, able to make use of the fact that $k_{\nu\ell} = k_\nu$ and $m_{\nu\ell} = m_\nu$ for $\ell = 1, \ldots, n_\nu$, speeding up the computations. However, for some of the scenarios the computation times required with the Meijer G function implementation still run on the couple of hundred thousand fold when compared with the computation times needed with the EGIG p.d.f. or c.d.f. implementation, as it may be seen from the computation times in Tables 4.8 and 4.9, with p.d.f.'s and c.d.f.'s computed at the z values in Table 4.7.

Some of the scenarios considered have quite similar parameters, in order to illustrate the fact that while the EGIG p.d.f. and c.d.f. implementations show very similar computing times for such scenarios, the Meijer G function implementation of Mathematica® may show quite different computing times. Specifically for scenarios II.1–II.6 the Meijer G function implementation shows a rather erratic behavior. These scenarios correspond to quite simple cases, all with $m^* = 1$ in Theorem 3.2 and Corollary 4.2, although with rather large values of m_ν. This is the main reason for the computation problems exhibited by the Meijer G function implementations, whose rather unpredictable behavior is indeed an additional disadvantage of this implementation, since we never really know what to expect in terms of computing times.

Scenarios II.1–II.6 correspond in fact to distributions of the likelihood ratio statistic in Sect. 5.1.1 of Chap. 5 for the values of p, q, and n in Table 4.10. We should note that this statistic is actually, in terms of distribution, one of the simplest ones among those addressed in Chap. 5.

It is useful to remark here that the computing times reported in Tables 4.8 and 4.9 refer to the CPU time used to compute the corresponding p.d.f. and c.d.f. values, which are reported by Mathematica® through the use of the command `Timing`. However, for the user it might have been in fact more useful to have

Table 4.8 Computing times (in seconds) for p.d.f. formulations based on the EGIG and on the Meijer G function implementations in Mathematica®, for the scenarios in Table 4.7, relative to the results in Corollary 4.2—computations done at the running values z in Table 4.7

		EGIG		Meijer G		
Scen.	Result	p.d.	c. time (a)	p.d.	c. time (b)	(b)/(a)
II.1	$4.5753687675\times10^{+0}$	93	0.019	95	2.730	143.7
II.2	$2.9893962061\times10^{+0}$	102	0.023	104	162.272	7055.3
II.3	$3.9356365854\times10^{+0}$	114	0.028	116	18.861	673.6
II.4	$3.0430407963\times10^{+0}$	146	0.042	148	124.364	2961.0
II.5	$3.9178108341\times10^{+0}$	249	0.110	252	246.934	2244.9
II.6	$6.7289686118\times10^{+0}$	260	0.118	260	(>14,400.000)[a]	(>122,033.9)
II.7	$9.5550214608\times10^{-2}$	30	0.008	32	119.918	14,989.8
II.8	$2.6692869224\times10^{+0}$	29	0.018	30	275.123	15,284.6
II.9	$1.3342185143\times10^{+0}$	30	0.023	29	717.620	31,200.9
II.10	$6.2631230424\times10^{+0}$	22	0.012	23	2713.528	226,127.3
II.11	$5.0890259974\times10^{-2}$	32	0.020	33	3695.554	184,777.7
II.12	$5.2076496397\times10^{+0}$	28	0.022	29	4069.426	186,201.2
II.13	$6.1818066639\times10^{+0}$	22	0.014	23	2960.571	211,469.4

[a] Fills up the memory after 1/2 h and still gives no result after 4 h of computing

Table 4.9 Computing times (in seconds) for c.d.f. formulations based on the EGIG and on the Meijer *G* function implementations in Mathematica®, for the scenarios in Table 4.7, relative to the results in Corollary 4.2—computations done at the running values *z* in Table 4.7

		EGIG		Meijer G		
Scen.	Result	p.d.	c. time (a)	p.d.	c. time (b)	(b)/(a)
II.1	$1.1162120385\times10^{-1}$	91	0.022	93	40.966	1862.1
II.2	$5.8322225017\times10^{-2}$	102	0.026	104	252.098	9696.1
II.3	$8.0070562298\times10^{-2}$	113	0.030	113	(>14,400.000)[a]	(>450,000.0)
II.4	$4.9664713456\times10^{-2}$	145	0.047	145	(>14,400.000)[b]	(>306,383.0)
II.5	$5.0602461411\times10^{-2}$	249	0.120	249	(>14,400.000)[c]	(>120,000.0)
II.6	$1.0197987382\times10^{-1}$	259	0.129	259	(>14,400.000)[c]	(>111,627.9)
II.7	$9.9735110008\times10^{-1}$	27	0.009	29	162.849	18,094.3
II.8	$9.4226935744\times10^{-1}$	28	0.018	29	343.748	19,097.1
II.9	$9.7669306666\times10^{-1}$	29	0.023	30	933.479	40,586.0
II.10	$4.7206366395\times10^{-1}$	21	0.012	22	3399.496	283,291.3
II.11	$9.9910171941\times10^{-1}$	30	0.021	31	4704.741	224,035.3
II.12	$8.7216366829\times10^{-1}$	27	0.022	28	5160.856	234,584.4
II.13	$6.3591995712\times10^{-1}$	22	0.014	23	3549.928	253,566.3

[a]Fills up the memory after around 20 min and still gives no result after 4 h of computing
[b]Fills up the memory after around 40 min and still gives no result after 4 h of computing
[c]Fills up the memory after around 30 min and still gives no result after 4 h of computing

Table 4.10 Values of p, q, and n for the test in Sect. 5.1.1, for scenarios II.1–II.6

Scen.	p	q	n
II.1	14	11	152
II.2	11	15	167
II.3	14	13	182
II.4	15	15	227
II.5	25	15	377
II.6	15	25	377

reported the "waiting times," that is, the time that the user really waits to obtain the results, and which are obtained in Mathematica® through the use of the command `AbsoluteTiming`. These later times are for the EGIG implementation in general exactly the same as the computing times reported, while for the Meijer *G* function implementation they are generally exceedingly higher than the computing times reported, being in the order of many hours for scenarios II.10–II.13 and showing an extremely large variability.

In Table 4.11 are listed the 0.05 and 0.01 quantiles for scenarios II.1–II.13 obtained with the module which requires the specification of a starting value and the number of precision digits to be used. The values for the quantiles are listed together with the corresponding computing times. We may see how extremely low computing times were obtained in most cases not even attaining one tenth of a second, which is remarkable, namely for scenario II.3, for which it was not even possible to obtain the computation of a single value of the c.d.f. when

Table 4.11 Computing times (in seconds) for quantiles computed using the formulation based on the EGIG implementation in Mathematica®, for the scenarios in Table 4.7, relative to the results in the second part of Corollary 4.1

			$\alpha = 0.05$			$\alpha = 0.01$		
Scen.	s.v.	eps	Quantile	p.d.	c. time	Quantile	p.d.	c. time
II.1	0.286	10^{-11}	$2.9549282334\times10^{-01}$	160	0.046	$2.6852114706\times10^{-01}$	158	0.051
II.2	0.291	10^{-11}	$3.0005091666\times10^{-01}$	182	0.055	$2.7415478889\times10^{-01}$	180	0.062
II.3	0.296	10^{-11}	$3.0397610520\times10^{-01}$	204	0.065	$2.7903696803\times10^{-01}$	202	0.078
II.4	0.305	10^{-11}	$3.1310989426\times10^{-01}$	269	0.103	$2.9050914630\times10^{-01}$	267	0.115
II.5	0.324	10^{-11}	$3.2984550494\times10^{-01}$	476	0.267	$3.1194362670\times10^{-01}$	473	0.295
II.6	0.314	10^{-11}	$3.2015288706\times10^{-01}$	494	0.284	$3.0255994937\times10^{-01}$	490	0.318
II.7	0.140	10^{-11}	$1.4986933646\times10^{-01}$	21	0.021	$1.1290555338\times10^{-01}$	19	0.022
II.8	0.015	10^{-12}	$1.7395779979\times10^{-02}$	23	0.035	$1.0873283513\times10^{-02}$	24	0.036
II.9	0.013	10^{-13}	$1.4393560742\times10^{-02}$	23	0.042	$8.9811831049\times10^{-03}$	23	0.044
II.10	0.100	10^{-12}	$1.1329703801\times10^{-01}$	20	0.025	$8.4663003248\times10^{-02}$	21	0.025
II.11	0.050	10^{-12}	$2.5834522290\times10^{-02}$	27	0.042	$1.7129386168\times10^{-02}$	26	0.048
II.12	0.020	10^{-12}	$2.5814916306\times10^{-02}$	27	0.041	$1.7115823476\times10^{-02}$	26	0.040
II.13	0.090	10^{-12}	$9.8493799132\times10^{-02}$	20	0.029	$7.3383680517\times10^{-02}$	20	0.030

Table 4.12 Computing times (in seconds) for quantiles computed using the formulation based on the EGIG implementation in Mathematica®, for the scenarios in Table 4.7, for the fully automated module

	Quantile			Quantile	
Scen.	0.05	0.01	Scen.	0.05	0.01
II.1	0.104	0.105	II.8	0.097	0.084
II.2	0.125	0.115	II.9	0.115	0.106
II.3	0.148	0.142	II.10	0.065	0.063
II.4	0.178	0.181	II.11	0.108	0.114
II.5	0.451	0.376	II.12	0.112	0.110
II.6	0.449	0.479	II.13	0.087	0.077
II.7	0.061	0.055			

using the Meijer *G* function formulation. Even for the most computationally heavy cases only one of the computing times barely exceeded three tenths of a second.

These computing times for quantiles do not grow that much higher when using the totally automated module, that is, the one that does not need the specification of a starting value nor a value for the number of precision digits to be used. These computing times are listed in Table 4.12.

All computing times still remain well below one second, with only the computing times for scenarios II.5 and II.6 almost reaching one half of a second.

In Table 4.13 we have the computing times obtained for the p.d.f., c.d.f., and quantiles for scenarios II.1–II.13 when we use a N280 CPU based laptop, with only 1GB of RAM. Still all computing times remain very low, with only the computation time for the 0.05 quantile for scenarios II.5 and II.6 almost reaching two and a half seconds and the 0.01 quantile for scenario II.6 barely exceeding this time.

Table 4.13 Computing times (in seconds) for the p.d.f., c.d.f. and quantiles computed with the formulation based on the EGIG implementation in Mathematica®, for the scenarios in Table 4.7, using an ASUS Eee PC, with 1GB of RAM, running on an Intel Atom CPU N280 at 1.66GHz with Windows 7 starter (service pack 1)

			Quantiles					
			Assisted				Fully automated	
Scen.	p.d.f.	c.d.f.	0.05	p.d.	0.01	p.d.	0.05	0.01
II.1	0.084	0.095	0.211	160	0.233	158	0.539	0.558
II.2	0.101	0.115	0.253	185	0.280	183	0.673	0.608
II.3	0.121	0.135	0.300	204	0.332	202	0.765	0.758
II.4	0.187	0.208	0.503	269	0.554	267	0.895	0.930
II.5	0.486	0.532	1.401	476	1.599	473	2.440	2.069
II.6	0.524	0.573	1.538	494	1.753	490	2.408	2.586
II.7	0.035	0.036	0.090	23	0.086	21	0.290	0.259
II.8	0.073	0.076	0.150	26	0.157	26	0.447	0.389
II.9	0.095	0.098	0.179	26	0.189	26	0.516	0.490
II.10	0.050	0.051	0.108	25	0.108	25	0.304	0.304
II.11	0.087	0.090	0.182	29	0.198	29	0.483	0.517
II.12	0.093	0.096	0.179	27	0.173	26	0.522	0.494
II.13	0.059	0.062	0.123	23	0.130	23	0.403	0.359

Of course, all the quantile computations are done with the GIG/EGIG based formulation of the p.d.f.'s and c.d.f.'s in Corollary 4.2. This is the only formulation that allows for the computation of quantiles for all cases in an easy manner and with extremely low computation times. In addition, it is the case that for a few of the scenarios the computation of quantiles would not be feasible at all if one tried to use the Meijer G function formulation.

For the distributions in Corollary 4.3, which refer to r.v.'s Z whose distribution is addressed in Theorem 3.3, fourteen different scenarios are considered. These are listed in Table 4.14.

For scenarios III.1–III.14 listed in Table 4.14 the computing times for the p.d.f. and c.d.f. computed at the running values z specified in Table 4.14 are reported in Tables 4.15 and 4.16 for both the Meijer G function and the EGIG implementations. Once again the EGIG implementation displays extremely low computing times, which for the p.d.f. of scenario III.10 is less than one millionth of the computing time exhibited by the Meijer G function implementation.

Also for these scenarios III.1–III.14 a similar remark to the one made concerning CPU computing times and "absolute computing times," made for scenarios II.1–II.13 applies, once more with the "absolute computing times" for the Meijer G function implementation displaying much larger values than the reported CPU computing times in Tables 4.15 and 4.16, and showing also a large variability.

Table 4.14 Parameters for the scenarios used to obtain computation times for distributions in Corollary 4.3 for r.v.'s in Theorem 3.2

Scen.	m^*	n_ν and a_ν	k_ν and s_ν	z
III.1	2	$n_\nu=\{2,2\},\ a_\nu=\{45.3,12.3\}$	$k_\nu=\{6,2\},\ s_\nu=\{2,3\}$	6×10^{-2}
III.2	2	$n_\nu=\{4,2\},\ a_\nu=\{45.3,12.3\}$	$k_\nu=\{3,5\},\ s_\nu=\{2,3\}$	1×10^{-1}
III.3	2	$n_\nu=\{4,3\},\ a_\nu=\{45.3,12.3\}$	$k_\nu=\{3,5\},\ s_\nu=\{2,3\}$	5×10^{-4}
III.4	2	$n_\nu=\{5,3\},\ a_\nu=\{45.3,12.3\}$	$k_\nu=\{3,5\},\ s_\nu=\{2,3\}$	45×10^{-5}
III.5	2	$n_\nu=\{2,2\},\ a_\nu=\{45.3,12.3\}$	$k_\nu=\{7,2\},\ s_\nu=\{2,3\}$	4×10^{-2}
III.6	2	$n_\nu=\{2,3\},\ a_\nu=\{45.3,12.3\}$	$k_\nu=\{6,2\},\ s_\nu=\{2,3\}$	6×10^{-2}
III.7	3	$n_\nu=\{2,3\},\ a_\nu=\{45.3,12.3\}$	$k_\nu=\{7,2\},\ s_\nu=\{2,3\}$	6×10^{-2}
III.8	3	$n_\nu=\{4,2,2\},\ a_\nu=\{45.3,12.3,22.3\}$	$k_\nu=\{3,4,5\},\ s_\nu=\{2,3,3\}$	1×10^{-2}
III.9	3	$n_\nu=\{3,3,2\},\ a_\nu=\{45.3,12.3,22.3\}$	$k_\nu=\{3,4,5\},\ s_\nu=\{2,3,3\}$	1×10^{-2}
III.10	3	$n_\nu=\{3,2,3\},\ a_\nu=\{45.3,12.3,12.3\}$	$k_\nu=\{2,3,3\},\ s_\nu=\{2,4,2\}$	4×10^{-3}
III.11	3	$n_\nu=\{5,4,3\},\ a_\nu=\{45.3,12.3,12.3\}$	$k_\nu=\{2,3,2\},\ s_\nu=\{2,4,2\}$	5×10^{-2}
III.12	3	$n_\nu=\{5,4,3\},\ a_\nu=\{45.3,12.3,12.3\}$	$k_\nu=\{2,3,3\},\ s_\nu=\{2,4,2\}$	5×10^{-2}
III.13	3	$n_\nu=\{4,5,7\},\ a_\nu=\{45.3,12.3,22.3\}$	$k_\nu=\{4,3,4\},\ s_\nu=\{3,4,4\}$	1×10^{-4}
III.14	2	$n_\nu=\{6,3\},\ a_\nu=\{45.3,12.3\}$	$k_\nu=\{3,5\},\ s_\nu=\{2,3\}$	4×10^{-4}

Table 4.15 Computing times (in seconds) for p.d.f. formulations based on the EGIG and on the Meijer G function implementations in Mathematica®, for the scenarios in Table 4.14, relative to the results in Corollary 4.3—computations done at the running values z in Table 4.14

		EGIG		Meijer G		
Scen.	Result	p.d.	c. time (a)	p.d.	c. time (b)	(b)/(a)
III.1	$4.6215723567\times10^{+0}$	40	0.022	42	10.592	481.5
III.2	$2.6185313990\times10^{-5}$	55	0.025	57	105.524	4221.0
III.3	$2.3579931214\times10^{+2}$	44	0.037	44	415.944	11,241.7
III.4	$2.7998792329\times10^{+2}$	43	0.043	45	3195.050	74,303.5
III.5	$6.6560841878\times10^{+0}$	54	0.032	57	59.920	1872.5
III.6	$1.1584854490\times10^{+1}$	37	0.025	39	3795.908	151,836.3
III.7	$1.7664441597\times10^{+1}$	59	0.038	59	(31,740.000)[a]	(835,263.2)
III.8	$1.3731344126\times10^{+1}$	44	0.047	44	2664.360	56,688.5
III.9	$2.4824437418\times10^{+1}$	52	0.051	54	4041.515	79,245.4
III.10	$3.4828513691\times10^{+1}$	44	0.019	45	20,005.288	1,052,909.9
III.11	$1.3099180455\times10^{-1}$	64	0.024	65	387.115	16129.8
III.12	$1.7864491270\times10^{-4}$	80	0.031	81	(12,120.000)[b]	(390,967.7)
III.13	$4.9394812751\times10^{+3}$	71	0.131	72	12,340.090	94,199.2
III.14	$3.2796506218\times10^{+2}$	43	0.046	45	40,786.420	886,661.3

[a]Fills up the memory after 20 min and issues an error message of "memory full," on average, around 8 h 49 min, halting the computation and giving no result
[b]Issues an error message of "memory full," on average, around 3 h 22 min, halting the computation and giving no result

In Table 4.17 we have the 0.05 and 0.01 quantiles for scenarios III.1–III.14, together with the computing times obtained when using the nonautomated module for their computation. Once again we may find very low computing times, with very

Table 4.16 Computing times(in seconds) for c.d.f. formulations based on the EGIG and on the Meijer G function implementations in Mathematica®, for the scenarios in Table 4.14, relative to the results in Corollary 4.3—computations done at the running values z in Table 4.14

		EGIG		Meijer G		
Scen.	Result	p.d.	c. time (a)	p.d.	c. time (b)	(b)/(a)
III.1	$5.6824913519\times 10^{-2}$	39	0.025	40	1688.594	67,543.8
III.2	$9.9999982013\times 10^{-1}$	49	0.028	51	131.992	4714.0
III.3	$4.8485955265\times 10^{-2}$	43	0.040	44	706.556	17,663.9
III.4	$5.2617549197\times 10^{-2}$	43	0.046	44	9887.419	214,943.9
III.5	$5.5707438142\times 10^{-2}$	53	0.035	53	(>36,000.000)[a]	(>1,028,571.4)
III.6	$2.0426172757\times 10^{-1}$	36	0.028	37	18,221.160[b]	650,755.7
III.7	$5.6661716862\times 10^{-1}$	59	0.041	59	(33,300.000)[c]	(812 195.1)
III.8	$9.6807870296\times 10^{-1}$	43	0.051	44	5791.760	113,563.9
III.9	$9.6762404987\times 10^{-1}$	51	0.054	52	8044.419	148,970.7
III.10	$4.7230128267\times 10^{-2}$	44	0.021	45	(46,800.000)[c]	(2,228,571.4)
III.11	$9.9913950805\times 10^{-1}$	62	0.025	62	2371.095	94,843.8
III.12	$9.9999922427\times 10^{-1}$	75	0.033	75	(12,120.000)[c]	(367,272.7)
III.13	$3.5842511344\times 10^{-1}$	70	0.135	71	15,227.686	112,797.7
III.14	$5.5366585700\times 10^{-2}$	43	0.050	43	(>54,000.000)[a]	(>1,080,000.0)

[a]Still no result after: (a) 10h00; (b) 15h00
[b]But may also take 10 h without giving any result
[c]Issues an error message of "memory full," halting the computation and giving no result, on average around: (a) 9h15; (b) 13h00; (c) 3h22

Table 4.17 Computing times (in seconds) for quantiles computed using the formulation based on the EGIG implementation in Mathematica®, for the scenarios in Table 4.14, relative to the results in the second part of Corollary 4.3

			$\alpha = 0.05$			$\alpha = 0.01$		
Scen.	s.v.	eps	Quantile	p.d.	c. time	Quantile	p.d.	c. time
III.1	5×10^{-2}	10^{-12}	$5.8461215401\times 10^{-02}$	55	0.068	$4.3487570435\times 10^{-02}$	55	0.049
III.2	2×10^{-3}	10^{-12}	$2.6264715749\times 10^{-03}$	44	0.051	$1.5549847134\times 10^{-03}$	44	0.051
III.3	3×10^{-4}	10^{-14}	$5.0637882156\times 10^{-04}$	68	0.076	$2.7870201200\times 10^{-04}$	67	0.071
III.4	34×10^{-5}	10^{-14}	$4.4055172126\times 10^{-04}$	68	0.090	$2.4221054914\times 10^{-04}$	69	0.085
III.5	3×10^{-2}	10^{-12}	$3.9112488619\times 10^{-02}$	86	0.083	$2.8915814051\times 10^{-02}$	85	0.070
III.6	35×10^{-3}	10^{-12}	$4.2208534712\times 10^{-02}$	40	0.054	$3.0604463704\times 10^{-02}$	40	0.053
III.7	25×10^{-3}	10^{-12}	$2.8289592149\times 10^{-02}$	79	0.076	$2.0395293233\times 10^{-02}$	79	0.083
III.8	1×10^{-3}	10^{-14}	$1.1923283803\times 10^{-03}$	44	0.082	$7.2987355762\times 10^{-04}$	43	0.087
III.9	35×10^{-5}	10^{-14}	$4.7350465705\times 10^{-04}$	64	0.100	$2.7597507420\times 10^{-04}$	63	0.095
III.10	3×10^{-3}	10^{-14}	$4.0783903146\times 10^{-03}$	72	0.046	$2.4850027036\times 10^{-03}$	69	0.042
III.11	5×10^{-3}	10^{-14}	$3.1863428996\times 10^{-03}$	85	0.053	$1.9307890713\times 10^{-03}$	83	0.054
III.12	6×10^{-4}	10^{-14}	$1.0275611968\times 10^{-03}$	93	0.067	$5.9220326135\times 10^{-04}$	90	0.067
III.13	28×10^{-6}	10^{-15}	$3.6419762710\times 10^{-05}$	106	0.204	$2.0580616508\times 10^{-05}$	103	0.215
III.14	3×10^{-4}	10^{-14}	$3.8329415617\times 10^{-04}$	70	0.085	$2.1050407474\times 10^{-04}$	70	0.092

stable values even for scenarios for which the Meijer G function implementation of the p.d.f. and c.d.f. display extremely variable computing times. None of the computing times exceeded one tenth of a second, except for scenario III.13 which shows computing times around two tenths of a second.

In Table 4.18 the computing times for quantiles obtained with the automated module, which does not require a starting value nor the indication of the precision to be used in the computation, show once again a larger computing time for scenario III.13, with a time around five tenths of a second for both quantiles, but with all computing times remaining well below one second.

In Table 4.19 we have the computing times for the p.d.f., c.d.f. and quantiles of scenarios III.1–III.14 obtained when using the EGIG formulation for the p.d.f.'s and c.d.f.'s with the less powerful laptop with a N280 CPU and 1GB of RAM. It is for

Table 4.18 Computing times (in seconds) for quantiles computed using the formulation based on the EGIG implementation in Mathematica®, for the scenarios in Table 4.14, for the fully automated module

	Quantile			Quantile	
Scen.	0.05	0.01	Scen.	0.05	0.01
III.1	0.107	0.103	III.8	0.205	0.190
III.2	0.131	0.132	III.9	0.216	0.223
III.3	0.189	0.180	III.10	0.106	0.101
III.4	0.197	0.200	III.11	0.115	0.124
III.5	0.150	0.182	III.12	0.159	0.153
III.6	0.124	0.122	III.13	0.509	0.515
III.7	0.151	0.166	III.14	0.215	0.207

Table 4.19 Computing times (in seconds) for the p.d.f., c.d.f., and quantiles computed with the formulation based on the EGIG implementation in Mathematica®, for the scenarios in Table 4.14, using an ASUS Eee PC, with 1GB of RAM, running on an Intel Atom CPU N280 at 1.66GHz with Windows 7 starter (service pack 1)

			Quantiles					
			Assisted				Fully automated	
Scen.	p.d.f.	c.d.f.	0.05	p.d.	0.01	p.d.	0.05	0.01
III.1	0.095	0.108	0.211	55	0.214	55	0.526	0.531
III.2	0.113	0.121	0.222	45	0.222	45	0.631	0.663
III.3	0.159	0.175	0.335	68	0.309	70	0.911	0.880
III.4	0.188	0.204	0.344	68	0.369	69	0.956	0.976
III.5	0.136	0.154	0.361	86	0.304	85	0.783	0.982
III.6	0.109	0.121	0.233	40	0.231	40	0.602	0.607
III.7	0.164	0.182	0.327	79	0.358	79	0.729	0.820
III.8	0.205	0.221	0.353	44	0.378	43	0.983	0.944
III.9	0.222	0.240	0.406	65	0.409	63	1.039	1.095
III.10	0.083	0.092	0.200	72	0.183	69	0.504	0.492
III.11	0.103	0.111	0.228	87	0.242	87	0.556	0.596
III.12	0.137	0.145	0.289	94	0.246	93	0.764	0.729
III.13	0.579	0.606	0.902	106	0.947	103	2.632	2.638
III.14	0.202	0.218	0.373	70	0.398	70	1.053	1.005

scenarios III.1–III.14 that the differences between the computing times reported for the more powerful laptop and the less powerful one are larger. Anyway, it is only for scenarios III.3, III.4, III.8, and III.9 that we find computing times for the automated computation of quantiles around one second, for the less powerful laptop, and a computing time a little over two and a half seconds for scenario III.13.

There is a Mathematica® module to implement the computation of Fox H function, available from Yilmaz and Alouini (2009). This module uses the equivalence between the Fox H function and the Meijer G function reported in Sect. 2.1 in Chap. 2 for all cases where the power parameters in the Fox H function are rational. Then in these cases it calls the native Mathematica® Meijer G function module to implement the computations. Since for all cases addressed in Corollaries 4.1–4.5 the power parameters in all Fox H functions end up being rational, in case we would use such module to compute the Fox H function implementation of the distributions involved, we would end up using the native Mathematica® Meijer G module and as such all computing times would be similar to the ones reported in this section for the use of the Meijer G formulation for p.d.f.'s and c.d.f.'s.

The huge gains in computation time, and the computation times themselves, obtained when using the GIG/EGIG formulation for the p.d.f.'s and c.d.f.'s, show the immense advantage in using this formulation in practical applications and as such its extreme usefulness when implementing the likelihood ratio tests described in the next chapter.

The results in this section, with the extremely low computing times for all formulations based on the GIG/EGIG p.d.f. and c.d.f. also show that there is indeed, both in terms of computing time and in terms of precision, a great advantage in using this formulation to compute exact quantiles instead of trying to obtain approximations through simulation.

For the code and details on the usage of the Mathematica® modules used in the computations reported in this section, see Appendix.

Appendix: Modules for the Computation of the p.d.f.'s, c.d.f.'s, and Quantiles for the Distributions in Corollaries 4.1–4.5

In this appendix we present the Mathematica® modules used to compute the GIG/EGIG versions of the p.d.f.'s and c.d.f.'s in Corollaries 4.1–4.5 as well as the corresponding quantiles, together with sets of commands that may be used to compute such values. Also the modules used to compute the p.d.f.'s and c.d.f.'s for the distributions in the same three corollaries, based on the Meijer G function are provided, together with a few examples of their implementation.

All modules in this appendix were programmed in Mathematica®, version 9.0.0, and are available at https://sites.google.com/site/exactdistributions/.

In Fig. 4.1 we have the Mathematica® modules `PDFCorEGIG` and `CDFCorEGIG` used to compute the p.d.f.'s and c.d.f.'s in Corollaries 4.1–4.5, based on the

```
PDFCorEGIG[cor_,a_,n_,k_,m_,z_,prec_:50,ind_:0]:=Module[
                                    {vec,vec1,vec2,precn},
  If[ToString[prec]=="Null",precn=50,precn=prec];
  If[cor==1,vec=ParCor1EGIG[a,k,m],
   If[cor==2,vec=ParCor2EGIG[a,n,k,m],
    If[cor==3,vec=ParCor3EGIG[a,k,n,m], If[cor==4,
      vec1=ParCor1EGIG[a[[1]],k[[1]],m[[1]]],
      vec1=ParCor2EGIG[a[[1]],n[[1]],k[[1]],m[[1]]]];
      vec2=ParCor3EGIG[a[[2]],k[[2]],n[[2]],m[[2]]];
      vec=Screen2[Flatten[{vec1[[1]],vec2[[1]]}],
                      Flatten[{vec1[[2]],vec2[[2]]}]]]]]];
  If[ind==0,EGIGpdffast[vec[[2]],vec[[1]],z,precn],
                      EGIGpdf[vec[[2]],vec[[1]],z,precn]]
 ]

CDFCorEGIG[cor_,a_,n_,k_,m_,z_,prec_:50,ind_:0]:=Module[
                                    {vec,vec1,vec2,precn},
  If[ToString[prec]=="Null",precn=50,precn=prec];
  If[cor==1,vec=ParCor1EGIG[a,k,m],
   If[cor==2,vec=ParCor2EGIG[a,n,k,m],
    If[cor==3,vec=ParCor3EGIG[a,k,n,m], If[cor==4,
      vec1=ParCor1EGIG[a[[1]],k[[1]],m[[1]]],
      vec1=ParCor2EGIG[a[[1]],n[[1]],k[[1]],m[[1]]]];
      vec2=ParCor3EGIG[a[[2]],k[[2]],n[[2]],m[[2]]];
      vec=Screen2[Flatten[{vec1[[1]],vec2[[1]]}],
                      Flatten[{vec1[[2]],vec2[[2]]}]]]]]];
  If[ind==0,EGIGcdffast[vec[[2]],vec[[1]],z,precn],
                      EGIGcdf[vec[[2]],vec[[1]],z,precn]]
 ]
```

Fig. 4.1 Mathematica® modules used to compute the p.d.f.'s and c.d.f.'s in Corollaries 4.1–4.5

EGIG distribution (see the examples of commands in Fig. 4.15). These modules have six mandatory arguments, which are

- `cor` – the number of the corollary we are referring to: 1, 2, 3, 4, or 5, respectively for Corollary 4.1, 4.2, 4.3, 4.4, or 4.5
- `a` – a list with the values of the parameters a_ν ($\nu = 1, \dots, m^*$)
- `n` – a list with the values of the parameters n_ν ($\nu = 1, \dots, m^*$) (irrelevant when carrying out computations for Corollary 4.1, in which case this argument may be given any value, like zero or 1, since in this case the n_ν are obtained from the length of the sub-lists in `k` below)
- `k` – a two-level list with the values of the parameters $k_{\nu\ell}$ ($\nu = 1, \dots, m^*$; $\ell = 1, \dots, n_\nu$), in case of Corollary 4.1 or a simple list with the parameters k_ν ($\nu = 1, \dots, m^*$) in case of Corollary 4.2 or 4.3
- `m` – a two-level list with the values of the parameters $m_{\nu\ell}$ ($\nu = 1, \dots, m^*$;

$\ell = 1, \ldots, n_\nu$), in case of Corollary 4.1 or a simple list with the parameters m_ν ($\nu = 1, \ldots, m^*$) in case of Corollary 4.2 or 4.3

`z` – the running value of the r.v. Z in Corollaries 4.1–4.5 to be used for the computation of the p.d.f. or the c.d.f.

which have to be given in the above order, and two optional arguments

`prec` – the precision (number of digits) to be used in the computations (default value: 50)

`ind` – an index which if given a value different from 0 (zero) will make the computations to be done with the more precise modules `EGIGpdf` and `EGIGcdf` in Fig. 4.8, instead of the faster modules `EGIGpdffast` and `EGIGcdffast`, also in Fig. 4.8 (default value: 0) (all computations of p.d.f.'s and c.d.f.'s shown in Sect. 4.4 were carried out with the default value of 0 (zero) for this argument, thus using the faster modules; these results are exactly the same that would be obtained would the more precise modules had been used).

For Corollaries 4.4 and 4.5

`a` and `n` – are double lists with the values of the parameters a_ν and a_ν^* or n_ν and n_ν^*, respectively

`k` and `m` – are double lists with the values of the parameters $k_{\nu\ell}$ and k_ν^* or $m_{\nu\ell}$ and m_ν^* in case of Corollary 4.4 or of the parameters k_ν and k_ν^* or m_ν and m_ν^* in case of Corollary 4.5.

Modules `ParCor1EGIG` and `ParCor2EGIG` in Fig. 4.2 are used to compute the rate and shape parameters for the EGIG versions of the distributions respectively in Corollaries 4.1 and 4.2, the first one based on the parameters $a_{\nu\ell}$, $k_{\nu\ell}$ and $m_{\nu\ell}$ ($\nu = 1, \ldots, m^*$; $\ell = 1, \ldots, n_\nu$), and the second one based on the parameters a_ν, n_ν, k_ν, and m_ν ($\nu = 1, \ldots, m^*$). In Fig. 4.3 we have module `ParCor3EGIG` which computes the rate and shape parameters for the EGIG based form of the distributions in Corollary 4.3, based on the parameters a_ν, k_ν, n_ν, and s_ν ($\nu = 1, \ldots, m^*$).

These modules are called by modules `PDFCorEGIG` and `CDFCorEGIG` and the user actually does not have to worry about the details on how to use them.

In Fig. 4.4 we have module `Rat`, used to rationalize real values. Module `Rat` is called by modules `ParCor1EGIG`, `ParCor2EGIG`, and `ParCor3EGIG`. Although Mathematica® has the native function/module `Rationalize`, the need for module `Rat` is made clear through the examples in Fig. 4.5, where it is shown that the Mathematica® native function/module does not display the necessary behavior, while the use of the module `Rat` gives the right results.

Modules `Screen` and `Screen2` in Fig. 4.6 do something similar to what the Mathematica® function/module `Tally` does. They scan a list of real values looking for possible repetitions and register these in a vector of integer values with the multiplicities of the corresponding real values, outputting then both sets of values.

```
ParCor1EGIG[ao_,k_,m_] := Module[{ms,n,lvf,a},
  a=Rat[ao];
  ms=Length[a];
  n=Map[Length,k];
  lvf=Flatten[Table[Table[Table[a[[nu]]+(-Sum[k[[nu,r]],
           {r,1,ell}]+i)/k[[nu,ell]],{i,0,m[[nu,ell]]-1}],
                              {ell,1,n[[nu]]}],{nu,1,ms}]];
  Screen[lvf]
  ]

ParCor2EGIG[ao_,n_,k_,m_]:=Module[{a,ms,lvf,p,hnj,rnj},
  a=Rat[ao]; ms=Length[a];
  lvf=Flatten[Table[Table[a[[nu]]-n[[nu]]+(j-1)/k[[nu]],
           {j,1,m[[nu]]+k[[nu]]*(n[[nu]]-1)}],{nu,1,ms}]];
  p=n*k;
  hnj=Table[Table[Count[{p[[nu]],m[[nu]]}-(j-1),_?Positive],
           {j,1,m[[nu]]+k[[nu]]*(n[[nu]]-1)}],{nu,1,ms}];
  hnj=hnj-1;
  rnj=hnj;
  Do[Do[rnj[[nu,j]]=hnj[[nu,j]]+rnj[[nu,j-k[[nu]]]],
     {j,1+k[[nu]],m[[nu]]+k[[nu]]*(n[[nu]]-1)}],{nu,1,ms}];
  rnj=Flatten[rnj];
  {lvf[[Flatten[Position[rnj,_?Positive]]]],Select[rnj,#>0&]}
  ]
```

Fig. 4.2 Mathematica® modules ParCor1EGIG and ParCor2EGIG called by the modules in Fig. 4.1 to compute the shape and rate parameters in the EGIG distributions in Corollaries 4.1 and 4.2, from the parameters a_ν, n_ν, $k_{\nu\ell}$, and $m_{\nu\ell}$

```
ParCor3EGIG[ao_,k_,n_,s_]:=Module[{a,ms,rj,lj},
  a=Rat[ao]; ms=Length[a];
  rj=Table[Table[k[[nu]],{j,1,n[[nu]]*k[[nu]]+s[[nu]]}],
                                                 {nu,1,ms}];
  Do[Do[rj[[nu,j]]=k[[nu]]+1+Floor[(s[[nu]]-j)/n[[nu]]],
      {j,s[[nu]]+1,n[[nu]]*k[[nu]]+s[[nu]]}],{nu,1,ms}];
  rj=Flatten[rj];
  lj=Flatten[Table[Table[a[[nu]]+(s[[nu]]-j)/n[[nu]],
       {j,1,n[[nu]]*k[[nu]]+s[[nu]]}],{nu,1,ms}]];
  If[Length[lj]!=Length[Union[lj]],Screen2[lj,rj],{lj,rj}]
  ]
```

Fig. 4.3 Mathematica® module ParCor3EGIG called by the modules in Fig. 4.1 to compute the shape and rate parameters in the EGIG distributions in Corollary 4.3, from the parameters a_ν, n_ν, $k_{\nu\ell}$, and $m_{\nu\ell}$

Their use and functioning are illustrated in Fig. 4.7. Module `Screen` accepts as input a list of real values and it just uses the native Mathematica® function/module `Tally`, to output then the set of real values, together with the set of their corre-

```
Rat[x_] := Module[{RD,nn},
  If[ToString[Head[x]]!="Real", x, RD=RealDigits[x];
      nn=Length[RD[[1]]]-RD[[2]]; Round[x*10^nn]/10^nn]]

SetAttributes[Rat, Listable];
```

Fig. 4.4 Mathematica® modules called by the modules in Figs. 4.2 and 4.3, used alternatively to the native Mathematica® function `Rationalize`, to rationalize values

```
SetPrecision[Rationalize[{.342222234567125, 56.7274657943653,
                                          23.23769439485495}], 20]
{0.34222223456712502765, 56.727465794365301122, 23.237694394854951696}
SetPrecision[Rationalize[{.342222234567125, 56.7274657943653,
                                       23.23769439485495}, 0], 20]
{0.34222223456712502846, 56.727465794365295731, 23.237694394854951103}
SetPrecision[Rat[{.342222234567125, 56.7274657943653, 23.23769439485495}], 20]
{0.34222223456712500000, 56.727465794365300000, 23.237694394854950000}
```

Fig. 4.5 Examples illustrating the need for the use of module `Rat` in Fig. 4.4 instead of the native Mathematica® function/module `Rationalize`

```
Screen[v_] := Module[{vec},
  vec = Tally[v];
  {vec[[All,1]], vec[[All,2]]}]

Screen2[v1_,v2_]:=Module[{v},
  v =Tally[Flatten[Table[Table[v1[[j]],{i,1,v2[[j]]}],
                                        {j,1,Length[v1]}]]];
  {v[[All,1]], v[[All,2]]}]
```

Fig. 4.6 Mathematica® modules called by the modules in Fig. 4.4 to screen the list of rate parameters in the EGIG and GIG distributions for duplicated values and give as output a list of rate without duplicated values and the corresponding list of shape parameters

sponding multiplicities, while module `Screen2` does a slightly more sophisticated thing which is to accept as input a first argument which is a list of real values, which may still have repeated values, and a second argument which is a list with the multiplicities of the corresponding real values, outputting then the set of real values, without repetitions, and the set of the corresponding multiplicities.

In Fig. 4.8 are listed modules `EGIGpdffast`, `EGIGcdffast`, `EGIGpdf`, and `EGIGcdf`, called by modules `PDFCorEGIG` and `CDFCorEGIG` to compute the EGIG p.d.f. and c.d.f. being the first two modules, as their names clearly indicate, the modules associated with the fast version.

The modules in Fig. 4.8 call in turn the modules in Figs. 4.9 and 4.10 to compute the EGIG p.d.f.'s and c.d.f.'s from the corresponding GIG p.d.f.'s and c.d.f.'s.

```
Screen[{1.23, 3.4, 1.23, 2.1, 3.4, 5.6, 4.7, 1.23, 2.1, 5.6, 3.4, 5.6, 3.4}]
{{1.23, 3.4, 2.1, 5.6, 4.7}, {3, 4, 2, 3, 1}}
Screen2[{1.23, 3.4, 2.1, 5.6, 4.7, 2.1, 5.6, 1.23}, {3, 4, 2, 3, 1, 2, 1, 4}]
{{1.23, 2.1, 3.4, 4.7, 5.6}, {7, 4, 4, 1, 4}}
```

Fig. 4.7 Examples illustrating the use and functioning of modules `Screen` and `Screen2` in Fig. 4.6

```
EGIGpdffast[r_,li_,z_,prec_:50]:=
                          GIGpdffast[r,li,-Log[z],prec]*1/z

EGIGcdffast[r_,li_,z_,prec_:50]:=
                          1-GIGcdffast[r,li,-Log[z],prec]

EGIGpdf[r_,li_,z_,prec_:50]:=
  If[z>0&&z<1,GIGpdf[r,li,-Log[Rat[z]],prec]*1/Rat[z],
   Print["Third argument must be a value between 0 and 1"]]

EGIGcdf[r_,li_,z_,prec_:50]:=
  If[z>0&&z<1,1-GIGcdf[r,li,-Log[Rat[z]],prec],
   Print["Third argument must be a value between 0 and 1"]]
```

Fig. 4.8 Mathematica® modules for the computation of the EGIG p.d.f. and c.d.f., based on the GIG distribution modules in Fig. 4.7

```
GIGpdffast[r_,li_,z_,prec_:50]:=Module[{p,l,c},
  p=Length[r]; c=Makec[r,l,p];
  SetPrecision[Product[l[[j]]^r[[j]],{j,1,p}]*
    Sum[Sum[c[[j]][[k]]*z^(k-1),{k,1,r[[j]]}]*Exp[-l[[j]]*z],
                                              {j,1,p}],prec]]

GIGcdffast[r_,li_,z_,prec_:50]:=Module[{p,l,c},
  p=Length[r]; c=Makec[r,l,p];
  1-Product[l[[j]]^r[[j]],{j,1,p}]*
     SetPrecision[Sum[Sum[c[[j]][[k]]*(k-1)!*
        Sum[z^i/(i!*l[[j]]^(k-i)),{i,0,k-1}],{k,1,r[[j]]}]*
                                  Exp[-l[[j]]*z],{j,1,p}],prec]]
```

Fig. 4.9 Mathematica® modules called by the modules in Fig. 4.6 to compute the p.d.f. and c.d.f. of the GIG distribution in a fast way

Module `Makec` in Fig. 4.11 is called by modules `GIGpdf` and `GIGcdf` in Figs. 4.9 and 4.10. It is used to compute the coefficients $c_{j,k}$ given by (2.20)–(2.22), which appear in the formulation of the GIG p.d.f. and c.d.f. in (2.18) and (2.19).

```
GIGpdf[r_,li_,zi_,prec_:50]:=Module[{p,l,z},
  If[Count[r, _Integer]==Length[r] && And@@NonNegative[r] &&
       And@@Positive[li] && Length[r]==Length[li],
   If[zi>0,
    p=Length[r]; l=Rat[li]; c=Makec[r,l,p]; z=Rat[zi];
    SetPrecision[Product[l[[j]]^r[[j]],{j,1,p}]*
      Sum[Sum[c[[j]][[k]]*z^(k-1),{k,1,r[[j]]}]*
        Exp[-l[[j]]*z],{j,1,p}],prec],
    Print["Third argument must be a positive value"]]]]

GIGcdf[r_,li_,zi_,prec_:50]:=Module[{p,l,c,z},
  If[Count[r,_Integer]==Length[r] && And@@NonNegative[r] &&
       And@@Positive[li] && Length[r]==Length[li],
   If[zi>0,
    p=Length[r]; l=Rat[li]; c=Makec[r, l, p]; z=Rat[zi];
    1-Product[l[[j]]^r[[j]],{j,1,p}]*
      SetPrecision[Sum[Sum[c[[j]][[k]]*(k-1)!*
         Sum[z^i/(i!*l[[j]]^(k-i)),{i,0,k-1}],{k,1,r[[j]]}]*
           Exp[-l[[j]]*z],{j,1,p}],prec],
    Print["Third argument must be a positive value"]]]]
```

Fig. 4.10 Mathematica® modules called by the modules in Fig. 4.6 to compute the p.d.f. and c.d.f. of the GIG distribution in a more precise way

```
Makec[r_,l_,p_]:=Module[{c},
  c=Table[Table[1,{j,1,Max[r]}],{i,1,p}];
  Do[c[[i,r[[i]]]]=Product[(l[[j]]-l[[i]])^(-r[[j]]),
                                                  {j,1,i-1}]*
     Product[(l[[j]]-l[[i]])^(-r[[j]]),{j,i+1,p}]/
                                         (r[[i]]-1)!,{i,1,p}];
  Do[Do[c[[i,r[[i]]-k]]=Sum[((r[[i]]-k+j-1)!*
       (Sum[r[[h]]/(l[[i]]-l[[h]])^j,{h,1,i-1}]+
        Sum[r[[h]]/(l[[i]]-l[[h]])^j,{h,i+1,p}])*
         c[[i]][[r[[i]]-(k-j)]])/(r[[i]]-k-1)!,{j,1,k}]/
                           k,{k,1,r[[i]]-1}],{i,1,p}];
  c]
```

Fig. 4.11 Mathematica® module used to compute the coefficients $c_{j,k}$ ($k = 1, \ldots, r_j$; $j = 1, \ldots, p$) in the GIG p.d.f. and c.d.f.

In Figs. 4.12, 4.13, and 4.14 are listed the Mathematica® modules used to compute the p.d.f.'s and c.d.f.'s for each of the Corollaries 4.1, 4.2, and 4.3, based on the Meijer G function formulation. Modules `PDFCor1Meijer` and `CDFCor1Meijer` call module `ParCor1Meijer` to setup, from the values of the parameters $a_{\nu\ell}$, $k_{\nu\ell}$, and $m_{\nu\ell}$ ($\nu = 1, \ldots, m^*$; $\ell = 1, \ldots, n_\nu$), the arguments for the Mathematica® native function/module `Meijer`, which is then used to make the computations. Modules `PDFCor1Meijer` and `CDFCor1Meijer` have four

```
PDFCor1Meijer[a_,k_,m_,z_]:=Module[{vec},
  vec=ParCor1Meijer[a,k,m];
  vec[[3]]*MeijerG[{{},vec[[1]]+vec[[2]]-1},
                                              {vec[[1]]-1,{}},z]
  ]

CDFCor1Meijer[a_,k_,m_,z_]:=Module[{vec},
  vec=ParCor1Meijer[a,k,m];
  vec[[3]]*MeijerG[{{1},vec[[1]]+vec[[2]]},{vec[[1]],{0}},z]
  ]

ParCor1Meijer[a_,k_,m_]:=Module[{ms,n,av,bv,K1},
  ms=Length[a];
  n=Map[Length,k];
  av=Table[Table[Table[Rat[a[[nu]]]-(j+Sum[k[[nu,r]],
            {r,1,ell-1}])/k[[nu,ell]],{j,1,k[[nu,ell]]}],
                              {ell,1,n[[nu]]}],{nu,1,ms}];
  bv=Table[Table[Table[m[[nu,ell]]/k[[nu,ell]],
            {j,1,k[[nu,ell]]}],{ell,1,n[[nu]]}],{nu,1,ms}];
  K1=Product[Product[Product[Gamma[av[[nu,ell,j]]+
                       bv[[nu,ell,j]]]/Gamma[av[[nu,ell,j]]],
            {j,1,k[[nu,ell]]}],{ell,1,n[[nu]]}],{nu,1,ms}];
  {Flatten[av],Flatten[bv],K1}
  ]
```

Fig. 4.12 Mathematica® modules used to implement the Meijer G functions for the p.d.f. and c.d.f. in Corollary 4.1

mandatory arguments which are, in this order, the list of the a_ν parameters, the two-level lists of the $k_{\nu\ell}$ and $m_{\nu\ell}$ parameters and the running value of the r.v. Z where the p.d.f. or the c.d.f. is to be computed.

Modules `PDFCor2Meijer` and `CDFCor2Meijer` have a similar functioning and usage, with the only difference that these have five mandatory arguments, which are, in this order, the lists of the a_ν, n_ν, k_ν, and m_ν ($\nu = 1, \ldots, m^*$) parameters and the running value of the r.v. Z, while modules `PDFCor3Meijer` and `CDFCor3Meijer` also have five mandatory arguments which are, in this order, the values for the a_ν, k_ν, n_ν, and s_ν ($\nu = 1, \ldots, m^*$) parameters and the running value for the r.v. Z.

In Fig. 4.15 it is exemplified how one can compute the values for the p.d.f. and the c.d.f. for scenarios I.1, II.1, and III.1 using the EGIG implementation based modules `PDFCorEGIG` and `CDFCorEGIG`, both with the fast and the more precise versions, and in Fig. 4.16 we have the set of commands to be used with modules `PDFCor1Meijer`, `PDFCor2Meijer`, and `PDFCor3Meijer` to compute the p.d.f. values for these same scenarios by using the Meijer G function based implementations. Similar commands which use the corresponding c.d.f. modules would be used to compute the c.d.f. values.

```
PDFCor2Meijer[a_,n_,k_,m_,z_]:=Module[{vec},
  vec=ParCor2Meijer[a,n,k,m];
  vec[[3]]*MeijerG[{{},vec[[1]]+vec[[2]]-1},
                                        {vec[[1]]-1,{}},z]
  ]

CDFCor2Meijer2[a_,n_,k_,m_,z_]:=Module[{vec},
  vec=ParCor2Meijer[a,n,k,m];
  vec[[3]]*MeijerG[{{1},vec[[1]]+vec[[2]]},{vec[[1]],{0}},z]
  ]

ParCor2Meijer[a_,n_,k_,m_]:=Module[{ms,av,bv,K1},
  ms=Length[a];
  av=Table[Table[Table[Rat[a[[nu]]]+1-ell-j/k[[nu]],
               {j,1,k[[nu]]}],{ell,1,n[[nu]]}],{nu,1,ms}];
  bv=Table[Table[Table[m[[nu]]/k[[nu]],{j,1,k[[nu]]}],
                           {ell,1,n[[nu]]}],{nu,1,ms}];
  K1=Product[Product[Product[Gamma[av[[nu,ell,j]]+
                 bv[[nu,ell,j]]]/Gamma[av[[nu,ell,j]]],
               {j,1,k[[nu]]}],{ell,1,n[[nu]]}],{nu,1,ms}];
  {Flatten[av],Flatten[bv],K1}
  ]
```

Fig. 4.13 Mathematica® modules used to implement the Meijer G functions for the p.d.f. and c.d.f. in Corollary 4.2

```
PDFCor3Meijer[a_,k_,n_,s_,z_]:=Module[{vec},
  vec=ParCor3Meijer[a,k,n,s];
  vec[[3]]*MeijerG[{{},vec[[1]]+vec[[2]]-1},
                                        {vec[[1]]-1,{}},z]
  ]

CDFCor3Meijer[a_,k_,n_,s_,z_]:=Module[{vec},
  vec=ParCor3Meijer[a,k,n,s];
  vec[[3]]*MeijerG[{{1},vec[[1]]+vec[[2]]},{vec[[1]],{0}},z]
  ]

ParCor3Meijer[a_,k_,n_,s_]:=Module[{ms,av,bv,K1},
  ms=Length[a];
  av=Flatten[Table[Table[Table[Rat[a[[nu]]]-
            ((ell-1)*k[[nu]]+j)/n[[nu]],{j,1,k[[nu]]}],
                              {ell,1,n[[nu]]}],{nu,1,ms}]];
  bv=Flatten[Table[Table[Table[(j+(ell-1)*k[[nu]]+ell+
       s[[nu]]-1)/n[[nu]],{j,1,k[[nu]]}],{ell,1,n[[nu]]}],
                                              {nu,1,ms}]];
  K1=Product[Gamma[av[[j]]+bv[[j]]]/Gamma[av[[j]]],
                                        {j,1,Length[av]}];
  {av,bv,K1}  ]
```

Fig. 4.14 Mathematica® modules used to implement the Meijer G functions for the p.d.f. and c.d.f. in Corollary 4.3

```
(* Computation of p.d.f. and c.d.f. for Scenario I.1 *)
a = {16.3, 18.6}; k = {{3, 1, 4}, {2, 5, 3, 2, 2}};
m = {{2, 2, 3}, {1, 5, 2, 4, 3}}; z = 1 / 10;
PDFCorEGIG[1, a, 0, k, m, z, 21]
PDFCorEGIG[1, a, 0, k, m, z, 21, 1]
CDFCorEGIG[1, a, 0, k, m, z, 20]
CDFCorEGIG[1, a, 0, k, m, z, 20, 1]
0.87650963635
0.87650963635
0.014972280165
0.014972280165
(* Computation of p.d.f. and c.d.f. for Scenario II.1 *)
a = {71}; n = {7}; k = {2}; m = {10};
z = Ratk[.313];
PDFCorEGIG[2, a, n, k, m, z, 93]
PDFCorEGIG[2, a, n, k, m, z, 93, 1]
CDFCorEGIG[2, a, n, k, m, z, 91]
CDFCorEGIG[2, a, n, k, m, z, 91, 1]
4.5753687675
4.5753687675
0.11162120385
0.11162120385
(* Computation of p.d.f. and c.d.f. for Scenario III.1 *)
a = {45.3, 12.3}; n = {2, 2}; k = {6, 2}; s = {2, 3};
z = Ratk[.06];
PDFCorEGIG[3, a, n, k, s, z, 40]
PDFCorEGIG[3, a, n, k, s, z, 40, 1]
CDFCorEGIG[3, a, n, k, s, z, 39]
CDFCorEGIG[3, a, n, k, s, z, 39, 1]
4.6215723567
4.6215723567
0.056824913519
0.056824913519
```

Fig. 4.15 Examples illustrating the use of modules PDFCorEGIG and CDFCorEGIG to compute values for the p.d.f. and c.d.f. for scenarios I.1, II.1, and III.1

Figure 4.17 bears the listing of the module QuantCorEGIG, used to compute the quantiles for the distributions in Corollaries 4.1–4.5, in an "assisted" way. We call this version of the computation of the quantiles "assisted" since when using this module we need to provide both a starting value and the precision for the computations. An adequate starting value may be easily obtained from a few computations of the corresponding c.d.f., carried out using module CDFCorEGIG,

```
a = {16.3, 18.6}; k = {{3, 1, 4}, {2, 5, 3, 2, 2}};
m = {{2, 2, 3}, {1, 5, 2, 4, 3}}; z = 1/10;
SetPrecision[PDFCor1Meijer[a, k, m, z], 22]
a = {71}; n = {7}; k = {2}; m = {10};
z = Ratk[.313];
SetPrecision[PDFCor2Meijer[a, n, k, m, z], 95]
a = {45.3, 12.3}; n = {2, 2}; k = {6, 2}; s = {2, 3};
z = Ratk[.06];
SetPrecision[PDFCor3Meijer[a, n, k, s, z], 42]
```

Fig. 4.16 Commands to be used to compute the values for the p.d.f.'s in scenarios I.1, II.1, and III.1, based on the Meijer *G* function formulation

```
QuantCorEGIG[cor_,a_,n_,k_,m_,quant_,xo_,eps_,prec_:40,
      ind_:0]:=Module[{nx,xa,vec,r,p,l,c,pr, F1, F2, quantn},
  quantn=Rat[quant];
  nx=Rat[xo];
  xa=0;
  If[cor==1,vec=ParCor1EGIG[a,k,m],
     If[cor==2,vec=ParCor2EGIG[a,n,k,m],
                    vec=ParCor3EGIG[a,k,n,m]]];
  r=vec[[2]];
  p=Length[r];
  l=Rat[vec[[1]]];
  c=Makec[r,l,p];
  pr=Product[l[[j]]^r[[j]],{j,1,p}];
  If[TrueQ[ind==0||ind==Null],
   {F1[x_]=EGIGpdfsimpfast[r,l,x,p,pr,c];
    F2[x_]=EGIGcdfsimpfast[r,l,x,p,pr,c];
    While[Abs[nx-xa]>eps,{xa=nx;
      nx=xa+(quantn-SetPrecision[F2[xa],prec])/
         SetPrecision[F1[xa],prec]}]},
   {F1[x_]=EGIGpdfsimp[r,l,x,p,pr,c,prec];
    F2[x_]=EGIGcdfsimp[r,l,x,p,pr,c,prec];
    While[Abs[nx-xa]>eps,{xa=nx;
      nx=xa+(quant-F2[Rat[xa]])/F1[Rat[xa]]}]}];
  nx  ]
```

Fig. 4.17 Mathematica® module used for the computation of quantiles for the distributions in Corollaries 4.1–4.3

while the minimal precision needed may then be obtained by starting with the same precision, that is, the same number of digits, used to compute the corresponding c.d.f., or something slightly higher, and, if necessary, adjusting it to slightly higher values, in case the module gives as output an error message or no output at all or yet if it does not give any result in a quite short time.

```
a = {16.3, 18.6}; k = {{3, 1, 4}, {2, 5, 3, 2, 2}};
m = {{2, 2, 3}, {1, 5, 2, 4, 3}};
QuantCorEGIG[1, a, 0, k, m, 0.05, .1, 10^-12, 23]
a = {71}; n = {7}; k = {2}; m = {10};
QuantCorEGIG[2, a, n, k, m, 0.05, .286, 10^-11, 160]
a = {45.3, 12.3}; n = {2, 2}; k = {6, 2}; s = {2, 3};
QuantCorEGIG[3, a, n, k, s, 0.05, .05, 10^-12, 55]
```

Fig. 4.18 Commands to be used to compute the 0.05 quantiles for scenarios I.1, II.1, and III.1 with the nonautomated module `QuantCorEGIG`

Module `QuantCorEGIG` uses a Newton type algorithm to compute successive approximations for the quantiles, starting from a given initial value. Such procedure reveals itself as much more efficient and reliable than the use of the generalist native Mathematica® module/function `FindRoot`.

In Fig. 4.18 we have the set of commands used to obtain the 0.05 quantiles for scenarios I.1, II.1, and III.1 which are listed in Tables 4.4, 4.11, and 4.17, using module `QuantCorEGIG`. This module has eight mandatory arguments and two optional ones. The first five mandatory arguments are exactly the same as those for modules `PDFCorEGIG` and `CDFCorEGIG`, while the remaining three ones are:

`quant` – the quantile value, a value between 0 (zero) and 1
`xo` – the initial value for the quantile computation
`eps` – the "epsilon" value, that is, an upper-bound for the difference between two consecutive approximations for the quantile

in this order. The two optional arguments are:

`prec` – the precision, that is, the number of digits to be used in computing the p.d.f. and the c.d.f. values
`ind` – see the argument with a similar name for modules `PDFCorEGIG` and `CDFCorEGIG`.

The modules in Fig. 4.19 are two sets of modules used to compute the EGIG p.d.f. and c.d.f. in repeated computations of these functions. These modules call the corresponding computing GIG modules in Fig. 4.20. These modules, instead of computing the $c_{j,k}$ coefficients in the GIG/EGIG distributions, take these as an argument list, which is denoted as the argument `c` in the modules in Figs. 4.19 and 4.20. This list is computed inside the module `QuantCorEGIG`, before the modules that compute the EGIG p.d.f. and c.d.f. are called. Anyway, these are computational details that allow huge improvements in the computing times of quantiles, but which are transparent to the end-user, that is, these are details about which the end-user does not have to worry about.

Module `QuantCor` in Fig. 4.21 is the module used to compute the "automated" or unassisted version of the quantiles. This module does not require either a starting value or an *a priori* setup precision value as arguments. The module has six mandatory arguments which are:

```
EGIGpdfsimpfast[r_,l_,z_,p_,pr_,c_,prec_:50]:=
    GIGpdfsimpfast[r,l,-Log[Rat[z]],p,pr,c,prec]*1/Rat[z]

EGIGcdfsimpfast[r_,l_,z_,p_,pr_,c_,prec_:50]:=
    1-GIGcdfsimpfast[r,l,-Log[Rat[z]],p,pr,c,prec]

EGIGpdfsimp[r_,l_,z_,p_,pr_,c_,prec_:50]:=
    GIGpdfsimp[r,l,-Log[Rat[z]],p,pr,c,prec]*1/Rat[z]

EGIGcdfsimp[r_,l_,z_,p_,pr_,c_,prec_:50]:=
    1-GIGcdfsimp[r,l,-Log[Rat[z]],p,pr,c,prec]
```

Fig. 4.19 Mathematica® modules called by module QuantCorEGIG in Fig. 4.13, used to compute the EGIG p.d.f. and c.d.f.

```
GIGpdfsimp[r_,l_,zi_,p_,pr_,c_,prec_:50]:=Module[{z},
  z=Rat[zi];
  SetPrecision[pr*Sum[Sum[c[[j]][[k]]*z^(k-1),{k,1,r[[j]]}]*
      Exp[-l[[j]]*z],{j,1,p}],prec]]

GIGcdfsimp[r_,l_,zi_,p_,pr_,c_,prec_:50]:= Module[{z},
  z=Rat[zi];
  1-pr*SetPrecision[Sum[Sum[c[[j]][[k]]*(k-1)!*
         Sum[z^i/(i!*l[[j]]^(k-i)),{i,0,k-1}],{k,1,r[[j]]}]*
                                    Exp[-l[[j]]*z],{j,1,p}],prec]]

GIGpdfsimpfast[r_,l_,z_,p_,pr_,c_,prec_:50]:=
  pr*Sum[Sum[c[[j]][[k]]*z^(k-1),{k,1,r[[j]]}]*Exp[-l[[j]]*z]
                                                      ,{j,1,p}]

GIGcdfsimpfast[r_,l_,z_,p_,pr_,c_,prec_:50]:=
  1-pr*Sum[Sum[c[[j]][[k]]*(k-1)!*
        Sum[z^i/(i!*l[[j]]^(k-i)),{i,0,k-1}],  {k,1,r[[j]]}]*
                                       Exp[-l[[j]]*z],{j,1,p}]
```

Fig. 4.20 Mathematica® modules called by the modules in Fig. 4.14, used to compute the GIG p.d.f. and c.d.f. avoiding the computation of the parameters c_{jk} in each step

- `cor` – the number of the corollary we are referring to: 1, 2, 3, 4, or 5, respectively for Corollary 4.1, 4.2, 4.3, 4.4, or 4.5
- `qquant` – the quantile value, a value between 0 (zero) and 1
- `a` – the same as the second mandatory argument of modules `PDFCorEGIG` and `CDFCorEGIG` (see pages 56–57)
- `n` – the same as the third mandatory argument of modules `PDFCorEGIG` and `CDFCorEGIG` (see pages 56–57)
- `k` – the same as the fourth mandatory argument of modules `PDFCorEGIG` and `CDFCorEGIG` (see pages 56–57)

```
QuantCor[cor_,qquant_,a_,n_,k_,m_,eps_:10^-11,ind_:0,pp_:11,
        preco_:50,xoo_:1/2]:=
 Module[{quant,xo,prec,vec,r,p,l,c,pr,xa,xb,res,nx,F,F1,F2},
  quant=Rat[qquant];
  If[TrueQ[xoo==0||xoo==Null],xo=1/2,xo=Rat[xoo]];
  prec=preco;
  If[cor==1,vec=ParCor1EGIG[a,k,m],
       If[cor==2,vec=ParCor2EGIG[a,n,k,m],
               vec=ParCor3EGIG[a,k,n,m]]];
  r=vec[[2]];
  p=Length[r];
  l=vec[[1]];
  c=Makec[r, l, p];
  pr=Product[l[[j]]^r[[j]],{j,1,p}];
  F[x_]=EGIGcdfsimpfast[r,l,x,p,pr,c];
  While[Length[RealDigits[res=SetPrecision[F[xo],prec]][[1]]]
                              <15*Max[11,pp],prec=3/2*prec];
  If[res>quant,xa=xo;xo=xo/2;
   While[SetPrecision[F[xo],prec]>quant,xa=xo;xo=xo/2],
   xa=xo;xo=xo*2;
        While[SetPrecision[F[xo],prec]<quant,xa=xo;xo=xo*2]];
  If[xo>xa,xb=xo,xb=xa;xa=xo];
  xo=(xa+xb)/2;
  res=SetPrecision[F[xo], prec];
  While[Abs[res-quant]>10^-3,xo=(xa+xb)/2;
   If[(res=SetPrecision[F[xo],prec])>quant,xb=xo,xa=xo]];
  nx=Rat[xo];
  xa=0;
  If[TrueQ[ind==0||ind==Null],
                        {F1[x_]=EGIGpdfsimpfast[r,l,x,p,pr,c];
    While[Abs[nx-xa]>eps,{xa=nx;
      nx=xa+(quant-SetPrecision[F[xa],prec])/
                               SetPrecision[F1[xa],prec]}]},
   {F1[x_]=EGIGpdfsimp[r,l,x,p,pr,c];
    F2[x_]=EGIGcdfsimp[r,l,x,p,pr,c];
    While[Abs[nx-xa]>eps,{xa=nx;
      nx=xa+(quant-SetPrecision[F2[Rat[xa]],prec])/
                         SetPrecision[F1[Rat[xa]],prec]}]}];
  SetPrecision[nx,pp]]
```

Fig. 4.21 Mathematica® module used for the unassisted computation of quantiles for the distributions in Corollaries 4.1–4.3

m – the same as the fifth mandatory argument of modules PDFCorEGIG and CDFCorEGIG (see pages 56–57)

in this order, and four optional arguments, which are

eps – the "epsilon" value, that is, an upper-bound for the difference between two consecutive approximations for the quantile (default value: 10^{-11})

```
QuantCor[1, 0.05, a, 0, k, m, 10^-12]
QuantCor[2, 0.05, a, n, k, m]
QuantCor[3, 0.05, a, n, k, s, 10^-12]
```

Fig. 4.22 Commands to be used to compute the 0.05 quantiles for scenarios I.1, II.1, and III.1 with the automated module `QuantCor`

`ind` – see the argument with a similar name for modules `PDFCorEGIG` and `CDFCorEGIG` (all computing times reported in Tables 4.4, 4.11, and 4.17 were obtained by using the default value of 0 (zero) for this argument)

`pp` – the number of precision digits to be used to print the quantile value (default value: 11)

`preco` – the number of digits to be used in the initial computations of the c.d.f. carried out by the module in order to determine a good starting value (default value: 50) (this value is internally adjusted by the module if necessary, even when a value is explicitly given for this argument, anyway it may have to be used for cases where the argument `a` has extremely large values)

`xo` – the original starting value for the computation of the quantile (default value: 1/2)

Module `QuantCor` uses a variant of the bisection algorithm in order to obtain a good starting value for the quantile computation. This starting value has to yield a value for the c.d.f. that falls within 10^{-3} of the required α-value specified in the mandatory argument `qquant`. Once this starting value is found, a Newton-type algorithm in all similar to the one used by module `QuantCorEGIG` is used till the final approximate value of the quantile is obtained.

In Fig. 4.22 we have the commands used to compute the 0.05 quantiles for scenarios I.1, II.1, and III.1 with this automated module using the corresponding parameter definitions in Fig. 4.18.

References

Erdélyi, A., Magnus, W., Oberhettinger, F., Tricomi, F.G. (eds.): Higher Transcendental Functions (The Bateman Manuscript Project), vol. I. McGraw-Hill, New York (1953)

Fox, C.: The *G* and *H* functions as symmetrical kernels. Trans. Amer. Math. Soc. **98** 395–429 (1961)

Meijer, C.S.: Über Whittakersche bzw. Besselsche Funktionen und deren Produkte. Nieuw Archief voor Wiskunde **18**(2), 10–29 (1936a)

Meijer, C.S.: Neue Integraldarstellungen aus der Theorie der Whittakerschen und Hankelschen Funktionen. Math. Annalen **112**(2), 469–489 (1936b)

Meijer, C.S.: Multiplikationstheoreme für die Funktion $G_{p,q}^{m,n}(z)$. Proc. Koninklijk Nederlandse Akademie van Weteenschappen **44**, 1062–1070 (1941)
Meijer, C.S.: On the G-function. Proc. Koninklijk Nederlandse Akademie van Weteenschappen **49**, 227–237, 344–356, 457–469, 632–641, 765–772, 936–943, 1063–1072, 1165–1175 (1946)
Yilmaz, F., Alouini, M.-S.: Product of the powers of generalized Nakagami-m variates and performance of cascaded fading channels. Proceedings of IEEE Global Telecommunications Conference, pp. 1–8 (2009)

Chapter 5
Application of the Finite Form Representations of Meijer *G* and Fox *H* Functions to the Distribution of Several Likelihood Ratio Test Statistics

Abstract In this chapter the authors present a number of l.r.t. statistics used in Multivariate Analysis, both for real and complex random variables, whose exact distributions are particular cases of the products of independent Beta r.v.'s in Theorems 3.1–3.3 in Chap. 3, and whose p.d.f.'s and c.d.f.'s have thus their expressions given by Corollaries 4.1–4.5 in the previous chapter.

Keywords Circularity test · Composition of tests · Covariance structure tests · Equality of mean vectors · Exact distribution · Exact quantiles · Independence test · Invariance · Parallelism test for profiles · Scalar block sphericity test · Tests for an expected value matrix

In this chapter several l.r.t. statistics are briefly described and studied, and it is shown how their exact distributions may be obtained from that of the products of Beta r.v.'s in Theorems 3.1–3.3 in Chap. 3, and how their exact p.d.f.'s and c.d.f.'s may thus be obtained from Corollaries 4.1–4.5 in Chap. 4.

To be more precise, in fact these statistics are not exactly l.r.t. statistics, but rather powers of these statistics, functions of the sample sizes, which will be clearly stated in each case. However, without any risk of confusion and without anyway harming the correctness of the results, for simplicity of writing, in most cases, after this specification is clearly made, the statistic will be addressed as a "l.r.t. statistic."

At the end of the title of each subsection is provided an acronym for the l.r.t. in that subsection. This acronym will be used as part of the name of the Mathematica®, MAXIMA, and R modules/functions provided to implement the computation of the p.d.f., c.d.f., quantiles and p-values for the corresponding test statistic.

For each l.r.t. five end-user modules/functions are provided: Pval<acronym>, PvalData<acronym>, Quant<acronym>, together with PDF<acronym> and CDF<acronym>. The first of these modules is used to compute the p-value corresponding to a given computed value of the statistic, while the second is used to compute the p-value corresponding to the value of the statistic obtained from a data set, given in a file. The third module is used to compute the α-quantile for any given value of α in the range (0,1) and the two last modules are used to compute values

C. A. Coelho, B. C. Arnold, *Finite Form Representations for Meijer G and Fox H Functions*, Lecture Notes in Statistics 223,
https://doi.org/10.1007/978-3-030-28790-0_5

or make plots respectively of the exact p.d.f. and c.d.f. of the statistic. We should remark that the Pval<acronym> and CDF<acronym> modules are mostly the same module, being the CDF<acronym> modules provided only to make things conformable with the existence of the PDF<acronym> modules.

Simple examples of the usage of these modules are provided in each subsection. These modules are also part of the Mathematica®, MAXIMA, and R packages which are described in more detail in Chap. 6. As already remarked and explained in Chap. 1, while all random sample matrices in the text are of the form $p{\times}n$, where p is the number of variables and n the sample size, all sample matrices used in the computation modules are of the form $n{\times}p$, since this is the most common setup for data matrices and datafiles.

Although for every test addressed, a module called Quant<acronym> is provided in all three softwares to compute quantiles, together with another module called Pval<acronym> which computes the p-value that corresponds to a given computed value of the test statistic, tables of quantiles are provided for each test for some selected values of the parameters in the book's supplementary material web site. However, it would be an impossible task to provide in these tables quantiles for all possible combinations of the parameters, particularly for tests as the ones in Sects. 5.1.11 and 5.1.12. As such, the authors strongly rather advise the use of the above-mentioned modules, as they are also able to provide more significant digits than the six significant digits supplied in the tables and even have a somewhat simpler usage than these tables.

We should also note here that all tests addressed in this chapter enjoy some form of invariance for some type of linear transformation, it being usually easy to discern from the context and the expressions for the l.r.t. statistics the type of invariance that each test enjoys. Anyway in each subsection it is expressly stated under which conditions the test is invariant, with all tests for complex random variables being invariant under similar conditions to those of the corresponding test for real random variables. Also, for each and every test, any linear transformation matrix, or any of its diagonal blocks, in case it is block-diagonal, may be replaced by any permutation matrix multiplied by any non-null real.

5.1 Likelihood Ratio Test Statistics Whose Distributions Correspond to the Products in Theorems 3.1 and 3.2 and That Have p.d.f. and c.d.f. Given by Corollary 4.1 or 4.2

The products of independent Beta r.v.'s in Theorems 3.1 and 3.2 yield the exact distribution of several l.r. (likelihood ratio) statistics used in Multivariate Analysis. Several representative examples are provided in the next subsections, with real data applications.

5.1.1 The Likelihood Ratio Statistic to Test the Equality of Mean Vectors for Real Random Variables [EqMeanVecR]

Let $\underline{X}_k \sim N_p(\underline{\mu}_k, \Sigma)$ $(k = 1, \ldots, q)$ be q random vectors with real p-variate Normal distributions, with expected value $\underline{\mu}_k$ and covariance matrix Σ. Let us suppose we have q independent samples, one from each $\underline{X}_k$, the k-th of which is of size n_k $(k = 1, \ldots, q)$, and let $n = \sum_{k=1}^{q} n_k$.

The $(2/n)$-th power of the l.r.t. statistic to test

$$H_0 : \underline{\mu}_1 = \underline{\mu}_2 = \cdots = \underline{\mu}_q \tag{5.1}$$

is (see for example Kshirsagar (1972, Chap. 9))

$$\Lambda = \frac{|A|}{|A + B|} \tag{5.2}$$

where

$$A = \sum_{k=1}^{q} (n_k - 1) S_k \quad \text{and} \quad B = \sum_{k=1}^{q} n_k (\overline{\underline{X}}_k - \overline{\underline{X}})(\overline{\underline{X}}_k - \overline{\underline{X}})' \tag{5.3}$$

are respectively the "within" and "between" sum of squares and sum of products matrices. In (5.3), S_k and $\overline{\underline{X}}_k$ are respectively the sample covariance matrix and mean vector of the k-th sample and

$$\overline{\underline{X}} = \frac{1}{n} \sum_{k=1}^{q} n_k \, \overline{\underline{X}}_k \,. \tag{5.4}$$

Given the independence between the sample covariance matrices S_k and the mean vectors $\overline{\underline{X}}_k$, matrices A and B in (5.2) and (5.3) are independent, and while the matrix A in (5.3) has, for $n > p + q - 1$, a Wishart $W_p(n - q, \Sigma)$ distribution, the matrix B, under H_0 in (5.1), has a Wishart or pseudo-Wishart $W_p(q - 1, \Sigma)$ distribution.

For references on the Wishart distribution, see Wishart (1928), Kshirsagar (1972, Chap. 3), Anderson (2003, Sect. 7.2), and Muirhead (2005, Sect. 3.2), and for the pseudo-Wishart distribution see Kshirsagar (1972, Sect. 3.6).

As such (see Kshirsagar (1972, Chap. 9) and Appendix 1), for $n > p + q - 1$, under H_0 in (5.1),

$$\Lambda \stackrel{d}{\equiv} \prod_{j=1}^{p} Y_j \stackrel{d}{\equiv} \prod_{k=1}^{q-1} Y_k^* \tag{5.5}$$

where, for $j = 1, \ldots, p$, and $k = 1, \ldots, q-1$,

$$Y_j \sim Beta\left(\frac{n-q+1-j}{2}, \frac{q-1}{2}\right) \quad \text{and} \quad Y_k^* \sim Beta\left(\frac{n-p-k}{2}, \frac{p}{2}\right) \tag{5.6}$$

represent respectively a set of p and a set of $q-1$ independent random variables. The exact p.d.f. and c.d.f. of Λ are thus given in terms of the Meijer G function as

$$f_\Lambda(z) = \left\{\prod_{j=1}^{p} \frac{\Gamma\left(\frac{n-j}{2}\right)}{\Gamma\left(\frac{n-q+1-j}{2}\right)}\right\} G_{p,p}^{p,0}\left(\left.\begin{array}{c}\left\{\frac{n-j}{2}-1\right\}_{j=1:p} \\ \left\{\frac{n-q+1-j}{2}-1\right\}_{j=1:p}\end{array}\right| z\right)$$

and

$$F_\Lambda(z) = \left\{\prod_{j=1}^{p} \frac{\Gamma\left(\frac{n-j}{2}\right)}{\Gamma\left(\frac{n-q+1-j}{2}\right)}\right\} G_{p+1,p+1}^{p,1}\left(\left.\begin{array}{c}\left\{1, \left\{\frac{n-j}{2}\right\}_{j=1:p}\right\} \\ \left\{\left\{\frac{n-q+1-j}{2}\right\}_{j=1:p}, 0\right\}\end{array}\right| z\right)$$

or

$$f_\Lambda(z) = \left\{\prod_{k=1}^{q-1} \frac{\Gamma\left(\frac{n-k}{2}\right)}{\Gamma\left(\frac{n-p-k}{2}\right)}\right\} G_{q-1,q-1}^{q-1,0}\left(\left.\begin{array}{c}\left\{\frac{n-k}{2}-1\right\}_{k=1:q-1} \\ \left\{\frac{n-p-k}{2}-1\right\}_{k=1:q-1}\end{array}\right| z\right)$$

and

$$F_\Lambda(z) = \left\{\prod_{k=1}^{q-1} \frac{\Gamma\left(\frac{n-k}{2}\right)}{\Gamma\left(\frac{n-p-k}{2}\right)}\right\} G_{q,q}^{q-1,1}\left(\left.\begin{array}{c}\left\{1, \left\{\frac{n-k}{2}\right\}_{k=1:q-1}\right\} \\ \left\{\left\{\frac{n-p-k}{2}\right\}_{k=1:q-1}, 0\right\}\end{array}\right| z\right).$$

But, for even p or odd q, the exact distribution of Λ is of the type of the products in Theorem 3.1 or 3.2, and therefore its exact p.d.f. and c.d.f. have very manageable expressions which are given, for even p, by Corollary 4.1 or 4.2, with

$$m^* = 1, \quad a_1 = \frac{n-q+1}{2}, \quad n_1 = \frac{p}{2}, \quad k_1 = 2 \quad \text{and} \quad m_1 = q-1, \tag{5.7}$$

while for odd q, the exact p.d.f. and c.d.f. of Λ are given by Corollary 4.1 or 4.2, with

$$m^* = 1, \quad a_1 = \frac{n-p}{2}, \quad n_1 = \frac{q-1}{2}, \quad k_1 = 2 \quad \text{and} \quad m_1 = p. \tag{5.8}$$

Then, the results in Theorem 3.1 or 3.2 and in Corollary 4.1 or 4.2, yield absolutely identical formulations for either the case in (5.7) or the case in (5.8), giving the exact p.d.f. and c.d.f. of Λ, expressed in terms of the EGIG distribution, respectively as (see Appendix of Chap. 2 for notation and definition of the EGIG distribution)

$$f_\Lambda(z) = f^{EGIG}\left(z \,\middle|\, \{r_j\}_{j=1:p+q-3};\left\{\frac{n-2-j}{2}\right\}_{j=1:p+q-3};p+q-3\right) \tag{5.9}$$

and

$$F_\Lambda(z) = F^{EGIG}\left(z \,\middle|\, \{r_j\}_{j=1:p+q-3};\left\{\frac{n-2-j}{2}\right\}_{j=1:p+q-3};p+q-3\right) \tag{5.10}$$

where

$$r_j = \begin{cases} h_j\,, & j=1,2 \\ r_{j-2}+h_j\,, & j=3,\dots,p+q-3 \end{cases}$$

with

$$h_j = (\#\text{ of elements in } \{p,q-1\} \geq j) - 1\,, \quad j=1,\dots,p+q-3\,,$$

result that confirms the ones obtained in Coelho et al. (2010), Marques et al. (2011), and Arnold et al. (2013).

Schatzoff (1966) and Lee (1972) obtained the skeletal form of the p.d.f. and c.d.f. of Λ in (5.2) in case p is even and q is odd, but, as they themselves say, did not explicitly work out the formulation for the parameters in the distribution, with Lee (1972) imposing some further restrictions on the relation between the values of p and q. Pillai and Gupta (1969) fully worked out just a few cases but only on a case by case basis, and only for a few small values of p.

For the cases where p is odd and q is even, near-exact distributions for the statistic Λ in (5.2) are available (Coelho et al., 2010; Marques et al., 2011) or the reader may also refer to the material in Chap. 7.

In order to carry out a classical level α test of the null hypothesis in (5.1), we should reject this hypothesis if the computed value of Λ in (5.2) is smaller than the α quantile of Λ, which may be computed using the module `QuantEqMeanVecR`. Otherwise, we may just compute the p-value that corresponds to the computed value of Λ, using module `PvalEqMeanVecR`.

The present test is invariant under linear transformations of the type $\underline{X}_k \rightarrow C\underline{X}_k+\underline{b}$, where C is any nonrandom, full-rank, $p\times p$ matrix and $\underline{b}$ is any nonrandom $p\times 1$ vector.

For two alternative manners to implement the present test see Sects. 5.1.4.2 and 5.1.9.4.

We should note here that the distribution studied in this subsection, for the l.r.t. statistic Λ in (5.2), remains valid, under the null hypothesis in (5.1), if the distribution of the random vectors $\underline{X}_k$ is not multivariate Normal, but rather some elliptically contoured or left orthogonal-invariant distribution—see Theorem 5.2.4 in Fang and Zhang (1990, Chap. V) and Sect. 8.11 in Anderson (2003), and also Jensen and Good (1981), Kariya (1981), Anderson et al. (1986), and Anderson and Fang (1990).

When the different populations (or samples) correspond to the different levels of a factor, the present test is the test of significance for a factor with q levels in a one-way MANOVA (Multivariate ANalysis Of VAriance) model. Indeed, for $p = 1$ we have, from (5.5) and (5.6)

$$\Lambda \stackrel{d}{\equiv} Y_1 \sim Beta\left(\frac{n-q}{2}, \frac{q-1}{2}\right) \implies F = \frac{n-q}{q-1}\frac{1-\Lambda}{\Lambda} \sim F_{q-1,n-q}\,,$$

where F is the usual F statistic used to test the significance of the effect of a factor with q levels in a univariate one-way ANOVA model.

See Sect. 5.1.1.3 for further details on the distribution of Λ for $p = 1$ or $q = 2$, and the relation between the Beta and F distributions.

The representation of the exact p.d.f. and c.d.f. of the statistic Λ in (5.2) through the EGIG p.d.f. and c.d.f. has the advantage of enabling a very quick computation of p-values and quantiles and also a fast implementation of plots of these functions as those in Figs. 5.1 and 5.2. These plots, which may be easily obtained with the software Mathematica®, using the EGIG formulation and a set of commands as the one in Fig. 5.3 would not be feasible, especially the ones for the larger values of p and q, if we would try to use the Meijer G function implementation of these p.d.f.'s and c.d.f.'s, since they need the computation of the p.d.f.'s or c.d.f.'s for several values of the statistic Λ.

Although for very small values of p and q the difference in terms of execution times between the Meijer G function formulation of the distribution and the EGIG finite representation may be only a couple of hundred times fold, this difference increases sharply as the values of either p or q increase, as it may be seen from scenarios II.1–II.6 in Sect. 4.4 of Chap. 4, which correspond to instances of the distribution of the statistic Λ in the present subsection, as noted in Table 4.10. We may note how for $p = 15$ and $q = 15$ it was not even possible to obtain a value for the c.d.f. of Λ when using the Meijer G function formulation, while modules `QuantEqMeanVecR` and `PvalEqMeanVecR` have absolutely no problem in computing quantiles and p-values even for larger values of p and q.

These modules may be used to easily compute p-values and quantiles for the statistic in this subsection for virtually any values of p and q, as long as p is even or q is odd. A simple set of commands used to compute the 0.05 quantile for $p = 25$, $q = 25$, and $n = 635$ and then to compute the p-value exactly at this quantile is available in Fig. 5.4, where the syntax used is for the software Mathematica®.

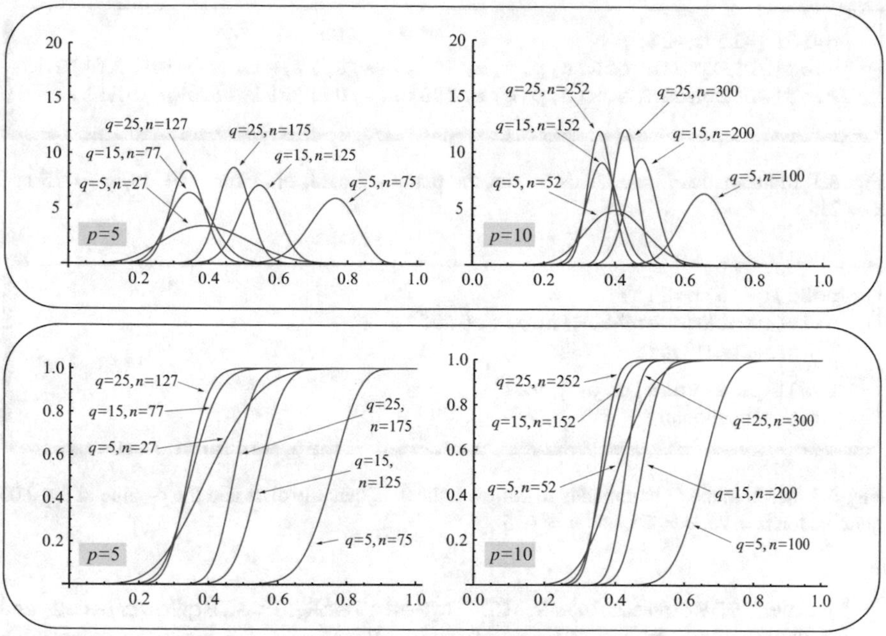

Fig. 5.1 Plots of p.d.f.'s and c.d.f.'s of Λ in (5.2) for $p = 5$ and $p = 10$, with $q = 5, 15, 25$ and $n = pq + 2$ and $n = pq + 50$

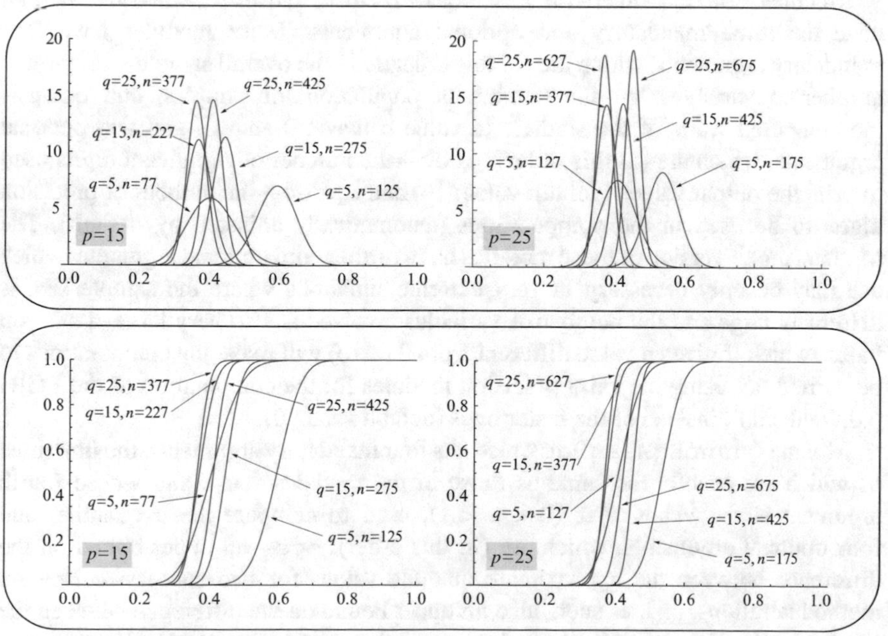

Fig. 5.2 Plots of p.d.f.'s and c.d.f.'s of Λ in (5.2) for $p = 15$ and $p = 25$, with $q = 5, 15, 25$ and $n = pq + 2$ and $n = pq + 50$

```
p=15;q=15;n=235;
Plot[PDFEqMeanVecR[n,p,q,z,200],{z,0,1},PlotRange->All]
Plot[CDFEqMeanVecR[n,p,q,z,200],{z,0,1},PlotRange->All]
```

Fig. 5.3 Mathematica® commands to plot the p.d.f. and c.d.f. of Λ for $p = 15$, $q = 15$ and $n = 235$

```
p=25;q=25;n=675;
val=QuantEqMeanVecR[n,p,q,0.05]
  0.36266964157933
PvalEqMeanVecR[n,p,q,val]
  0.050000000000
```

Fig. 5.4 Mathematica® commands to compute the 0.05 quantile of Λ and the p-value at the 0.05 quantile for $p = 25$, $q = 25$, and $n = 675$

Modules `PDFEqMeanVecR`, `CDFEqMeanVecR`, `PvalEqMeanVecR`, and `QuantEqMeanVecR` are able to automatically choose the necessary precision for the computations, and module `QuantEqMeanVecR` is also able to choose an adequate starting value for the Newton method, used to compute the approximation to the desired quantile with the required precision.

Modules `PDFEqMeanVecR`, `CDFEqMeanVecR`, and `PvalEqMeanVecR` all have the same mandatory and optional arguments. These modules have four mandatory arguments, which are (in this order): `n`—the overall sample size, `p`—the number of variables, `q`—the number of populations or samples, and `comp`—the computed value of the statistic (a value between 0 and 1), and two optional arguments, which are (in this order): `prec`—the number of significant digits used to print the output values (default value: 10) and `sprec`—the number of precision digits to be used in the computations (automatically adjusted by default). The Mathematica® version of these modules has a further third optional argument, which use may be only necessary in very extreme situations where the sample size is extremely large and the number of variables involved is also very large, `ind`—an index, which if given a value different from 0 (zero) will make the computations to be carried out using the extra precision modules for the computation of the EGIG p.d.f and c.d.f., instead of the faster ones (default value: 0).

Module `QuantEqMeanVecR` also has four mandatory arguments, the first three of which are exactly the same as those of the modules above and whose fourth argument is the value of α $(0 < \alpha < 1)$, used to compute the α-quantile, and four optional arguments, which are (in this order): `eps`—an upper bound on the difference between the approximate quantile values for two successive Newton method iterations, and, as such, also an upper bound on the difference between the reported value for the quantile and its real value (default value: 10^{-11}), `prec`—

starting number of precision digits used in the quantile computations (default value: maximum between 50 and $-2\log_{10}(\texttt{eps})$); if given a value smaller than $-2\log_{10}(\texttt{eps})$, it will take this latter value, which will be further adjusted if it is found to be too low, sv—a starting value for the Newton method (automatically determined if not given), pp—the number of digits used to print the quantile (default value: $-\log_{10}(\texttt{eps})$). Also the Mathematica® version of QuantEqMeanVecR has a fifth optional argument ind, which is exactly the same index used as the third optional argument of the three previously described modules, and which use, once again, may be only necessary in very extreme situations of extremely large sample sizes or extremely small values of eps.

Although usually the use of the last three optional arguments is not necessary, the combined use of the second and third optional arguments of QuantEqMeanVecR, if given adequate values, may lead to a faster computation of quantiles, as illustrated in Fig. 5.5. In this figure is also illustrated how the use of the first optional argument of PvalEqMeanVecR may also help to speed up the computations.

The use of the first optional argument for QuantEqMeanVecR, together with the use of the fifth optional argument which is the number of digits to be used in printing the quantile, is illustrated in Fig. 5.6. In this figure is also illustrated the use of the third optional argument of PvalEqMeanVecR, to set the number of digits to print the p-value. The result of the use of the default values for these arguments may be analyzed in Figs. 5.4 and 5.5.

Since the actual value for the difference between the value obtained for the exact quantile and the real value of this quantile is indeed in general much smaller than the default value of 10^{-11} for the optional argument eps of QuantEqMeanVecR,

```
p=25;q=25;n=675;
sv=0.2; prec=950;
Timing[QuantEqMeanVecR[n,p,q,0.05]]
Timing[QuantEqMeanVecR[n,p,q,0.05,,prec]]
Timing[QuantEqMeanVecR[n,p,q,0.05,,prec,sv]]

  {2.386815, 0.36266964157933}
  {1.341609, 0.36266964157933}
  {1.185608, 0.36266964157933}

val=QuantEqMeanVecR[n,p,q,0.05]
Timing[PvalEqMeanVecR[n,p,q,val]]
Timing[PvalEqMeanVecR[n,p,q,val,prec]]

  {1.216808, 0.050000000000}
  {0.421203, 0.050000000000}
```

Fig. 5.5 Mathematica® commands to compute the 0.05 quantile of Λ and the p-value at the 0.05 quantile for $p = 25$, $q = 25$, and $n = 675$, giving the number of precision digits to be used in the computations (prec) and the starting value for the computation of the quantile (sv)

```
p=25;q=25;n=675;
eps=10^-25; pp=45;
val1=QuantEqMeanVecR[n,p,q,0.05]
val2=QuantEqMeanVecR[n,p,q,0.05,,,,pp]
val3=QuantEqMeanVecR[n,p,q,0.05,eps,,,pp]
  0.36266964158
  0.362669641579325547819131393317593905757629942
  0.362669641579325547819131393317593905755236950

PvalEqMeanVecR[n,p,q,val1,pp]
PvalEqMeanVecR[n,p,q,val2,pp]
PvalEqMeanVecR[n,p,q,val3,pp]
  0.0500000000003377877258034116496618213723018850З
  0.0500000000000000000000000000000000000119848867
  0.0499999999999999999999999999999999999999999995
```

Fig. 5.6 Mathematica® commands to compute the 0.05 quantile of Λ and the p-value at the 0.05 quantile for $p = 25$, $q = 25$, and $n = 675$, giving specifications for the optional arguments for the upper bound on the difference between the exact value of the quantile and the output value (eps) and the number of digits to be used in outputting the values for the quantile and the p-values (pp)

in many cases the simple use of the optional argument pp, which sets the number of digits to be used in printing the quantile, with a value larger than its default value may be enough to obtain a value for the quantile that gives a much better precision, being not necessary to give the optional argument eps a lower value as it is illustrated in Fig. 5.6.

We have decided to undertake in this subsection a more detailed assessment of the functioning of these modules because the corresponding modules for the l.r.t. statistics studied in all subsections ahead do have an absolutely similar operating mode, with exactly the same optional arguments. Only the non-optional arguments of these modules will be directly related to the particular test being addressed. This is also the reason for the existence of the next subsection, where the module addressed is able to handle data files with different structures. These data files are the type of files used by similar modules which refer to tests that use several samples.

5.1.1.1 The Module PvalDataEqMeanVecR

The module PvalDataEqMeanVecR computes the value of the statistic Λ from data stored in a data file and the corresponding p-value. This module uses a module called ReadFileR, whose functionality is described in more detail in Chap. 6, to read a data file where the data for the q samples are stored. The data file may be organized in one of several ways. One possible way is as a set of q samples, the k-th of which arranged as an $n_k \times p$ table ($k = 1, \ldots, q$), that is, a table with the n_k observations organized by rows and the p variables defining the columns, with

```
 Names of variables
  X1      X2      X3      X4

        1st sample
23.4    56.7    21.2    36.5
34.5    23.4    28.9    25.4
21.2    33.3    56.7    25.6

        2nd sample
56.8    56.4    25.6    43.3
23.4    56.8    23.4    52.2
33.5    45.8    23.6    42.2
33.7    38.9    56.8    53.2

        3rd sample
43.2    45.6    34.5    32.2
56.4    24.7    78.3    34.3
74.5    83.5    84.3    52.3
24.5    25.7    37.4    52.3
35.7    74.5    84.6    83.3
```

```
        sample 1
  X1      X2      X3      X4
23.4    56.7    21.2    36.5
34.5    23.4    28.9    25.4
21.2    33.3    56.7    25.6
        sample 2
  X1      X2      X3      X4
56.8    56.4    25.6    43.3
23.4    56.8    23.4    52.2
33.5    45.8    23.6    42.2
33.7    38.9    56.8    53.2
        sample 3
  X1      X2      X3      X4
43.2    45.6    34.5    32.2
56.4    24.7    78.3    34.3
74.5    83.5    84.3    52.3
24.5    25.7    37.4    52.3
35.7    74.5    84.6    83.3
```

```
        sample 1
23.4    56.7    21.2    36.5
34.5    23.4    28.9    25.4
21.2    33.3    56.7    25.6
        sample 2
56.8    56.4    25.6    43.3
23.4    56.8    23.4    52.2
33.5    45.8    23.6    42.2
33.7    38.9    56.8    53.2
        sample 3
43.2    45.6    34.5    32.2
56.4    24.7    78.3    34.3
74.5    83.5    84.3    52.3
24.5    25.7    37.4    52.3
35.7    74.5    84.6    83.3
```

```
23.4    56.7    21.2    36.5
34.5    23.4    28.9    25.4
21.2    33.3    56.7    25.6

56.8    56.4    25.6    43.3
23.4    56.8    23.4    52.2
33.5    45.8    23.6    42.2
33.7    38.9    56.8    53.2

43.2    45.6    34.5    32.2
56.4    24.7    78.3    34.3
74.5    83.5    84.3    52.3
24.5    25.7    37.4    52.3
35.7    74.5    84.6    83.3
```

Fig. 5.7 Examples of different acceptable layouts with a multi-sample setup for the data file for module `PvalDataEqMeanVecR`

each sample separated from the following one by one or more empty rows or rows with any text. Some possible layouts for such a file are illustrated in Fig. 5.7. Such a file may also have any number of empty or text rows at its beginning. This set of rows may for example describe the data and have the indication of the names of the variables.

If a file with this structure is found by module `PvalDataEqMeanVecR` no further questions are placed to the user concerning the reading of the file contents and each chunk of consecutive data rows will be taken as corresponding to one of the q samples. After reading the file an informative line of text is printed, indicating how many samples were found in the file and how many variables were read.

Alternatively, the set of q samples may be organized as a single table with $n = \sum_{k=1}^{q} n_k$ rows corresponding to the whole set of n observations and p columns corresponding to the p variables observed. In this case this set of n rows should have no empty or text rows in between, although the file may have a set of any number of empty or text rows at its beginning. A file with this structure may also have a column which has the sample assignments. In this case the sample observations do not have to be consecutively listed in the data file. See Fig. 5.8a for an example of such a file. When a file with a single table structure is detected by module `PvalDataEqMeanVecR`, a question is issued to the user asking whether there is or not a column in the file with the sample assignments. If the answer is "Yes," then the module asks the user to provide the number of the column in the file with the sample assignments, since this may be indeed any column, although, for the sake of human readability of the file, it is strongly advisable to make it the first or the last column in the file. This column will then be taken out of the data samples, not appearing anymore as a variable, in the analysis. If the user says that there is no column in the file with the sample assignments, then he/she will be asked to provide a list of integer values with the sample sizes n_k $(k = 1, \ldots, q)$. These have to be integer values whose sum matches the overall number of data rows in the file and the length of this list will be used to set the number of samples.

As such, any of the file layouts in both Figs. 5.7 and 5.8 will produce exactly the same set of samples and as such the same computed value for the statistic Λ in (5.2).

Module `PvalDataEqMeanVecR` has one mandatory argument which is the name of the data file and which is given as a string, enclosed in double quotes and eventually preceded by the drive letter and file "path." The module also has other four optional arguments. These are (in this order): (i) `index1`—an index which if given the value 1 gives the user the possibility to select and/or reorder samples and variables (default value: 0), (ii) `index2`—a second index which if given the value 1 allows the user to select for each sample the sample observations to enter the analysis (default value: 0), (iii) `prec1`—an argument which sets the number of digits to be used in printing the p-value and which has a default value of 10, and (iv) `prec2`—a fourth optional argument which sets the number of digits used in printing the computed value of the statistic Λ and which also has the default value of 10. Although these arguments are optional, in case we want or need to use any one of them, then all the previous ones, even if not intended to be used, have to be mentioned, being eventually left blank or empty or given its default value. A simple example of the use of these arguments is given below in this subsection in connection with the contents of Fig. 5.10. See also Sect. 5.1.1.5, where a few real data sets are analyzed.

For the Mathematica® version of the module `PvalDataEqMeanVecR` there are two other optional arguments which are called by their names: `cov` and `norm`. If either one of them is given the value 1, this will entail the testing of preliminary hypotheses which may be important to assure that all conditions are verified in

a)

Names of variables

X1	X2	X3	X4	ind
23.4	56.7	21.2	36.5	1
34.5	23.4	28.9	25.4	1
21.2	33.3	56.7	25.6	1
56.8	56.4	25.6	43.3	2
23.4	56.8	23.4	52.2	2
33.5	45.8	23.6	42.2	2
33.7	38.9	56.8	53.2	2
43.2	45.6	34.5	32.2	3
56.4	24.7	78.3	34.3	3
74.5	83.5	84.3	52.3	3
24.5	25.7	37.4	52.3	3
35.7	74.5	84.6	83.3	3

b)

Names of variables

X1	X2	X3	X4
23.4	56.7	21.2	36.5
34.5	23.4	28.9	25.4
21.2	33.3	56.7	25.6
56.8	56.4	25.6	43.3
23.4	56.8	23.4	52.2
33.5	45.8	23.6	42.2
33.7	38.9	56.8	53.2
43.2	45.6	34.5	32.2
56.4	24.7	78.3	34.3
74.5	83.5	84.3	52.3
24.5	25.7	37.4	52.3
35.7	74.5	84.6	83.3

c)

Names of variables

X1	X2	X3	X4	ind
35.7	74.5	84.6	83.3	3
23.4	56.7	21.2	36.5	1
21.2	33.3	56.7	25.6	1
56.8	56.4	25.6	43.3	2
34.5	23.4	28.9	25.4	1
23.4	56.8	23.4	52.2	2
33.7	38.9	56.8	53.2	2
43.2	45.6	34.5	32.2	3
56.4	24.7	78.3	34.3	3
33.5	45.8	23.6	42.2	2
74.5	83.5	84.3	52.3	3
24.5	25.7	37.4	52.3	3

d)

23.4	56.7	21.2	36.5
34.5	23.4	28.9	25.4
21.2	33.3	56.7	25.6
56.8	56.4	25.6	43.3
23.4	56.8	23.4	52.2
33.5	45.8	23.6	42.2
33.7	38.9	56.8	53.2
43.2	45.6	34.5	32.2
56.4	24.7	78.3	34.3
74.5	83.5	84.3	52.3
24.5	25.7	37.4	52.3
35.7	74.5	84.6	83.3

Fig. 5.8 Examples of different acceptable layouts with a single table setup for the same data file in Fig. 5.7, for module `PvalDataEqMeanVecR`: (**a**, **c**) setup with a column with the sample assignments; (**b**, **d**) setup without column with the sample assignments

order to apply the test under all its assumptions. If `cov` is given the value 1, by indicating the option `cov- >1` in the arguments of the Mathematica® version of `PvalDataEqMeanVecR`, a l.r. test of equality of the covariance matrices will be performed and a near-exact p-value for this test will be computed, based on the near-exact distribution developed by Coelho et al. (2010) for the l.r.t. statistic of this test. If `norm` is given the value 1, by indicating the option `norm- >1`, the whole set of multivariate normality tests available in Mathematica®, versions 9 and later, will be implemented. These multivariate normality tests will only be carried out for samples for which $n_k > p$.

A third optional argument which may be used by calling its name is `neprec`, acronym for "near-exact precision," which sets the number of precision digits to be used in the computations of the near-exact distributions for the test of equality of covariance matrices. This argument may have to be used in cases where p and/or q are quite large or n is very large, but its use is not necessary for most common values of p, q, and n.

The reason why these further options are only implemented in the Mathematica® version of `PvalDataEqMeanVecR` is that this is the only software for which the near-exact distributions for the l.r.t. test of equality of covariance matrices are implemented and also the only software that has the multivariate normality tests implemented. We may note here that the near-exact distribution used for the l.r.t. statistic of the test of equality of covariance matrices is the one that matches four exact moments of the logarithm of the l.r.t. statistic, which is very manageable and displays a very good performance as may be seen in Coelho et al. (2010).

Examples of the use of these named optional arguments are considered in Fig. 5.11 and also in Sect. 5.1.1.5 and in Chap. 6.

In Fig. 5.9 we use the MAXIMA version of the module `PvalDataEqMeanVecR` to illustrate the use of the data files in Figs. 5.7 and 5.8. Details on the interface

```
PvalDataEqMeanVecR("<Path>/ex1_1.dat")$
   There are 3 samples, with sizes: [3,4,5] and 4 variables
Computed value of Lambda: 0.2099507759                              a)
p-value: 0.1788101398
______________________________________________________________

PvalDataEqMeanVecR("<Path>/ex1_2.dat")$
  Is there a column in the data file with the sample
                                        assignments? (1-Yes) 1;
 Which is the column in the data file with the sample assignments? 5;
   There are 3 samples, with sizes: [3,4,5] and 4 variables
Computed value of Lambda: 0.2099507759                              b)
p-value: 0.1788101398
______________________________________________________________

PvalDataEqMeanVecR("<Path>/ex1_3.dat")$
  Is there a column in the data file with the sample
                                        assignments? (1-Yes) 0;
 Please give the sample sizes for the q samples as a list with
                                         [n1,n2,...,nq]: [3,4,5];
   There are 3 samples, with sizes: [3,4,5] and 4 variables
Computed value of Lambda: 0.2099507759                              c)
p-value: 0.1788101398
```

Fig. 5.9 Using module `PvalDataEqMeanVecR` to compute the value of Λ and corresponding p-value from data in data files with different structures but equivalent contents: (**a**) for a data file with a structure as one of those displayed in Fig. 5.7; (**b**) for a file with a structure as one of those in Fig. 5.8a; (**c**) for a file with a structure as one of those in Fig. 5.8b

```
PvalDataEqMeanVecR("<Path>/ex1_1.dat",1)$
 Do you want to select samples? (1-Yes) 1;
 Please insert a list, inside square brackets,                         a)
   with the numbers of the samples you want to keep,
                                      separated by commas: [1,3];
 Do you want to select variables? (1-Yes) 1;
 Please insert a list, inside square brackets,
   with the numbers of the variables you want to keep,
                                      separated by commas: [1,4];
   There are 2 samples, with sizes: [3,5] and 2 variables
   Original samples [1,3]
   Original variables [1,4]
Computed value of Lambda: 0.2099507759
p-value: 0.1788101398
```

```
PvalDataEqMeanVecR("<Path>/ex1_1.dat",[],1,15,17)$
  There are 3 samples, with sizes: [3,4,5] and 4 variables            b)
 Please give a double list with the elements to be kept in each
                            sample: [[1,2,3],[1,2,3],[1,2,3]];
  There are now 3 samples, with sizes: [3,3,3] and 4 variables
Computed value of Lambda: 0.07189452966586930
p-value: 0.199123970795265
```

Fig. 5.10 Illustration of the use of the three non-named optional arguments for module `PvalDataEqMeanVecR`: (**a**) use of the first optional index which allows for the selection of samples and/or variables; (**b**) use of the second optional index which allows for the selection of observations in each sample and the third optional argument which allows for the specification of the number of digits used in printing the p-value

of the other two softwares may be analyzed in Chap. 6. Concerning the data files used in the examples in Fig. 5.9, the contents of file `ex1_1.dat` are similar to the first example in Fig. 5.7, while those of file `ex1_2.dat` are similar to the second example in Fig. 5.8a and file `ex1_3.dat` has a content similar to that of the first example in Fig. 5.8b.

In Fig. 5.10 is illustrated the use of the optional arguments of the module `PvalDataEqMeanVecR`. Once again it is used the MAXIMA version of this module. When using this version of the module, if one wants to use one of the optional arguments that is not the first one, all previous optional arguments will have to be given a value or left empty. In Fig. 5.10a is illustrated the use of the first optional index of `PvalDataEqMeanVecR`, which if given the value 1 enables the user to select samples and/or variables. In Fig. 5.10b is illustrated the use of the second optional index of `PvalDataEqMeanVecR`, which if given the value 1 enables the user to select the observations to be used in each sample, and also of the third and fourth optional arguments which establish the number of digits to be used in printing the p-value and the computed value of the statistic. Just note how now the p-value was printed with 15 decimal digits and the computed value of the statistic with 17 digits. In giving the numbers of the observations to be kept in each sample we might had used `seq(1,3)` instead of `[1,2,3]`. The MAXIMA, Mathematica®, and R functions `seq` are programmed to generate a sequence of integers between their first and second arguments. They may use an optional third argument specifying a step different from 1.

Figure 5.11 illustrates the use of the two optional arguments `cov` and `norm` for the Mathematica® version of module `PvalDataEqMeanVecR`, or rather, it illustrates the consequence of giving these two arguments the value 1. We may note how samples 1 and 2 are too small to carry out the normality tests and also to allow for the implementation of the test of equality of covariance matrices. Sample 3 is also too small for the Cramér-von Mises and the Jarque-Bera normality tests to be carried out. Further examples of the use of these two arguments may be found in Sect. 5.1.1.5 ahead and in Chap. 6.

```
PvalDataEqMeanVecR["<Path>/ex1_1.dat", , , , , norm → 1, cov → 1]

 There are 3 samples, with sizes {3, 4, 5} and 4 variables

Sample 1 is too small to carry out a normality test

Sample 2 is too small to carry out a normality test
```

		Statistic	P-Value
Sample 3	Anderson-Darling	3.14059	0.977271
	Kolmogorov-Smirnov	3.13401	0.976567
	Kuiper	1.83129	0.389023
	Mardia Combined	15.1707	0.182143
	Mardia Kurtosis	−2.22051	0.182927
	Mardia Skewness	5.12	0.241667
	Pearson χ^2	2.49228	0.79576
	Shapiro-Wilk	0.552182	0.000126531
	Watson U^2	2.45792	0.778769

```
There are 2 samples which are too small
   to carry out a test of equality of covariance matrices

Computed value of Λ: 0.2099507759

p-value: 0.1788101398
```

Fig. 5.11 Illustration of the use of the optional arguments `cov` and `norm` for the Mathematica® module `PvalDataEqMeanVecR`

5.1.1.2 On the Availability of Quantile Tables and the Pressing Need for the Present Book

Quantiles for Λ, rounded to six decimal places, are available for $\alpha = 0.05$ and $\alpha = 0.01$ and for $p = 1$ through $p = 8$ and $q = 2, \ldots, 16, 19, 22, 25, 28, 31, 41, 61, 81, 101, 121$ as Table 1 in Kres (1983), which, as the author notes, is taken from Wall (1967). Although this table refers in fact to the test statistic in Sect. 5.1.4, we may use it for the present test by taking q as our $q - 1$, n as our $n - q$ and p equivalent to our p. A remark to be made is that the first $p - 1$ rows for each value of p are spurious since we need to have a sample of size $n > p + q - 1$ for the test statistic to be well defined, because if on a first look we need to have the degrees of freedom of the Wishart distributions of the matrices A and B in (5.3) larger than zero, which would require just $n > q$, on a second, more attentive look, we see that we indeed need to have $n > p + q - 1$ in order for the first parameters in the Beta r.v.'s in (5.6) to be all positive. Another remark is that not always all the six decimal digits displayed are correct. It is not uncommon that the two last digits are not correct, at least for the smaller sample sizes, as for example occurs in the case of the 0.05 quantile for $p = 3$, $q = 7$ and $n = 21$ (our q and n) where the value listed is 0.140775 when it should be 0.140762. This problem gets worse for larger values of p and q, and for the 0.01 quantiles. For example, the 0.05 quantile listed for $p = 5$, $q = 61$, and $n = 101$ (our q and n) is 0.004417, when it should be 0.004025, while the 0.01 quantile listed is 0.003378, when it should be 0.002942. This situation improves anyway for much larger values of n, with both the 0.05 and 0.01 quantiles for the same values of p and q displaying all six decimal digits correct for $n = 501$ (our n). However, for very small sample sizes it may happen that none of the significant digits shown is correct, as it happens for example for $n = 13$, $p = 6$, and $q = 7$.

An abbreviated version of this table only for 0.05 quantiles and for $p = 1, \ldots, 8$ but only for $q = 2, \ldots, 13$ and with quantiles rounded to three decimal places may be found as Table A.9 in Rencher (2002) or Rencher and Christensen (2012).

Approximate quantiles for Λ may also be obtained from Table 9 in Muirhead (2005) and Table B.1 in Anderson (2003). These tables list correction factors to be applied to the asymptotic chi-square quantiles in order to obtain the corresponding quantiles of Λ or a function of its logarithm. While Muirhead (2005) lists correction factors for $\alpha = 0.1, 0.05, 0.025$ and 0.01, for values of $q = 4, \ldots, 13$ and several different values of $p \geq q - 1$ and different sample sizes, Anderson (2003) only lists correction factors for $\alpha = 0.05$ and 0.01, for $p = 3, \ldots, 10$ and even values of $q - 1 \geq 2$. Once again, although these tables were indeed constructed for the test statistic in Sect. 5.1.4, they may be used for the test statistic in the present subsection. Table 9 in Muirhead (2005), where r and m are interchangeable, is taken, as the author remarks, from the tables in Schatzoff (1966), Pillai and Gupta (1969), Lee (1972), and Davis (1979) and may be used for the test in this subsection by taking m as our $q - 1$, r as our p and M as our $n + 1 - p - q$. A general result for l.r.t. statistics asserts that for the statistic Λ in (5.2), as n (the sample size) tends to infinity, $-n \log \Lambda$ converges in distribution to a $\chi^2_{p(q-1)}$ r.v. It is based on this limiting distribution that these tables are constructed. They list factors which multiplied by the corresponding $1-\alpha$ quantile of the $\chi^2_{p(q-1)}$ distribution will yield approximations to the exact quantile of $\{n-1-(p+q)/2\} \log \Lambda$. That is, calling "$c$" to the "correction" factors listed in those tables and taking m and r in Muirhead (2005) respectively as our $q-1$ and p we may obtain approximations for the α quantile of Λ in (5.2) by taking

$$e^{-c\,\chi^2_{p(q-1)}(1-\alpha)/(n-1-(p+q)/2)}, \tag{5.11}$$

where n, p, and q stand for these quantities in our notation and $\chi^2_{p(q-1)}(1-\alpha)$ stands for the $1-\alpha$ quantile of a chi-square distribution with $p(q-1)$ degrees of freedom. These approximations may actually yield, for smaller sample sizes, better values than the ones obtained from the tables in Wall (1967) and Kres (1983). For example, for $n = 13$, $p = 6$, and $q = 7$ the value listed in Table 1 of Kres (1983) is 0.000012 while the correct value is 7.558736×10^{-7} and the value obtained from the application of the correction factors in Table 9 of Muirhead (2005) is 7.568355×10^{-7}. For $n = 17$ the value listed in Table 1 of Kres (1983) is 0.003035 while the correct value is 0.002914 and the value obtained from the application of the correction factor in Table 9 of Muirhead (2005) together with (5.11) is 0.002907. Also for $n = 22$ the value in Table 1 of Kres (1983) is 0.026433 while the correct value is 0.026384 and the value from Table 9 in Muirhead (2005) is 0.026339. Then, for larger values of the sample size, the values in Table 1 in Kres (1983) regain a better precision, with the value listed in this table for $n = 36$ being 0.164629 while the exact value is 0.164623 and the value obtained from Table 9 in Muirhead (2005) being 0.164683. A similar table to Table 9 in Muirhead (2005) is published as Table D13 in Seber (1984), where d is to be taken as our $q-1$, m_H as our p, and, as in Table 9 in Muirhead (2005), M is to be taken as our $n+1-p-q$.

Table B.1 in Anderson (2003) is indeed, as the author notes, Table 47 in Pearson and Hartley (1972), where we should take p as our p, m as our $q-1$, and M as our $n+1-p-q$.

Anyway, although all these tables exist, inevitably they will always be restricted to a few values of α and also to just a few values of n, p, and q. As such, in case p is even or q is odd, the authors strongly recommend the use of the modules `QuantEqMeanVecR` or `PvalEqMeanVecR` which will readily give the exact quantiles or p-values for any sample size and any combination of values of p and q, and with as many decimal places as one wishes, with the use of the `PvalEqMeanVecR` module giving, for a computed value of the statistic, a better basis and a wider view towards a decision, since in fact the use of a p-value to take that decision will in general be preferable to the use of a fixed α quantile; and for the computation of these p-values one will always need a computer module to obtain the value.

The use of p-values in Multivariate Statistics is not that common since till now their computation was usually seen as too hard and time consuming, even with the possible availability of computer modules, themselves usually seen as not easy to be programmed. But, this is one of the aims, or it may be the major aim, of this book: to make easily available the computation of exact p-values, and also quantiles, for many of the likelihood ratio tests used in Multivariate Analysis.

Moreover, with the availability of the module `PvalDataEqMeanVecR` which gives the computed value of the statistic Λ in (5.2) as well as the corresponding p-value for a set of q samples stored in a data file the use of quantile tables is virtually not any longer necessary.

5.1.1.3 The Exact Distribution of Λ in (5.2) for $p = 1$ or 2 and for $q = 2$ or 3

For $p = 1$ or 2 or for $q = 2$ or 3 it is easy to relate the exact distribution of Λ in (5.2) with that of an F distributed r.v. From (5.5) and (5.6) it is easy to see that for $p = 1$ and $q = 2$ the exact distribution of Λ is a Beta distribution. More precisely, from (5.5) and (5.6) we have that for $p = 1$

$$\Lambda \sim Beta\left(\frac{n-q}{2}, \frac{q-1}{2}\right)$$

and for $q = 2$,

$$\Lambda \sim Beta\left(\frac{n-p-1}{2}, \frac{p}{2}\right),$$

so that, given the relation between the Beta and F distributions, which may be stated as

$$Y \sim Beta(a, b) \implies \frac{a}{b}\,\frac{1-Y}{Y} \sim F_{2b,2a}\,, \tag{5.12}$$

we see that for $p = 1$,

$$\frac{n-q}{q-1}\,\frac{1-\Lambda}{\Lambda} \sim F_{q-1,n-q}\,,$$

which is the statistic commonly used in the one-way ANOVA model for a factor with q levels, while for $q = 2$,

$$\frac{n-p-1}{p}\,\frac{1-\Lambda}{\Lambda} \sim F_{p,n-p-1}\,,$$

which bears a direct relation with the Hotelling T^2 statistic, commonly used in this setting.

In fact, for $q = 2$, that is, when we only have two mean vectors and we want to test the null hypothesis

$$H_0 : \underline{\mu}_1 = \underline{\mu}_2 , \tag{5.13}$$

the statistic Λ in (5.2) has a simple relation with the Hotelling T^2 statistic, which is commonly used to test the null hypothesis in (5.13).

If we take the Hotelling T^2 statistic defined as (see Sect. 4.2 of Morrison (2005) and Sect. 5.2 of Anderson (2003))

$$T^2 = \frac{n_1 n_2}{n_1 + n_2} \left(\underline{X}_1 - \underline{X}_2\right)' A^{-1} \left(\underline{X}_1 - \underline{X}_2\right), \tag{5.14}$$

where A is the matrix in (5.3), then we have (see Appendix 2)

$$T^2 = \frac{1-\Lambda}{\Lambda},$$

so that

$$\frac{n-p-1}{p} T^2 \sim F_{p,n-p-1},$$

and, for a test of size α, we would reject the null hypothesis in (5.13) if the computed value of $\frac{n-p-1}{p} T^2$ exceeds the $1-\alpha$ quantile of an $F_{p,n-p-1}$ distribution, yielding the usual T^2 test for H_0 in (5.13).

For $p = 2$, once again from (5.5) and (5.6), we may write the h-th moment of $\Lambda^{1/2}$ as

$$E\left(\Lambda^{h/2}\right) = \frac{\Gamma\left(\frac{n-1}{2}\right)\Gamma\left(\frac{n-q}{2}+\frac{h}{2}\right)}{\Gamma\left(\frac{n-q}{2}\right)\Gamma\left(\frac{n-1}{2}+\frac{h}{2}\right)} \frac{\Gamma\left(\frac{n-2}{2}\right)\Gamma\left(\frac{n-q-1}{2}+\frac{h}{2}\right)}{\Gamma\left(\frac{n-q-1}{2}\right)\Gamma\left(\frac{n-2}{2}+\frac{h}{2}\right)}$$

which, using the Gamma duplication formula in (5.583) may be written as

$$E\left(\Lambda^{h/2}\right) = \frac{\Gamma(n-2)\,\Gamma(n-q-1+h)}{\Gamma(n-q-1)\,\Gamma(n-2+h)},$$

which in turn shows that, for $p = 2$,

$$\Lambda^{1/2} \sim Beta(n-q-1, q-1)$$

so that

$$\frac{n-q-1}{q-1}\,\frac{1-\Lambda^{1/2}}{\Lambda^{1/2}} \sim F_{2(q-1),2(n-q-1)}\,.$$

Then, for $q = 3$, from (5.5) and (5.6), using once again (5.583), we have the h-th moment of $\Lambda^{1/2}$ given by

$$E\left(\Lambda^{h/2}\right) = \frac{\Gamma\left(\frac{n-1}{2}\right)\Gamma\left(\frac{n-p-1}{2}+\frac{h}{2}\right)\Gamma\left(\frac{n-2}{2}\right)\Gamma\left(\frac{n-p-2}{2}+\frac{h}{2}\right)}{\Gamma\left(\frac{n-p-1}{2}\right)\Gamma\left(\frac{n-1}{2}+\frac{h}{2}\right)\Gamma\left(\frac{n-p-2}{2}\right)\Gamma\left(\frac{n-1}{2}+\frac{h}{2}\right)}$$
$$= \frac{\Gamma(n-2)\,\Gamma(n-p-2+h)}{\Gamma(n-p-2)\,\Gamma(n-2+h)},$$

which shows that in this case,

$$\Lambda^{1/2} \sim Beta(n-p-2, p)$$

and as such, for $q = 3$,

$$\frac{n-p-2}{p}\frac{1-\Lambda^{1/2}}{\Lambda^{1/2}} \sim F_{2p,2(n-p-2)}\,. \tag{5.15}$$

Anyway, for the cases $p = 2$ and $q = 3$ we have always the exact distribution of Λ as an EGIG distribution, with p.d.f. and c.d.f. given by (5.9) and (5.10).

5.1.1.4 On the Test for Equality of Covariance Matrices

As it may be seen from the first sentence in Sect. 5.1.1, in fact, the correct implementation of the test of equality of mean vectors requires the equality of the covariance matrices of the q population vectors $\underline{X}_k$ $(k = 1, \ldots, q)$. As such, and although some authors as Rencher (1998, Sect. 4.2; 2002, Sect. 6.2) and Rencher and Christensen (2012, Sect. 6.2) refer that the l.r. test of equality of mean vectors, namely if the normality assumption holds and samples are large and equal in size, may be somewhat robust to the violation of the assumption of equal covariance matrices (see Ito and Schull (1964) and also Sect. 3.7 of Rencher (1998) for a more thorough discussion and other references for the case of only two populations), before performing the test of equality of mean vectors one should test the hypothesis of equality of the covariance matrices, also because not all authors agree with the above assertion (Olson, 1974, Sect. 5). It happens though that the l.r.t. statistic for this test has a distribution which does not fit within the framework of the results in Theorems 3.1–3.3 (see Coelho and Marques 2012). Anyway, implementing the test of equality of covariance matrices becomes quite simple if one uses one of the near-exact distributions developed for the l.r.t. statistic in Coelho et al. (2010) or in Coelho and Marques (2012). This is exactly what the Mathematica® module `PvalDataEqMeanVecR` does when the optional argument named `cov` is given the value 1, making the module to compute a near-exact p-value for the test of equality of covariance matrices, based on a near-exact distribution that matches the first four exact moments of the logarithm of the l.r.t. statistic, following the developments in Coelho and Marques (2012, Sec. 3) for the case of different sample sizes.

However, in case one wants to carry out the test of equality of covariance matrices there is one further restriction that we have to be aware of. This is the need for each sample size to be larger than p. In case there is at least one sample that has a size which is smaller than p and the optional argument `cov` is given the value 1, a message is issued telling the user that the test will not be carried out because of the existence of at least one sample which is too small.

This option is only implemented in the Mathematica® version of the `PvalData` module because this is the only software where the near-exact distribution for the l.r.t. statistic of equality of covariance matrices is at present implemented.

5.1.1.5 Real Data Examples

As a first example with a real data set we carry out a test of equality of mean vectors for the three species of *Iris* in the well-known iris data set used by Fisher (1936), where $p = 4$ variables (sepal length, sepal width, petal length, and petal width) were measured on 50 flowers of each one of the three species *Iris setosa*, *Iris versicolor*, and *Iris virginica*. This data set is available as Table I in Fisher (1936), as Table 11.5 in Johnson and Wichern (2014), or even in R just by typing "iris," and we have decided to use it also to point out some commonly overlooked features about it.

The data file used has 5 columns, the first 4 of which correspond to the four variables measured, with the 150 observations for the three species of *Iris* piled on top of each other and a fifth column with a variable with the values 1, 2, or 3 (respectively for *Iris setosa*, *Iris versicolor*, and *Iris virginica*), indicating the species of iris to which the observation belongs.

A plot of the sample means for the four variables and the three species of iris may be observed in Fig. 5.12. Such a plot is also commonly called a plot of the "profiles" for the three species—see Sect. 5.1.3 for the definition of "profile."

We are interested in testing the equality of the population mean vectors, each with dimension $p = 4$, for the three species of iris. The command used with the Mathematica® version of module `PvalDataEqMeanVecR` and the output obtained are shown in Fig. 5.13.

The module used is the module available in Mathematica® because, as already noted, this module provides the opportunity to carry out the multivariate normality tests available in Mathematica® and also the l.r.t. for equality of covariance matrices.

By looking at the output in Fig. 5.13 we may see that the p-value for the test of equality of covariance matrices is extremely low, showing that this hypothesis should be rejected for any sensible α-value, which is a fact usually not taken into account. This p-value is indeed so low that the decision of rejecting the equality of the covariance matrices for the three species of iris may be taken with some assurance although this test is quite sensitive to the normality assumption and the

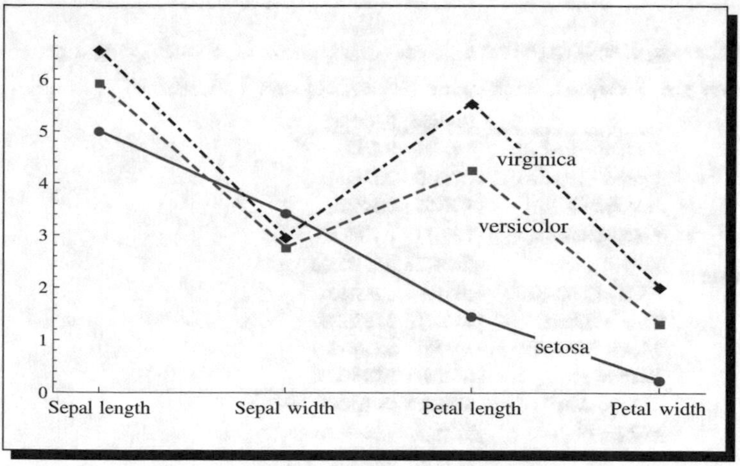

Fig. 5.12 Sample means (or "profiles") for the four variables measured on the three species of iris

multivariate normality tests give somewhat mixed results, with the Kuiper, Pearson chi-square, and Shapiro-Wilk tests giving rather low p-values.

Although, as noted before, the test of equality of mean vectors is somewhat robust to the inequality of covariance matrices when the normality assumption is plausible and the samples have equal sizes, we should anyway be careful when looking at the p-value associated with the test of equality of mean vectors. But since this p-value is so small, there should not be much doubt about rejecting the hypothesis of equality of the mean vectors for the three species of iris. By looking at the plot in Fig. 5.12, the very low p-value for the test of equality of mean vectors should not come as much of a surprise, since the means for the four variables differ much for the three species, mainly for petal length, at the same time that the samples are rather large. Nevertheless, we should be aware of the fact that especially in multivariate settings, our judgements may be easily fooled by only looking at graphical representations.

In case one may be a bit concerned with the extremely low p-value for the test of equality of mean vectors, since in this case we have $q = 3$, one direct way we have to check the correctness of this p-value is to use the relation in (5.15), although we will then need a software that is able to give the necessary precision in handling the c.d.f. for the F distribution. We may look at the Mathematica® commands in Fig. 5.14, used to compute the p-value based on the relation in (5.15). We should note how in fact the p-value computations carried out in module `PvalDataEqMeanVecR` use many more precision digits than just the 10 displayed by default for the computed value of the Λ statistic.

In case one may still be a bit concerned with the very low p-values in Fig. 5.13, mainly the one associated with the test of equality of covariance matrices, one may become assured of the correct functioning of module `PvalDataEqMeanVecR` by executing a set of commands such as the ones in Fig. 5.15, where three independent pseudo-random samples of size 50 are obtained from a 4-variate Normal distribution

```
PvalDataEqMeanVecR["<Path>/iris.dat", , , , , cov → 1, norm → 1]
```

```
There are 3 samples, with sizes {50, 50, 50} and 4 variables
```

Sample 1

	Statistic	P-Value
Anderson-Darling	1.95751	0.4717
Cramér-von Mises	1.76619	0.348016
Jarque-Bera ALM	1.32681	0.127227
Kolmogorov-Smirnov	1.29178	0.114815
Kuiper	0.424729	0.00135593
Mardia Combined	26.797	0.259892
Mardia Kurtosis	0.758712	0.0572239
Mardia Skewness	24.1551	0.258778
Pearson χ^2	0.472811	0.00208228
Shapiro-Wilk	0.958785	0.0790587
Watson U^2	1.01826	0.0447941

Sample 2

	Statistic	P-Value
Anderson-Darling	2.60025	0.844303
Cramér-von Mises	2.39015	0.743198
Jarque-Bera ALM	1.97845	0.485637
Kolmogorov-Smirnov	1.75782	0.342853
Kuiper	1.17169	0.0783851
Mardia Combined	26.8012	0.25966
Mardia Kurtosis	−1.03422	0.992936
Mardia Skewness	23.7039	0.283674
Pearson χ^2	0.969761	0.0368509
Shapiro-Wilk	0.930434	0.00573895
Watson U^2	1.45954	0.181652

```
DistributionFitTest::ties:
  Ties exist in the data and will be ignored for the {KolmogorovSmirnov,
    Kuiper, WatsonU2} tests which assume unique values >>
```

Sample 3

	Statistic	P-Value
Anderson-Darling	2.46405	0.781852
Cramér-von Mises	2.33799	0.714086
Jarque-Bera ALM	1.88888	0.426361
Kolmogorov-Smirnov	1.71819	0.318799
Kuiper	0.818529	0.0187036
Mardia Combined	26.9553	0.251195
Mardia Kurtosis	−0.338428	0.397885
Mardia Skewness	24.7257	0.230495
Pearson χ^2	0.877296	0.0246816
Shapiro-Wilk	0.934143	0.00795539
Watson U^2	1.27767	0.110044

(p-value for the test of equality of covariance matrices: $3.597085734 \times 10^{-20}$)

Computed value of Λ: 0.02343863065

p-value: $1.365005833 \times 10^{-112}$

Fig. 5.13 Command used and corresponding output obtained with the Mathematica® module `PvalDataEqMeanVecR` to analyze the iris data set

```
lambda = 0.02343863065;
val = (150 - 4 - 2) / 4 * SetPrecision[(1 - Sqrt[lambda]) / Sqrt[lambda], 20];
SetPrecision[1 - CDF[FRatioDistribution[2 * 4, 2 * (150 - 4 - 2)], val], 20]

1.365005828920680883 6 × 10^-112

PvalDataEqMeanVecR["<Path>/iris_3.dat", 0, 0, 22, 20]

 There are 3 samples, with sizes {50, 50, 50} and 4 variables

Computed value of Λ: 0.02343863065087836172727

p-value: 1.3650058325899569273 × 10^-112

lambda = 0.02343863065087836172727;
val = (150 - 4 - 2) / 4 * SetPrecision[(1 - Sqrt[lambda]) / Sqrt[lambda], 20];
SetPrecision[1 - CDF[FRatioDistribution[2 * 4, 2 * (150 - 4 - 2)], val], 20]

1.3650058325899569273 × 10^-112
```

Fig. 5.14 Computation of the p-value for the test of equality of mean vectors for the iris data set using the F-distribution given by (5.15)

with a null mean vector and a covariance matrix which was chosen to be equal to the sample covariance matrix for the *Iris setosa* sample, rounded to four decimal places. The output obtained is also shown in Fig. 5.15, although, given the generation of pseudo-random samples, different outputs would be obtained for different runs of the commands in that figure. We expect, on the long run, that 95% of the executions will show p-values above 0.05 for the test of equality of covariance matrices and for the test of equality of mean vectors, as well as for the multivariate normality tests, which is exactly what happens, with the exception of the Shapiro-Wilk test which persists in giving quite low p-values in most cases.

One other example is taken from the data in Table 9.12 of Johnson and Wichern (2014), where we have the data for three variables that were measured on 50 sales people to assess their performance. These variables are: "growth of sales," "profitability of sales," and "new account-sales," which were then transformed to a scale where the value 100 indicates "average." Also, each one of the 50 people were scored on each of four tests measuring creativity, mechanical reasoning, abstract reasoning and mathematical ability. The data file used mimics exactly Table 9.12 in Johnson and Wichern (2014), with 50 rows and 8 columns. As such we will have to extract the variables corresponding to columns 2, 3, and 4 for the analysis and as such we need to use the optional argument in module `PvalDataEqMeanVecR` that comes right after the file name, with a value of 1, in order to be able to select variables and samples from the file.

In a first analysis we are interested in showing the importance of abstract reasoning and as such we decided to test the equality of the mean vectors of the three sales performance variables for the 4th, 7th, and 8th classes of the abstract reasoning test, which correspond respectively to scores of 8, 11, and 12 for this test and for which we have respectively samples of size 5, 15, and 10, which are the

```
CovMat = {{0.1242, 0.0992, 0.0164, 0.0103}, {0.0992, 0.1437, 0.0117, 0.0093},
    {0.0164, 0.0117, 0.0302, 0.0061}, {0.0103, 0.0093, 0.0061, 0.0111}};
x1 = RandomReal[MultinormalDistribution[{0, 0, 0, 0}, CovMat], 50];
x2 = RandomReal[MultinormalDistribution[{0, 0, 0, 0}, CovMat], 50];
x3 = RandomReal[MultinormalDistribution[{0, 0, 0, 0}, CovMat], 50];
Export["<Path>/Random.dat", Join[x1, x2, x3]];
PvalDataEqMeanVecR["<Path>/Random.dat", , , , , cov → 1, norm → 1]

There are 3 samples, with sizes {50, 50, 50} and 4 variables
```

Sample 1

	Statistic	P-Value
Anderson-Darling	3.74762	0.999831
Cramér-von Mises	3.62535	0.999179
Jarque-Bera ALM	2.00711	0.504742
Kolmogorov-Smirnov	3.66246	0.999459
Kuiper	3.25216	0.986967
Mardia Combined	18.7523	0.89502
Mardia Kurtosis	−1.92766	0.16477
Mardia Skewness	13.8516	0.620043
Pearson χ^2	2.31124	0.698614
Shapiro-Wilk	0.972893	0.302279
Watson U^2	3.20531	0.983382

Sample 2

	Statistic	P-Value
Anderson-Darling	3.37528	0.993653
Cramér-von Mises	3.26574	0.987889
Jarque-Bera ALM	1.31239	0.12202
Kolmogorov-Smirnov	3.10513	0.973281
Kuiper	3.16382	0.97963
Mardia Combined	20.7093	0.818722
Mardia Kurtosis	−0.626014	0.608311
Mardia Skewness	18.7163	0.714017
Pearson χ^2	2.67295	0.872684
Shapiro-Wilk	0.968837	0.207489
Watson U^2	2.79605	0.912744

Sample 3

	Statistic	P-Value
Anderson-Darling	3.57551	0.998647
Cramér-von Mises	3.32538	0.991369
Jarque-Bera ALM	2.44996	0.774729
Kolmogorov-Smirnov	3.35198	0.992653
Kuiper	2.96605	0.952381
Mardia Combined	13.2957	0.17976
Mardia Kurtosis	−1.73505	0.28943
Mardia Skewness	9.47482	0.133131
Pearson χ^2	2.92794	0.944966
Shapiro-Wilk	0.970052	0.232585
Watson U^2	2.66541	0.869903

Fig. 5.15 Commands used and corresponding output obtained with the Mathematica® module `PvalDataEqMeanVecR` to analyze a set of three pseudo-random samples drawn from the same multivariate Normal distribution

numbers of sales people with each of these scores. We have chosen these classes since the other classes have too low of a number of individuals in order to carry out the multivariate normality and the equality of covariance matrices tests, with only three of them showing a sample of size 4 and all the other five classes showing even smaller sample sizes. A plot of the sample means or profiles of the three classes of abstract reasoning for the three performance variables is shown in Fig. 5.16, and we are interested in testing the equality of the mean vectors for these three classes of abstract reasoning scores.

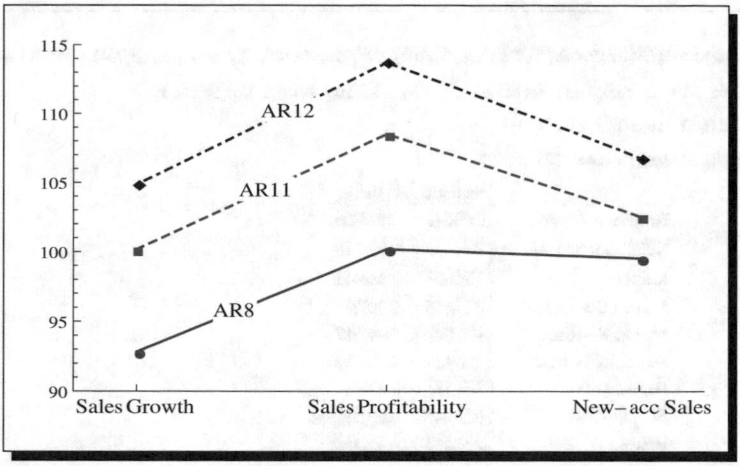

Fig. 5.16 Sample means (or "profiles") for the three performance variables (Sales growth, Sales profitability, New-account sales) for the three classes defined by scores of 8, 11, and 12 on the abstract reasoning test

Once again, in order to be able to carry out the l.r.t. for equality of covariance matrices and the normality tests, we use the Mathematica® `PvalDataEqMeanVecR` module. The command used and the output obtained may be analyzed in Fig. 5.17.

We may see how all multivariate normality tests, except the Shapiro-Wilk test, give quite high p-values, showing that the normality assumption may be assumed with quite large confidence and also how the test of equality of covariance matrices gives a somewhat large p-value which gives us some assurance that the we may be not making that much of an error in assuming the equality of the covariance matrices. Then the test of equality of mean vectors gives a quite low p-value, showing that this hypothesis should be rejected. This is in conformity with what we feel when looking at the plot in Fig. 5.16, although, as pointed out before, we have to be extra careful when making judgements based on the analysis of plots, especially in Multivariate Analysis.

The Cramér-von Mises and Jarque-Bera normality tests are not carried out for the first sample because the first of these two tests only gives a valid p-value for samples with a size larger than 6 and the second test for samples with a size larger than 9.

As we will see in Sect. 5.1.3, the parallelism of the three profiles will not be rejected, showing that although the differences for the mean values of the three performance variables are significant among the three profiles defined by the three different abstract reasoning scores, with higher scores in the abstract reasoning test associated with higher values for the sales performance variables, these differences are similar in magnitude for all three performance variables, when we compare any two profiles.

```
PvalDataEqMeanVecR["<Path>/Tab9_12_JW.dat", 1, , , , cov → 1, norm → 1]
There are 3 samples, with sizes {5, 15, 10} and 3 variables
Original samples {4, 7, 8}
Original variables {2, 3, 4}
```

Sample 1

	Statistic	P-Value
Anderson-Darling	2.79042	0.998466
Kolmogorov-Smirnov	2.66579	0.993778
Kuiper	2.26363	0.933451
Mardia Combined	8.26656	0.99887
Mardia Kurtosis	−1.76521	0.946987
Mardia Skewness	2.41434	0.986766
Pearson χ^2	1.59585	0.57159
Shapiro-Wilk	0.586433	0.000397958
Watson U^2	2.63832	0.992115

Sample 2

	Statistic	P-Value
Anderson-Darling	2.66941	0.993978
Cramér-von Mises	2.59522	0.988947
Jarque-Bera ALM	1.58355	0.562466
Kolmogorov-Smirnov	2.55977	0.985781
Kuiper	1.96327	0.814308
Mardia Combined	10.0295	0.835011
Mardia Kurtosis	−1.28343	0.927197
Mardia Skewness	6.33038	0.807382
Pearson χ^2	2.34166	0.952444
Shapiro-Wilk	0.916741	0.171768
Watson U^2	2.22656	0.922886

Sample 3

	Statistic	P-Value
Anderson-Darling	2.88213	0.999727
Cramér-von Mises	2.76774	0.997912
Jarque-Bera ALM	2.17066	0.904929
Kolmogorov-Smirnov	2.56801	0.986564
Kuiper	2.2326	0.924681
Mardia Combined	10.4571	0.600744
Mardia Kurtosis	−1.45832	0.903236
Mardia Skewness	5.53416	0.649187
Pearson χ^2	2.67532	0.994296
Shapiro-Wilk	0.780381	0.00785047
Watson U^2	2.60983	0.990101

```
(p-value for the test of equality of covariance matrices: 0.1457299624)
Computed value of Λ: 0.3859344456
p-value: 0.0003872679905
```

Fig. 5.17 Command used and corresponding output obtained with the Mathematica® module `PvalDataEqMeanVecR` to test the equality of mean vectors for the three sales performance variables in Table 9.12 of Johnson and Wichern (2014), for the 4th, 7th, and 8th classes of abstract reasoning, corresponding to scores of 8, 11, and 12 in the abstract reasoning test

One third example is drawn from the data in Table 4.9.4 of Timm (2002) where base grade-equivalent scores for reading (ZR) and mathematics (ZM) were obtained for 15 males and 15 females, which were then randomly assigned in groups of 5 to the current learning program (CL) and to two new experimental programs (E1 and E2). Then, 6 months later new grade-equivalent scores for reading (YR) and mathematics (YM) were obtained for these 30 people. In order to carry out a one-way MANOVA we will only consider one of the sexes. The male data is more adequate to help us illustrate the features we have in mind and as such we will only use the data for males.

A plot of the sample means for the four variables YR, YM, ZR, and ZM for each of the three treatments CL, E1, and E2, that is, a plot of the profiles for these three treatments, is available in Fig. 5.18.

The results for the test of equality of the mean vectors for these three treatments are displayed in Fig. 5.19, where we may see how the normality tests in general give quite large p-values. However, there are some strange values such as the case of the Shapiro-Wilk test with equal values for all three samples. Also, the Mardia combined test gives a p-value of 1 for the third sample and the Mardia kurtosis test a p-value of zero. Anyway, this is an issue on which we will not further elaborate. Then, the test of equality of covariance matrices gives a quite large p-value which would lead us to not reject this hypothesis, while the p-value for the test of equality of mean vectors is so low that we would for any reasonable α-value reject this hypothesis, a fact that is in good agreement with the idea we get from looking at the plot in Fig. 5.18.

We will see how the results from the test of parallelism of the profiles carried out in Sect. 5.1.3 will help in giving a good interpretation to the analysis of these data.

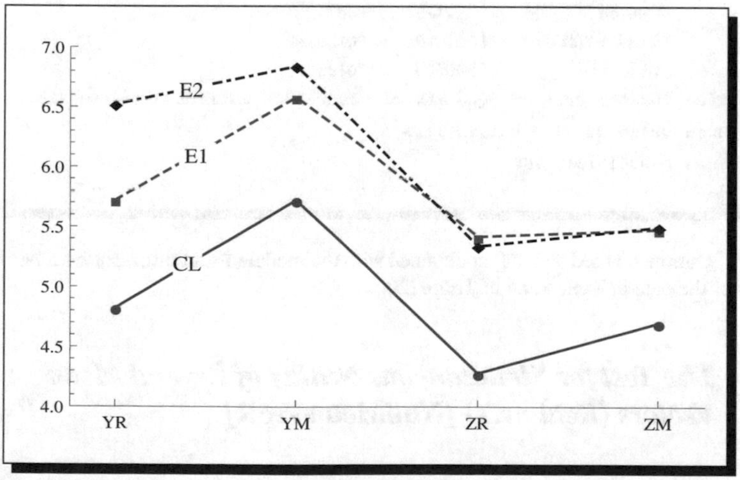

Fig. 5.18 Sample means (or "profiles") for variables YR, YM, ZR, and ZM (new and former grade-equivalent scores for reading and math) for the current learning (CL) program and two new experimental programs (E1 and E2)

```
PvalDataEqMeanVecR ["<Path>/Timm_Tab494_m.dat", , , , , cov → 1, norm → 1]
 There are 3 samples, with sizes {5, 5, 5} and 4 variables
```

		Statistic	P-Value
Sample 1	Anderson-Darling	3.54137	0.998157
	Kolmogorov-Smirnov	3.40091	0.994633
	Kuiper	2.89236	0.937305
	Mardia Combined	15.1707	0.196429
	Mardia Kurtosis	□2.22051	0.275
	Mardia Skewness	5.12	0.376667
	Pearson □^{2}	2.49228	0.79576
	Shapiro-Wilk	0.552182	0.000126531
	Watson U^2	3.09809	0.972429

		Statistic	P-Value
Sample 2	Anderson-Darling	3.52197	0.997824
	Kolmogorov-Smirnov	3.3062	0.990345
	Kuiper	2.85362	0.928115
	Mardia Combined	15.1707	0.164286
	Mardia Kurtosis	□2.22051	0.164634
	Mardia Skewness	5.12	0.163402
	Pearson □^{2}	2.49228	0.79576
	Shapiro-Wilk	0.552182	0.000126531
	Watson U^2	3.03254	0.963497

		Statistic	P-Value
Sample 3	Anderson-Darling	3.49801	0.997354
	Kolmogorov-Smirnov	3.55143	0.998313
	Kuiper	2.43733	0.768243
	Mardia Combined	15.1707	1.
	Mardia Kurtosis	□2.22051	0.
	Mardia Skewness	5.12	0.114706
	Pearson □^{2}	3.22125	0.984675
	Shapiro-Wilk	0.552182	0.000126531
	Watson U^2	3.13009	0.976139

```
(p-value for the test of equality of covariance matrices: 0.3084731442)
Computed value of Λ: 0.05107285711
p-value: 0.0001713473557
```

Fig. 5.19 Command used and output obtained with the module `PvalDataEqMeanVecR` for the analysis of the data in Table 4.9.4 of Timm (2002)

5.1.2 *The Test for Simultaneous Nullity of Several Mean Vectors (Real r.v.'s) [NullMeanVecR]*

Under a similar setup to the one in the previous subsection, let us assume that we are interested in testing the null hypothesis

$$H_0 : \underline{\mu}_1 = \underline{\mu}_2 = \cdots = \underline{\mu}_q = \underline{0}, \tag{5.16}$$

that is the hypothesis of simultaneous nullity of the q mean vectors $\underline{\mu}_k$ $(k = 1, \ldots, q)$.

Then it is not hard to determine that for $n = n_1 + n_2 + \cdots + n_q$, where n_k is the size of the sample from $\underline{X}_k$ $(k = 1, \ldots, q)$, the $(2/n)$-th power of the l.r.t. statistic to test H_0 in (5.16) is

$$\Lambda = \frac{|A|}{|A + B^*|} \tag{5.17}$$

where A is the matrix in (5.3) and

$$B^* = \sum_{k=1}^{q} n_k \overline{\underline{X}}_k \overline{\underline{X}}_k'$$

where, as in the previous subsection, $\overline{\underline{X}}_k$ is the sample mean vector for the k-th sample, and where B^* has a Wishart or pseudo-Wishart $W_p(q, \Sigma)$ distribution.

Then, from Appendix 1 it follows that for $n > p + q - 1$ we have

$$\Lambda \stackrel{d}{\equiv} \prod_{j=1}^{p} Y_j \stackrel{d}{\equiv} \prod_{k=1}^{q} Y_k^* \tag{5.18}$$

where, for $j = 1, \ldots, p$ and $k = 1, \ldots, q$,

$$Y_j \sim Beta\left(\frac{n-q+1-j}{2}, \frac{q}{2}\right) \quad \text{and} \quad Y_k^* \sim Beta\left(\frac{n-p+1-k}{2}, \frac{p}{2}\right) \tag{5.19}$$

represent respectively a set of p and a set of q independent r.v.'s.

The exact p.d.f. and c.d.f. of Λ are thus given in terms of the Meijer G function as

$$f_\Lambda(z) = \left\{\prod_{j=1}^{p} \frac{\Gamma\left(\frac{n+1-j}{2}\right)}{\Gamma\left(\frac{n-q+1-j}{2}\right)}\right\} G_{p,p}^{p,0}\left(\begin{array}{c} \left\{\frac{n+1-j}{2}-1\right\}_{j=1:p} \\ \left\{\frac{n-q+1-j}{2}-1\right\}_{j=1:p} \end{array} \middle| \, z\right)$$

and

$$F_\Lambda(z) = \left\{\prod_{j=1}^{p} \frac{\Gamma\left(\frac{n+1-j}{2}\right)}{\Gamma\left(\frac{n-q+1-j}{2}\right)}\right\} G_{p+1,p+1}^{p,1}\left(\begin{array}{c} \left\{1, \left\{\frac{n+1-j}{2}\right\}_{j=1:p}\right\} \\ \left\{\left\{\frac{n-q+1-j}{2}\right\}_{j=1:p}, 0\right\} \end{array} \middle| \, z\right)$$

or

$$f_\Lambda(z)=\left\{\prod_{k=1}^{q}\frac{\Gamma\left(\frac{n+1-k}{2}\right)}{\Gamma\left(\frac{n-p+1-k}{2}\right)}\right\}G_{q,q}^{q,0}\left(\begin{array}{c}\left\{\frac{n+1-k}{2}-1\right\}_{k=1:q}\\\left\{\frac{n-p+1-k}{2}-1\right\}_{k=1:q}\end{array}\middle|\,z\right)$$

and

$$F_\Lambda(z)=\left\{\prod_{k=1}^{q}\frac{\Gamma\left(\frac{n+1-k}{2}\right)}{\Gamma\left(\frac{n-p+1-k}{2}\right)}\right\}G_{q+1,q+1}^{q,1}\left(\begin{array}{c}\left\{1,\left\{\frac{n+1-k}{2}\right\}_{k=1:q}\right\}\\\left\{\left\{\frac{n-p+1-k}{2}\right\}_{k=1:q},0\right\}\end{array}\middle|\,z\right).$$

For even p, the distribution of Λ in (5.17) is given by Theorem 3.1 or 3.2 and its p.d.f. and c.d.f. by Corollary 4.1 or 4.2 with

$$m^*=1,\quad a_1=\frac{n-q+1}{2},\quad n_1=\frac{p}{2},\quad k_1=2\quad\text{and}\quad m_1=q\,,$$

or with

$$m^*=1,\quad a_1=\frac{n-p+1}{2},\quad n_1=\frac{q}{2},\quad k_1=2\quad\text{and}\quad m_1=p\,,$$

for even q, yielding for either of these cases the exact p.d.f. and c.d.f. of Λ, through the EGIG p.d.f. and c.d.f., in the form

$$f_\Lambda(z)=f^{EGIG}\left(z\,\middle|\,\{r_j\}_{j=1:p+q-2};\left\{\frac{n-1-j}{2}\right\}_{j=1:p+q-2};p+q-2\right)$$

and

$$F_\Lambda(z)=F^{EGIG}\left(z\,\middle|\,\{r_j\}_{j=1:p+q-2};\left\{\frac{n-1-j}{2}\right\}_{j=1:p+q-2};p+q-2\right)$$

where

$$r_j=\begin{cases}h_j\,, & j=1,2\\ r_{j-2}+h_j\,, & j=3,\dots,p+q-2\end{cases}$$

with

$$h_j=(\#\text{ of elements in }\{p,q\}\geq j)-1\,,\quad j=1,\dots,p+q-2\,.$$

This test is invariant under linear transformations of the form $\underline{X}_k \rightarrow C\underline{X}_K$, where C is any $p\times p$ full-rank matrix.

For the implementation of this test modules `PDFNullMeanVecR`, `CDFNullMeanVecR`, `PvalNullMeanVecR`, `QuantNullMeanVecR`, and `PvalDataNullMeanVecR` are made available. These modules have similar arguments and similar usage to the corresponding modules in the previous subsection.

5.1.3 *The Parallelism Test for Profiles (Real r.v.'s) [ProfParR]*

Let us suppose a similar set-up to the one in the two previous subsections where we have q populations, the k-th of which has a p-variate Normal distribution with mean vector $\underline{\mu}_k$ and covariance matrix Σ $(k = 1, \ldots, q)$, and that we obtain from the k-th population a sample of size n_k, where the q samples are independent.

We will assume that we consider p variables in each population and we will represent the k-th population random vector by $\underline{X}_k = [X_{1k}, \ldots, X_{jk}, \ldots, X_{pk}]'$, so that we will write $\underline{X}_k \sim N_p(\underline{\mu}_k, \Sigma)$, where $\underline{\mu}_k = [\mu_{1k}, \ldots, \mu_{jk}, \ldots, \mu_{pk}]'$.

We will call the "profile of the k-th population," the line drawn by passing through the p points with coordinates (X_{jk}, μ_{jk}), where the X_{jk} are represented on a common horizontal axis, using the same common arbitrary scale for all q samples. We may see in Fig. 5.20 what may be a graphical representation of some sample profiles, with dots having ordinates that correspond to the sample averages $\overline{X}_{jk} = \frac{1}{n_k}\sum_{\ell=1}^{n_k} X_{jk\ell}$ for the j-th variable in the sample from the k-th population.

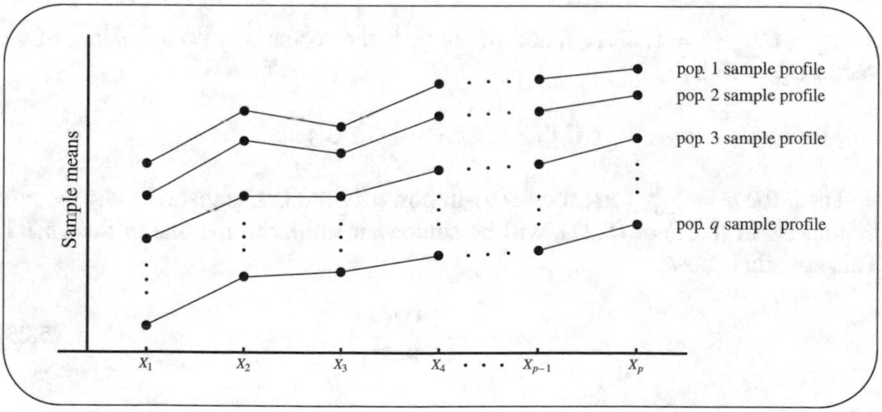

Fig. 5.20 Plot of sample profiles for populations 1 through q

In this subsection we will be concerned with the test to the null hypothesis of parallelism of the profiles. If we consider the matrix

$$\underset{(p-1)\times p}{C} = \begin{bmatrix} 1 & -1 & 0 & 0 & \cdots & 0 & 0 \\ 0 & 1 & -1 & 0 & \cdots & 0 & 0 \\ 0 & 0 & 1 & -1 & \cdots & 0 & 0 \\ \vdots & \vdots & \vdots & \vdots & & \vdots & \vdots \\ 0 & 0 & 0 & 0 & \cdots & 1 & -1 \end{bmatrix}, \tag{5.20}$$

then, taking into account that

$$C\underline{\mu}_k = \begin{bmatrix} \mu_{1k} - \mu_{2k} \\ \mu_{2k} - \mu_{3k} \\ \vdots \\ \mu_{p-1,k} - \mu_{pk} \end{bmatrix}$$

where μ_{jk} represents the j-th component of $\underline{\mu}_k$ $(j = 1, \ldots, p : k = 1, \ldots, q)$, the hypothesis of parallelism of the q profiles may be written as

$$H_0 : C\underline{\mu}_1 = \cdots = C\underline{\mu}_k = \cdots = C\underline{\mu}_q\,. \tag{5.21}$$

This hypothesis may be equivalently written as

$$H_0 : \underline{\nu}_1 = \cdots = \underline{\nu}_k = \cdots = \underline{\nu}_q \tag{5.22}$$

for $\underline{\nu}_k = C\underline{\mu}_k$ $(k = 1, \ldots, q)$, and where $\underline{\nu}_k$ is the vector of expected values of the vector $C\underline{X}_k$, with

$$C\underline{X}_k \sim N_{p-1}(\underline{\nu}_k, C\Sigma C')\,.$$

Then, for $n = \sum_{k=1}^{q} n_k$, the $(2/n)$-th power of the l.r.t. statistic to test the null hypothesis in (5.21) or (5.22), will be somewhat similar to the one in Sect. 5.1.1. This statistic is now

$$\Lambda = \frac{|A^*|}{|A^* + B^*|}\,, \tag{5.23}$$

where

$$A^* = \sum_{k=1}^{q} (n_k - 1) C S_k C' = CAC' \quad \text{and} \quad B^* = \sum_{k=1}^{q} n_k C(\overline{\underline{X}}_k - \overline{\underline{X}})(\overline{\underline{X}}_k - \overline{\underline{X}})' C' = CBC'$$

where A and B are the matrices defined in (5.3) and, as in (5.3), S_k and $\overline{X}_k$ are respectively the sample covariance matrix and mean vector for the k-th population, and $\overline{X}$ is the overall mean vector in (5.4).

The distribution of the statistic Λ in (5.23) is similar to that of the statistic Λ in Sect. 5.1.1, with p replaced by $p-1$, since while the matrix A^* has, for $n > p+q-2$, either under H_0 in (5.21) or (5.22) or even under the alternative hypothesis, a Wishart $W_{p-1}(n-q, \Sigma)$ distribution, the matrix B^*, has, under H_0, a $W_{p-1}(q-1, \Sigma)$ distribution, independent of A^*. As such, under H_0 in (5.21) or (5.22), for $n > p+q-2$, (see Appendix 1)

$$\Lambda \stackrel{d}{\equiv} \prod_{j=1}^{p-1} Y_j \stackrel{d}{\equiv} \prod_{k=1}^{q-1} Y_k^* \tag{5.24}$$

where, for $j = 1, \ldots, p-1$, and $k = 1, \ldots, q-1$,

$$Y_j \sim Beta\left(\frac{n-q+1-j}{2}, \frac{q-1}{2}\right) \quad \text{and} \quad Y_k^* \sim Beta\left(\frac{n-p+1-k}{2}, \frac{p-1}{2}\right) \tag{5.25}$$

represent respectively a set of $p-1$ and a set of $q-1$ independent random variables. Therefore, the exact p.d.f. and c.d.f. of Λ are given in terms of the Meijer G function as

$$f_\Lambda(z) = \left\{\prod_{j=1}^{p-1} \frac{\Gamma\left(\frac{n-j}{2}\right)}{\Gamma\left(\frac{n-q+1-j}{2}\right)}\right\} G^{p-1,0}_{p-1,p-1}\left(\begin{array}{c} \left\{\frac{n-j}{2}-1\right\}_{j=1:p-1} \\ \left\{\frac{n-q+1-j}{2}-1\right\}_{j=1:p-1} \end{array} \middle| \, z \right)$$

and

$$F_\Lambda(z) = \left\{\prod_{j=1}^{p-1} \frac{\Gamma\left(\frac{n-j}{2}\right)}{\Gamma\left(\frac{n-q+1-j}{2}\right)}\right\} G^{p-1,1}_{p,p}\left(\begin{array}{c} \left\{1, \left\{\frac{n-j}{2}\right\}_{j=1:p-1}\right\} \\ \left\{\left\{\frac{n-q+1-j}{2}\right\}_{j=1:p-1}, 0\right\} \end{array} \middle| \, z \right)$$

or as

$$f_\Lambda(z) = \left\{\prod_{k=1}^{q-1} \frac{\Gamma\left(\frac{n-k}{2}\right)}{\Gamma\left(\frac{n-p+1-k}{2}\right)}\right\} G^{q-1,0}_{q-1,q-1}\left(\begin{array}{c} \left\{\frac{n-k}{2}-1\right\}_{k=1:q-1} \\ \left\{\frac{n-p+1-k}{2}-1\right\}_{k=1:q-1} \end{array} \middle| \, z \right)$$

and

$$F_\Lambda(z)=\left\{\prod_{k=1}^{q-1}\frac{\Gamma\left(\frac{n-k}{2}\right)}{\Gamma\left(\frac{n-p+1-k}{2}\right)}\right\}G_{q,q}^{q-1,1}\left(\begin{matrix}1,\left\{\frac{n-k}{2}\right\}_{k=1:q-1}\\ \left\{\frac{n-p+1-k}{2}\right\}_{k=1:q-1},0\end{matrix}\,\middle|\,z\right).$$

For odd p or odd q, the exact distribution of Λ is of the type of the product in Theorem 3.2 and thus its exact p.d.f. and c.d.f. will be given in a finite highly manageable form by Corollary 4.2, with

$$m^*=1,\quad a_1=\frac{n-q+1}{2},\quad n_1=\frac{p-1}{2},\quad k_1=2\quad\text{and}\quad m_1=q-1,$$

for odd p, and with

$$m^*=1,\quad a_1=\frac{n-p+1}{2},\quad n_1=\frac{q-1}{2},\quad k_1=2\quad\text{and}\quad m_1=p-1.$$

for odd q.

For either of these two cases Corollary 4.2 gives the representation of the exact p.d.f. and c.d.f. of Λ, in terms of the EGIG p.d.f. and c.d.f., respectively as

$$f_\Lambda(z)=f^{EGIG}\left(z\,\middle|\,\{r_j\}_{j=1:p+q-4};\left\{\frac{n-2-j}{2}\right\}_{j=1:p+q-4};p+q-4\right)\tag{5.26}$$

and

$$F_\Lambda(z)=F^{EGIG}\left(z\,\middle|\,\{r_j\}_{j=1:p+q-4};\left\{\frac{n-2-j}{2}\right\}_{j=1:p+q-4};p+q-4\right),\tag{5.27}$$

where

$$r_j=\begin{cases}h_j, & j=1,2\\ r_{j-2}+h_j, & j=3,\dots,p+q-4\end{cases}$$

with

$$h_j=(\#\text{ of elements in }\{p-1,q-1\}\geq j)-1,\quad j=1,\dots,p+q-4.$$

As in the previous subsection, the representation of the exact form of this distribution through the EGIG p.d.f. and c.d.f. has the advantage of enabling a very quick computation of p-values and quantiles and also allows the easy implementation of plots of these functions.

As for all l.r. tests, also for the test in this subsection, in order to carry out an α-level test we should reject H_0 in (5.21) or (5.22) if the computed value of Λ in (5.23) is smaller than the α quantile of Λ, or equivalently, if the corresponding p-value is smaller than our value of α. These quantiles or p-values may be easily obtained by using respectively modules `QuantProfParR` or `PvalProfParR`. Also modules `PDFProfParR` and `CDFProfParR` are made available to compute and plot the p.d.f. and c.d.f. of Λ in (5.23) and the module `PvalDataProPar` to compute the p-value corresponding to the value of Λ calculated from a data set in a file. All these modules have similar arguments and a similar usage to the corresponding modules in Sect. 5.1.1. Namely, module `PvalDataProPar` uses exactly the same type of data files as module `PvalDataEqMeanVecR`, described in detail in Sect. 5.1.1, and it also uses exactly the same arguments as this module.

The test for profile parallelism is invariant under linear transformations of the type $\underline{X}_k \to C^*\underline{X}_k + \underline{b}$, where C^* is any nonrandom full-rank $p\times p$ circulant matrix (see Sect. 5.1.22 for the definition of a circulant or circular matrix) or $C^* = aI_p^*$ where I_p^* is any $p\times p$ permutation matrix and a any non-null real value and where $\underline{b} = bE_{p1}$, with b any real.

We may note here that the matrix C in (5.20) may be replaced by any matrix $C^+ = I_{p-1}^* C$, where I_{p-1}^* is any permutation matrix of order $p-1$.

For an alternative implementation of the test in this subsection we refer the reader to Sect. 5.1.4.3, where the test is implemented through the l.r.t. for the linear model.

5.1.3.1 On the Availability of Quantile Tables

The quantile tables described in Sect. 5.1.1.2 may be used to obtain "exact" or approximate quantiles also for the statistic Λ in (5.23), using now the value of $p-1$ where in Sect. 5.1.1.2 one would use the value of p and where we write "exact" because of the problems already reported in Sect. 5.1.1.2 about these supposedly exact quantiles listed in Table 1 of Kres (1983) and Table A.9 of Rencher (2002) and Rencher and Christensen (2012). That is, in Table 1 of Kres (1983) we would replace p by our $p-1$, q by our $q-1$, and n by our $n-q$, while in Table 9 of Muirhead (2005) we would replace m by our $q-1$, r by our $p-1$, and M by our $n-(p-1)-(q-1)$. For example, to obtain the "exact" 0.01 quantile of Λ in (5.23) for $n=10$, $p=5$, and $q=5$, one would use Table 1 in Kres (1983) with $p=4$, $q=4$, and $n=5$, obtaining the value 0.000250, instead of the real exact value of 0.000161, which may be obtained with the module `QuantProfParR`. To obtain the approximate value for this quantile from Table 9 in Muirhead (2005), one would use $m=4$, $r=4$, and $M=2$, obtaining the correction factor $c=1.229$, which once inserted in expression (5.11), with p replaced by $p-1$, would give the value 0.000160 for the approximate quantile, and which indeed is a much better result.

However, as already pointed out in Sect. 5.1.1.2, the availability of modules for the computation of quantiles, namely, for the present test the availability of the module QuantProfParR makes the use of these tables, for odd p or odd q, unnecessary and obsolete, giving the user a much simpler and more precise way to obtain such quantiles with virtually any desired precision. This is an even stronger argument nowadays, given the ubiquitous use and availability of all sorts of personal computers, laptops, and notebooks. Modules PvalProfParR and CDFProfParR give an even more useful related quantity, the p-value associated with a given computed value of the statistic Λ. Module PvalDataProfParR, similar to module PvalDataEqMeanVecR, and using similar arguments and a similar syntax, is able to compute the value of the statistic Λ in (5.23) and the corresponding p-value for a data set stored in a file. Examples of the use of this module may be found in Sect. 5.1.3.4.

5.1.3.2 The Exact Distribution of Λ in (5.23) for $p = 2$ or 3 and $q = 2$ or 3

Similar to what happens with the statistic Λ in (5.2), also for the statistic Λ in (5.23) is possible to establish a direct relation of its exact distribution and that of an F distributed r.v. for a few values of p or q. This is actually also true for all l.r.t. statistics presented in this book whenever their distributions reduce to just a Beta distribution and in most cases where their distribution is that of a product of only two independent Beta r.v.'s.

For $p = 2$ we have, from (5.24) and (5.25),

$$\Lambda \sim Beta\left(\frac{n-q}{2}, \frac{q-1}{2}\right)$$

so that using the relation in (5.12) we have

$$\frac{n-q}{q-1}\frac{1-\Lambda}{\Lambda} \sim F_{q-1,n-q}\,, \tag{5.28}$$

while for $q = 2$, also from (5.24) and (5.25),

$$\Lambda \sim Beta\left(\frac{n-p}{2}, \frac{p-1}{2}\right)$$

and as such, in this case,

$$\frac{n-p}{p-1}\frac{1-\Lambda}{\Lambda} \sim F_{p-1,n-p}\,, \tag{5.29}$$

where $\frac{1-\Lambda}{\Lambda}$ is equivalent to the T^2 statistic commonly used for $q = 2$.

For $p=3$, once again from (5.24) and (5.25), and using the Gamma duplication formula in (5.583), we may write the h-th moment of $\Lambda^{1/2}$ as

$$E\left(\Lambda^{h/2}\right)=\frac{\Gamma\left(\frac{n-1}{2}\right)\Gamma\left(\frac{n-q}{2}+\frac{h}{2}\right)\Gamma\left(\frac{n-2}{2}\right)\Gamma\left(\frac{n-q-1}{2}+\frac{h}{2}\right)}{\Gamma\left(\frac{n-q}{2}\right)\Gamma\left(\frac{n-1}{2}+\frac{h}{2}\right)\Gamma\left(\frac{n-q-1}{2}\right)\Gamma\left(\frac{n-2}{2}+\frac{h}{2}\right)}=\frac{\Gamma(n-2)\,\Gamma(n-q-1+h)}{\Gamma(n-q-1)\,\Gamma(n-2+h)}$$

which shows that in this case

$$\Lambda^{1/2}\sim Beta(n-q-1,q-1)$$

so that, using (5.12),

$$\frac{n-q-1}{q-1}\,\frac{1-\Lambda^{1/2}}{\Lambda^{1/2}}\sim F_{2(q-1),2(n-q-1)}\,, \tag{5.30}$$

while for $q=3$ we have the h-th moment of $\Lambda^{1/2}$ given by

$$E\left(\Lambda^{h/2}\right)=\frac{\Gamma\left(\frac{n-1}{2}\right)\Gamma\left(\frac{n-p}{2}+\frac{h}{2}\right)\Gamma\left(\frac{n-2}{2}\right)\Gamma\left(\frac{n-p-1}{2}+\frac{h}{2}\right)}{\Gamma\left(\frac{n-p}{2}\right)\Gamma\left(\frac{n-1}{2}+\frac{h}{2}\right)\Gamma\left(\frac{n-p-1}{2}\right)\Gamma\left(\frac{n-2}{2}+\frac{h}{2}\right)}=\frac{\Gamma(n-2)\,\Gamma(n-p-1+h)}{\Gamma(n-p-1)\,\Gamma(n-2+h)}$$

which shows that

$$\Lambda^{1/2}\sim Beta(n-p-1,p-1)$$

or that

$$\frac{n-p-1}{p-1}\,\frac{1-\Lambda^{1/2}}{\Lambda^{1/2}}\sim F_{2(p-1),2(n-p-1)}\,. \tag{5.31}$$

In all the above cases, for a test of size α, one will reject the null hypothesis of profile parallelism if the computed value of the statistics in (5.28), (5.29), (5.30), or (5.31) exceeds the $1-\alpha$ quantile of the corresponding F distribution, while for either $p=3$ or $q=3$ we also have the exact distribution of Λ as an EGIG distribution, with p.d.f. and c.d.f. given by (5.26) and (5.27).

5.1.3.3 On the Test for Equality of Covariance Matrices and the Tests of Multivariate Normality

The tests of multivariate normality and the l.r.t. for equality of covariance matrices are implemented, similar to what happens with the l.r.t. in Sect. 5.1.1, in the Mathematica® version of the module `PvalDataProfParR` and may be carried out by giving respectively the named optional arguments `norm` and `cov` the value 1, as illustrated in Sect. 5.1.1 for module `PvalDataEqMeanVecR`. The l.r.t. for equality of covariance matrices uses exactly the same near-exact distribution described in Sect. 5.1.1.4.

For a brief discussion on the importance of these tests we refer the reader to Sect. 5.1.1, since similar considerations apply to the present test of profile parallelism.

5.1.3.4 Real Data Examples

We will use in this subsection exactly the same data sets used in Sect. 5.1.1.5 to illustrate tests of equality of mean vectors, to illustrate the test in the present subsection and also to help us complete some of the analyses started in that subsection. As such, in most cases no normality tests and no l.r.t. for equality of covariance matrices will be carried out, since these would give exactly the same results as those in Sect. 5.1.1.5.

We start with the iris data set and consider testing the parallelism of the profiles in Fig. 5.12, for the three iris species. It is with no surprise that a computed value for Λ in (5.23) of 0.04115316581, with an associated p-value of $2.3958320350\times10^{-97}$, leads us to reject the hypothesis of parallelism of the three profiles.

For the three profiles in Fig. 5.16, concerning the sales professionals with scores of 8, 11, and 12 in the abstract reasoning test, for the data reported in Table 9.12 of Johnson and Wichern (2014), it is also with not much surprise that we obtain a computed value of Λ equal to 0.7687678499, with a corresponding p-value of 0.1376999357 which leads us to not reject the null hypothesis of profile parallelism, taking us to conclude that although there are significant differences among the means for the three sales performance variables for people with scores of 8, 11, and 12 in the abstract reasoning test, these differences are similar for all these three variables when we compare any two profiles. In Fig. 5.21 we may analyze the command used and the output obtained, noticing that the first unnamed optional argument of module `PvalDataProfParR` was used since the data file, which mimics Table 9.12 in Johnson and Wichern (2014), has a number of columns, from which we want to select only columns 2, 3, and 4, which correspond to the three sales performance variables, it being the case that we also want to select only samples 4, 7, and 8, which correspond to people with scores of 8, 11, and 12 in the abstract reasoning test.

Concerning the three profiles in Fig. 5.18, which originated from the data in Table 4.9.4 of Timm (2002) on the four variables measuring reading and math

```
PvalDataProfParR["<Path>/Tab9_12_JW.dat", 1]
 There are 3 samples, with sizes {5, 15, 10} and 3 variables
 Original samples {4, 7, 8}
 Original variables {2, 3, 4}
Computed value of Λ: 0.7687678499
p-value: 0.1376999357
```

Fig. 5.21 Command used and output obtained for the profile parallelism test of the three profiles whose sample means are plotted in Fig. 5.16

```
PvalDataProfParR["<Path>/Timm_Tab494_m.dat"]
 There are 3 samples, with sizes {5, 5, 5} and 4 variables
Computed value of Λ: 0.1667263313
p-value: 0.003371690331
PvalDataProfParR["<Path>/Timm_Tab494_m.dat", 1]
 There are 3 samples, with sizes {5, 5, 5} and 3 variables
 Original variables {2, 3, 4}
Computed value of Λ: 0.7605378588
p-value: 0.5341288545
PvalDataProfParR["<Path>/Timm_Tab494_m.dat", 1]
 There are 2 samples, with sizes {5, 5} and 4 variables
 Original samples {1, 2}
Computed value of Λ: 0.7835305862
p-value: 0.664981500
```

Fig. 5.22 Commands used and output obtained for the profile parallelism test of the three profiles whose sample means are plotted in Fig. 5.18

capabilities, described in Sect. 5.1.1.5, and corresponding to the three learning programs (current: CL, experimental 1: E1, and experimental 2: E2) administered to males, we are interested in testing the parallelism of these three profiles. As it may be seen from the first output in Fig. 5.22, when considering the four variables YR, YM, ZR, and ZM, this hypothesis should be rejected for most common values of α, but when we consider only YM, ZR, and ZM, then the parallelism hypothesis should not be rejected, which is in agreement with what we grasp from looking at Fig. 5.18.

These results, together with the results obtained in Sect. 5.1.1.5 where we rejected the equality of the three profiles, would lead us to conclude that although we would not consider the profiles for the current learning program (CL) and for the first experimental program (E1) to be equal (the test of equality of these two profiles gives a computed value of 0.1647248198 for the statistic in (5.2), with an associated p-value of 0.034009585) they should be considered to be parallel, showing that the males subject to the first experimental program (E1) had higher original reading and math capabilities, that is, higher values for ZR and ZM, than those subject to the current learning program (CL), and that although there might have been some improvement of these capabilities when they were measured 6 months later (and this is a different type of analysis, addressed in Sects. 5.1.4.6 and 5.1.9.6), this change in the reading and math capabilities was similar for the two groups, CL and E1, while the second experimental program (E2) seems to lead to a significant improvement in the reading ability measured six months later (YR).

We display in Fig. 5.23 the results from the tests of equality and parallelism for the pairs of profiles CL-E1 and E1-E2, with CL, E1, and E2 corresponding respectively to samples 1, 2, and 3 in the data file.

In case we may be concerned with the level of the overall testing procedure, given that multiple tests were carried out on the same data, we may say that we are somewhat safeguarded against this issue by the use of p-values, which had in all occasions values that lead to clear decisions.

5.1.4 *The Likelihood Ratio Statistic to Test Hypotheses on an Expected Value Matrix (Real r.v.'s) [MatEVR]*

In this subsection we will address a test which has as particular cases what is usually called the "test of the general linear hypothesis" and as such also the overall test for the Multivariate Analysis of Variance (MANOVA) model (see for example

```
PvalDataEqMeanVecR["<Path>/Timm_Tab494_m.dat", 1]
 There are 2 samples, with sizes {5, 5} and 4 variables
 Original samples {1, 3}
Computed value of Λ: 0.09939850675
p-value: 0.010128238
PvalDataEqMeanVecR["<Path>/Timm_Tab494_m.dat", 1]
 There are 2 samples, with sizes {5, 5} and 4 variables
 Original samples {2, 3}
Computed value of Λ: 0.1543319820
p-value: 0.029139504
PvalDataProfParR["<Path>/Timm_Tab494_m.dat", 1]
 There are 2 samples, with sizes {5, 5} and 4 variables
 Original samples {1, 3}
Computed value of Λ: 0.1048619678
p-value: 0.002420968
PvalDataProfParR["<Path>/Timm_Tab494_m.dat", 1]
 There are 2 samples, with sizes {5, 5} and 4 variables
 Original samples {2, 3}
Computed value of Λ: 0.1560197516
p-value: 0.007805502
```

Fig. 5.23 Commands used and output obtained for the tests of equality and parallelism of pairs of profiles CL-E2 and E1-E2, corresponding to samples 1–3 and 2–3 (see Fig. 5.18 for sample plots)

Anderson (2003, Chap. 8)), which thus includes the test for equality of mean vectors in Sect. 5.1.1. Furthermore, as it is shown, this test is also able, under a slightly extended version, to include as a particular case the test for profile parallelism addressed in the previous subsection.

Let us start by considering the $p\times n$ matrix X with a real multivariate Normal distribution with expected value $\boldsymbol{\mu} M$ and variance $Var(vec(X)) = I_n \otimes \Sigma$, where $\boldsymbol{\mu}$ is a $p\times q$ real matrix and M is $q\times n$ of rank q $(\leq n)$. We will denote this fact by

$$\underset{p\times n}{X} \sim N_{p\times n}(\boldsymbol{\mu} M, I_n \otimes \Sigma)\,. \tag{5.32}$$

Let us then suppose that we want to test the hypothesis

$$H_0 : \boldsymbol{\mu}_{(p\times q)} = \beta^*_{(p\times q)}\,, \tag{5.33}$$

where β^* is a given $p\times q$ real matrix (for the case of complex matrices and complex r.v.'s, see Sect. 5.1.8).

Then, the $(2/n)$-th power of the l.r.t. statistic to test H_0 is (the derivation is in all ways similar to the one used in Khatri (1965) for the complex case)

$$\Lambda = \frac{|A|}{\left|A + \frac{1}{n}(\beta - \beta^*)(MM')(\beta - \beta^*)'\right|} \tag{5.34}$$

where

$$A = \tfrac{1}{n} X \left(I_n - M' \left(MM'\right)^{-1} M \right) X' \tag{5.35}$$

and

$$\beta = XM' \left(MM'\right)^{-1} \tag{5.36}$$

are respectively the m.l.e.'s of Σ and μ, and as such independent.

But, since $\left(I_n - M' \left(MM'\right)^{-1} M\right)$, the projector on the null space of the columns of M, is idempotent with

$$rank \left(I_n - M' \left(MM'\right)^{-1} M \right) = tr \left(I_n - M' \left(MM'\right)^{-1} M \right) = n - q\,,$$

from the distribution of X in (5.32), we have

$$A = \tfrac{1}{n} X \left(I_n - M' \left(MM'\right)^{-1} M \right) X' \sim W_p \left(n - q, \tfrac{1}{n}\Sigma \right). \tag{5.37}$$

From (5.32) and the results in Appendix 3 we may also easily see that

$$\beta = XM'\left(MM'\right)^{-1} \sim N_{p\times q}\left(\boldsymbol{\mu},\, \left(MM'\right)^{-1} \otimes \Sigma\right),$$

so that

$$(\beta-\beta^*)\left(MM'\right)^{1/2} \sim N_{p\times q}\left((\boldsymbol{\mu}-\beta^*)\left(MM'\right)^{1/2},\, I_q \otimes \Sigma\right),$$

where, under H_0 in (5.33), $(\boldsymbol{\mu}-\beta^*)\left(MM'\right)^{1/2} = 0$. As such, under this null hypothesis,

$$(\beta-\beta^*)\left(MM'\right)(\beta-\beta^*)' \sim W_p(q,\, \Sigma),$$

independent of A, and thus, under H_0 in (5.33),

$$A + \tfrac{1}{n}(\beta-\beta^*)(MM')(\beta-\beta^*)' \sim W_p\left(n,\, \tfrac{1}{n}\Sigma\right).$$

Therefore, from the results in Appendix 1, we may say that, for $n > p+q-1$, under H_0 in (5.33),

$$\Lambda \stackrel{\text{d}}{\equiv} \prod_{j=1}^{p} Y_j \stackrel{\text{d}}{\equiv} \prod_{k=1}^{q} Y_k^*, \tag{5.38}$$

where, for $j=1,\dots,p$ and $k=1,\dots,q$

$$Y_j \sim Beta\left(\frac{n-q+1-j}{2}, \frac{q}{2}\right) \quad \text{and} \quad Y_k^* \sim Beta\left(\frac{n-p+1-k}{2}, \frac{p}{2}\right) \tag{5.39}$$

form two sets of independent r.v.'s.

The exact p.d.f. and c.d.f. of Λ in (5.34) are thus given in terms of the Meijer G function by

$$f_\Lambda(z) = \left\{\prod_{j=1}^{p} \frac{\Gamma\left(\frac{n+1-j}{2}\right)}{\Gamma\left(\frac{n-q+1-j}{2}\right)}\right\} G_{p,p}^{p,0}\left(\left.\begin{array}{c}\left\{\frac{n+1-j}{2}-1\right\}_{j=1:p}\\ \left\{\frac{n-q+1-j}{2}-1\right\}_{j=1:p}\end{array}\right| z\right)$$

and

$$F_\Lambda(z)=\left\{\prod_{j=1}^{p}\frac{\Gamma\left(\frac{n+1-j}{2}\right)}{\Gamma\left(\frac{n-q+1-j}{2}\right)}\right\}G^{p,1}_{p+1,p+1}\left(\begin{array}{c}\left\{1,\left\{\frac{n+1-j}{2}\right\}_{j=1:p}\right\}\\ \left\{\left\{\frac{n-q+1-j}{2}\right\}_{j=1:p},0\right\}\end{array}\middle|\, z\right)$$

or

$$f_\Lambda(z)=\left\{\prod_{k=1}^{q}\frac{\Gamma\left(\frac{n+1-k}{2}\right)}{\Gamma\left(\frac{n-p+1-k}{2}\right)}\right\}G^{q,0}_{q,q}\left(\begin{array}{c}\left\{\frac{n+1-k}{2}-1\right\}_{k=1:q}\\ \left\{\frac{n-p+1-k}{2}-1\right\}_{k=1:q}\end{array}\middle|\, z\right)$$

and

$$F_\Lambda(z)=\left\{\prod_{k=1}^{q}\frac{\Gamma\left(\frac{n+1-k}{2}\right)}{\Gamma\left(\frac{n-p+1-k}{2}\right)}\right\}G^{q,1}_{q+1,q+1}\left(\begin{array}{c}\left\{1,\left\{\frac{n+1-k}{2}\right\}_{k=1:q}\right\}\\ \left\{\left\{\frac{n-p+1-k}{2}\right\}_{k=1:q},0\right\}\end{array}\middle|\, z\right).$$

Therefore, for either p or q even, the exact distribution of Λ in (5.34) is an EGIG distribution. More precisely, for even p or even q, the exact distribution of Λ in (5.34) is of the type of the product in Theorem 3.2 and thus its exact p.d.f. and c.d.f. are given by Corollary 4.2, with

$$m^*=1,\quad a_1=\frac{n-q+1}{2},\quad n_1=\frac{p}{2},\quad k_1=2\ \text{ and }\ m_1=q\,, \tag{5.40}$$

for even p, or

$$m^*=1,\quad a_1=\frac{n-p+1}{2},\quad n_1=\frac{q}{2},\quad k_1=2\ \text{ and }\ m_1=p\,, \tag{5.41}$$

for even q.

In terms of the EGIG distribution representation the exact p.d.f. and c.d.f. of Λ in (5.34) will be, for either one of the two cases above, respectively given by

$$f_\Lambda(z)=f^{EGIG}\left(z\,\middle|\,\{r_j\}_{j=1:p+q-2};\left\{\frac{n-1-j}{2}\right\}_{j=1:p+q-2};p+q-2\right) \tag{5.42}$$

and

$$F_\Lambda(z)=F^{EGIG}\left(z\,\middle|\,\{r_j\}_{j=1:p+q-2};\left\{\frac{n-1-j}{2}\right\}_{j=1:p+q-2};p+q-2\right), \tag{5.43}$$

where

$$r_j = \begin{cases} h_j\,, & j = 1, 2 \\ r_{j-2} + h_j\,, & j = 3, \ldots, p+q-2 \end{cases} \tag{5.44}$$

with

$$h_j = (\#\text{ of elements in } \{p, q\} \geq j) - 1\,, \quad j = 1, \ldots, p+q-2\,. \tag{5.45}$$

Since from (5.32) we may write

$$E(X) = \mu M\,, \tag{5.46}$$

a Multivariate Linear Model for X may be written as

$$X = \mu M + \mathcal{E}\,, \tag{5.47}$$

with

$$\mathcal{E}_{p\times n} \sim N_{p\times n}(0_{p\times n}, I_n \otimes \Sigma)\,. \tag{5.48}$$

As such, an interesting and useful way to look at the present test, for the case $\beta^* = 0$ in (5.33), is to see it as the test of the nullity of the matrix of parameters, μ, in a Multivariate Linear Model, where the variables in X act as the response variables and the variables in M as the explanatory variables, or, equivalently, as the test of the fit of the model in (5.46) or (5.47)–(5.48), since a rejection of the null hypothesis in (5.33) for $\beta^* = 0$ is equivalent to saying that "the model fits," in the sense that the variables, or the "information" in M explains or models "significantly well" the values of the variables in X and their variability.

This is a very useful way to look at this test, which will enable us to give practical meanings to several particular cases of the extension of this test which will be presented next.

We should note that for $\beta^* = 0_{p\times q}$ the present test is strictly equivalent to the test in Sect. 5.1.2. Indeed, for $\beta^* = 0$ the statistic in (5.34) and its distribution in (5.38)–(5.39) are equivalent to the statistic in (5.17) and its distribution given by (5.18)–(5.19). In addition, the distribution of Λ in (5.34) remains valid if the distribution of X is some left orthogonal-invariant distribution (Anderson, 2003, Sect. 8.11) or even any elliptically contoured distribution (see Jensen and Good (1981), Kariya (1981), Anderson et al. (1986), and Anderson and Fang (1990)).

This l.r. test also enjoys some invariance properties under linear transformations. Whenever it is being used under a generalized form that reproduces another l.r. test as for example those in Sect. 5.1.1 or 5.1.3, or even some of the tests in Sect. 5.1.9, it will enjoy the same invariance properties of those tests—see Sects. 5.1.4.2, 5.1.4.3, and 5.1.4.6.

5.1.4.1 Extending the Test

We may extend the test addressed in this subsection in such a way that the tests in Sects. 5.1.1 and 5.1.3 may be viewed as particular cases of this test. This is so because indeed the likelihood ratio tests in these two subsections may also be viewed as likelihood ratio tests for the nullity of an expected value matrix, and as such they may be viewed as particular cases of an extended version of the test being addressed in this subsection.

If we take C to be a nonrandom real $p^*\times p$ matrix, with $rank(C) = p^* \leq p$, and D to be a nonrandom real $q\times q$ matrix with $rank(D) = q^* \leq q$, then, by the invariance property of m.l.e.'s, $C\beta D$ is the m.l.e. of $C\mu D$. This property permits an almost immediate derivation of the l.r.t. statistic to test the null hypothesis

$$H_0 : C\mu D = \beta^*_{(p^*\times q)} \tag{5.49}$$

although the derivation and study of this l.r.t. statistic has to be done with some extra care in this more general setup.

First we have to notice that the matrix C yields a linear combination of the p variables in X, so that now we have

$$\underset{(p^*\times n)}{CX} \sim N_{p^*\times n}(C\mu M, I_n \otimes C\Sigma C'),$$

where the m.l.e. of $C\Sigma C'$ is, for A in (5.35),

$$A^* = CAC' = \frac{1}{n}CX\left(I_n - M'\left(MM'\right)^{-1}M\right)X'C' \sim W_{p^*}\left(n-q, \frac{1}{n}C\Sigma C'\right) \tag{5.50}$$

and the m.l.e. of $C\mu$ is, for β in (5.36) (see Appendix 3)

$$C\beta = CXM'\left(MM'\right)^{-1} \sim N_{p^*\times q}\left(C\boldsymbol{\mu}, \left(MM'\right)^{-1} \otimes C\Sigma C'\right), \tag{5.51}$$

so that if one wants to test the null hypothesis

$$H_0 : C\mu = \beta^*_{(p^*\times q)} \tag{5.52}$$

we have under this null hypothesis,

$$(C\beta - \beta^*)\left(MM'\right)^{1/2} \sim N_{p^*\times q}\left((C\boldsymbol{\mu} - \beta^*)\left(MM'\right)^{1/2}, I_q \otimes C\Sigma C'\right),$$

where, under H_0 in (5.52), $(C\boldsymbol{\mu} - \beta^*)\left(MM'\right)^{1/2} = 0$. As such, under this null hypothesis,

$$(C\beta - \beta^*)\left(MM'\right)(C\beta - \beta^*)' \sim W_{p^*}(q, C\Sigma C'),$$

independent of A^*, so that, under H_0 in (5.52),

$$A^* + \frac{1}{n}(C\beta - \beta^*)(MM')(C\beta - \beta^*)' \sim W_{p^*}\left(n, \frac{1}{n}C\Sigma C'\right).$$

Thus the l.r.t. statistic to test H_0 in (5.52) is

$$\Lambda^* = \frac{|A^*|}{\left|A^* + \frac{1}{n}(C\beta - \beta^*)(MM')(C\beta - \beta^*)'\right|} \tag{5.53}$$

with

$$\Lambda^* \stackrel{d}{\equiv} \prod_{j=1}^{p^*} Y_j^* \stackrel{d}{\equiv} \prod_{k=1}^{q} Y_k^{**}, \tag{5.54}$$

where, for $n > p^* + q - 1$, $j = 1, \ldots, p^*$, and $k = 1, \ldots, q$

$$Y_j^* \sim Beta\left(\frac{n-q+1-j}{2}, \frac{q}{2}\right) \quad \text{and} \quad Y_k^{**} \sim Beta\left(\frac{n-p^*+1-k}{2}, \frac{p^*}{2}\right)$$

represent two sets of independent r.v.'s.

As such, for either p^* or q even, the exact distribution of Λ^* in (5.53) is an EGIG distribution. More precisely, for even p^* or even q, the exact distribution of Λ^* in (5.53) is of the type of the product in Theorem 3.2 and thus its exact p.d.f. and c.d.f. are given, for even p^*, by Corollary 4.2, with

$$m^* = 1, \quad a_1 = \frac{n-q}{2}, \quad n_1 = \frac{p^*}{2}, \quad k_1 = 2 \quad \text{and} \quad m_1 = q, \tag{5.55}$$

while for even q, the exact distribution of Λ^* will be given by Theorem 3.2, with

$$m^* = 1, \quad a_1 = \frac{n-p^*}{2}, \quad n_1 = \frac{q}{2}, \quad k_1 = 2 \quad \text{and} \quad m_1 = p^*, \tag{5.56}$$

with the exact p.d.f. and c.d.f. of Λ^* being given, either for even p^* or even q, by (5.42)–(5.45) with p replaced by p^*.

Then, if we are interested in testing the null hypothesis in (5.49), we may do it just by using an l.r.t. statistic which is similar to Λ^* in (5.53), with $C\beta$ replaced by

$C\beta D$. However, in doing this replacement one has to be careful, since from (5.51) and Appendix 3,

$$(C\beta D - \beta^*) \sim N_{p^* \times q^{**}}\left(C\mu D - \beta^*,\, D'(MM')^{-1}D \otimes C\Sigma C'\right),$$

where, under H_0 in (5.49), $(C\boldsymbol{\mu} D - \beta^*) = 0$, so that we only need to post-multiply $C\mu D - \beta^*$ by a full-rank $q \times q$ matrix, let us call it H, so that $Var((C\beta D - \beta^*)H) = H'D'(MM')^{-1}DH$ is an idempotent matrix of rank $q^* \leq q$. Since this matrix is clearly symmetric we only need it to be idempotent in order to be a projector, so that

$$B = \big((C\mu D - \beta^*)H\big)\big((C\mu D - \beta^*)H\big)' \sim W_{p^*}(q^*, C\Sigma C'). \tag{5.57}$$

Therefore, the l.r.t. statistic to test H_0 in (5.49) is the statistic

$$\Lambda^{**} = \frac{|A^*|}{\left|A^* + \frac{1}{n}(C\beta D - \beta^*)(HH')(C\beta D - \beta^*)'\right|} \tag{5.58}$$

where A^* remains defined as in (5.50), and where H has to be defined in such a way that $H'D'(MM')^{-1}DH$ is an idempotent matrix, that is, in such a way that

$$H'D'(MM')^{-1}DHH'D'(MM')^{-1}DH = H'D'(MM')^{-1}DH\,, \tag{5.59}$$

where the $q \times q$ matrix H has to be full-rank in order to have

$$rank(D'(MM')^{-1}D) = rank(H'D'(MM')^{-1}DH)\,.$$

But then, since the full-rank property of H implies that HH' is invertible, if we left-multiply both sides of (5.59) by $(HH')^{-1}H$, we get

$$D'(MM')^{-1}DHH'D'(MM')^{-1}D = D'(MM')^{-1}D\,, \tag{5.60}$$

which shows that HH' has to be defined in such a way that it is a weak inverse (Seber, 2008, Sects. 7.1, 7.2) of $D'(MM')^{-1}D$. In practice, given the general availability of computer modules to compute generalized inverses, we will take HH' as a generalized inverse (Golub and Loan, 1996, Sect. 5.5.4) of $D'(MM')^{-1}D$, which is a stronger requirement than that imposed by (5.59) and (5.60). However, in many useful practical applications of this generalization of the test the matrix H may have a simple structure as will be illustrated in the next Sects. 5.1.4.2 and 5.1.4.3.

We may note that since the matrix B in (5.57) is independent of A^*, we have, under H_0 in (5.49),

$$A^* + \frac{1}{n}B = A^* + \frac{1}{n}(C\beta D - \beta^*)(HH')(C\beta D - \beta^*)' \sim W_{p^*}\left(n - q + q^*, \frac{1}{n}C\Sigma C'\right),$$

and we also thus have, for $n > p^* + q - 1$, (see Appendix 1)

$$\Lambda^{**} \stackrel{d}{\equiv} \prod_{j=1}^{p^*} Y_j^{**} \stackrel{d}{\equiv} \prod_{k=1}^{q^*} Y_k^{***}, \tag{5.61}$$

where, for $j = 1, \ldots, p^*$ and $k = 1, \ldots, q^*$,

$$Y_j^{**} \sim Beta\left(\frac{n-q+1-j}{2}, \frac{q^*}{2}\right) \quad \text{and} \quad Y_k^{***} \sim Beta\left(\frac{n-q+q^*-p^*+1-k}{2}, \frac{p^*}{2}\right) \tag{5.62}$$

are two sets of independent r.v.'s.

Hence, for either p^* or q^* even, the exact distribution of Λ^{**} in (5.58) is an EGIG distribution. Indeed, for even p^* or even q^*, the exact distribution of Λ^{**} in (5.58) is of the type of the product in Theorem 3.2 and thus its exact p.d.f. and c.d.f. are given, for even p^*, by Corollary 4.2, with

$$m^* = 1, \quad a_1 = \frac{n-q+1}{2}, \quad n_1 = \frac{p^*}{2}, \quad k_1 = 2 \quad \text{and} \quad m_1 = q^*, \tag{5.63}$$

while for even q^*, the exact p.d.f. and c.d.f. of Λ^{**} will be given by Corollary 4.2, with

$$m^* = 1, \quad a_1 = \frac{n-q+q^*-p^*+1}{2}, \quad n_1 = \frac{q^*}{2}, \quad k_1 = 2 \quad \text{and} \quad m_1 = p^*. \tag{5.64}$$

When p^* and q^* are both even, their interchangeability in (5.63) and (5.64) becomes clear only if we write the first parameter of the Y_j^{**} Beta r.v.'s in (5.62), added by $j/2$, or a_1 in (5.63) as

$$\frac{n-q+q^*-q^*+1}{2}$$

and we keep unchanged the "structural" part $\frac{n-q+q^*}{2}$, which "replaces" the "structural" part $\frac{n}{2}$ in (5.39)–(5.43), and then replace in the remaining part q^* by p^*. Therefore, for either even p^* or even q^*, the exact p.d.f. and c.d.f. of Λ^{**} in (5.58) are given by (5.42)–(5.45) with p replaced by p^*, q by q^* and n replaced by $n - q + q^*$, that is, respectively as

$$f_\Lambda(z) = f^{EGIG}\left(z \,\middle|\, \{r_j\}_{j=1:p^*+q^*-2}; \left\{\frac{n-q+q^*-1-j}{2}\right\}_{j=1:p^*+q^*-2}; p^*+q^*-2\right)$$

and

$$F_\Lambda(z) = F^{EGIG}\left(z \,\middle|\, \{r_j\}_{j=1:p^*+q^*-2};\, \left\{\frac{n-q+q^*-1-j}{2}\right\}_{j=1:p^*+q^*-2};\, p^*+q^*-2\right),$$

where

$$r_j = \begin{cases} h_j, & j=1,2 \\ r_{j-2}+h_j, & j=3,\ldots,p^*+q^*-2 \end{cases}$$

with

$$h_j = (\#\text{ of elements in } \{p^*, q^*\} \geq j) - 1\,, \qquad j=1,\ldots,p^*+q^*-2\,,$$

and for general p^* and q^*, in terms of the Meijer G function as

$$f_\Lambda(z) = \left\{\prod_{j=1}^{p^*} \frac{\Gamma\left(\frac{n-q+q^*+1-j}{2}\right)}{\Gamma\left(\frac{n-q+1-j}{2}\right)}\right\} G^{\,p^*,0}_{\,p^*,p^*}\left(\begin{matrix} \left\{\frac{n-q+q^*+1-j}{2}-1\right\}_{j=1:p^*} \\ \left\{\frac{n-q+1-j}{2}-1\right\}_{j=1:p^*} \end{matrix} \,\middle|\, z \right)$$

and

$$F_\Lambda(z) = \left\{\prod_{j=1}^{p^*} \frac{\Gamma\left(\frac{n-q+q^*+1-j}{2}\right)}{\Gamma\left(\frac{n-q+1-j}{2}\right)}\right\} G^{\,p^*,1}_{\,p^*+1,p^*+1}\left(\begin{matrix} \left\{1, \left\{\frac{n-q+q^*+1-j}{2}\right\}_{j=1:p^*}\right\} \\ \left\{\left\{\frac{n-q+1-j}{2}\right\}_{j=1:p^*}, 0\right\} \end{matrix} \,\middle|\, z \right)$$

or

$$f_\Lambda(z) = \left\{\prod_{k=1}^{q^*} \frac{\Gamma\left(\frac{n-q+q^*+1-k}{2}\right)}{\Gamma\left(\frac{n-q+q^*-p^*+1-k}{2}\right)}\right\} G^{\,q^*,0}_{\,q^*,q^*}\left(\begin{matrix} \left\{\frac{n-q+q^*+1-k}{2}-1\right\}_{k=1:q^*} \\ \left\{\frac{n-q+q^*-p^*+1-k}{2}-1\right\}_{k=1:q^*} \end{matrix} \,\middle|\, z \right)$$

and

$$F_\Lambda(z) = \left\{\prod_{k=1}^{q^*} \frac{\Gamma\left(\frac{n-q+q^*+1-k}{2}\right)}{\Gamma\left(\frac{n-q+q^*-p^*+1-k}{2}\right)}\right\} G^{\,q^*,1}_{\,q^*+1,q^*+1}\left(\begin{matrix} \left\{1, \left\{\frac{n-q+q^*+1-k}{2}\right\}_{k=1:q^*}\right\} \\ \left\{\left\{\frac{n-q+q^*-p^*+1-k}{2}\right\}_{k=1:q^*}, 0\right\} \end{matrix} \,\middle|\, z \right).$$

As for all l.r. tests, also for the test in this subsection, in order to carry out an α-level test we should reject H_0 in (5.49) if the computed value of Λ^{**} in (5.58) is smaller than the α quantile of Λ, or equivalently, if the corresponding p-value is smaller than our value of α. Quantiles and p-values may be easily obtained for this statistic by using modules `QuantMatEVR` and `PvalMatEVR`. Modules `PDFMatEVR` and `CDFMatEVR` which are also made available may be used to compute and plot the p.d.f. and c.d.f. of Λ^{**} in (5.58) and module `PvalDataMatEVR` may be used to compute the p-value corresponding to the value of Λ^{**} calculated from a data set in a file. The first four of these modules have an usage similar to the corresponding modules in the previous subsections, with the only difference being that the first three mandatory arguments are now replaced by four mandatory arguments which are (in this order): `n`—the overall sample size, or, equivalently, the number of columns of the matrix X, `p*`—the number of rows of the matrix C `q`—the number of columns of the matrices μ and D, or the number of rows of the matrices M and D, and `q*`—the rank of the matrix D (which will be equal to q in case for example the matrix D is the identity matrix).

Module `PvalDataMatEVR` has five "mandatory" arguments, which we call "mandatory" more in order to make a parallel with the mandatory argument of the `PvalData` modules of the previous Sects. 5.1.1–5.1.3, since indeed the last four of these arguments may, in some cases, be left empty or just not given, as will be illustrated in Sect. 5.1.4.8. These "mandatory" arguments are the names of the data files with the matrices X, M, C, D, and β, in this order. The first one of these is the only one that is indeed absolutely mandatory. The data file with the matrix X has to be a file with a similar structure to any of the files accepted by module `PvalDataEqMeanVecR`, with the variables defining the columns and the sample units the rows. See Sect. 5.1.1.1 and Figs. 5.7 and 5.8 for details. In case this file is arranged as one of the files in Fig. 5.7, that is, with a "multi-sample" layout, then the matrix M will be taken as the "design matrix" corresponding to the samples (see next Sect. 5.1.4.2 for further details), that is, as the matrix formed by rows by the indicator variables for the samples, and in this case no file should be supplied with the matrix M, thus leaving the second mandatory argument empty or just specifying it as " ". In case this data file for the matrix X is arranged as a "single sample" file (see Fig. 5.8) and the second argument for `PvalDataMatEVR` is not left empty or given as " ", that is, if a file name is provided for a file with the matrix M, the matrix in that file is taken as the design matrix and no further questions are asked to the user, concerning the sample structure of the data file, unless the first optional argument (`index1`) is given the value 2, in which case the usual questions concerning the sample structure of the data file are asked to the user and also he/she will be given the possibility to select samples and/or variables. In case this argument is left empty or given as " ", then the user will be asked if there is a column in the data file defining the sample assignments. If the answer is positive, then he/she will be asked to give the column number with those sample assignments and if the answer is negative he/she will be asked to provide a list with the sample sizes. In case one really wants to treat the whole data file as a "single sample," then one has to provide a file with the matrix M.

The third argument, which is the name of the file with the matrix C, may be left empty or given as " " in case this matrix is intended to be the identity matrix of order p. Also the fourth and fifth arguments may be left empty or given as " ", in case, respectively the matrix D is intended to be the identity matrix of order q, or the matrix β^* in (5.49) is intended to be a null matrix.

There are five optional arguments for module `PvalDataMatEVR`. The first four of these are exactly the same as the optional arguments used for the module `PvalDataEqMeanVecR`, described in Sect. 5.1.1.1, with the only difference that the optional argument `index1` has now a slightly different behavior. If a file with the matrix M is provided, then if `index1` is given the value 1, it will only allow the user to select variables, and not samples. If the user also wants or needs to select samples, and/or needs to indicate a column with the sample assignments, then he/she will have to give `index1` the value 2. See Sect. 5.1.4.8 for an example. The fifth optional argument is an argument which we may call `test`, and which if given the value 1 or 2 will make the module to carry out respectively the test of equality of mean vectors in Sect. 5.1.1, or the profile parallelism test in Sect. 5.1.3. In these cases the user only needs to supply the first mandatory argument with the name of the data file for the matrix X, with the other four following arguments being left empty or given as " ". For a couple of examples on the use of module `PvalDataMatEVR` see Sect. 5.1.4.8. We decided to place this optional argument `test` as the last optional argument since although we think it may be used quite often, the most convenient way to carry out a test of equality of mean vectors or profile parallelism is to use the `PvalData` modules respectively in Sect. 5.1.1 or 5.1.3.

5.1.4.2 Specializing the Present Test to the Test in Sect. 5.1.1

An adequate choice of the matrices X, M, C, and D may then allow us to implement the tests in Sects. 5.1.1 and 5.1.3 as tests to the hypothesis in (5.49).

Let us consider the $p\times q$ matrix

$$\underset{p\times q}{X^*} = \left[\underline{X}_1, \dots, \underline{X}_k, \dots, \underline{X}_q\right] \tag{5.65}$$

where $\underline{X}_k \sim N_p(\underline{\mu}_k, \Sigma)$ are the random vectors in any of the previous subsections.

Then we will have

$$\underset{p\times q}{\mu} = E(X^*) = \left[\underline{\mu}_1, \dots, \underline{\mu}_k, \dots, \underline{\mu}_q\right] = [\mu_{jk}]_{\substack{j=1,\dots,p\\ k=1,\dots,q}}, \tag{5.66}$$

with $\mu_{jk} = E(X_{jk})$ $(j = 1, \dots, p; k = 1, \dots, q)$ representing the expected value of the j-th variable in the k-th population or set of variables, that is, the j-th component of $\underline{X}_k$.

If we now take X to be the $p\times n$ overall sample matrix, that is, the matrix obtained by combining the q samples, the k-th of them with size n_k, with $n=\sum_{k=1}^{q} n_k$, then we have

$$\underset{(p\times n)}{X} = \begin{bmatrix} \overbrace{X_{111} \dots X_{11n_1}}^{n_1} & \overbrace{X_{211} \dots X_{21n_2}}^{n_2} & \dots & \overbrace{X_{k11} \dots X_{k1n_k}}^{n_k} & \dots & \overbrace{X_{q11} \dots X_{q1n_q}}^{n_q} \\ \vdots \quad\quad \vdots & \vdots \quad\quad \vdots & & \vdots \quad\quad \vdots & & \vdots \quad\quad \vdots \\ X_{1j1} \dots X_{1jn_1} & X_{2j1} \dots X_{2jn_2} & \dots & X_{kj1} \dots X_{kjn_k} & \dots & X_{qj1} \dots X_{qjn_q} \\ \vdots \quad\quad \vdots & \vdots \quad\quad \vdots & & \vdots \quad\quad \vdots & & \vdots \quad\quad \vdots \\ X_{1p1} \dots X_{1pn_1} & X_{2p1} \dots X_{2pn_2} & \dots & X_{kp1} \dots X_{kpn_k} & \dots & X_{qp1} \dots X_{qpn_q} \end{bmatrix} \tag{5.67}$$

where $X_{kj\ell}$ represents the ℓ-th observation for the j-th variable in the k-th population, for $k=1,\dots,q$, $j=1,\dots,p$ and $\ell=1,\dots,n_k$. Then we take M to be the corresponding $q\times n$ *design* matrix,

$$\underset{(q\times n)}{M} = \begin{bmatrix} \overbrace{1 \;\; 1 \dots 1}^{n_1} & \overbrace{0 \;\; 0 \dots 0}^{n_2} & \dots & \overbrace{0 \;\; 0 \dots 0}^{n_k} & \dots & \overbrace{0 \;\; 0 \dots 0}^{n_q} \\ 0 \;\; 0 \dots 0 & 1 \;\; 1 \dots 1 & \dots & 0 \;\; 0 \dots 0 & \dots & 0 \;\; 0 \dots 0 \\ \vdots \;\; \vdots \quad \vdots & \vdots \;\; \vdots \quad \vdots & & \vdots \;\; \vdots \quad \vdots & & \vdots \;\; \vdots \quad \vdots \\ 0 \;\; 0 \dots 0 & 0 \;\; 0 \dots 0 & \dots & 1 \;\; 1 \dots 1 & \dots & 0 \;\; 0 \dots 0 \\ \vdots \;\; \vdots \quad \vdots & \vdots \;\; \vdots \quad \vdots & & \vdots \;\; \vdots \quad \vdots & & \vdots \;\; \vdots \quad \vdots \\ 0 \;\; 0 \dots 0 & 0 \;\; 0 \dots 0 & \dots & 0 \;\; 0 \dots 0 & \dots & 1 \;\; 1 \dots 1 \end{bmatrix} \tag{5.68}$$

In this way we have

$$X \sim N_{p\times n}(\mu M, I_n \otimes \Sigma)\,,$$

where μ is the matrix in (5.66), or

$$CX \sim N_{p\times n}(C\mu M,\ I_n \otimes C\Sigma C')\,.$$

Then, if we take $C=I_p$ and

$$\underset{q\times q}{D} = I_q - \frac{1}{n} MM'E_{qq} = \begin{bmatrix} 1-\frac{n_1}{n} & -\frac{n_1}{n} & \dots & -\frac{n_1}{n} \\ -\frac{n_2}{n} & 1-\frac{n_2}{n} & \dots & -\frac{n_2}{n} \\ \vdots & \vdots & \ddots & \vdots \\ -\frac{n_q}{n} & -\frac{n_q}{n} & \dots & 1-\frac{n_q}{n} \end{bmatrix}, \tag{5.69}$$

where E_{qq} is a $n\times n$ matrix of 1's, clearly with $rank(D) = q-1$, we will have

$$\underset{p\times q}{C\mu D} = \begin{bmatrix} \mu_{11}-\mu_1 & \dots & \mu_{1k}-\mu_1 & \dots & \mu_{1q}-\mu_1 \\ \vdots & & \vdots & & \vdots \\ \mu_{j1}-\mu_j & \dots & \mu_{jk}-\mu_j & \dots & \mu_{jq}-\mu_j \\ \vdots & & \vdots & & \vdots \\ \mu_{p1}-\mu_p & \dots & \mu_{pk}-\mu_p & \dots & \mu_{pq}-\mu_p \end{bmatrix}, \tag{5.70}$$

where

$$\mu_j = \frac{1}{n}\sum_{k=1}^{q} n_k\, \mu_{jk} \tag{5.71}$$

with

$$\underline{\mu} = \left[\mu_j\right]_{j=1,\dots,p} = \frac{1}{n}\sum_{k=1}^{q} n_k\, \underline{\mu}_k \,.$$

Then, to test the null hypothesis of equality of the q mean vectors

$$\underline{\mu}_k = \left[\mu_{jk}\right]_{j=1,\dots,p} \qquad (k=1,\dots,q)$$

which is the null hypothesis (5.1) in Sect. 5.1.1, will be equivalent to testing the null hypothesis

$$H_0 : C\mu D = \mathbf{0}_{p\times q}\,.$$

But then, from the definition of the matrix D in (5.69) and its properties, shown in Appendix 4, we may quite easily see that the choice $H = M$ will satisfy both relations (5.59) and (5.60), so that the l.r.t. statistic Λ^{**} in (5.58) may be written, for $\beta^* = 0_{p\times q}$, as

$$\Lambda^{**} = \frac{|A|}{\left|A + \frac{1}{n}(\beta D)(MM')(\beta D)'\right|} \tag{5.72}$$

where A, given by (5.37), is equal to $1/n$ times the matrix A in (5.3) and where $(\beta D)(MM')(\beta D)'$, with β given by (5.36), is exactly the matrix B in (5.3), since

$$(\beta D)(MM')(\beta D)' = (\beta DM)(\beta DM)' = \sum_{k=1}^{q} n_k \left(\underline{\overline{X}}_k - \underline{\overline{X}}\right)\left(\underline{\overline{X}}_k - \underline{\overline{X}}\right)' \tag{5.73}$$

with

$$
\underset{(p\times n)}{\beta DM} = XM'\left(MM'\right)^{-1} DM
$$

$$
= \left[\begin{array}{c}
\overbrace{\overline{X}_{11}-\overline{X}_1 \;\ldots\; \overline{X}_{11}-\overline{X}_1}^{n_1} \;\;\ldots\;\; \overbrace{\overline{X}_{k1}-\overline{X}_1 \;\ldots\; \overline{X}_{k1}-\overline{X}_1}^{n_k} \;\;\ldots \;\; \overbrace{\overline{X}_{q1}-\overline{X}_1 \;\ldots\; \overline{X}_{q1}-\overline{X}_1}^{n_q}\\
\vdots \\
\overline{X}_{1j}-\overline{X}_j \;\ldots\; \overline{X}_{1j}-\overline{X}_j \;\;\ldots\;\; \overline{X}_{kj}-\overline{X}_j \;\ldots\; \overline{X}_{kj}-\overline{X}_j \;\;\ldots\;\; \overline{X}_{qj}-\overline{X}_j \;\ldots\; \overline{X}_{qj}-\overline{X}_j\\
\vdots \\
\overline{X}_{1p}-\overline{X}_p \;\ldots\; \overline{X}_{1p}-\overline{X}_p \;\;\ldots\;\; \overline{X}_{kp}-\overline{X}_p \;\ldots\; \overline{X}_{kp}-\overline{X}_p \;\;\ldots\;\; \overline{X}_{qp}-\overline{X}_p \;\ldots\; \overline{X}_{qp}-\overline{X}_p
\end{array}\right]
$$

where

$$
\overline{X}_{kj} = \frac{1}{n_k}\sum_{\ell=1}^{n_k} X_{kj\ell} \qquad k=1,\ldots,q;\; j=1,\ldots,p
$$

represents the sample average for the j-th variable in the k-th sample, and

$$
\overline{X}_j = \frac{1}{n}\sum_{k=1}^{q} n_k\,\overline{X}_{kj} = \frac{1}{n}\sum_{k=1}^{q}\sum_{\ell=1}^{n_k} X_{kj\ell} \qquad j=1,\ldots,p
$$

with

$$
\underline{\overline{X}_k} = \left[\overline{X}_{kj}\right]_{j=1,\ldots,p} \qquad \text{and} \qquad \underline{\overline{X}} = \left[\overline{X}_j\right]_{j=1,\ldots,p},
$$

which are the same $\underline{\overline{X}_k}$ and $\underline{\overline{X}}$ vectors as in (5.3) and (5.4), that is, the sample mean vector for the k-th sample and the overall sample mean vector, where $\overline{X}_j$ represents the sample of the j-th variable in the overall or global sample.

As such, the statistic Λ^{**} in (5.72) is the same as the statistic Λ in (5.2). Its exact distribution may be obtained from the results on the distribution of Λ^{**} in (5.61) and (5.62), in this case with $p^* = p$ and $q^* = q-1$. These results confirm that the l.r.t. statistic in (5.72) has the same distribution as the l.r.t. Λ in Sect. 5.1.1, which had to be the case, since they are indeed the same l.r.t. statistic.

As such, this test may also be implemented on a set of samples organized in a data file with the same structure as the one used in Sect. 5.1.1, by using module `PvalDataMatEVR` with the appropriate C and D matrices. See Sect. 5.1.4.8 and Fig. 5.31 on how to do this.

In fact, the use of the matrix D in (5.69) has an effect similar to that of inducing the centering of the matrix M, since DM is just a centered version of the matrix M for row means. See Sect. 5.1.9.4 for a thorough explanation of this.

Also, with the setup of the matrix X^* in (5.65) we may now understand better what $C\mu D$ represents. It is the expected value of the random matrix CX^*D, where C is the matrix of linear combinations of the p variables and D the matrix of linear combinations of the q populations or random vectors $\underline{X}_k$.

5.1.4.3 Specializing the Present Test to the Test in Sect. 5.1.3

The test in Sect. 5.1.3 may be implemented as a particular case of the generalized test in Sect. 5.1.4.1, that is, the test to the null hypothesis in (5.49), by taking matrices X, M, and D similar to the ones in the previous subsection $\beta^* = 0$, and a $(p-1)\times p$ matrix C as the one in (5.20). We may note that in this case we will have

$$\underset{(p-1)\times q}{C\mu D} = \left[\begin{array}{ccccc} (\mu_{11}-\mu_{21})-(\mu_1-\mu_2) & \dots & (\mu_{1k}-\mu_{2k})-(\mu_1-\mu_2) & \dots & (\mu_{1q}-\mu_{2q})-(\mu_1-\mu_2) \\ \vdots & & \vdots & & \vdots \\ (\mu_{j1}-\mu_{j+1,1})-(\mu_j-\mu_{j+1}) & \dots & (\mu_{jk}-\mu_{j+1,k})-(\mu_j-\mu_{j+1}) & \dots & (\mu_{jq}-\mu_{j+1,q})-(\mu_j-\mu_{j+1}) \\ \vdots & & \vdots & & \vdots \\ (\mu_{p-1,1}-\mu_{p1})-(\mu_{p-1}-\mu_p) & \dots & (\mu_{p-1,k}-\mu_{pk})-(\mu_{p-1}-\mu_p) & \dots & (\mu_{p-1,q}-\mu_{pq})-(\mu_{p-1}-\mu_p) \end{array}\right],$$

where the μ_j $(j = 1, \dots, p)$ are defined as in (5.71), and where, for $j = 1, \dots, p-1$,

$$\bigwedge_{k=1}^{q} (\mu_{jk}-\mu_{j+1,k})-(\mu_j-\mu_{j+1}) = 0$$
$$\iff \mu_{j1}-\mu_{j+1,1} = \dots = \mu_{jk}-\mu_{j+1,k} = \dots = \mu_{jq}-\mu_{j+1,q}\,,$$

so that indeed to test H_0 in (5.21) is the same as to test the null hypothesis

$$C\mu D = 0_{(p-1)\times q}\,.$$

The l.r. statistic to be used is the statistic Λ^{**} in (5.58), which, for the present choice for the matrices X, M, C, and D, and taking $\beta^* = 0$, is indeed the same as the l.r.t. statistic Λ in (5.23) and which, from (5.61) and (5.62), with $p^* = p - 1$ and $q^* = q - 1$, has the same distribution as that of Λ in Sect. 5.1.3, given by (5.24) and (5.25).

To see how we can implement this test on a set of q independent samples, by using the approach outlined in this sub-subsection and module `PvalDataMatEVR`, refer to Sect. 5.1.4.8 and Fig. 5.32.

5.1.4.4 A Test Regarding a Submatrix of Parameters

Let, for $q = q_1 + q_2$,

$$\underset{(p\times q)}{\mu} = \Big[\underset{(p\times q_1)}{\mu_1} \;\Big|\; \underset{(p\times q_2)}{\mu_2} \Big] \tag{5.74}$$

and let us suppose we want to test the null hypothesis

$$H_0 : \mu_2 = \beta^*_{2\,(p\times q_2)}\,, \tag{5.75}$$

allowing μ_1 to be any $p\times q_1$ matrix.

Then, if we take

$$C = I_p \qquad \text{and} \qquad D = \begin{bmatrix} 0_{(q_1\times q_1)} & 0_{(q_1\times q_2)} \\ 0_{(q_2\times q_1)} & I_{q_2} \end{bmatrix}, \tag{5.76}$$

we have $C\mu D = [\,0 \mid \mu_2\,]$, and testing the null hypothesis in (5.75) is equivalent to testing the null hypothesis $H_0 : C\mu D = \big[0|\beta_2^*\big]$, with Λ^{**} given by (5.58), where

$$HH' = \begin{bmatrix} 0_{(q_1\times q_1)} & 0_{(q_1\times q_2)} \\ 0_{(q_2\times q_1)} & M_{22.1} \end{bmatrix}, \tag{5.77}$$

with

$$M_{22.1} = M_{22} - M_{21}M_{11}^{-1}M_{12}\,, \tag{5.78}$$

where M_{22}, M_{21}, M_{11}, and M_{12} are defined as in (5.500), which is indeed equivalent to taking HH' as the generalized inverse of $D'(MM')^{-1}D$. Then, the distribution of Λ^{**} in (5.58), with $A^* = \frac{1}{n}X(I_n - M'(MM')^{-1}M)X'$, $\beta = XM'(MM')^{-1}$, and D and HH' given by (5.76) and (5.77), is given by (5.61)–(5.62) or, for even p or even q_2, by Theorem 3.2 and Corollary 4.2 with the parameters given by (5.63) or (5.64), with $p^* = p$ and $q^* = q_2$.

This shows how the definition of the extended test in Sect. 5.1.4.1 permits a very simple approach to testing the null hypothesis in (5.75). It is also easy to see that the definition of the matrix HH' in (5.77) satisfies the relations in (5.59) and (5.60).

The test of the hypothesis in (5.75) for $\beta_2^* = 0$ is indeed the multivariate generalization of the well-known partial F test and we will call it the partial Wilks' Lambda test, given its relation with both the partial F test and the Wilks' Lambda statistic. In fact, it only makes sense to test this null hypothesis once the hypothesis $\mu = 0$ for the whole parameter matrix μ has been rejected.

Since for D in (5.76),

$$\mu D = \left[\, \mu_1 \mid \mu_2 \,\right] \begin{bmatrix} 0 & 0 \\ 0 & I_{p_2} \end{bmatrix} = \left[\, 0_{(p\times q_1)} \mid \mu_{2\,(p\times q_2)} \,\right],$$

and, by splitting the matrix M according to the split in μ as

$$M = \begin{bmatrix} \underset{(q_1\times n)}{M_1} \\ \underset{(q_2\times n)}{M_2} \end{bmatrix}, \tag{5.79}$$

we have

$$\mu M = \left[\, \mu_1 \mid \mu_2 \,\right] \begin{bmatrix} M_1 \\ M_2 \end{bmatrix} = \mu_1 M_1 + \mu_2 M_2 \,,$$

we may write H_0 and the alternative hypothesis H_1, in terms of models, as

$$H_0 : E(X) = \mu_1 M_1 + \beta_2^* M_2 \quad \text{vs.} \quad H_1 : E(X) = \mu_1 M_1 + \mu_2 M_2 \tag{5.80}$$

or

$$H_0 : X = \mu_1 M_1 + \beta_2^* M_2 + \mathcal{E} \quad \text{vs.} \quad H_1 : X = \mu_1 M_1 + \mu_2 M_2 + \mathcal{E} \tag{5.81}$$

with $\mathcal{E} \sim N(0, I_n \otimes \Sigma)$. This clearly shows that the test of H_0 in (5.75) for $\beta_2^* = 0$ is indeed the multivariate generalization of the partial F test, since we are indeed testing if the variables represented by the rows of M_2 are or are not important in explaining the variables, or rather, the variability among the variables, represented in the rows of X, in a model where the variables represented by the rows of M_1 are already acting as explanatory variables. In this case the model associated with H_1

may be taken as the full or complete model and the model associated with H_0 as the submodel.

Another useful way to approach this test is to see the l.r.t. statistic for this test as the ratio of two l.r.t. statistics. With this in view, let us start by splitting the $p{\times}q$ matrix β^* according to the split of μ in (5.74), as

$$\underset{(p\times q)}{\beta^*} = \Big[\underset{(p\times q_1)}{\beta_1^*} \;\Big|\; \underset{(p\times q_2)}{\beta_2^*} \Big]$$

where β_2^* is the matrix in (5.75) and β_1^* is any $p{\times}q_1$ matrix, which may be taken to be the null matrix, given that the "final" test between H_0 and H_1 in (5.80) or (5.81) will not be a function of β_1^*. Then let us consider the models associated with the hypotheses H_1 and H_0 in (5.80) or (5.81) and the model

$$H_0^* : X = \beta_1^* M_1 + \beta_2^* M_2 + \mathcal{E}\,,$$

which will be the null or empty model in case we take both β_1^* and β_2^* as null matrices.

Then let L_{H_1}, L_{H_0}, and $L_{H_0^*}$ be the suprema of the likelihood functions respectively under H_1, H_0, and H_0^*, and take Λ_{H_a,H_b} to denote the l.r.t. statistic to test the null hypothesis H_b versus the alternative hypothesis H_a. Then we have

$$\Lambda_{H_1,H_0^*} = \frac{L_{H_0^*}}{L_{H_1}} \qquad \text{and} \qquad \Lambda_{H_0,H_0^*} = \frac{L_{H_0^*}}{L_{H_0}}\,,$$

which in case we take β_1^* and β_2^* as null matrices will just be respectively the l.r.t statistic to test the fit of the model in H_1 (the complete model) and the fit of the model in H_0 (the submodel), since in this case, as noted above, the model associated with H_0^* will be the null or empty model. In any case, the l.r.t. statistic corresponding to the test of the null hypothesis in (5.75) may then be written as

$$\Lambda_{H_1,H_0} = \frac{\Lambda_{H_1,H_0^*}}{\Lambda_{H_0,H_0^*}} = \frac{L_{H_0^*}}{L_{H_1}}\,\frac{L_{H_0}}{L_{H_0^*}} = \frac{L_{H_0}}{L_{H_1}}\,, \tag{5.82}$$

leading us to the decision of sticking with the original or complete model in case we reject H_0 and leading us to use the submodel in case we do not reject H_0.

Building a bridge between the relation in (5.82) and Lemma 10.3.1 in Anderson (2003) we may note that the relation in (5.82) may be equivalently written as

$$\Lambda_{H_1,H_0^*} = \Lambda_{H_1,H_0}\,\Lambda_{H_0,H_0^*}\,,$$

where

$$\Omega_{H_0^*} \subseteq \Omega_{H_0} \subseteq \Omega_{H_1}$$

with Ω_{H_a} denoting the parameter space for H_a.

In our case we would have,

$$\Lambda_{H_1,H_0^*}=\frac{|A_1|}{|A_1+B_1|}\,,\qquad \Lambda_{H_0,H_0^*}=\frac{|A_0|}{|A_0+B_0|}$$

and

$$\Lambda_{H_1,H_0}=\frac{|A_1|}{|A_1+B_1|}\bigg/\frac{|A_0|}{|A_0+B_0|}=\frac{|A_1|}{|A_0|}\,, \tag{5.83}$$

where, from Sect. 5.1.4,

$$A_1=\tfrac{1}{n}X(I_n-M'(MM')^{-1}M)X'\,,\qquad B_1=\tfrac{1}{n}XM'(MM')^{-1}MX'\,, \tag{5.84}$$

and

$$A_0=\frac{1}{n}X(I_n-M_1'(M_1M_1')^{-1}M_1)X'\,,\qquad B_0=\frac{1}{n}XM_1'(M_1M_1')^{-1}M_1X'\,, \tag{5.85}$$

clearly with $A_1+B_1 = A_0+B_0$, thus giving rise to the second expression for Λ_{H_1,H_0} in (5.83). Since A_1 is clearly the same as A^* in (5.50) and (5.58), to show that the statistic Λ_{H_1,H_0} in (5.83) is indeed the same as the statistic Λ^{**} in (5.58) it remains to show that $A_0 = A_1 + \frac{1}{n}\beta DHH'D'\beta'$. That this is indeed the case is shown in Appendix 5.

The formulation developed in Appendix 5 regarding the matrix $\beta DHH'D'\beta'$ also helps one to see that the approach followed in this subsection, based on the extension of the test presented in Sect. 5.1.4.1 is indeed equivalent to the expositions in Sect. 8.5 of Kshirsagar (1972) and Sects. 8.3 and 8.7 in Anderson (2003) for the l.r.t.'s for the matrix of parameters in the Multivariate Linear Model.

The test in this subsection may be implemented with the `PvalDataMatEVR` module by giving the value 3 to the fifth optional argument of this module, which was called `test` in Sect. 5.1.4.1 and at the same time give to the fourth "mandatory" argument of `PvalDataMatEVR`, which usually would carry the name of the file with the matrix D, the value of q_2. One further detail is that in case a non-null matrix β_2^* is used, then a file with the whole β^* matrix or just the matrix β_2^* has to be provided as the fifth "mandatory" argument. In case only the β_2^* matrix is provided, the matrix β_1^* will be taken as a null matrix. Alternatively, the test may be implemented by providing a file with the appropriate D matrix as a fourth argument and a file with the full $\beta_{p\times q}^*$ matrix as the fifth argument. See Sect. 5.1.4.8 for an example using a real data set.

In more general setups one may need to use a matrix C other than I_p. Also, in a more general setup where the test of the whole $p\times q$ expected value matrix $\underline{\mu}$ would involve a $q\times q$ matrix

$$D=\begin{bmatrix} \underset{q_1\times q_1}{D_{11}} & \underset{q_1\times q_2}{D_{12}} \\ \underset{q_2\times q_1}{D_{21}} & \underset{q_2\times q_2}{D_{22}} \end{bmatrix} \tag{5.86}$$

we would have to consider instead of the matrix D in (5.76) a matrix

$$D=\begin{bmatrix} \underset{(q_1\times q_1)}{0} & \underset{(q_1\times q_2)}{0} \\ \underset{(q_2\times q_1)}{0} & \underset{(q_2\times q_2)}{D_{22}} \end{bmatrix} \tag{5.87}$$

where D_{22} is the submatrix of dimensions $q_2\times q_2$ at the lower right-hand corner in (5.86)—see Sect. 5.1.4.6.

5.1.4.5 The Test for a Mean Vector

Let us suppose that $\underline{X} \sim N_p(\underline{\mu}, \Sigma)$ and that we are interested in testing the null hypothesis

$$H_0 : \underline{\mu} = \underline{\beta}^*_{(p\times 1)} \tag{5.88}$$

where $\underline{\mu}$ and $\underline{\beta}^*$ are both $p\times 1$ real vectors, with $\underline{\beta}^*$ being possibly a null vector.

The question is: can we use the test in Sect. 5.1.4.1, or more precisely, the statistic Λ^{**} in (5.58) to test the null hypothesis in (5.88)? The answer is "Yes." And it can be done in a quite straightforward way by taking the $p\times n$ matrix X in (5.32) as the matrix of the sample of size n from $\underline{X}$. Then, under the present distributional assumption for $\underline{X}$, the matrix X will have the distribution in (5.32) with μ being a $p\times 1$ vector and $M = E_{1n}$. As such, all we have to do is to implement the test in Sect. 5.1.4.1 with $q = 1$, $M = E_{1n}$, $C = I_p$, and $D = I_1 = 1$. Then HH' is "automatically" taken as the inverse of

$$D'(MM')^{-1}D = (MM')^{-1} = \frac{1}{n}\,,$$

that is, we have

$$HH' = n\,,$$

and the statistic Λ^{**} in (5.58) reduces in the present case to

$$\Lambda^{**} = \frac{|A|}{|A + (\underline{X} - \beta^*)(\underline{X} - \beta^*)'|},$$

where β^* is now a $p \times 1$ vector, and, given that $M = E_{1n}$, we have, from (5.36),

$$\beta = XM'(MM')^{-1} = \frac{1}{n} X E_{n1} = \underline{X},$$

where $\underline{X}$ is the sample mean vector.

One further question is: the test to the hypothesis in (5.88) is commonly carried out by using the statistic

$$T^2 = (n-1)\left(\underline{X} - \underline{\beta}^*\right)' A^{-1} \left(\underline{X} - \underline{\beta}^*\right), \tag{5.89}$$

with

$$\frac{n-p}{p} \frac{T^2}{n-1} \sim F_{p,n-p},$$

where A is the m.l.e. of Σ; as such, what is the relation between the two tests?

The answer is: they are absolutely equivalent. In fact, from (5.61) and (5.62), since in the present case we have $p^* = p$ and $q^* = q = 1$, the distribution of Λ^{**} in (5.58) is now the same as that of a $Beta\left(\frac{n-p}{2}, \frac{p}{2}\right)$ r.v., so that using the relation between the Beta and the F distributions—see for example (5.12)—we have

$$\frac{n-p}{p} \frac{1-\Lambda^{**}}{\Lambda^{**}} \sim F_{p,n-p},$$

where, for the Hotelling T^2 statistic in (5.89),

$$\frac{T^2}{n-1} = \frac{1-\Lambda^{**}}{\Lambda^{**}}.$$

This equality is easy to prove since we may write

$$\begin{aligned} \frac{1-\Lambda^{**}}{\Lambda^{**}} = \frac{1}{\Lambda^{**}} - 1 &= \frac{|A + (\underline{X} - \beta^*)(\underline{X} - \beta^*)'|}{|A|} - 1 \\ &= (\underline{X} - \beta^*)' A^{-1} (\underline{X} - \beta^*). \end{aligned} \tag{5.90}$$

This last equality may be obtained in several different ways. One of them is by considering the matrix

$$\widetilde{A} = \left[\begin{array}{c|c} 1 & -(\underline{\overline{X}}-\beta^*)' \\ \hline \underline{\overline{X}}-\beta^* & A \end{array}\right],$$

and then writing

$$|\widetilde{A}| = |A| \times \left(1 + (\underline{\overline{X}}-\beta^*)'A^{-1}(\underline{\overline{X}}-\beta^*)\right)$$

or

$$|\widetilde{A}| = 1 \times |A + (\underline{\overline{X}}-\beta^*)(\underline{\overline{X}}-\beta^*)'|\,,$$

so that we have

$$|A + (\underline{\overline{X}}-\beta^*)(\underline{\overline{X}}-\beta^*)'| / |A| = 1 + (\underline{\overline{X}}-\beta^*)'A^{-1}(\underline{\overline{X}}-\beta^*)\,,$$

which is just another version of (5.90).

An example of application of this test is addressed in the Subsect. 5.1.4.8.

5.1.4.6 Paired Samples, Repeated Measures, Longitudinal Studies, Random-Block Designs, and Growth Curve Models

By using the extended version of the present test in Sect. 5.1.4.1, it is very easy to implement the test for the case of two paired samples, that is, for longitudinal studies with two dates or occasions of data collection, or equivalently, of MANOVA block designs with one factor with two levels. However, for cases where there are more than two paired samples, or equivalently, for longitudinal studies with more than two dates or instances of data collection, or for block designs with more than one factor or with one factor with more than two levels, the test is easier to implement through the partial Wilks' Lambda test in Sect. 5.1.9.5, and we refer the reader to Sect. 5.1.9.6.

As an example of application of the test of an expected value matrix to the case of two paired samples we will use the bone mineral content data in Tables 1.8 and 6.16 of Johnson and Wichern (2014), where the bone mineral content was measured on six bones (dominant and nondominant radius, humerus, and ulna) at the beginning of the study (Table 1.8) and 1 year later (Table 6.16) and where we take the data for the first 24 individuals in Table 1.8 and the data in Table 6.16 as two paired samples, since they are indeed taken on the same individuals, 1 year apart. We are interested in testing if there are differences in the mean bone mineral content for the six bones between the two dates of measurement.

To implement this test we take a data matrix

$$\underset{(12\times24)}{X} = \begin{bmatrix} \underset{(6\times24)}{X_1} \\ \hline \underset{(6\times24)}{X_2} \end{bmatrix} \tag{5.91}$$

where X_1 and X_2 are the 6×24 data matrices respectively for the first and second measurement dates, that is, respectively the transpose of the first 24 rows in Table 1.8 and the transpose of Table 6.16 of Johnson and Wichern (2014), so that a given column corresponds to a given individual.

We will also then take

$$\underset{(6\times12)}{C} = \left[\, I_6 \,\middle|\, -I_6 \,\right], \quad M = E_{1,24}, \quad D = I_1 = 1, \tag{5.92}$$

and we will be interested in testing the hypothesis

$$H_0 : C\mu D = \underset{(6\times1)}{\underline{0}}, \tag{5.93}$$

with

$$\underset{(12\times1)}{\mu} = \left[\, \mu_{11} \ldots \mu_{16} \middle| \mu_{21} \ldots \mu_{26} \,\right]'$$

where μ_{ij} represents the mean bone mineral content for the j-th bone ($j = 1, \ldots, 6$) on the i-th measurement date ($i = 1, 2$). Since we have

$$C\mu D = \begin{bmatrix} \mu_{11} - \mu_{21} \\ \mu_{12} - \mu_{22} \\ \mu_{13} - \mu_{23} \\ \mu_{14} - \mu_{24} \\ \mu_{15} - \mu_{25} \\ \mu_{16} - \mu_{26} \end{bmatrix}$$

we will be testing

$$H_0 : \bigwedge_{j=1}^{6} \mu_{1j} - \mu_{2j} = 0\,.$$

The command used to carry out this test with the Mathematica® version of the `PvalDataMatEVR` module and the output obtained may be found in Fig. 5.24,

```
PvalDataMatEVR[path <> "X_2ps.dat", path <> "M_2ps.dat", path <> "C_2ps.dat"]

Computed value of Λ: 0.7182600770

p-value: 0.3616271845
```

Fig. 5.24 Command used and output obtained with the Mathematica® module `PvalDataMatEVR` to implement the test of equality of mean vectors, based on two paired samples, using the bone mineral data in Tables 1.8 and 6.16 of Johnson and Wichern (2014)

where the contents of files `X_2ps.dat`, `M_2ps.dat`, and `C_2ps.dat` are respectively the transpose of matrix X in (5.91), and matrices M and C in (5.92), and where the file names may eventually have to be preceded by the path indication. The p-value obtained shows that we should not reject the null hypothesis in (5.93).

We may note that we may avoid using the files with the specifications for the D matrix and the vector β^*, which will be respectively taken as an identity matrix of order 1 and a null vector of dimension 6.

One other choice would be to compute the data matrix $X^* = X_1 - X_2$, to store it in a file, say called `Xstar.dat` and then, using the same M and D matrices as before, and now a matrix $C = I_6$, thus with $C\mu D = \mu$, to implement the test of the null hypothesis

$$H_0 : \mu = \underset{(6\times 1)}{0} \tag{5.94}$$

where now

$$\mu = \left[\, \mu_1^* \ldots \mu_6^* \,\right]'$$

represents the mean vector for the differences of bone mineral content between the two dates, with

$$\mu_j^* = \mu_{1j} - \mu_{2j} \quad (j = 1, \ldots, 6)$$

where μ_{ij} has the same meaning as above. This test is actually just a test of a mean vector, which was addressed in the previous subsection. The command used to carry out this test with the Mathematica® version of the `PvalDataMatEVR` module and the output obtained are shown in Fig. 5.25, and, of course, as expected, both the computed value of the statistic and the p-value obtained are exactly the same as those in Fig. 5.24. In the command in Fig. 5.25 no file is used with the specification of the matrix C, since this is now an identity matrix. These results are also the same as those obtained in Sect. 5.1.9.6 for a different implementation of the same test, using the partial Wilks' Lambda test approach.

However, if we want to carry out a test involving more than two paired samples, we will need to take a different, slightly more elaborate, approach, which, of course,

```
PvalDataMatEVR[path <> "Xstar.dat", path <> "M_2ps.dat"]

Computed value of Λ: 0.7182600770

p-value: 0.3616271845
```

Fig. 5.25 Command used and output obtained with the Mathematica® module `PvalDataMatEVR` to implement the test of equality of mean vectors, based on two paired samples, using a data matrix with the differences of the bone mineral values in Tables 1.8 (first 24 observations) and 6.16 of Johnson and Wichern (2014)

may also be adopted in the case of only two paired samples. We will first consider this more general implementation of the paired samples or longitudinal study using the bone density data. Since our data matrix is now a 6×48 matrix formed by X_1 placed side by side with X_2. We will need to define a 25×24 matrix

$$M = \begin{bmatrix} I_{24} & I_{24} \\ 0_{1\times 24} & E_{1,24} \end{bmatrix} \tag{5.95}$$

where the first 24 rows represent the indicator variables for the pairings and the last row is the indicator variable for the second measurement date, that is, the indicator variable that remained after the one for the first level or first date of measurement was removed.

Then we will need to define first, for $q = 25$ and $n = 48$, a matrix

$$D = I_q - \frac{1}{n} MM'E_{qq} - E_{qq}\,, \tag{5.96}$$

which is split as in (5.86) with $q_1 = 24$ and $q_2 = 1$. Subsequently, since what we really want to do is to test the nullity of the parameter associated with the indicator variable for the second date of measurement, represented in the last row of M, while considering in the model the other indicator variables in M, which are the indicator variables for the pairings, we need, according to Sect. 5.1.4.4, to re-define the matrix D as in (5.87), with $q_1 = 24$ and $q_2 = 1$. All we need to do now is to carry out a test for a sub-hypothesis such as the one in (5.75) with $q_1 = 24$ and $q_2 = 1$, using $\beta_2^* = 0_{(6\times 1)}$. This is done with the command in Fig. 5.26, where file `X_2psn.dat` has the transpose of the 6×48 data matrix and files `MM_2ps.dat` and `D_2ps.dat` have contents respectively equal to matrices M in (5.95) and D in (5.96), using the setup in (5.87) for $q_1 = 24$ and $q_2 = 1$. We may see that both the computed value of the statistic and the corresponding p-value exactly match the ones previously obtained.

As a second example, this one with more than two paired samples, we will use the five forged paired samples used in Sect. 5.1.9.6 as part of the first example with five paired samples in that subsection. The data matrix used has the structure of

```
PvalDataMatEVR[path <> "X_2psn.dat", path <> "MM_2ps.dat", , path <> "D_2ps.dat"]
Computed value of Λ: 0.7182600770
p-value: 0.3616271845
```

Fig. 5.26 Command used and output obtained with the Mathematica® module `PvalDataMatEVR` to implement the more general version of the test of equality of mean vectors, based on two paired samples, using the bone mineral data in Tables 1.8 and 6.16 of Johnson and Wichern (2014)

```
PvalDataMatEVR[path <> "X_5ps.dat", path <> "M_5ps.dat", , path <> "D_5ps.dat"]
Computed value of Λ: 0.7345352070
p-value: 0.2639833687
```

Fig. 5.27 Command used and output obtained with the Mathematica® module `PvalDataMatEVR` to implement the more general version of the test of equality of mean vectors, based on five paired samples (to be compared with the result from the first command in Fig. 5.53

matrix X_1 in (5.215) and (5.216), with the corresponding data file `X_5ps.dat` yielding a transposed matrix, exactly equal to the first six columns of the data file `bones_gc_1.dat` in Fig. 5.54t1 and t2 of the book supplementary material, with its five samples obtained from Table 6.16 in Johnson and Wichern (2014) by adding to the values in this table independent random quantities respectively with $N(0.1, 1)$, $N(0.2, 1)$, $N(0.3, 1)$, $N(0.4, 1)$ and $N(0.5, 1)$ distributions. The matrix M used is the matrix

$$M = \begin{bmatrix} I_{24} & I_{24} & I_{24} & I_{24} & I_{24} \\ 0_{1\times 24} & E_{1,24} & 0_{1\times 24} & 0_{1\times 24} & 0_{1\times 24} \\ 0_{1\times 24} & 0_{1\times 24} & E_{1,24} & 0_{1\times 24} & 0_{1\times 24} \\ 0_{1\times 24} & 0_{1\times 24} & 0_{1\times 24} & E_{1,24} & 0_{1\times 24} \\ 0_{1\times 24} & 0_{1\times 24} & 0_{1\times 24} & 0_{1\times 24} & E_{1,24} \end{bmatrix} \tag{5.97}$$

that is, equal to matrix X_2 in (5.216), with a first row of an I_{24} matrix added on top of each of the five I^*_{24} matrices. Then the matrix D is first given by (5.96), which is then split as in (5.86) with $q_1 = 24$ and $q_2 = 4$, to be then re-defined as in (5.87).

The command used and the result obtained with the Mathematica® version of module `PvalDataMatEVR` are shown in Fig. 5.27.

The contents of file `M_5ps.dat` is the matrix M in (5.97) and the contents of file `D_5ps.dat` is the matrix D obtained through the process described above.We may see that the computed value of the statistic and the corresponding p-value exactly match those obtained from the first command in Fig. 5.53.

We should note here that although all examples presented deal with balanced designs, that is, situations where the individuals were the same in each sample and even appeared always in the same order, which allows the matrix M to have a very nice and simple structure, when we take the more general implementation this does not have to be the case. We may indeed have all kinds of situations, where samples, although being paired or taken along time, may have different numbers of observations, with a few different individuals being present in different samples taken along time and even showing in a different order in each sample. In this case the matrix M, which is the design matrix, has to be adequately defined. That is, samples do not need to have the same length, nor need to have the individuals present in the same order, only the pairings of the observations and their bond to a given sample need to be reflected in the matrix M. This is one more advantage of this more general approach. One other side note is that the initial definition of the matrix D may indeed be taken as in (5.69) or otherwise as in (5.96), which although giving different D matrices will work well in all situations unless we have $q_2 = 1$. In this situation only the definition in (5.96) works well.

Another set of models adequate for modeling repeated measures and longitudinal studies are the Growth Curve models. Although the setup in the present Sect. 5.1.4 was indeed the one originally used by several authors to implement the growth curve models—see for example Potthoff and Roy (1964) and Khatri (1964, 1966, 1973)—we will prefer to use the setup in Sect. 5.1.9.6 for the implementation of these models. This decision is taken since although the setup in the present subsection allows for the testing of any growth curve model this would take us into the recurring definition of a new D matrix for each test to be carried out.

5.1.4.7 On the Availability of Quantile Tables

The tables we made reference to in Sects. 5.1.1.2 and 5.1.3.1 were indeed built for the non-extended version of the present test, that is, for the statistic Λ in (5.34), as remarked at the beginning of Sect. 5.1.1.2. "Exact" quantiles for this statistic may thus be obtained from Table 1 in Kres (1983), taking p in this table as our p, q as our q and n as our $n - q$, and where we write "exact" given the limitations, already remarked in Sect. 5.1.1.2, of the quantiles listed in this table. For example, for $n = 20$, $p = 4$, and $q = 12$ the 0.05 exact quantile for Λ in (5.34), rounded to six decimal places is 0.001517 (see the first command in Fig. 5.28 for the MAXIMA command used to obtain the exact value of this quantile rounded to its first ten decimal places, where, given that the `QuantMatEVR` module is programmed for the extended version of the test, one has to use $q^* = q$), while the listed value in those tables is 0.001766. A better approximation for this exact quantile may actually be obtained from Table 9 in Muirhead (2005), with r as our p, m as our q and $M = n + 1 - p - q = 5$, where the correction factor $c = 1.145$ is to be used in an expression similar to expression (5.11), which now is

$$e^{-c\,\chi^2_{pq}(1-\alpha)/(n-(p+q+1)/2)},$$

```
QuantMatEVR(20,4,12,12,.05)$
1.517172223e-3
QuantMatEVR(21,4,13,12,.05)$
QuantEqMeanVecR(21,4,13,.05)$
QuantProfParR(21,5,13,.05)$
1.517172223e-3
1.517172223e-3
1.517172223e-3
```

Fig. 5.28 Using module QuantMatEVR to compute the 0.05 quantile for Λ in (5.34) for $n = 20$, $p = 4$, and $q = 12$

in order to obtain the approximate value of 0.001520.

As such, the authors once again strongly instead recommend the use of module QuantMatEVR, which, as already mentioned in Sect. 5.1.4.1, uses as mandatory arguments the values for n, p^*, q, and q^* and the α-value for the quantile, in this order. Its optional arguments being exactly the same as those for module QuantEqMeanVecR described in Sect. 5.1.1. The use of this module, QuantMatEVR, has all the advantages already noted for similar modules in Sects. 5.1.1.2 and 5.1.3.1, namely in terms of flexibility of arguments and also in terms of precision, while it also allows for the computation of quantiles for the extended version of the test in Sect. 5.1.4.1.

Even more useful may be the use of module PvalMatEVR to provide the p-value associated with a computed value of the statistic Λ in (5.34) or of the statistic Λ^{**} in (5.58).

It comes in handy here to note that the distribution of Λ in (5.34), for given $n = n^+$, $p = p^+$ and $q = q^+$ is the same as that of Λ^{**} in (5.58) for $n = n^+ + 1$, $p^* = p^+$, $q = q^+ + 1$ and $q^* = q^+$, which is also the same as that of Λ in (5.2) for $n = n^+ + 1$, $p = p^+$, and $q = q^+ + 1$ and the same as that of Λ in (5.23) for $n = n^+ + 1$, $p = p^+ + 1$, and $q = q^+ + 1$. This is the reason why the last three commands in Fig. 5.28 all give exactly the same result as that of the first command in that same figure. Therefore, modules QuantMatEVR and PvalMatEVR, through the use of the appropriate set of arguments for these functions, may also be used to obtain quantiles and p-values for the tests in Sects. 5.1.1 and 5.1.3 (as already mentioned, see Fig. 5.28 for an example relating to these tests). However, for the tests in Sects. 5.1.1 and 5.1.3 the modules dedicated to these tests are clearly much easier to use.

5.1.4.8 A Few Examples of Application of Several Variants of the Test with Real Data Sets

As a first application of the present test we will test if the expected value for all four variables in the Iris data set may be considered to be null, for all three species of

```
PvalDataMatEVR("<path>/iris.dat")$
 Is there a column in the data file with the sample assignments? (1-Yes)"1;
  Which is the column in the data file with the sample assignments ?5;
      There are 3 samples, with sizes: [50,50,50] and 4 variables
Computed value of Lambda: 1.7876652436e-4
p-value: 1.8478860210e-261
```

Fig. 5.29 Using module `PvalDataMatEVR` to test the nullity of the expected value for the four variables in the Iris data set, for the three iris species

iris. The data file used is exactly the same file that was used in Sect. 5.1.1.5 and the command used and corresponding output obtained are shown in Fig. 5.29.

By not using any file for the second to fifth mandatory arguments we are using matrices $M = I_3 \otimes \underline{1}'_{50}$, which is exactly the matrix of indicator variables for the three samples (where $\underline{1}_r$ is a vector of 1's of dimension $r\times 1$), $C = I_4$, $D = I_3$, and β^* a null matrix of dimensions 4×3. As expected, the p-value obtained is really low.

To reassure those who may be a bit concerned about the extremely small p-value obtained for this test and to show that the test really works well, let us carry out a similar test, where now the matrix β^* is the matrix

$$\beta^* = \begin{bmatrix} 5.0 & 5.9 & 6.6 \\ 3.4 & 2.8 & 3.0 \\ 1.5 & 4.3 & 5.6 \\ 0.2 & 1.3 & 2.0 \end{bmatrix} . \tag{5.98}$$

The value in row j and column k ($j = 1, \ldots, 4; k = 1, \ldots, 3$) in matrix β^* is indeed the value of the sample mean of the j-th variable (sepal length, sepal width, petal length, petal width) for the k-th iris species (setosa, versicolor, virginica), rounded to its first decimal place. We will then expect a rather high p-value, which is exactly what happens in Fig. 5.30, where the contents of the file `BetaIris.dat` is exactly the matrix β^* in (5.98).

In Fig. 5.31 we show a couple of equivalent ways to carry out the test of equality of mean vectors for the three iris species by using the present test, where the use of the fifth and first optional arguments of `PvalDataMatEVR` is illustrated.

We may see how the computed value of the statistic and the p-value obtained exactly match the ones in Fig. 5.13, obtained with module `PvalDataEqMeanVecR`. While the first command uses the more convenient way to implement the test, which uses the fifth optional argument of `PvalDataMatEVR` with a value of 1, the second command implements the

```
fileX:"<path>/iris.dat"$
fileBeta:"<path>/BetaIris.dat"$

PvalDataMatEVR(fileX,"","","",fileBeta)$
Is there a column in the data file with the sample assignments? (1-Yes) 1;
 Which is the column in the data file with the sample assignments? 5;
    There are 3 samples, with sizes: [50,50,50] and 4 variables
Computed value of Lambda: 8.8691209947e-1
p-value: 1.3106691442e-1
```

Fig. 5.30 Using module `PvalDataMatEVR` to test the hypothesis that the expected value for the four variables in the Iris data set, for the three iris species, is equal to the matrix β^* in (5.98)

```
fileX:"<path>/iris.dat"$
fileM:"<path>/MIris.dat"$
fileD:"<path>/DIris.dat"$

PvalDataMatEVR(fileX,"","","","",[],[],[],[],1)$
Is there a column in the data file with the sample assignments? (1-Yes) 1;
 Which is the column in the data file with the sample assignments? 5;
    There are 3 samples, with sizes: [50,50,50] and 4 variables
Computed value of Lambda: 2.3438630651e-2
p-value: 1.3650058326e-112

PvalDataMatEVR(fileX,fileM,"",fileD,"",1)$
Do you want to select variables? (1-Yes) " 1;
Please insert a list, inside square brackets,
 with the numbers of the variables you want to keep, separated by commas:[1,2,3,4];
    Original variables [1,2,3,4]
Computed value of Lambda: 2.3438630651e-2
p-value: 1.3650058326e-112
```

Fig. 5.31 Using module `PvalDataMatEVR` to test the hypothesis of equality of mean vectors for the three Iris species

same test by using as a second argument a file with a design matrix $M = I_3 \otimes \underline{1}'_{50}$, and as a fourth argument a file with a matrix

$$D = \begin{bmatrix} \frac{100}{150} & -\frac{50}{150} & -\frac{50}{150} \\ -\frac{50}{150} & \frac{100}{150} & -\frac{50}{150} \\ -\frac{50}{150} & -\frac{50}{150} & \frac{100}{150} \end{bmatrix} = \begin{bmatrix} 2/3 & -1/3 & -1/3 \\ -1/3 & 2/3 & -1/3 \\ -1/3 & -1/3 & 2/3 \end{bmatrix},$$

in accordance with the definition of the matrix D in (5.69).

A test for the profile parallelism is easily obtained with a command similar to the first command in Fig. 5.31, giving the fifth optional argument a value of 2, instead

```
fileX:"<path>/Timm_Tab494.dat"$
PvalDataMatEVR(fileX,"","","","",[],[],[],[],2)$
Is there a column in the data file with the sample assignments? (1-Yes) 1;
Which is the column in the data file with the sample assignments? 5;
     There are 3 samples, with sizes: [5,5,5] and 4 variables
Computed value of Lambda: 1.6672633130e-1
p-value: 3.3716903308e-3
```

Fig. 5.32 Using module `PvalDataMatEVR` to test the hypothesis of profile parallelism for the three profiles in Fig. 5.18

of 1. See Fig. 5.32 for the test of profile parallelism of the profiles in Fig. 5.18 and check how the computed value of the statistic and the p-value perfectly match the ones in Fig. 5.22, just with one more significant digit.

We next illustrate the implementation of the partial Wilks' Lambda test in Sect. 5.1.4.4 by using the data set in Table 9.12 of Johnson and Wichern (2014). We want to test whether in a model where the 4th, 5th, 6th, and 9th score classes for the abstract reasoning test are included in the model, the expected value for the three sales performance variables, sales growth, sales profitability, and new account-sales, may be taken as being equal to 98, 106, and 101 for the 7th score class and as 107, 116, and 109 for the 8th score class, that is, we will take a matrix

$$\beta_2^* = \begin{bmatrix} 98 & 107 \\ 106 & 116 \\ 101 & 109 \end{bmatrix},$$

which is the contents of the file `Beta2.dat`. We may note that these values lie not too far from the sample means for the variables in each score class. To carry out this test we used the command in Fig. 5.33, and obtained a p-value of 0.117777 which shows that our null hypothesis may be taken as plausible.

We may note the ordering of the samples given in the first command in Fig. 5.33. The "automatic" implementation of the test in Sect. 5.1.4.4 requires that the samples that correspond to the populations being "conditionally" tested appear as the q_2 last ones in this list, and the use of the first optional argument with a value of 1 not only enables the user to select samples but also to reorder them as indicated and where the ordering of the first four might be any other ordering.

In case we take $\beta_2^* = 0$ it is possible to compute the statistic for the test in Sect. 5.1.4.4 through two implementations of the `PvalDataMatEVR` module, by carrying out with one of them the test between the H_1 and H_0^* models and with the other the test between the H_0 and H_0^* models. This may be done using for β_1^* any $p\times q_1$ matrix. The first use of module `PvalDataMatEVR` in part (a) of Fig. 5.34 uses $\beta_1^* = 0$ and carries out the test to the "conditional" hypothesis $\beta_2^* = 0$ for the 7th and 8th abstract reasoning score classes, in the presence of the 4th, 5th, 6th, and

```
fileX:"<path>/Tab9_12_JW.dat"$
fileB:"<path>/Beta2.dat"$

PvalDataMatEVR(fileX,"","",2,fileB,1,[],[],[],3)$
Is there a column in the data file with the sample assignments? (1-Yes) 1;
 Which is the column in the data file with the sample assignments? 7;
Do you want to select samples? (1-Yes) 1;
Please insert a list inside square brackets
 with the numbers of the samples you want to keep, separated by commas:[4,5,6,9,7,8];
Do you want to select variables? (1-Yes)" 1;
Please insert a list, inside square brackets,
 with the numbers of the variables you want to keep, separated by commas:[2,3,4];
    There are 6 samples, with sizes: [5,4,4,4,15,10] and 3 variables
    Original samples [4,5,6,9,7,8]
    Original variables [2,3,4]
Computed value of Lambda: 7.4775740774e-1
p-value: 1.1777734689e-1
```

Fig. 5.33 Using module `PvalDataMatEVR` to implement the partial Wilks' Lambda test in Sect. 5.1.4.4

9th ones. Then the two uses of this module in part (b) of the same figure enable us to obtain the computed value of Λ in part (a) as the ratio of the two computed values of the statistics, as shown. In part (c) of the same figure a matrix

$$\beta_1^* = \begin{bmatrix} 21 & 51 & 31 & 71 \\ 25 & 45 & 44 & 26 \\ 34 & 26 & 68 & 35 \end{bmatrix}$$

is used as the contents of the file `Beta1star.dat`, being the contents of the file `Betastar.dat` the matrix $\beta^* = [\beta_1^* \,|\, \beta_2^*]$, with β_2^* being a $3{\times}2$ null matrix.

As shown, the two uses of module `PvalDataMatEVR` in part (c) of Fig. 5.34 lead exactly to the same value for the ratio of the computed values of the statistics, which once again exactly matches the value obtained for Λ in part (a) of the same figure, showing that we may indeed use for β_1^* any $p{\times}q_1$ matrix.

The contents of the file `M1T912.dat` is the matrix

$$M_1 = \begin{bmatrix} \underline{1}_5' & \underline{0}_4' & \underline{0}_4' & \underline{0}_4 & \underline{0}_{15}' & \underline{0}_{10}' \\ \underline{0}_5' & \underline{1}_4' & \underline{0}_4' & \underline{0}_4 & \underline{0}_{15}' & \underline{0}_{10}' \\ \underline{0}_5' & \underline{0}_4' & \underline{1}_4' & \underline{0}_4 & \underline{0}_{15}' & \underline{0}_{10}' \\ \underline{0}_5' & \underline{0}_4' & \underline{0}_4' & \underline{1}_4 & \underline{0}_{15}' & \underline{0}_{10}' \end{bmatrix}$$

```
fileX:"<path>/Tab9_12_JW.dat"$
fileM:"<path>/M1T912.dat"$
fileB:"<path>/Betastar.dat"$
fileB1:"<path>/Beta1star.dat"$
PvalDataMatEVR(fileX,"","",2,"",1,[],[],[],3)$

    [... same interaction with user as in part a) of Fig.5.33 ...]

Computed value of Lambda: 9.3134486735e-4
p-value: 1.7704359240e-49                                        a)
______________________________________________________
PvalDataMatEVR(fileX,"","","","",1)$

    [... same interaction with user as in part a) of Fig.5.33 ...]

Computed value of Lambda: 3.5063489698e-4
p-value: 9.2260492679e-51                                        b)

PvalDataMatEVR(fileX,fileM,"","","",2)$

    [... same interaction with user as in part a) of Fig.5.33 ...]

Computed value of Lambda: 3.7648234212e-1
p-value: 2.1981840984e-4

3.5063489698e-4/3.7648234212e-1;
 9.3134486735e-4
______________________________________________________
PvalDataMatEVR(fileX,"","","",fileB,1)$

    [... same interaction with user as in part a) of Fig.5.33 ...]

Computed value of Lambda: 2.4937308274e-6
p-value: 3.5738295204e-87                                        c)

PvalDataMatEVR(fileX,fileM,"","",fileB1,2)$

    [... same interaction with user as in part a) of Fig.5.33 ...]

Computed value of Lambda: 2.6775589954e-3
p-value: 5.9881650425e-41

2.4937308274e-6/2.6775589954e-3;
 9.3134486735e-4
```

Fig. 5.34 Using module PvalDataMatEVR to implement the partial Wilks' Lambda test in Sect. 5.1.4.4 with $\beta_2^* = 0$: (**a**) single implementation of module with $\beta_1^* = 0$; (**b**) double implementation of module with $\beta_1^* = 0$ and (**c**) double implementation of module with $\beta_1^* \neq 0$

which is the part of the whole design matrix M formed by the rows corresponding to the samples from the 4th, 5th, 6th, and 9th score classes of the abstract reasoning test.

Although the implementations of module PvalDataMatEVR in parts (b) and (c) of Fig. 5.34 are useful to show that the statistic Λ in Sect. 5.1.4.4 does not depend on the choice of β_1^*, of course the implementation in part (a) is the most convenient way to carry out the test and obtain the corresponding p-value.

In each part (b) and (c) in Fig. 5.34 two other important features are illustrated. One of them, already brought to our attention in Fig. 5.33 is that the samples which correspond to the populations to be "conditionally" tested have to be placed last, by reordering, if necessary, the original samples. This is allowed to the user by giving the first optional argument of the module the value 1. Also, notice how when we provide a file with an M matrix, as second argument, as it is the case with the second implementations of module `PvalDataMatEVR` in parts (b) and (c) of Fig. 5.34 we need to give this first optional argument the value 2, instead of 1, in order to be able to still select and/or reorder samples and variables from the data file. Also, notice how in the first implementation of module `PvalDataMatEVR` in parts (b) and (c) of Fig. 5.34 when using MAXIMA one has to give explicitly the value [] (nill) to the nonused optional arguments.

5.1.5 The Likelihood Ratio Statistic to Test the Equality of Mean Vectors for Complex Random Variables [EqMeanVecC]

Let us now assume that the random vectors $\underline{X}_k$ have complex p-variate Normal distributions, with expected value $\underline{\mu}_k$ and covariance matrix Σ. That is, let $\underline{X}_k \sim CN_p(\underline{\mu}_k, \Sigma)$, and let us assume that we are interested in testing the hypothesis

$$H_0 : \underline{\mu}_1 = \underline{\mu}_2 = \cdots = \underline{\mu}_q\,, \tag{5.99}$$

where $\underline{\mu}_k \in \mathbb{C}^p$, $k = 1, \ldots, q$.

The complex multivariate Normal distribution we use in this subsection and throughout this book is the same as the one used in Wooding (1956), Goodman (1957, 1963a,b), James (1964, Sect. 8), Khatri (1965), Gupta (1971), Krishnaiah et al. (1976), Fang et al. (1982), Brillinger (2001, Sect. 4.2), and Anderson (2003, probl. 2.64), where, for $k = 1, \ldots, q$,

$$\underline{X}_k = \underline{Y}_k + \mathrm{i}\underline{Z}_k\,,$$

with

$$Var(Re(\underline{X}_k)) = Var(Im(\underline{X}_k)) = Var(\underline{Y}_k) = Var(\underline{Z}_k) = \Gamma\,,$$

$$Cov(Re(\underline{X}_k), Im(\underline{X}_k)) = Cov(\underline{Y}_k, \underline{Z}_k) = \Phi,$$

and

$$Cov(Im(\underline{X}_k), Re(\underline{X}_k)) = Cov(\underline{Z}_k, \underline{Y}_k) = \Phi' = -\Phi$$

where Γ is a positive-definite matrix and Φ is a skew-symmetric matrix, so that

$$\Sigma = Var(\underline{X}_k) = 2\Gamma - 2\mathrm{i}\Phi$$

is an Hermitian positive-definite matrix. See Appendix 7 and the references above for the expression of the p.d.f. of $\underline{X}_k$ and the derivation of the expression for $Var(\underline{X}_k)$.

Let us suppose we have a set of q independent samples, the k-th one from $\underline{X}_k$, with size n_k $(k = 1, \ldots, q)$. Then, for $n = \sum_{k=1}^{q} n_k$, the $(1/n)$-th power of the l.r.t. statistic used to test the null hypothesis in (5.99) will still have a similar formulation to the one in (5.2), that is,

$$\Lambda = \frac{|A|}{|A + B|} \tag{5.100}$$

now with

$$A = \sum_{k=1}^{q} \left(X_k - \overline{\underline{X}}_k E_{1n_k}\right)\left(X_k - \overline{\underline{X}}_k E_{1n_k}\right)^{\#} \quad \text{and} \quad B = \sum_{k=1}^{q} n_k \left(\overline{\underline{X}}_k - \overline{\underline{X}}\right)\left(\overline{\underline{X}}_k - \overline{\underline{X}}\right)^{\#} \tag{5.101}$$

where $(\cdot)^{\#}$ denotes the transpose of the complex conjugate of $(\cdot)$, X_k is the $p{\times}n_k$ matrix of the random sample of size n_k from $\underline{X}_k$, E_{nm} represents a matrix of ones with dimensions $n{\times}m$, and

$$\overline{\underline{X}}_k = \frac{1}{n_k} X_k E_{n_k 1}$$

is the vector of sample means for the k-th sample, with $\overline{\underline{X}}$, defined in a similar manner of that in (5.4), as the overall sample mean vector.

The $p{\times}p$ matrix A has what is called a complex Wishart distribution (Goodman, 1963a,b) with $n - q$ degrees of freedom and parameter matrix Σ. We will denote this fact by

$$A \sim CW_p(n - q, \Sigma)\,.$$

Under H_0 in (5.1),

$$B \sim CW_p(q - 1, \Sigma)\,.$$

But then, given the independence, for Normal r.v.'s, of the m.l.e.'s of the mean and variance, and since the matrix A in (5.101) is the m.l.e. of Σ (see Goodman (1963a) and Anderson (2003, problem 3.11) for references on the m.l.e. of Σ for the complex multivariate Normal distribution) and the matrix B is only built based on the vectors $\underline{\overline{X}}_k$, m.l.e.'s of $\underline{\mu}_k$, the matrices A and B are independent and thus

$$A + B \sim CW_p(n-1, \Sigma)\,.$$

As such (see Appendix 8), under H_0 in (5.99), for $n > p+q-1$,

$$\Lambda \stackrel{d}{\equiv} \prod_{j=1}^{p} Y_j \quad \text{where} \quad \begin{array}{l} Y_j \sim Beta\,(n-q-j+1, q-1)\;,\;\; j=1,\dots,p \\ \text{are } p \text{ independent r.v.'s.} \end{array} \tag{5.102}$$

This means that the exact distribution of Λ in (5.100) is given by Theorem 3.2 and its exact p.d.f. and c.d.f. by Corollary 4.2, for

$$m^* = 1\,, \quad a_1 = n-q+1\,, \quad k_1 = 1\,, \quad n_1 = p\,, \quad m_1 = q-1\,. \tag{5.103}$$

Based on the results in Theorem 3.2 and also on the results in Appendix 8, it is easy to show that we may interchange p and $q-1$ in (5.102), and thus also in (5.103), yielding as an alternative possible choice for the parameters in (5.103),

$$m^* = 1\,, \quad a_1 = n-p\,, \quad k_1 = 1\,, \quad n_1 = q-1\,, \quad m_1 = p\,. \tag{5.104}$$

This way, in terms of the EGIG distribution representation, we have the exact p.d.f. and c.d.f. of Λ in (5.100), for any p and any q, given by

$$f_\Lambda(z) = f^{EGIG}\left(z \,\middle|\, \{r_j\}_{j=1:p+q-2}; \{n-1-j\}_{j=1:p+q-2}\,; p+q-2\right)$$

and

$$F_\Lambda(z) = F^{EGIG}\left(z \,\middle|\, \{r_j\}_{j=1:p+q-2}; \{n-1-j\}_{j=1:p+q-2}\,; p+q-2\right),$$

where

$$r_j = \begin{cases} h_j\,, & j=1 \\ r_{j-1}+h_j\,, & j=2,\dots,p+q-2 \end{cases}$$

with

$$h_j = (\#\text{ of elements in } \{p, q-1\} \geq j) - 1\,, \quad j=1,\dots,p+q-2\,,$$

a result that confirms the one obtained in Coelho et al. (2015).

In Fig. 5.35 we may analyze some plots of p.d.f.'s and c.d.f.'s of Λ in (5.100) for the same values of p, q, and n as in Fig. 5.1, for the case of real r.v.'s, and we may

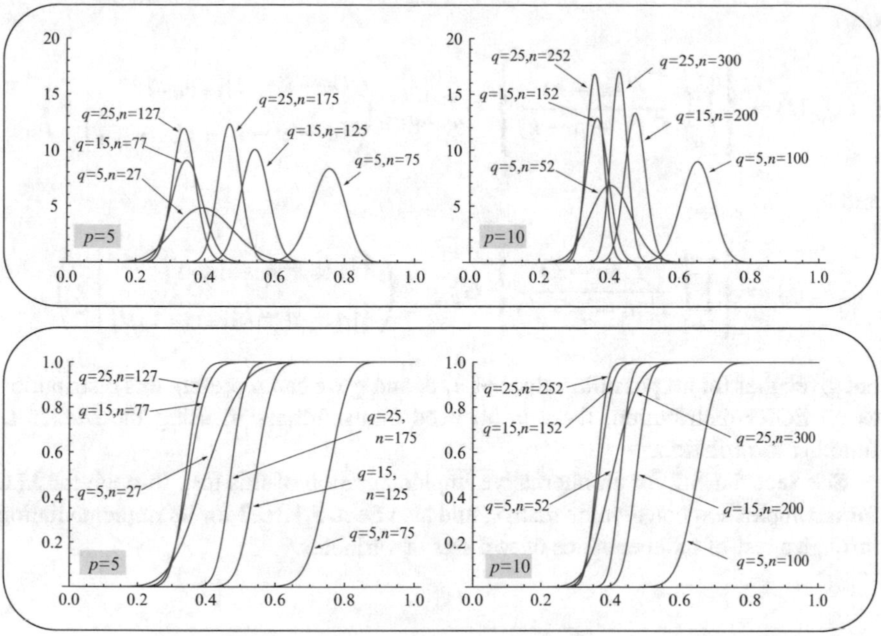

Fig. 5.35 Plots of p.d.f.'s and c.d.f.'s for Λ in (5.100), for $p = 5$ and $p = 10$, $q = 5$, 15 and 25 and $n = pq + 2$ and $n = pq + 50$

see how now, in the complex case, the distributions are quite a bit more concentrated around the mean.

Also for the statistic in this subsection modules are made available to compute the p.d.f., the c.d.f., p-values, and quantiles. These are the modules `PDFEqMeanVecC`, `CDFEqMeanVecC`, `PvalEqMeanVecC`, `PvalDataEqMeanVecC`, and `QuantEqMeanVecC`, which have a very similar usage to that of the corresponding modules in Sect. 5.1.1. The plots in Fig. 5.35 were obtained by using the first two of these modules with a set of commands similar to the ones in Fig. 5.3.

In terms of the Meijer G function, the exact p.d.f. and c.d.f of Λ in (5.100) are given by

$$f_\Lambda(z) = \left\{ \prod_{j=1}^{p} \frac{\Gamma(n-j)}{\Gamma(n-q+1-j)} \right\} G_{p,p}^{p,0} \left(\begin{array}{c} \{n-j-1\}_{j=1:p} \\ \{n-q+1-j-1\}_{j=1:p} \end{array} \middle| \, z \right)$$

and

$$F_\Lambda(z) = \left\{ \prod_{j=1}^{p} \frac{\Gamma(n-j)}{\Gamma(n-q+1-j)} \right\} G_{p+1,p+1}^{p,1} \left(\begin{array}{c} \{1, \{n-j\}_{j=1:p}\} \\ \{\{n-q+1-j\}_{j=1:p}, 0\} \end{array} \middle| \, z \right)$$

or

$$f_\Lambda(z) = \left\{ \prod_{k=1}^{q-1} \frac{\Gamma(n-k)}{\Gamma(n-p-k)} \right\} G^{q-1,0}_{q-1,q-1}\left(\begin{array}{c} \{n-k-1\}_{k=1:q-1} \\ \{n-p-k-1\}_{k=1:q-1} \end{array} \middle| \, z \right)$$

and

$$F_\Lambda(z) = \left\{ \prod_{k=1}^{q-1} \frac{\Gamma(n-k)}{\Gamma(n-p-k)} \right\} G^{q-1,1}_{q,q}\left(\begin{array}{c} \{1, \{n-k\}_{k=1:q-1}\} \\ \{\{n-p-k\}_{k=1:q-1}, 0\} \end{array} \middle| \, z \right),$$

but given that for all possible values of n, p, and q we can represent this distribution as an EGIG distribution, there is no need or usefulness in using the Meijer G function formulation.

See Sect. 5.1.8.2 for an alternative implementation of this test, through the l.r.t. for a complex expected value matrix, and also Sect. 5.1.10.2 for its implementation through a test of independence of two sets of variables.

5.1.6 The Test for Simultaneous Nullity of Several Mean Vectors (Complex r.v.'s) [NullMeanVecC]

Under a similar setup to the one in the previous subsection, let us assume that we are interested in testing the null hypothesis

$$H_0 : \underline{\mu}_1 = \underline{\mu}_2 = \cdots = \underline{\mu}_q = \underline{0}\,, \tag{5.105}$$

that is the hypothesis of simultaneous nullity of the q mean vectors $\underline{\mu}_k$ ($k = 1, \ldots, q$).

Then, for $n = n_1 + n_2 + \cdots + n_q$, where n_k is the size of the sample from $\underline{X}_k$ ($k = 1, \ldots, q$), the $(1/n)$-th power of the l.r.t. statistic to test H_0 in (5.105) is

$$\Lambda = \frac{|A|}{|A + B^*|} \tag{5.106}$$

where A is the matrix in (5.101) and

$$B^* = \sum_{k=1}^{q} n_k \overline{\underline{X}}_k \overline{\underline{X}}_k^{\#}$$

where, as in the previous subsection, $\underline{\overline{X}}_k$ is the sample mean vector for the k-th sample and $\underline{\overline{X}}_k^{\#}$ denotes the transpose of its complex conjugate, and where B^* has a complex Wishart or pseudo-Wishart $CW_p(q, \Sigma)$ distribution.

Then, from Appendix 8 it is easy to see that for Λ in (5.106) we have

$$\Lambda \stackrel{d}{\equiv} \prod_{j=1}^{p} Y_j \stackrel{d}{\equiv} \prod_{k=0}^{q} Y_k^*$$

where, for $n > p+q-1$, $j = 1, \ldots, p$ and $k = 1, \ldots, q$,

$$Y_j \sim Beta\,(n-q+1-j, q) \quad \text{and} \quad Y_k^* \sim Beta\,(n-p+1-k, p)$$

represent respectively a set of p and a set of q independent r.v.'s.

The exact p.d.f. and c.d.f. of Λ are thus given in terms of the Meijer G function as

$$f_\Lambda(z) = \left\{\prod_{j=1}^{p} \frac{\Gamma(n+1-j)}{\Gamma(n-q+1-j)}\right\} G_{p,p}^{p,0}\left(\left.\begin{array}{c}\{n-j\}_{j=1:p}\\ \{n-q-j\}_{j=1:p}\end{array}\right| z\right)$$

and

$$F_\Lambda(z) = \left\{\prod_{j=1}^{p} \frac{\Gamma(n+1-j)}{\Gamma(n-q+1-j)}\right\} G_{p+1,p+1}^{p,1}\left(\left.\begin{array}{c}\{1, \{n+1-j\}_{j=1:p}\}\\ \{\{n-q+1-j\}_{j=1:p}, 0\}\end{array}\right| z\right)$$

or

$$f_\Lambda(z) = \left\{\prod_{k=1}^{q} \frac{\Gamma(n+1-k)}{\Gamma(n-p+1-k)}\right\} G_{q,q}^{q,0}\left(\left.\begin{array}{c}\{n-k\}_{k=1:q}\\ \{n-p-k\}_{k=1:q}\end{array}\right| z\right)$$

and

$$F_\Lambda(z) = \left\{\prod_{k=1}^{q} \frac{\Gamma(n+1-k)}{\Gamma(n-p+1-k)}\right\} G_{q+1,q+1}^{q,1}\left(\left.\begin{array}{c}\{1, \{n+1-k\}_{k=1:q}\}\\ \{\{n-p+1-k\}_{k=1:q}, 0\}\end{array}\right| z\right),$$

or, alternatively, given by Theorem 3.1 or 3.2 and Corollary 4.1 or 4.2 with

$$m^* = 1, \quad a_1 = n-q+1, \quad n_1 = p, \quad k_1 = 1 \quad \text{and} \quad m_1 = q,$$

or,

$$m^* = 1, \quad a_1 = n-p+1, \quad n_1 = q, \quad k_1 = 1 \quad \text{and} \quad m_1 = p,$$

yielding for either of these cases the exact p.d.f. and c.d.f. of Λ, through the EGIG p.d.f. and c.d.f., in the form

$$f_\Lambda(z) = f^{EGIG}\left(z \,\middle|\, \{r_j\}_{j=1:p+q-1}; \{n-j\}_{j=1:p+q-1}; p+q-1\right)$$

and

$$F_\Lambda(z) = F^{EGIG}\left(z \,\middle|\, \{r_j\}_{j=1:p+q-1}; \{n-j\}_{j=1:p+q-1}; p+q-1\right)$$

where

$$r_j = \begin{cases} h_j\,, & j=1 \\ r_{j-1}+h_j\,, & j=2,\ldots,p+q-1 \end{cases}$$

with

$$h_j = (\#\text{ of elements in } \{p,q\} \geq j) - 1\,, \quad j=1,\ldots,p+q-1\,.$$

For the implementation of this test modules PDFNullMeanVecC, CDFNullMeanVecC, PvalNullMeanVecC, QuantNullMeanVecC, and PvalDataNullMeanVecC are made available. These modules have similar arguments and similar usage to the corresponding modules in Sect. 5.1.1.

5.1.7 The Parallelism Test for Profiles (Complex r.v.'s) [ProfParC]

Using a similar setup to the one in the previous subsection and also to the setup in Sect. 5.1.3, where real r.v.'s were used, we have q populations, the k-th of which has a p-variate complex Normal distribution with mean vector $\underline{\mu}_k$ and covariance matrix Σ $(k=1,\ldots,q)$, that is, $\underline{X}_k \sim CN(\underline{\mu}_k, \Sigma)$ for $k=1,\ldots,q$. Let us then suppose that we obtain a sample of size n_k from the k-th population, where the q samples are independent, and let $n=\sum_{k=1}^{q} n_k$.

Suppose that then we are interested in testing the null hypothesis of parallelism of the profiles, that is, a null hypothesis similar to the one in (5.21), but now for the complex valued vectors $\underline{\mu}_k$. This will be the null hypothesis

$$H_0 : C\underline{\mu}_1 = \cdots = C\underline{\mu}_k = \cdots = C\underline{\mu}_q\,, \tag{5.107}$$

where C is the matrix in (5.20).

Then we may establish between this test and the test in the previous subsection a relation parallel to the one that was established between the test in Sect. 5.1.3 and the one in Sect. 5.1.1. All we need in order to test the null hypothesis in (5.107) is to use a l.r.t. statistic similar to the one in (5.100), where the matrices A and B have to be pre-multiplied by C and post-multiplied by C'. More precisely, we will use the l.r.t. statistic

$$\Lambda = \frac{|\,A^*|}{|\,A^* + B^*|} \tag{5.108}$$

where

$$A^* = CAC' \qquad \text{and} \qquad B^* = CBC'$$

with A and B given by (5.101).

But then, the two matrices A^* and B^* are independent, and, under H_0 in (5.107),

$$A^* \sim CW_{p-1}(n-q, C\Sigma C') \qquad \text{and} \qquad B^* \sim CW_{p-1}(q-1, C\Sigma C'),$$

so that, for $n > p+q-2$, under H_0 in (5.107), we may write for the statistic Λ in (5.108) (see Appendix 8)

$$\Lambda \stackrel{d}{\equiv} \prod_{j=1}^{p-1} Y_j \stackrel{d}{\equiv} \prod_{k=1}^{q-1} Y_k^* \tag{5.109}$$

where for $n > p+q-2$, $j = 1, \ldots, p-1$, and $k = 1, \ldots, q-1$,

$$Y_j \sim Beta\,(n-q-j+1, q-1) \qquad \text{and} \qquad Y_k^* \sim Beta\,(n-p-k+1, p-1)$$

are two sets of independent r.v.'s.

As such, for any p and q, the exact distribution of Λ in (5.108) is given by Theorem 3.2 and its exact p.d.f. and c.d.f. by Corollary 4.2, for

$$m^* = 1\,, \quad a_1 = n-q+1\,, \quad k_1 = 1\,, \quad n_1 = p-1\,, \quad m_1 = q-1\,, \tag{5.110}$$

or

$$m^* = 1\,, \quad a_1 = n-p+1\,, \quad k_1 = 1\,, \quad n_1 = q-1\,, \quad m_1 = p-1\,, \tag{5.111}$$

yielding the exact p.d.f. and c.d.f. of Λ in (5.100) given, in terms of the EGIG p.d.f. and c.d.f. representation, by

$$f_\Lambda(z) = f^{EGIG}\left(z \,\middle|\, \{r_j\}_{j=1:p+q-3}; \{n-1-j\}_{j=1:p+q-3}; p+q-3\right)$$

and

$$F_\Lambda(z) = F^{EGIG}\left(z \,\Big|\, \{r_j\}_{j=1:p+q-3}; \{n-1-j\}_{j=1:p+q-3}; p+q-3\right),$$

where

$$r_j = \begin{cases} h_j\,, & j=1 \\ r_{j-1}+h_j\,, & j=2,\ldots,p+q-3 \end{cases}$$

with

$$h_j = (\#\text{ of elements in } \{p-1, q-1\} \geq j) - 1\,, \quad j=1,\ldots,p+q-3\,,$$

for any p and any q.

The exact p.d.f. and c.d.f. of Λ can be given in terms of the Meijer G function as

$$f_\Lambda(z) = \left\{\prod_{j=1}^{p-1} \frac{\Gamma(n-j)}{\Gamma(n-q+1-j)}\right\} G^{p-1,0}_{p-1,p-1}\left(\begin{array}{c} \{n-j-1\}_{j=1:p-1} \\ \{n-q+1-j-1\}_{j=1:p-1} \end{array} \Bigg|\, z\right)$$

and

$$F_\Lambda(z) = \left\{\prod_{j=1}^{p-1} \frac{\Gamma(n-j)}{\Gamma(n-q+1-j)}\right\} G^{p-1,1}_{p,p}\left(\begin{array}{c} \{1, \{n-j\}_{j=1:p-1}\} \\ \{\{n-q+1-j\}_{j=1:p-1}, 0\} \end{array} \Bigg|\, z\right)$$

or

$$f_\Lambda(z) = \left\{\prod_{k=1}^{q-1} \frac{\Gamma(n-k)}{\Gamma(n-p+1-k)}\right\} G^{q-1,0}_{q-1,q-1}\left(\begin{array}{c} \{n-k-1\}_{k=1:q-1} \\ \{n-p+1-k-1\}_{k=1:q-1} \end{array} \Bigg|\, z\right)$$

and

$$F_\Lambda(z) = \left\{\prod_{k=1}^{q-1} \frac{\Gamma(n-k)}{\Gamma(n-p+1-k)}\right\} G^{q-1,1}_{q,q}\left(\begin{array}{c} \{1, \{n-k\}_{k=1:q-1}\} \\ \{\{n-p+1-k\}_{k=1:q-1}, 0\} \end{array} \Bigg|\, z\right),$$

but, as with the test statistic in the previous subsection, there is no need to use this formulation since the EGIG formulation, which is easily computable is available for all values of n, p, and q.

Modules `PDFProfParC`, `CDFProfParC`, `PvalProfParC`, `QuantProfParC`, and `PvalDataProfParC` are made available to compute the p.d.f., c.d.f., p-values and quantiles. These modules are similar in usage to the corresponding modules in Sects. 5.1.3 and 5.1.1.

5.1.8 The Likelihood Ratio Statistic to Test Hypotheses on an Expected Value Matrix (Complex r.v.'s) [MatEVC]

In this subsection we will first address the test developed in Sect. 3.2 of Khatri (1965), which is a similar test to the one addressed in Sect. 5.1.4, but now for complex r.v.'s, and then we will formulate a generalization of this test which includes as particular cases the tests in Sects. 5.1.5 and 5.1.7.

Let $X_{(p\times n)}$ be a matrix with a complex multivariate Normal distribution with expected value $\boldsymbol{\mu}M$ and variance $Var(vec(X)) = I_n \otimes \Sigma$, where $\boldsymbol{\mu}$ is a $p\times q$ complex matrix and M is a real or complex $q\times n$ matrix of rank q ($\leq n$). We will denote this fact by

$$X_{(p\times n)} \sim CN_{p\times n}(\boldsymbol{\mu}M, I_n \otimes \Sigma)\,. \tag{5.112}$$

Let us then suppose that we want to test the hypothesis

$$H_0 : \boldsymbol{\mu}_{(p\times q)} = 0_{(p\times q)}\,. \tag{5.113}$$

Then, according to Khatri (1965), the $(1/n)$-th power of the l.r.t. statistic to test H_0 is

$$\Lambda = \frac{|\Psi|}{\left|\Psi + \frac{1}{n}\beta(MM^\#)\beta^\#\right|} \tag{5.114}$$

where $(\,\cdot\,)^\#$ denotes the transpose of the complex conjugate, and where

$$\Psi = \frac{1}{n}X\left(I_n - M^\#\left(MM^\#\right)^{-1}M\right)X^\# = \frac{1}{n}\left(XX^\# - \beta\left(MM^\#\right)\beta^\#\right) \tag{5.115}$$

and

$$\beta = XM^\#\left(MM^\#\right)^{-1} \tag{5.116}$$

are respectively the m.l.e.'s of Σ and μ, and as such independent.

But, since $\left(I_n - M^\#(MM^\#)^{-1}M\right)$ is the projector on the null space of the columns of M, it is idempotent with

$$rank\left(I_n - M^\#\left(MM^\#\right)^{-1}M\right) = tr\left(I_n - M^\#\left(MM^\#\right)^{-1}M\right) = n - q\,,$$

so that, from the distribution of X in (5.112), we have

$$\Psi = \frac{1}{n} X \left(I_n - M^{\#} \left(M M^{\#} \right)^{-1} M \right) X^{\#} \sim CW_p \left(n - q, \frac{1}{n} \Sigma \right) . \tag{5.117}$$

From (5.112) we may easily see that

$$\beta = X M^{\#} \left(M M^{\#} \right)^{-1} \sim CN_{p \times q} \left(\boldsymbol{\mu}, \left(M M^{\#} \right)^{-1} \otimes \Sigma \right) , \tag{5.118}$$

so that

$$\beta \left(M M^{\#} \right)^{1/2} \sim CN_{p \times q} \left(\boldsymbol{\mu} \left(M M^{\#} \right)^{1/2}, I_q \otimes \Sigma \right) ,$$

where, under H_0 in (5.113), $\boldsymbol{\mu}\left(MM^{\#}\right)^{1/2} = 0$. Therefore, under H_0 in (5.113),

$$\beta \left(M M^{\#} \right) \beta^{\#} \sim CW_p(q, \Sigma) ,$$

which is independent of Ψ, and hence, under H_0 in (5.113),

$$\Psi + \frac{1}{n} \beta (M M^{\#}) \beta^{\#} = \frac{1}{n} X X^{\#} \sim CW_p \left(n, \frac{1}{n} \Sigma \right) .$$

Thus, from the results in Appendix 8, we may say that, for $n > q + p - 1$,

$$\Lambda \stackrel{d}{\equiv} \prod_{j=1}^{p} Y_j \quad \text{where} \quad \begin{array}{l} Y_j \sim Beta\,(n - q - j + 1, q) \;\; (j = 1, \ldots, p) \\ \text{are } p \text{ independent r.v.'s.} \end{array} \tag{5.119}$$

When we compare (5.119) with (5.2.2) and (5.2.3) in Khatri (1965), we see that there is a small mistake in Khatri's paper, in that q has to be subtracted from the first argument of the Beta r.v.'s in (5.2.2) and (5.2.3) and as such also from the arguments of all Gamma functions in (5.3.1) of Khatri (1965).

In case that, instead of the hypothesis in (5.113), we want to test the null hypothesis

$$H_0 : \mu = \beta^*_{p \times q}$$

where β^* is some $p{\times}q$ given matrix, we only have to replace β by $\beta - \beta^*$ in the formulation of Λ in (5.114).

The exact distribution of Λ in (5.114) is therefore given by Theorem 3.2 and its exact p.d.f. and c.d.f. by Corollary 4.2, for

$$m^* = 1\,, \quad a_1 = n - q + 1\,, \quad k_1 = 1\,, \quad n_1 = p\,, \quad m_1 = q\,. \tag{5.120}$$

Based on the results in Appendix 8 and also on the results in Theorem 3.2, it is easy to show that we may interchange p and q in (5.119), and thus also in (5.120), yielding as alternative possible choices for the parameters in (5.120),

$$m^* = 1\,, \quad a_1 = n - p + 1\,, \quad k_1 = 1\,, \quad n_1 = q\,, \quad m_1 = p\,. \tag{5.121}$$

The p.d.f. and c.d.f. of Λ in (5.114), is thus, for any p and any q, given, in terms of the EGIG distribution representation, respectively by

$$f_\Lambda(z) = f^{EGIG}\left(z \,\middle|\, \{r_j\}_{j=1:p+q-1}; \{n-1-j\}_{j=1:p+q-1}\,; p+q-1\right) \tag{5.122}$$

and

$$F_\Lambda(z) = F^{EGIG}\left(z \,\middle|\, \{r_j\}_{j=1:p+q-1}; \{n-1-j\}_{j=1:p+q-1}\,; p+q-1\right), \tag{5.123}$$

where

$$r_j = \begin{cases} h_j\,, & j = 1 \\ r_{j-1} + h_j\,, & j = 2, \ldots, p+q-1 \end{cases} \tag{5.124}$$

with

$$h_j = (\#\text{ of elements in } \{p, q\} \geq j) - 1\,, \quad j = 1, \ldots, p+q-1\,, \tag{5.125}$$

which confirms the results in Sect. 2 of Coelho et al. (2015).

In terms of the Meijer G function, the exact p.d.f. and c.d.f. of Λ in (5.114) may be written as

$$f_\Lambda(z) = \left\{\prod_{j=1}^{p} \frac{\Gamma(n+1-j)}{\Gamma(n-q+1-j)}\right\} G^{p,0}_{p,p}\left(\begin{array}{c} \{n+1-j-1\}_{j=1:p} \\ \{n-q+1-j-1\}_{j=1:p} \end{array} \,\middle|\, z\right)$$

and

$$F_\Lambda(z) = \left\{\prod_{j=1}^{p} \frac{\Gamma(n+1-j)}{\Gamma(n-q+1-j)}\right\} G^{p,1}_{p+1,p+1}\left(\begin{array}{c} \{1, \{n+1-j\}_{j=1:p}\} \\ \{\{n-q+1-j\}_{j=1:p}\,, 0\} \end{array} \,\middle|\, z\right)$$

or

$$f_{\Lambda}(z) = \left\{ \prod_{k=1}^{q} \frac{\Gamma(n+1-k)}{\Gamma(n-p+1-k)} \right\} G_{q,q}^{q,0} \left(\begin{array}{c} \{n+1-k-1\}_{k=1:q} \\ \{n-p+1-k-1\}_{k=1:q} \end{array} \middle| z \right)$$

and

$$F_{\Lambda}(z) = \left\{ \prod_{k=1}^{q} \frac{\Gamma(n+1-k)}{\Gamma(n-p+1-k)} \right\} G_{q+1,q+1}^{q,1} \left(\begin{array}{c} \{1, \{n+1-k\}_{k=1:q}\} \\ \{\{n-p+1-k\}_{k=1:q}, 0\} \end{array} \middle| z \right),$$

but once again, as in the two previous subsections, there is no need to use this formulation since the EGIG formulation covers all cases.

An alternative finite form for the p.d.f. and c.d.f. of the l.r.t statistic to test H_0 in (5.113) may be found in Gupta (1971), where the author only provides the complete expressions for $p = 2$ and $p = 3$, leaving for the general case several parameters in the distribution without explicit expressions.

5.1.8.1 Extending the Test

Following similar steps to the ones in Sect. 5.1.4.1, we will first extend the present test in order to address the test of null hypotheses of the type

$$H_0 : C\mu = \beta^*_{(p^* \times q)} \tag{5.126}$$

where C is a nonrandom real $p^* \times p$ matrix, with $rank(C) = p^* \leq p$.

As it happened in Sect. 5.1.4.1, the matrix C gives rise to a linear combination of the p variables in X, so that now we have

$$\underset{(p^* \times n)}{CX} \sim CN_{p^* \times n}(C\mu M, I_n \otimes C\Sigma C'),$$

where, by the invariance property of m.l.e.'s, the m.l.e. of $C\Sigma C'$ is, for Ψ in (5.115),

$$\Psi^* = C\Psi C' = \frac{1}{n} CX \left(I_n - M^{\#} \left(MM^{\#} \right)^{-1} M \right) X^{\#} C' \sim CW_{p^*} \left(n-q, \frac{1}{n} C\Sigma C' \right) \tag{5.127}$$

and the m.l.e. of $C\mu$ is, for β in (5.116)

$$C\beta = CXM^{\#} \left(MM^{\#} \right)^{-1} \sim CN_{p^* \times q} \left(C\boldsymbol{\mu}, \left(MM^{\#} \right)^{-1} \otimes C\Sigma C' \right),$$

so that under H_0 in (5.126),

$$(C\beta-\beta^*)\left(MM^\#\right)^{1/2}\sim CN_{p^*\times q}\left(0_{p^*\times q},\,I_q\otimes C\Sigma C'\right).$$

Consequently, under this null hypothesis,

$$(C\beta-\beta^*)\left(MM^\#\right)(C\beta-\beta^*)^\#\sim CW_{p^*}(q,\,C\Sigma C'),$$

independent of Ψ^*, which implies, under H_0 in (5.126),

$$\Psi^*+\frac{1}{n}(C\beta-\beta^*)(MM^\#)(C\beta-\beta^*)^\#\sim CW_{p^*}\left(n,\frac{1}{n}C\Sigma C'\right).$$

Thus the l.r.t. statistic to test H_0 in (5.126) is

$$\Lambda^*=\frac{|\Psi^*|}{\left|\Psi^*+\frac{1}{n}(C\beta-\beta^*)(MM^\#)(C\beta-\beta^*)^\#\right|}\tag{5.128}$$

with (see Appendix 8)

$$\Lambda^*\stackrel{\mathrm{d}}{\equiv}\prod_{j=1}^{p^*}Y_j^*\stackrel{\mathrm{d}}{\equiv}\prod_{k=1}^{q}Y_k^{**},\tag{5.129}$$

where, for $n>p^*+q-1$, $j=1,\dots,p^*$, and $k=1,\dots,q$

$$Y_j^*\sim Beta\,(n-q+1-j,q)\quad\text{and}\quad Y_k^{**}\sim Beta\left(n-p^*+1-k,p^*\right)$$

represent two sets of independent r.v.'s.

The exact distribution of Λ^* in (5.128) has thus p.d.f. and c.d.f. given by Corollary 4.2, with

$$m^*=1,\quad a_1=n-q,\quad n_1=p^*,\quad k_1=1\ \text{ and }\ m_1=q\,,\tag{5.130}$$

or

$$m^*=1,\quad a_1=n-p^*,\quad n_1=q\,,\quad k_1=1\ \text{ and }\ m_1=p^*\,.\tag{5.131}$$

This exact p.d.f. and c.d.f. of Λ^* in (5.128) are given by (5.122)–(5.125) with p replaced by p^*.

Then, if one is interested in testing the null hypothesis

$$H_0: C\mu D=\beta^*_{(p^*\times q)}\,,\tag{5.132}$$

where D is a nonrandom real $q \times q$ matrix with $rank(D) = q^* \leq q$, by the invariance property of m.l.e.'s, and the induced invariance on the l.r.t. statistic, we may test such hypothesis just by using an l.r.t. statistic which is similar to Λ^* in (5.128), with $C\beta$ replaced by $C\beta D$, which is the m.l.e. of $C\mu D$. This amounts to using the l.r.t. statistic

$$\Lambda^{**} = \frac{|\Psi^*|}{\left|\Psi^* + \frac{1}{n}(C\beta D - \beta^*)(MM^\#)(C\beta D - \beta^*)^\#\right|} \tag{5.133}$$

where Ψ^* remains defined as in (5.127), and where

$$(C\beta D - \beta^*)\left(MM^\#\right)^{1/2} \sim CN_{p^* \times q}\left((C\mu D - \beta^*)\left(MM^\#\right)^{1/2}, D'D \otimes C\Sigma C'\right),$$

with $(C\mu D - \beta^*)(MM^\#)^{1/2} = 0$, under H_0 in (5.132), so that, under this null hypothesis, $(C\beta D - \beta^*)(MM^\#)(C\beta D - \beta^*)^\#$ will have a complex Wishart distribution with a number of degrees of freedom equal to $rank(D'D) = rank(D) = q^*$, that is,

$$(C\beta D - \beta^*)\left(MM^\#\right)(C\beta D - \beta^*)^\# \sim CW_{p^*}(q^*, C\Sigma C'),$$

independent of Ψ^*. Consequently, under H_0 in (5.132),

$$\Psi^* + \frac{1}{n}(C\beta D - \beta^*)(MM^\#)(C\beta D - \beta^*)^\# \sim CW_{p^*}\left(n - q + q^*, \frac{1}{n}C\Sigma C'\right),$$

and as such (see Appendix 8), for $n > p^* + q - 1$,

$$\Lambda^{**} \stackrel{d}{\equiv} \prod_{j=1}^{p^*} Y_j^{**} \stackrel{d}{\equiv} \prod_{k=1}^{q^*} Y_k^{***}, \tag{5.134}$$

where, for $j = 1, \ldots, p^*$ and $k = 1, \ldots, q^*$,

$$Y_j^{**} \sim Beta\left(n - q + 1 - j, q^*\right) \quad \text{and} \quad Y_k^{***} \sim Beta\left(n - q + q^* - p^* + 1 - k, p^*\right) \tag{5.135}$$

represent two sets of independent r.v.'s.

Therefore, the exact distribution of Λ^{**} in (5.133) is an EGIG distribution, with p.d.f. and c.d.f. given by Corollary 4.2, with

$$m^* = 1, \quad a_1 = n - q + 1, \quad n_1 = p^*, \quad k_1 = 1 \quad \text{and} \quad m_1 = q^*, \tag{5.136}$$

or,

$$m^* = 1, \quad a_1 = n - q + q^* - p^* + 1, \quad n_1 = q^*, \quad k_1 = 1 \quad \text{and} \quad m_1 = p^*, \tag{5.137}$$

and thus the exact p.d.f. and c.d.f. of Λ^{**} in (5.133) are given by (5.122)–(5.125) with n replaced by $n - q + q^*$, p replaced by p^* and q replaced by q^*.

As for all l.r.t.'s, also for the test in this subsection, in order to carry out an α-level test we should reject H_0 in (5.132) if the computed value of Λ^{**} in (5.133) is smaller than the α quantile of Λ^{**}, or equivalently, if the corresponding p-value is smaller than our value of α.

For the statistics in this subsection modules `PDFMatEVC`, `CDFMatEVC`, `QuantMatEVC`, `PvalMatEVC`, and `PvalDataMatEVC` are made available to compute the p.d.f., the c.d.f., quantiles, and p-values. These modules are to be used in a similar manner to that of the corresponding modules in Sect. 5.1.4.

5.1.8.2 Specializing the Present Test to the Test of Equality of Mean Vectors in Sect. 5.1.5

As it happens in the real case, an appropriate choice of the matrices X, M, C, and D will allow us to implement the tests in Sects. 5.1.5 and 5.1.7 as tests of the hypothesis in (5.132).

Similar to what happens in the real case, if we consider X and M matrices analogous to the X and M matrices in Sect. 5.1.4.2, that is, the sample and the *design* matrices, we will have

$$X \sim CN_{p\times n}(\mu M, I_n \otimes \Sigma),$$

where μ is now a complex matrix, with a similar structure to that of the matrix μ in (5.66).

Then, if we take $C = I_p$ and D as defined in (5.69), $C\mu D$ will be a matrix with a similar structure to the matrix $C\mu D$ in (5.70), now with complex-valued entries, and then to test the null hypothesis of equality of the q mean vectors $\underline{\mu}_k$, $k = 1, \ldots, q$, that is, the null hypothesis (5.99) in Sect. 5.1.5, will be equivalent to testing the null hypothesis

$$H_0 : C\mu D = \mathbf{0}_{p\times q}.$$

From the previous Sect. 5.1.8.1, the l.r.t. statistic will be

$$\Lambda = \frac{|\Psi|}{\left|\Psi + \frac{1}{n}(\beta D)(MM^{\#})(\beta D)^{\#}\right|} \tag{5.138}$$

where Ψ, given by (5.117), is equal to $1/n$ times the matrix A in (5.101) and where $(\beta D)(MM^{\#})(\beta D)^{\#}$, with β given by (5.118), is exactly the matrix B in (5.101), since

$$(\beta D)(MM^{\#})(\beta D)^{\#} = (\beta DM)(\beta DM)^{\#}$$

where the matrix $\beta DM = XM^{\#}(MM^{\#})^{-1}DM$ has a structure similar to that of the matrix βDM in Sect. 5.1.4.2.

As such, the exact distribution of Λ in (5.138) may be obtained from the results on the distribution of Λ^{**} in the previous Sect. 5.1.8.1, with $p^* = p$ and $q^* = q - 1$. Hence, from (5.134) and (5.135), the l.r.t. statistic in (5.138) has the same distribution as the l.r.t. Λ in Sect. 5.1.5, as given by (5.102), and it is indeed exactly the same l.r.t. statistic.

This test may therefore be implemented by using module `PvalDataMatEVC` in a similar manner to the one used with module `PvalDataMatEVR` to implement the test in Sect. 5.1.1. The command necessary to do this implementation will thus be similar to the first command in Fig. 5.31.

5.1.8.3 Specializing the Present Test to the Test in Sect. 5.1.7

The test in Sect. 5.1.7, for the parallelism of complex profiles, may be implemented as a particular case of the generalized test in Sect. 5.1.8.1, that is, the test to the null hypothesis in (5.132), in much the same way as the generalization of the test in Sect. 5.1.4 may be used to test the hypothesis of parallelism of the profiles of real r.v.'s, by taking matrices X, M, and D similar to the ones in the previous subsection, $\beta^* = 0$, and a $(p-1)\times p$ matrix C as the one in (5.20). We will then have a matrix $C\mu D$ similar to the one in Sect. 5.1.4.3 so that to test the null hypothesis in (5.107) will be the same as to test the null hypothesis

$$C\mu D = 0_{(p-1)\times q}\,.$$

The l.r. statistic to test this hypothesis will then be the statistic Λ^{**} in (5.133), which, for the present choice for the matrices X, M, C, and D, and taking $\beta^* = 0$, is indeed the same as the l.r.t. statistic Λ in (5.108) and which, from (5.134) and (5.135), taking $p^* = p - 1$ and $q^* = q - 1$, has the same distribution as that of Λ in Sect. 5.1.7.

Similar to what happens in the real case, this test may be implemented with module `PvalDataMatEVC` by using a command analogous to the command in Fig. 5.32.

5.1.9 The Likelihood Ratio Statistic to Test the Independence of Two Groups of Real Random Variables [Ind2R]

The test in this subsection is rather simple in nature, and its implementation is also quite simple. Nevertheless, as we will see, it plays an extremely important role in Multivariate Analysis, mainly in testing in the framework of the Multivariate Linear Model.

Let $\underline{X} \sim N_p(\underline{\mu}, \Sigma)$ be a real p-dimensional random vector, which is split into two subvectors, $\underline{X}_1$ and $\underline{X}_2$ of dimensions $p_1 \times 1$ and $p_2 \times 1$, with $p = p_1 + p_2$, and let Σ be accordingly partitioned as

$$\Sigma = \begin{bmatrix} \Sigma_{11} & \Sigma_{12} \\ \Sigma_{21} & \Sigma_{22} \end{bmatrix}. \tag{5.139}$$

Let us then suppose we are interested in testing the hypothesis of independence of $\underline{X}_1$ and $\underline{X}_2$. This hypothesis may be written as

$$H_0 : \Sigma = bdiag(\Sigma_{11}, \Sigma_{22}) \tag{5.140}$$

where $\Sigma_{11} = Var(\underline{X}_1)$ and $\Sigma_{22} = Var(\underline{X}_2)$. Alternatively, we may also write this null hypothesis as

$$H_0 : \Sigma_{12} = 0\,, \tag{5.141}$$

where $\Sigma_{12} = Cov(\underline{X}_1, \underline{X}_2) = \Sigma_{21}'$.

The m.l.e. of Σ is then the matrix

$$A = \begin{bmatrix} A_{11} & A_{12} \\ A_{21} & A_{22} \end{bmatrix} = \frac{1}{n} \begin{bmatrix} X_1 \\ X_2 \end{bmatrix} \left(I_n - \tfrac{1}{n} E_{nn} \right) \begin{bmatrix} X_1 \\ X_2 \end{bmatrix}' = \tfrac{1}{n} X^* X^{*\prime} \tag{5.142}$$

where

$$X = \begin{bmatrix} X_1 \\ X_2 \end{bmatrix} \quad \text{and} \quad X^* = \begin{bmatrix} X_1^* \\ X_2^* \end{bmatrix} = X \left(I_n - \tfrac{1}{n} E_{nn} \right) = X - \underbrace{\tfrac{1}{n}\, X E_{n1}}_{=\overline{X}} E_{1n}$$

are the $p \times n$ original and centered sample data matrices, with $n > p_1 + p_2$, and where $\overline{X}$ is the $p \times 1$ vector of sample means and $I_n - \frac{1}{n} E_{nn}$ is a symmetric idempotent matrix. The submatrices

$$A_{11} = \tfrac{1}{n} X_1 (I_n - \tfrac{1}{n} E_{nn}) X_1' = \tfrac{1}{n} X_1^* X_1^{*\prime}\,, \quad A_{22} = \tfrac{1}{n} X_2 (I_n - \tfrac{1}{n} E_{nn}) X_2' = \tfrac{1}{n} X_2^* X_2^{*\prime} \tag{5.143}$$

and

$$A_{12} = \tfrac{1}{n} X_1 (I_n - \tfrac{1}{n} E_{nn}) X_2' = \tfrac{1}{n} X_1^* X_2^{*\prime} = A_{21}' \tag{5.144}$$

in (5.142) are then the m.l.e.'s respectively of Σ_{11}, Σ_{22}, and Σ_{12}.

Under this setup, the $(2/n)$-th power of the l.r.t. statistic to test H_0 in (5.140) or (5.141) is (Anderson, 2003, Sects. 9.2, 9.8; Kshirsagar, 1972, Sect. 8.4; Muirhead, 2005, Sect. 11.2)

$$\Lambda = \frac{|A|}{|A_{11}|\,|A_{22}|}. \tag{5.145}$$

One optimal property of the statistic Λ in (5.145) is that it is invariant under linear transformations within each one of the sets $\underline{X}_1$ and $\underline{X}_2$, that is, under the class of transformations

$$\underline{X} = \left[\frac{\underline{X}_1}{\underline{X}_2}\right] \longrightarrow \left[\frac{C_1\underline{X}_1 + \underline{b}_1}{C_2\underline{X}_2 + \underline{b}_2}\right] \quad \text{or} \quad \underline{X} \longrightarrow C\underline{X} + \underline{b},$$

with $C = bdiag(C_1, C_2)$ and $\underline{b} = \left[\underline{b}_1', \underline{b}_2'\right]'$ where C_1 and C_2 are respectively $p_1{\times}p_1$ and $p_2{\times}p_2$ nonrandom full-rank matrices and $\underline{b}_1$ and $\underline{b}_2$ are nonrandom $p_1{\times}1$ and $p_2{\times}1$ vectors (see Anderson, 2003, Sect. 9.1; Bartlett, 1951; Muirhead, 2005, Sect. 11.2).

We may give Λ in (5.145) a similar expression to that of Λ in (5.2). By considering for the matrix A the partition in (5.142), similar to the partition of the matrix Σ in (5.139), we may write

$$|A| = |A_{22}|\,|A_{11} - A_{12}A_{22}^{-1}A_{21}| = |A_{11}|\,|A_{22} - A_{21}A_{11}^{-1}A_{12}| \tag{5.146}$$

and then write Λ in (5.145) as

$$\Lambda = \frac{|A|}{|A_{11}|\,|A_{22}|} = \frac{|A_{11} - A_{12}A_{22}^{-1}A_{21}|}{|A_{11}|} = \frac{|A^*|}{|A^* + B^*|} \tag{5.147}$$

where $A^* = A_{11.2} = A_{11} - A_{12}A_{22}^{-1}A_{21}$ and $B^* = A_{12}A_{22}^{-1}A_{21}$. Under H_0 in (5.141), A^* and B^* are independent, with $A^* \sim W_{p_1}(n-1-p_2, \frac{1}{n}\Sigma_{11})$ and $B^* \sim W_{p_1}(p_2, \frac{1}{n}\Sigma_{11})$ (see Theorem 3.2.10 in Muirhead (2005)). Alternatively, we may write

$$\Lambda = \frac{|A_{22} - A_{21}A_{11}^{-1}A_{12}|}{|A_{22}|} = \frac{|A^{**}|}{|A^{**} + B^{**}|}, \tag{5.148}$$

where $A^{**}=A_{22}-A_{21}A_{11}^{-1}A_{12}$ and $B^{**}=A_{21}A_{11}^{-1}A_{12}$, under H_0 in (5.141), are also independent, with $A^{**}\sim W_{p_2}(n-1-p_1,\frac{1}{n}\Sigma_{22})$ and $B^{**}\sim W_{p_2}(p_1,\frac{1}{n}\Sigma_{22})$.

As such (see Appendix 1), we have, for $n>p_1+p_2$,

$$\Lambda \stackrel{d}{\equiv} \prod_{j=1}^{p_1} Y_j \stackrel{d}{\equiv} \prod_{j=1}^{p_2} Y_j^*\,, \tag{5.149}$$

where

$$Y_j \sim Beta\left(\frac{n-p_2-j}{2},\frac{p_2}{2}\right), \quad \text{and} \quad Y_j^* \sim Beta\left(\frac{n-p_1-j}{2},\frac{p_1}{2}\right) \tag{5.150}$$

form two sets respectively of p_1 and p_2 independent r.v.'s.

Consequently, the exact p.d.f. and c.d.f. of Λ in (5.145), (5.147), or (5.148) are thus given in terms of the Meijer G function by

$$f_\Lambda(z)=\left\{\prod_{j=1}^{p_1}\frac{\Gamma\left(\frac{n-j}{2}\right)}{\Gamma\left(\frac{n-p_2-j}{2}\right)}\right\} G^{p_1,0}_{p_1,p_1}\left(\left.\begin{array}{c}\left\{\frac{n-j}{2}-1\right\}_{j=1:p_1}\\ \left\{\frac{n-p_2-j}{2}-1\right\}_{j=1:p_1}\end{array}\right| z\right)$$

and

$$F_\Lambda(z)=\left\{\prod_{j=1}^{p_1}\frac{\Gamma\left(\frac{n-j}{2}\right)}{\Gamma\left(\frac{n-p_2-j}{2}\right)}\right\} G^{p_1,1}_{p_1+1,p_1+1}\left(\left.\begin{array}{c}\left\{1,\left\{\frac{n-j}{2}\right\}_{j=1:p_1}\right\}\\ \left\{\left\{\frac{n-p_2-j}{2}\right\}_{j=1:p_1},0\right\}\end{array}\right| z\right)$$

or

$$f_\Lambda(z)=\left\{\prod_{j=1}^{p_2}\frac{\Gamma\left(\frac{n-j}{2}\right)}{\Gamma\left(\frac{n-p_1-j}{2}\right)}\right\} G^{p_2,0}_{p_2,p_2}\left(\left.\begin{array}{c}\left\{\frac{n-j}{2}-1\right\}_{j=1:p_2}\\ \left\{\frac{n-p_1-j}{2}-1\right\}_{j=1:p_2}\end{array}\right| z\right)$$

and

$$F_\Lambda(z)=\left\{\prod_{j=1}^{p_2}\frac{\Gamma\left(\frac{n-j}{2}\right)}{\Gamma\left(\frac{n-p_1-j}{2}\right)}\right\} G^{p_2,1}_{p_2+1,p_2+1}\left(\left.\begin{array}{c}\left\{1,\left\{\frac{n-j}{2}\right\}_{j=1:p_2}\right\}\\ \left\{\left\{\frac{n-p_1-j}{2}\right\}_{j=1:p_2},0\right\}\end{array}\right| z\right).$$

But then the distribution of Λ is, for even p_1, given by Theorem 3.2 and its p.d.f. and c.d.f. by Corollary 4.2, with

$$m^*=1\,,\quad k_1=2\,,\quad a_1=\frac{n-p_2}{2}\,,\quad n_1=\frac{p_1}{2}\,,\quad m_1=p_2 \tag{5.151}$$

while, for even p_2 the exact distribution and the exact p.d.f. and c.d.f. of Λ are given by the same Theorem and Corollary, now with

$$m^* = 1\,, \quad k_1 = 2\,, \quad a_1 = \frac{n-p_1}{2}\,, \quad n_1 = \frac{p_2}{2}\,, \quad m_1 = p_1\,. \tag{5.152}$$

The exact p.d.f. and c.d.f. of Λ in (5.147) or (5.148) are thus given, in terms of the EGIG distribution representation (see Appendix in Chap. 2), either for even p_1 or even p_2, respectively by

$$f_\Lambda(z) = f^{EGIG}\left(z \,\middle|\, \{r_j\}_{j=1:p_1+p_2-2};\, \left\{\frac{n-2-j}{2}\right\}_{j=1:p_1+p_2-2};\, p_1+p_2-2 \right)$$

and

$$F_\Lambda(z) = F^{EGIG}\left(z \,\middle|\, \{r_j\}_{j=1:p_1+p_2-2};\, \left\{\frac{n-2-j}{2}\right\}_{j=1:p_1+p_2-2};\, p_1+p_2-2 \right),$$

where

$$r_j = \begin{cases} h_j\,, & j=1,2 \\ r_{j-2}+h_j\,, & j=3,\ldots,p_1+p_2-2 \end{cases}$$

with

$$h_j = (\#\text{ of elements in } \{p_1,p_2\} \geq j) - 1\,, \quad j=1,\ldots,p_1+p_2-2\,. \tag{5.153}$$

This result confirms the results in Coelho (1998, 1999), Coelho et al. (2010), Marques et al. (2011), and Arnold et al. (2013).

When both p_1 and p_2 are odd, near-exact distributions for the statistic Λ in (5.145), (5.147) or (5.148) are available in Coelho (2004), Coelho et al. (2010) and Marques et al. (2011). For this case the reader may also refer to the material in Chap. 7.

The l.r.t. we are considering in this subsection has actually a similar distribution to that of the l.r.t. statistic we considered in Sect. 5.1.1. This may be readily seen from the distributions of the Beta r.v.'s in (5.5) and (5.149), taking p_1 in this subsection as p in Sect. 5.1.1 and p_2 as $q-1$, or vice versa.

As it happens in Sects. 5.1.1 and 5.1.4 and indeed with all other l.r.t. statistics so far addressed, the distribution here studied for the l.r.t. statistic Λ in (5.145) remains valid if the joint distribution of the random subvectors $\underline{X}_1$ and $\underline{X}_2$ is not multivariate Normal, but instead is some elliptically contoured or left orthogonal-invariant distribution—see Theorem 5.3.3 in Fang and Zhang (1990, Chap. V), Sect. 9.11 in Anderson (2003) and also Jensen and Good (1981), Kariya (1981), Anderson et al. (1986) and Anderson and Fang (1990).

The test of the hypothesis (5.140) or (5.141) is of interest in the Multivariate Linear Model or Multivariate Regression setting. In this context, one of the subvectors $\underline{X}_1$ or $\underline{X}_2$ will be viewed as the vector of response variables while the other one is viewed as the vector of explanatory variables. The statistic Λ in (5.145), (5.147), or (5.148) arises then as the statistic used to test the fit of the Multivariate Regression model, that is, the statistic used to determine whether the explanatory variables "belong" in the model, in the sense that, if the null hypothesis in (5.140) or (5.141) is rejected then $\underline{X}_1$ and $\underline{X}_2$ are not independent and values of the explanatory variables can be effectively used to predict the values of the response variables, as is typically done in linear regression. We then say that the Multivariate Regression model "fits" if we reject H_0 in (5.140) or (5.141). If that null hypothesis is not rejected, then the explanatory variables are not judged to be useful for prediction. In the present setting, the roles of the explanatory and response variables are somewhat arbitrarily defined and $\underline{X}_1$ or $\underline{X}_2$ can switch roles without any change in the test.

When $p_1 = 1$ or $p_2 = 1$ we have the usual Multiple Regression model and the statistic Λ may be transformed to an F statistic, with

$$\frac{n-p_2-1}{p_2}\,\frac{1-\Lambda}{\Lambda} \sim F_{p_2,n-p_2-1}$$

if $p_1 = 1$, or

$$\frac{n-p_1-1}{p_1}\,\frac{1-\Lambda}{\Lambda} \sim F_{p_1,n-p_1-1}$$

if $p_2 = 1$, which are the usual F statistics used to test the fit of the Multiple Regression model.

Of course, in case we have $p_1 = p_2 = 1$ we then have the well-known Simple Regression model, with

$$(n-2)\frac{1-\Lambda}{\Lambda} \sim F_{1,n-2}\,.$$

Testing the fit of the Multivariate Regression model is indeed the same as testing the nullity of the model parameter matrix, so that it is possible to establish a relationship between the test in the present subsection and the test in Sect. 5.1.4. Some further comments on this relation and on the issue of using centered or non-centered data matrices may be found in Sect. 5.1.9.3.

5.1.9.1 Equivalence Between the Test of Independence of $\underline{X}_1$ and $\underline{X}_2$ and the Test of Nullity of the Canonical Correlations

It is possible to prove that to test H_0 in (5.140) or (5.141) is equivalent to testing the simultaneous nullity of the so-called canonical correlations between the random vectors $\underline{X}_1$ and $\underline{X}_2$. In fact there is a matrix $U = bdiag(U_1', U_2')$ such that

$$\Sigma^* = U\Sigma U' = \begin{bmatrix} I_k & R \\ R & I_k \end{bmatrix}$$

where $k = \min(p_1, p_2)$ and

$$R = diag(\rho_1, \ldots, \rho_k)\,, \quad k = \min(p_1, p_2)$$

where $\rho_1 \geq \rho_2 \geq \cdots \geq \rho_k$ are the population canonical correlations between $\underline{X}_1$ and $\underline{X}_2$ (see Muirhead (2005, Sect. 11.3.2)).

Then, to test H_0 in (5.140) or (5.141) is equivalent to testing

$$H_0 : R = 0 \iff H_0 : \rho_1 = \cdots = \rho_k = 0\,, \tag{5.154}$$

that is, to test the independence of $\underline{X}_1$ and $\underline{X}_2$ is equivalent to test the simultaneous nullity of all population canonical correlations.

Matrices U_1 and U_2 are respectively defined as the matrices of eigenvectors of $\Sigma_{11}^{-1}\Sigma_{12}\Sigma_{22}^{-1}\Sigma_{21}$ and $\Sigma_{22}^{-1}\Sigma_{21}\Sigma_{11}^{-1}\Sigma_{12}$ associated with the non-null eigenvalues of these two matrices. More precisely, the matrix

$$U_1 = \left[\, \underline{u}_{11} \,|\, \underline{u}_{12} \,|\, \ldots \,|\, \underline{u}_{1k} \,\right]$$

is the matrix with column vectors $\underline{u}_{1\alpha}$ $(\alpha = 1, \ldots, k)$ which are the eigenvectors of $\Sigma_{11}^{-1}\Sigma_{12}\Sigma_{22}^{-1}\Sigma_{21}$, associated with the eigenvalues or latent roots ρ_α^2 $(\alpha = 1, \ldots, k)$, which are the population canonical correlations between $\underline{X}_1$ and $\underline{X}_2$ and are normalized to yield $||\underline{u}_{1\alpha}||^2_{\Sigma_{11}} = \underline{u}_{1\alpha}'\Sigma_{11}\underline{u}_{1\alpha} = 1$ (where $||\underline{u}_{1\alpha}||^2_{\Sigma_{11}}$ represents the square norm of $\underline{u}_{1\alpha}$ with respect to the metric Σ_{11}), and matrix

$$U_2 = \left[\, \underline{u}_{21} \,|\, \underline{u}_{22} \,|\, \ldots \,|\, \underline{u}_{2k} \,\right]$$

is the matrix with column vectors $\underline{u}_{2\alpha}$ $(\alpha = 1, \ldots, k)$ which are the eigenvectors of $\Sigma_{22}^{-1}\Sigma_{21}\Sigma_{11}^{-1}\Sigma_{12}$, associated with the non-null eigenvalues ρ_α^2 $(\alpha = 1, \ldots, k)$ and are normalized to yield $||\underline{u}_{2\alpha}||^2_{\Sigma_{22}} = \underline{u}_{2\alpha}'\Sigma_{22}\underline{u}_{2\alpha} = 1$. This implies that

$$U_1'\Sigma_{11}U_1 = I_k\,, \quad U_2'\Sigma_{22}U_2 = I_k \quad \text{and} \quad U_1'\Sigma_{12}U_2 = R\,.$$

For details see Appendix 9.

We may see that while Σ is the variance-covariance matrix of the vector $\underline{X} = [\underline{X}_1', \underline{X}_2']'$, with

$$\Sigma = Var\left(\begin{bmatrix}\underline{X}_1\\ \underline{X}_2\end{bmatrix}\right) = \begin{bmatrix}\Sigma_{11} & \Sigma_{12}\\ \Sigma_{21} & \Sigma_{22}\end{bmatrix},$$

where

$$Var(\underline{X}_1) = \Sigma_{11}\,, \quad Var(\underline{X}_2) = \Sigma_{22} \quad \text{and} \quad Cov(\underline{X}_1, \underline{X}_2) = \Sigma_{12}\,,$$

Σ^* is the variance-covariance matrix of the vector $\underline{Y} = [\underline{Y}_1', \underline{Y}_2']'$, where $\underline{Y}_1 = U_1'\underline{X}_1$ and $\underline{Y}_2 = U_2'\underline{X}_2$, with

$$\Sigma^* = Var\left(\begin{bmatrix}\underline{Y}_1\\ \underline{Y}_2\end{bmatrix}\right) = \begin{bmatrix}I_k & R\\ R & I_k\end{bmatrix}, \tag{5.155}$$

where

$$\begin{aligned} Var(\underline{Y}_1) &= U_1'\,Var(\underline{X}_1)U_1 = U_1'\Sigma_{11}U_1 = I_k\,,\\ Var(\underline{Y}_2) &= U_2'\,Var(\underline{X}_2)U_2 = U_2'\Sigma_{22}U_2 = I_k \end{aligned}$$

and

$$Cov(\underline{Y}_1, \underline{Y}_2) = U_1'\,Cov(\underline{X}_1, \underline{X}_2)U_2 = U_1'\Sigma_{12}U_2 = R\,. \tag{5.156}$$

The vectors $\underline{Y}_1$ and $\underline{Y}_2$ are the vectors of canonical variables, with

$$\underline{Y}_1 = \left[Y_{11}, \ldots, Y_{1k}\right]' \quad \text{and} \quad \underline{Y}_2 = \left[Y_{21}, \ldots, Y_{2k}\right]',$$

where from (5.155) and (5.156) it is clear that each pair $(Y_{1\alpha}, Y_{2,\alpha})$ has correlation equal to ρ_α, with the first pair (Y_{11}, Y_{21}) maximizing $Corr(Y_{11}, Y_{21}) = Cov(Y_{11}, Y_{21}) = \rho_1$, the second pair (Y_{12}, Y_{22}) maximizing $Corr(Y_{12}, Y_{22}) = Cov(Y_{12}, Y_{22}) = \rho_2$, subject to the conditions

$$Cov(Y_{11}, Y_{12}) = 0 \quad \text{and} \quad Cov(Y_{21}, Y_{22}) = 0\,,$$

the third pair (Y_{13}, Y_{23}) maximizing $Corr(Y_{13}, Y_{23}) = Cov(Y_{13}, Y_{23}) = \rho_3$, subject to the conditions

$$Cov(Y_{11}, Y_{12}) = Cov(Y_{11}, Y_{13}) = Cov(Y_{12}, Y_{13}) = 0$$

and

$$Cov(Y_{21}, Y_{22}) = Cov(Y_{21}, Y_{23}) = Cov(Y_{22}, Y_{23}) = 0\,,$$

and so on, for $\alpha = 1, \ldots, \min(p_1, p_2)$, so that we may indeed write

$$Cov(\underline{Y}_1, \underline{Y}_2) = Corr(\underline{Y}_1, \underline{Y}_2) = R\,,$$

which helps clarify why $\rho_1, \ldots, \rho_k$ $(k = \min(p_1, p_2))$ are called the canonical correlations between $\underline{X}_1$ and $\underline{X}_2$.

As Muirhead (2005, Sect. 11.3.2) remarks, Σ^* is called the canonical form of Σ for the group of transformations $\Sigma \longrightarrow C\Sigma C'$ with $C = bdiag(C_1, C_2)$ where C_1 and C_2 are full-rank matrices.

We may also see that we have

$$|\Sigma^*| = |I_k|\,|I_k - R^{*\prime} I_k^{-1} R| = \prod_{\alpha=1}^{k} (1 - \rho_\alpha^2)\,.$$

There is a parallel relation between the statistic Λ and the sample canonical correlations. These are defined as the sample counterparts of the population canonical correlations ρ_α, that is, as the quantities $\widehat{\rho}_\alpha$, which are the positive square roots of the non-null $k = \min(p_1, p_2)$ eigenvalues of either one of the two matrices

$$A_{11}^{-1/2} A_{12} A_{22}^{-1} A_{21} A_{11}^{-1/2} \quad \text{or} \quad A_{22}^{-1/2} A_{21} A_{11}^{-1} A_{12} A_{22}^{-1/2} \tag{5.157}$$

which are the same eigenvalues as those of the matrices

$$A_{11}^{-1} A_{12} A_{22}^{-1} A_{21} \quad \text{and} \quad A_{22}^{-1} A_{21} A_{11}^{-1} A_{12}\,, \tag{5.158}$$

since if for example $\underline{\hat{u}}_{1\alpha}^*$ are the unitary eigenvectors of the first matrix in (5.157), associated with the eigenvalues $\widehat{\rho}_\alpha^2$, then $\underline{\hat{u}}_{1\alpha} = A_{11}^{-1/2} \underline{\hat{u}}_{1\alpha}^*$ will be the eigenvectors of the first matrix in (5.158), associated with the same eigenvalues, which may be easily verified by left multiplying

$$A_{11}^{-1/2} A_{12} A_{22}^{-1} A_{21} A_{11}^{-1/2} \underline{\hat{u}}_{1\alpha}^* = \widehat{\rho}_\alpha^2 \underline{\hat{u}}_{1\alpha}^*$$

by $A_{11}^{-1/2}$. The existence of the matrices $A_{11}^{-1/2}$ and $A_{22}^{-1/2}$ is assured by the fact that both A_{11} and A_{22} are symmetric positive-definite matrices.

But then from (5.147) we may write

$$\Lambda = |A_{11}|^{-1}\,|A_{11} - A_{12} A_{22}^{-1} A_{21}| = |I_{p_1} - A_{11}^{-1} A_{12} A_{22}^{-1} A_{21}| = \prod_{\alpha=1}^{k} (1 - \widehat{\rho}_\alpha^2)$$

while from (5.148) we may write a similar expression by switching the indexes 1 and 2, or instead, from (5.145) write

$$\begin{aligned}\Lambda &= \left|\begin{bmatrix} A_{11} & A_{12} \\ A_{21} & A_{22} \end{bmatrix}\begin{bmatrix} A_{11}^{-1} & 0 \\ 0 & A_{22}^{-1} \end{bmatrix}\right| = \begin{vmatrix} I_{p_1} & A_{12}A_{22}^{-1} \\ A_{21}A_{11}^{-1} & I_{p_2} \end{vmatrix} \\ &= |I_{p_1}|\,|I_{p_2} - A_{21}A_{11}^{-1}I_{p_1}^{-1}A_{12}A_{22}^{-1}| = \prod_{\alpha=1}^{k}(1-\widehat{\rho}_\alpha^2)\,.\end{aligned}$$

We may note that the set of ordered eigenvalues (from larger to smaller) of the matrix

$$\left[\begin{array}{c|c} I_{p_1} & A_{12}A_{22}^{-1} \\ \hline A_{21}A_{11}^{-1} & I_{p_2} \end{array}\right]$$

are the values

$$\lambda_j = \begin{cases} 1+\widehat{\rho}_j\,, & j=1,\ldots,\min(p_1,p_2) \\ 1\,, & j=1+\min(p_1,p_2),\ldots,\max(p_1,p_2) \\ 1-\widehat{\rho}_{p_1+p_2-j+1}\,, & j=1+\max(p_1,p_2),\ldots,p_1+p_2 \end{cases} \tag{5.159}$$

yielding once again

$$\Lambda = \prod_{j=1}^{p_1+p_2}\lambda_j = \prod_{\alpha=1}^{k}(1-\widehat{\rho}_\alpha^2)\,.$$

We may also notice the interesting parallel between the definition of the eigenvalues λ_j in (5.159) and that of the parameters h_j in (5.153), which may alternatively be written as

$$h_j = \begin{cases} 1, & j=1,\ldots,\min(p_1,p_2) \\ 0\,, & j=1+\min(p_1,p_2),\ldots,\max(p_1,p_2) \\ -1\,, & j=1+\max(p_1,p_2),\ldots,p_1+p_2-2\,. \end{cases}$$

The equivalence of the test of the null hypotheses in (5.140) or (5.141) and the test of the null hypothesis in (5.154) is the reason why the test of independence of two sets of variables is commonly known as the test of fit of the Canonical Analysis or Canonical Correlation Analysis model, which may be seen as an all-embracing linear model since as Kshirsagar (1972, Sect. 7.8) states "Most of the practical problems arising in statistics can be translated, in some form or the other, as the problems of measurement of association between two vector variates $\underline{x}$ and

$\underline{y}$," and also as Knapp (1978) remarks, "virtually all of the commonly encountered parametric tests of significance can be treated as special cases of canonical-correlation analysis, which is the general procedure for investigating the relationship between two sets of variables," where we would only remark that "parametric tests" should refer to those parametric tests that can be translated in terms of a linear model. This is the reason why we will dedicate detailed attention to the present test.

5.1.9.2 On the Availability of Computer Modules

For the present test, that is for the test of the hypothesis in (5.140), (5.141), or even (5.154), computational modules are made available to compute the p.d.f., the c.d.f., quantiles and p-values for the statistic Λ in (5.145), (5.147) or (5.148). These are the modules `PDFInd2R`, `CDFInd2R`, `PvalInd2R`, `QuantInd2R` and `PvalDataInd2R`. The first four of these modules have four mandatory arguments. These are: a first argument which is the sample size, a second and a third arguments which are respectively the number of variables in $\underline{X}_1$ and $\underline{X}_2$, that is, p_1 and p_2, and a fourth argument which is for the first three modules the computed value of the statistic and for the fourth module the α value for the computation of the α-quantile. These four modules have exactly the same optional arguments as the corresponding modules in the previous subsections, namely the ones in Sect. 5.1.1.

The plots in Figs. 5.36 and 5.37 were obtained with the modules `PDFInd2R` and `CDFInd2R` and they show how the distributions, for a given difference between the sample size n and the global number of variables involved p, get more concentrated around their mean value as the value of p grows large, and also how the distributions get much and much more squeezed towards zero for small differences between n and p as p grows large. These are the reasons, mainly this latter one, why common asymptotic distributions for l.r.t. statistics in Multivariate Analysis have so much trouble dealing with increasing numbers of variables.

In what concerns module `PvalDataInd2R`, it has as first arguments the names of the two data files where the data matrices for the samples of $\underline{X}_1$ and $\underline{X}_2$ are stored, and where the second argument is left empty in case there is only one data file with the combined sample from $\underline{X}_1$ and $\underline{X}_2$. This module also has seven optional arguments, which, as in all other modules, follow these two "mandatory" arguments. The first of these optional arguments is an argument that should be left empty or given the value 0 (zero) in case there are two data files, one with the sample from $\underline{X}_1$ and another with the sample from $\underline{X}_2$, or to which should be given the value of p_1, the number of variables in $\underline{X}_1$, in case there is one only data file with the combined sample of $\underline{X}_1$ and $\underline{X}_2$ (default value: 0). The next four optional arguments are exactly the same as the optional arguments of module `PvalDataEqMeanVecR` in Sect. 5.1.1 and the last two are (in this order): `transp1`—an argument that has to be given the value 1 in case the data matrix with the sample of $\underline{X}_1$ or the data matrix with the combined sample of $\underline{X}_1$ and $\underline{X}_2$ is given in the form of a $p_1{\times}n$ or a $p{\times}n$ matrix (default value: 0); `transp2`—an argument that has to be given the value 1 in case the data matrix with the sample of $\underline{X}_2$ is given in the form of a $p_2{\times}n$

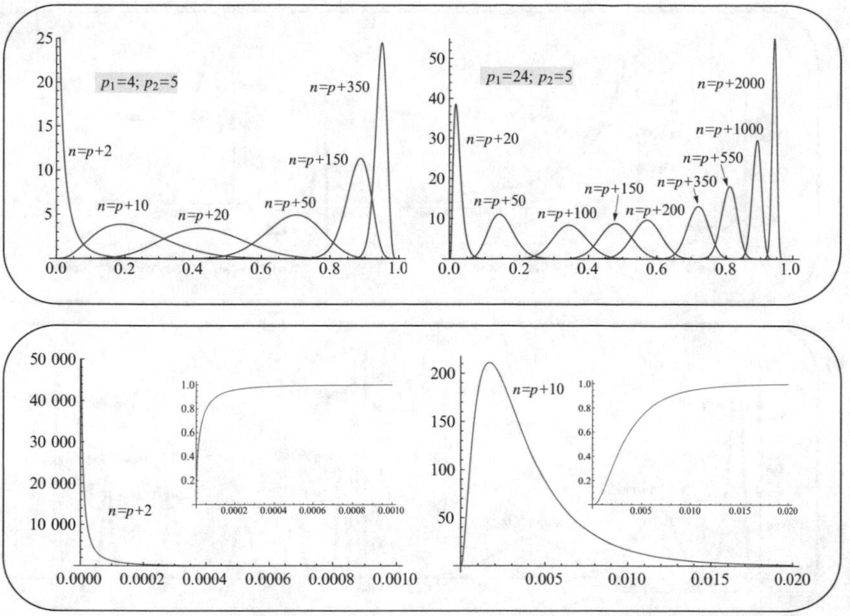

Fig. 5.36 Plots of p.d.f.'s of Λ in (5.145) for different values of n, p_1 and p_2, with $p = p_1 + p_2$: (**a**) p.d.f.'s for $p_1 = 4$, $p_2 = 5$ and $p_1 = 24$, $p_2 = 5$; (**b**) p.d.f.'s for $p_1 = 24$, $p_2 = 5$, with $n = p+2$ and $n = p+10$, with the corresponding c.d.f.'s in the insets. Note that the p.d.f. plots in (**a**) have different vertical scales and that the horizontal and vertical scales of the p.d.f. plots in (**b**) are also different (with the corresponding p.d.f. and c.d.f. plots having the same horizontal scales)

matrix (default value: 0). This latter argument should be left with its default value of 0 (zero) in case the data matrix with the sample of $\underline{X}_2$ is given in the form of an $n\times p_2$ matrix or in case there is one only data matrix with the combined sample of $\underline{X}_1$ and $\underline{X}_2$.

Although in our formulation and algebraic expressions the matrices X_1 and X_2 are denoted as $p_1\times n$ and $p_2\times n$ matrices, the default position assumed for these matrices in the data files, by the module `PvalDataInd2R` is in its transposed form, as $n\times p_1$ and $n\times p_2$ matrices. This is justified not only by the fact that since we need to have $n > p_1 + p_2$, this transposed position is usually seen as a more convenient way to store and display the data, but also because, as it will be seen in Sect. 5.1.9.2, this test may be seen also as a test to the equality of mean vectors, that is, the test addressed in Sects. 5.1.1 and 5.1.4.2, where the data matrix used in the `PvalData` modules was taken in the form of an $n\times p$ matrix. Anyway, there is the possibility of using data files with the data matrices in the form of $p_1\times n$ and $p_2\times n$ matrices, making then use of the optional arguments named `transp1` and `transp2` with a value of 1, to indicate that the data matrices are given in this form.

The Mathematica® version of module `PvalDataInd2R` has a further optional argument `norm` which has to be called by its name and which if given the value

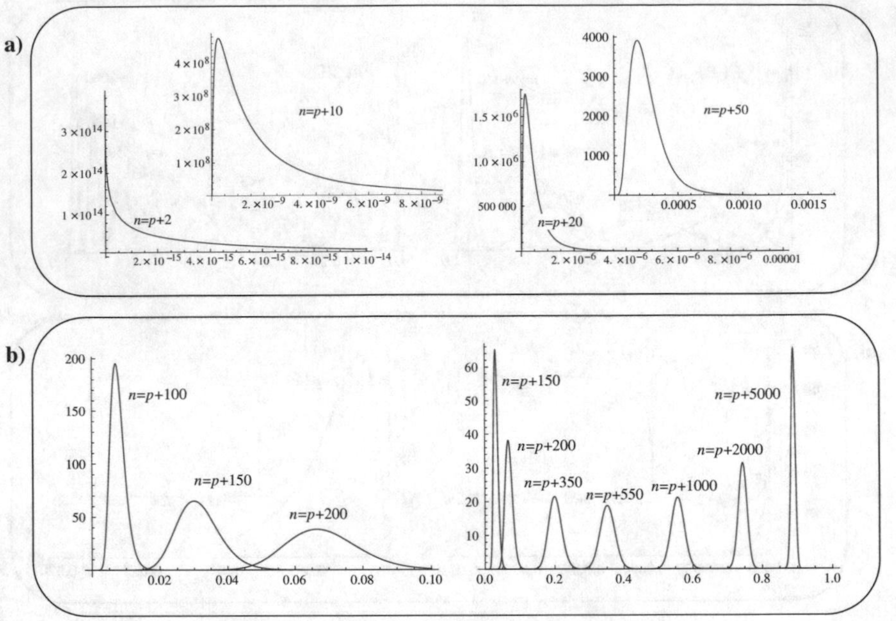

Fig. 5.37 Plots of p.d.f.'s of Λ in (5.145) for $p_1 = 24$ and $p_2 = 25$, with $p = p_1 + p_2$: (**a**) p.d.f.'s for the smaller values of n; (**b**) p.d.f.'s for the intermediate and larger values of n. Note that the plots have different vertical and horizontal scales

1 will trigger the implementation of the multivariate normality tests available in Mathematica®, for the overall vector $\underline{X}$ as well as for the subvectors $\underline{X}_1$ and $\underline{X}_2$.

We use the data in Table 9.12 of Johnson and Wichern (2014) to illustrate how one may carry out a test of independence between the set of three sales performance variables (sales growth, sales profitability, and new-account sales) and the four normalized test-scores intended to measure creativity, mechanical and abstract reasoning, and mathematical ability.

In Fig. 5.38 are shown the commands used and the output obtained when the Mathematica® version of module `PvalDataInd2R` is used. The data matrix stored in the data file `Tab9_12_JW.dat` reproduces exactly Table 9.12 in Johnson and Wichern (2014) and, as such, when reading the file, we will have to select the variables in columns 2 through 8 as corresponding to the variables to be used in the analysis, given that the first column has the observation indexes. Since we are using a single data file with the combined sample of $\underline{X}_1$ and $\underline{X}_2$, in Fig. 5.38 is also illustrated the use of the first optional argument of module `PvalDataInd2R`, to which a value of 3 is given, since we have $p_1 = 3$ variables in $\underline{X}_1$, the set of the sales performance variables.

We note that the optional argument `norm`, which is only available in the Mathematica® version of `PvalDataInd2R` and which has to be called by its name, has to appear after all other optional arguments, as illustrated.

```
PvalDataInd2R[path <> "Tab9_12_JW.dat", , 3, 1, , , , , , norm → 1]
 Original variables {2, 3, 4, 5, 6, 7, 8}
```

Combined sample of X1 and X2

	Statistic	P-Value
Anderson-Darling	4.33759	0.860357
Cramér-von Mises	4.13182	0.791936
Jarque-Bera ALM	2.93891	0.235331
Kolmogorov-Smirnov	3.4748	0.487124
Kuiper	1.97375	0.0220006
Mardia Combined	142.927	0.0000858494
Mardia Kurtosis	1.64619	0.0997256
Mardia Skewness	130.335	0.000116997
Pearson χ^2	2.08851	0.0318751
Shapiro-Wilk	0.824382	3.355×10^{-6}
Watson U^2	2.57334	0.115207

Sample of X1

	Statistic	P-Value
Anderson-Darling	1.90617	0.782289
Cramér-von Mises	1.92721	0.794419
Jarque-Bera ALM	0.951045	0.143368
Kolmogorov-Smirnov	1.94655	0.805233
Kuiper	1.16473	0.261111
Mardia Combined	22.1217	0.0495809
Mardia Kurtosis	−0.172187	0.417608
Mardia Skewness	20.2286	0.0489301
Pearson χ^2	1.37054	0.40363
Shapiro-Wilk	0.945336	0.0220207
Watson U^2	1.28346	0.340978

Sample of X2

	Statistic	P-Value
Anderson-Darling	2.43143	0.765182
Cramér-von Mises	2.20461	0.633769
Jarque-Bera ALM	1.98787	0.491911
Kolmogorov-Smirnov	1.52824	0.214302
Kuiper	0.80902	0.0178495
Mardia Combined	9.71903	0.0393178
Mardia Kurtosis	−1.48967	0.506244
Mardia Skewness	6.90891	0.0400908
Pearson χ^2	0.717972	0.0110718
Shapiro-Wilk	0.977853	0.465383
Watson U^2	1.28988	0.114164

```
Computed value of Λ: 0.002148472296
p-value: 1.254086916×10^-51

PvalDataInd2R[path <> "Tab9_12_JW.dat", , 3, 1]
 Original variables {2, 3, 4, 5, 6}
Computed value of Λ: 0.2875810886
p-value: 1.622881606×10^-10
```

Fig. 5.38 Command used and output obtained with the Mathematica® module `PvalDataInd2R` to test the independence between the set of sales performance variables and the normalized test-scores for creativity, mechanical and abstract reasoning, and mathematical ability considered in Problem 9.19 in Johnson and Wichern (2014) and which data is displayed in Table 9.12 of this same reference

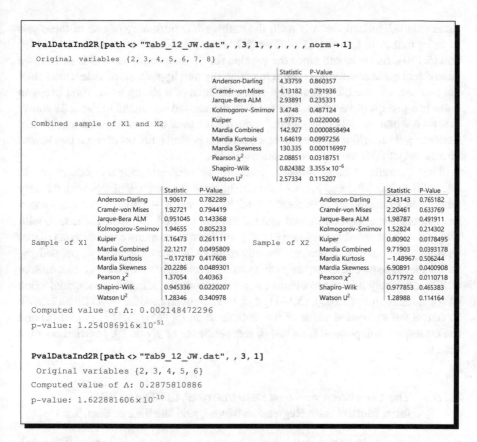

When looking at the first command and its result in Fig. 5.38 we see an extremely low p-value for the test of the null hypothesis of independence of the set of the three sales performance variables and the set of the four test scores, showing good evidence of a strong association between these two groups of variables. Although when looking at the p-values for the multivariate normality tests applied to the whole set of seven variables, which from the assumptions stated at the beginning of Sect. 5.1.9 seems to be the appropriate test to carry out, one finds some very low p-values, we think there is still no enough evidence to be concerned about the multivariate normality assumption, since when these tests are carried out separately for each one of the two sets of variables most tests give rather large p-values, and as it is more thoroughly discussed in Sect. 5.1.9.3 on page 179 and in Sect. 5.1.9.4, on page 185, right before the heading "Other choices for the matrix *M* which lead to the same results," indeed we only need the multivariate normality of one of these sets in order that Λ in (5.145), (5.147) or (5.148) has the distribution stated in (5.149) and (5.150). As such, and since the p-value for the independence test is so low, we should feel quite comfortable about rejecting the null hypothesis of independence of the two sets of variables, concluding the existence of a strong association between these two groups of variables. The result of the second command in Fig. 5.38 shows that even when we only consider the score for the tests of creativity and mechanical reasoning, there still is a very strong association between the set of these two scores and the set of three sales performance variables.

The quantile tables already referred to and thoroughly described in Sects. 5.1.1.2, 5.1.3.1 and 5.1.4.7 may also be used for the present test, with the only adaptation of replacing p in Sect. 5.1.1.2 by p_1 and $q-1$ by p_2, but once again the authors strongly recommend instead the use of module `QuantInd2R`, with which it will be possible to obtain exact quantiles with as many exact decimal places as it may be required or desired, for any α-value and any values of n, p_1, and p_2, as long as at least one of p_1 or p_2 is even. Alternatively modules `PvalInd2R` or `CDFInd2R` may be used to obtain the p-value corresponding to a computed value of the statistic Λ in (5.145), (5.147) or (5.148), or even module `PvalDataInd2R` to obtain the computed value of the statistic Λ in (5.145), (5.147) or (5.148) and the corresponding p-value for a test of independence of $\underline{X}_1$ and $\underline{X}_2$ carried out on a data set.

5.1.9.3 The Use of Non-centered Data Matrices, the Test of Fit for a Multivariate Regression Model, and the Test in Sect. 5.1.4

The statistic Λ in (5.145) is defined with the matrix A given by (5.142), that is, using a centered data matrix X, which implies that the matrices A_{11}, A_{22}, and A_{12} are also defined with the use of centered data matrices.

However, there are situations in which Λ in (5.147) or (5.148), defined with one of the data matrices or even both data matrices not centered still is a l.r.t. statistic. These situations are addressed next.

An Equivalent Way to Carry Out the Independence Test

Indeed by adding to one of the data sub-matrices X_1 or X_2 a row of 1's and then not centering this matrix, we may obtain an exactly equivalent test of independence, as long as we then correct for the degrees of freedom of the Wishart matrices involved. However, we cannot do this simultaneously to both data sub-matrices since then we would end up with an overall data matrix X with two equal rows of 1's, which would make the matrix A in (5.142) to have two equal rows and two equal columns, and consequently it would only be positive semi-definite, with one null eigenvalue and as such would have a null determinant. Without any loss of generality, let X_2 be the matrix to which we add a row of 1's and which we will not center. Again, without any loss of generality let this row be the first row of the new X_2^+ matrix. That is, let

$$X_2^+ = \left[\begin{array}{c} E_{1,n} \\ \hline X_2 \end{array}\right]\begin{array}{l} 1 \\ p_2 \end{array} \tag{5.160}$$

and then let A_{11} be defined as in (5.143), but let now

$$A_{22} = X_2^+ X_2^{+\prime}, \quad \text{and} \quad A_{12} = X_1\big(I_n - \tfrac{1}{n}E_{nn}\big)X_2^{+\prime}.$$

If we take the statistic Λ defined as in (5.147) or (5.148), with this new definition of the matrices A_{22} and A_{12}, then

$$\begin{aligned} A_{11.2} &= A_{11} - A_{12}A_{22}^{-1}A_{21} \\ &= X_1\Big(\big(I_n - \tfrac{1}{n}E_{nn}\big) - \underbrace{\big(I_n - \tfrac{1}{n}E_{nn}\big)X_2^{+\prime}(X_2^+X_2^{+\prime})^{-1}X_2^+\big(I_n - \tfrac{1}{n}E_{nn}\big)}_{H_1}\Big)X_1', \end{aligned} \tag{5.161}$$

and Λ will be the very same statistic as the one we would obtain by defining A_{22} and A_{12} as in (5.143) and (5.144).

In order to show this we will have to show that the matrix $A_{11.2}$ in (5.161) and the matrix $A_{11.2}$ in (5.147), which is, for A_{11}, A_{22}, and A_{12} in (5.143) and (5.144),

$$\begin{aligned} A_{11.2} &= A_{11} - A_{12}A_{22}^{-1}A_{21} \\ &= X_1\Big(\big(I_n - \tfrac{1}{n}E_{nn}\big) - \underbrace{\big(I_n - \tfrac{1}{n}E_{nn}\big)X_2'\big(X_2\big(I_n - \tfrac{1}{n}E_{nn}\big)X_2'\big)^{-1}X_2\big(I_n - \tfrac{1}{n}E_{nn}\big)}_{H_2}\Big)X_1' \end{aligned} \tag{5.162}$$

are the same matrix. This is, in turn, equivalent to showing that the matrices H_1 in (5.161) and H_2 in (5.162) are the same matrix. This may be easily done by using Result 2 on inverse matrices in Appendix 6.

From the definition of X_2^+ in (5.160) we have

$$X_2^+X_2^{+\prime} = \left[\begin{array}{c|c} n & E_{1n}X_2' \\ \hline X_2E_{n1} & X_2X_2' \end{array}\right]$$

and

$$\begin{aligned} \big(I_n - \tfrac{1}{n}E_{nn}\big)X_2^{+\prime} &= \big(I_n - \tfrac{1}{n}E_{nn}\big)\Big[\,E_{n1}\,\Big|\,X_2^{+\prime}\Big] \\ &= \Big[\,0_{n\times 1}\,\Big|\,\big(I_n - \tfrac{1}{n}E_{nn}\big)X_2'\Big], \end{aligned}$$

so that, using (5.505)–(5.507) in Appendix 6, we may write

$$(X_2^+ X_2^{+\prime})^{-1} = \left[\begin{array}{c|c} * & (**)' \\ \hline (**) & *** \end{array}\right],$$

where we are indeed only interested in the matrix

$$*** = (X_2 X_2' - X_2 E_{n1} \tfrac{1}{n} E_{1n} X_2')^{-1} = \left(X_2\left(I_n - \tfrac{1}{n} E_{nn}\right) X_2'\right)^{-1}.$$

But this shows the equality between H_1 in (5.161) and H_2 in (5.162) since we thus have

$$\begin{aligned}
&\left(I_n - \tfrac{1}{n}E_{nn}\right) X_2^{+\prime} (X_2^+ X_2^{+\prime})^{-1} X_2^+ \left(I_n - \tfrac{1}{n}E_{nn}\right) \\
&= \left[\, 0_{n\times 1} \,\middle|\, \left(I_n - \tfrac{1}{n}E_{nn}\right) X_2' \right] \left[\begin{array}{c|c} * & (**)' \\ \hline (**) & \left(X_2\left(I_n - \tfrac{1}{n}E_{nn}\right)X_2'\right)^{-1} \end{array}\right] \left[\begin{array}{c} 0_{1\times n} \\ \hline X_2\left(I_n - \tfrac{1}{n}E_{nn}\right) \end{array}\right] \\
&= \left[\left(I_n - \tfrac{1}{n}E_{nn}\right) X_2' (**) \,\middle|\, \left(I_n - \tfrac{1}{n}E_{nn}\right) X_2' \left(X_2\left(I_n - \tfrac{1}{n}E_{nn}\right)X_2'\right)^{-1}\right] \left[\begin{array}{c} 0_{1\times n} \\ \hline X_2\left(I_n - \tfrac{1}{n}E_{nn}\right) \end{array}\right] \\
&= \left(I_n - \tfrac{1}{n}E_{nn}\right) X_2' \left(X_2\left(I_n - \tfrac{1}{n}E_{nn}\right)X_2'\right)^{-1} X_2 \left(I_n - \tfrac{1}{n}E_{nn}\right).
\end{aligned}$$

This shows that the two approaches of either using X_2 centered, say, as usual, or having X_2 with an extra line of 1's and not centering it, give exactly the same result, as long as, for the parameters of the distribution of Λ we take in this latter case p_2 as the number of rows of X_2^+ minus 1.

We should note that while the approach of using the matrix X_2 centered may be viewed as corresponding to a Multivariate Regression through the origin, the approach of using the matrix X_2^+ corresponds to a model with an intercept parameter which corresponds to the row of 1's. If, without any loss of generality, we see the set of variables in $\underline{X}_1$ as the response variables and the set of variables in $\underline{X}_2$ as the explanatory variables, then Λ is the statistic for the test of the fit of the model

$$E(X_1^* | X_2^+) = \mu X_2^+ \tag{5.163}$$

where $X_1^* = X_1\left(I_n - \frac{1}{n}E_{nn}\right)$ here represents the centered sample matrix of $\underline{X}_1$, that is, a matrix with a $N_{p_1\times n}(\mu X_2^+, I_n \otimes \Sigma)$ multivariate Normal distribution, and where X_2^+ is the matrix in (5.160) and μ is a real $p_1 \times p_2$ matrix regarding which we are indeed conducting the test $H_0 : \mu = 0$.

Instead of looking at our model, whose "fit" we are testing, as the model in (5.163), we may indeed look at it as being the model

$$E(X_2^+ | X_1^*) = \mu' X_1^*$$

where we conduct the test of the same null hypothesis $H_0 : \mu = 0$. This shows that we may look at any of the vectors $\underline{X}_1$ or $\underline{X}_2$ as being the vector of response variables, while the other is the vector of explanatory variables. This actually also shows that we only need the multivariate normality of one of the sets, while the other may be seen as formed by nonrandom variables, or by r.v.'s with any distribution. See a note on this topic inserted in the next subsection, on page 185, right before the heading "Other choices for the matrix M which lead to the same results."

Module `PvalDataInd2R` detects if a data matrix has a first column of 1's and in this case it will not center this matrix. In case both data submatrices X_1 and X_2 have a first column of 1's module `PvalDataInd2R` will issue a warning and cease the computations, since this would cause the overall data matrix X to have two similar columns, which would make the matrix A to have a null determinant because of the existence of one null eigenvalue, indicating that between $\underline{X}_1$ and $\underline{X}_2$ there exists one coincident variable, or, if we want to think of the spaces spanned by the variables in $\underline{X}_1$ and $\underline{X}_2$, the existence of a coincident axis or "dimension," between these two spaces, which would directly entail the rejection of the null hypothesis of independence of $\underline{X}_1$ and $\underline{X}_2$, giving a null computed value for Λ.

However, there are cases when we may have both data matrices X_1 and X_2 not centered, as we will see next.

Implementing the Test in Sect. 5.1.4 Through a Test of Independence

In fact if we take the submatrix X_1 as the matrix X in Sect. 5.1.4 and the matrix X_2 as the matrix M, also in Sect. 5.1.4, and we do not center them, then the statistic Λ in (5.147), for

$$A_{11} = \tfrac{1}{n}XX' , \quad A_{22} = \tfrac{1}{n}MM' , \text{ and } A_{12} = \tfrac{1}{n}XM' ,$$

will be

$$\Lambda = \frac{\left|\frac{1}{n}XX' - \frac{1}{n}XM'(MM')^{-1}MX'\right|}{\left|\frac{1}{n}XX'\right|} ,$$

which is exactly the same as the statistic Λ in (5.34), used to test H_0 in (5.33), for $\beta^* = 0$.

5.1.9.4 The Test of Equality of Mean Vectors as a Test of Independence: The MANOVA Model

In this sub-subsection we will present the test of equality of mean vectors, already addressed in Sects. 5.1.1 and 5.1.4.2 as a test of independence between the random vector $\underline{X} \sim N_p(\underline{\mu}, \Sigma)$, from which we have an overall random sample of size $n = \sum_{k=1}^{q} n_k$, originating from combining q independent samples, the k-th of which with size n_k $(k = 1, \ldots, q)$, and a nonrandom vector, say $\underline{M}$, which is the vector of subpopulation indicator variables, the k-th of which will have the value 1 or 0 (zero) according to the fact that the corresponding observation belongs or does not belong to the k-th population $(k = 1, \ldots, q)$. The random vector $\underline{X}$ here is like a "multiple" random vector which "incorporates" all q random vectors $\underline{X}_k \sim N_p(\underline{\mu}_k, \Sigma)$ $(k = 1, \ldots, q)$ in Sect. 5.1.1.

Let then X be the $p\times n$ matrix of the sample of size n from $\underline{X}$, that is, the matrix X in (5.67) in Sect. 5.1.4.2 and let M be the $q\times n$ matrix with the values of the q indicator variables, that is, the matrix M in (5.68) in Sect. 5.1.4.2.

Then, to test the null hypothesis of equality of the q mean vectors $\underline{\mu}_k$ will be the same as to test the hypothesis of independence of the vectors $\underline{X}$ and $\underline{M}$, where $\underline{M}$ is the set of q indicator variables for the q populations. We will prove this by showing that the l.r.t. statistic for these two tests is indeed the same.

With the present setup the m.l.e.'s A_{11}, A_{22}, and $A_{12} = A_{21}'$ are respectively given by

$$A_{11} = \tfrac{1}{n} X\big(I_n - \tfrac{1}{n} E_{nn}\big)X', \qquad A_{22} = \tfrac{1}{n} M\big(I_n - \tfrac{1}{n} E_{nn}\big)M' \tag{5.164}$$

and

$$A_{12} = \tfrac{1}{n} X\big(I_n - \tfrac{1}{n} E_{nn}\big)M', \tag{5.165}$$

so that the l.r.t. statistic would be given by (5.147) with $A^* = A_{11.2} = A_{11} - A_{12}A_{22}^{-1}A_{12}$.

However, it happens that the matrix A_{22} in (5.164) is not of full-rank and as such not invertible. Indeed, the q indicator variables in $\underline{M}$ form a "one more than necessary" set of indicator variables, since in all cases if we know the values of $q-1$ of them we will then know the value of the remaining indicator variable.

As such, we should rather consider the vector $\underline{M}^*$ formed by any $q-1$ of the q components of $\underline{M}$, and then the matrix M^*, obtained from the matrix M by deleting the row corresponding to the indicator variable that was taken away from $\underline{M}$ to obtain $\underline{M}^*$. Only in order to make things more precise and a bit easier to handle, both in terms of notation and in terms of simplicity, let us consider that $\underline{M}^*$ is the vector obtained from $\underline{M}$ by removing the first indicator variable and that, accordingly, M^* is the matrix obtained from M by removing its first row. If we denote this row by M_1, then we have

$$\underset{(q\times n)}{M} = \left[\begin{array}{c} M_1 \;\; (1\times n) \\ \hline M^* \;\; ((q-1)\times n) \end{array}\right]\begin{matrix} 1 \\ q-1 \end{matrix} = \left[\begin{array}{c|c} E_{1n_1} & 0 \;\; (1\times(n-n_1)) \\ \hline \multicolumn{2}{c}{M^* \;\; ((q-1)\times n)} \end{array}\right]\begin{matrix} 1 \\ q-1 \end{matrix}. \tag{5.166}$$

In this way we will reformulate the matrices A_{22} and A_{12} as

$$A_{22} = \frac{1}{n} M^*\big(I_n - \tfrac{1}{n}E_{nn}\big)M^{*\prime} \quad \text{and} \quad A_{12} = \frac{1}{n} X\big(I_n - \tfrac{1}{n}E_{nn}\big)M^{*\prime} \tag{5.167}$$

so that the inverse A_{22}^{-1} exists and $A^* = A_{11.2}$ may be given by $A_{11} - A_{12}A_{22}^{-1}A_{21}$.

Now in order to prove that the l.r.t. statistic for the test of independence of $\underline{X}$ and $\underline{M}^*$ is the same as the l.r.t. statistic in Sects. 5.1.1 and 5.1.4.2 we have to show that the matrix

$$\begin{aligned} A_{11.2} = A_{11} - A_{12}A_{22}^{-1}A_{21} &= \tfrac{1}{n}X\big(I_n - \tfrac{1}{n}E_{nn}\big)X' \\ &\quad -\tfrac{1}{n}X\big(I_n - \tfrac{1}{n}E_{nn}\big)M^{*\prime}\left(M^*\big(I_n - \tfrac{1}{n}E_{nn}\big)M^{*\prime}\right)^{-1}M^*\big(I_n - \tfrac{1}{n}E_{nn}\big)X' \\ &= \tfrac{1}{n}X\left(\big(I_n - \tfrac{1}{n}E_{nn}\big) - \big(I_n - \tfrac{1}{n}E_{nn}\big)M^{*\prime}\left(M^*\big(I_n - \tfrac{1}{n}E_{nn}\big)M^{*\prime}\right)^{-1}M^*\big(I_n - \tfrac{1}{n}E_{nn}\big)\right)X' \end{aligned} \tag{5.168}$$

is the same as the matrix A in (5.72), which is the same as the matrix A in (5.37), and which is equal to $1/n$ times the matrix A in (5.3), and that

$$A_{11.2} + A_{12}A_{22}^{-1}A_{21} = A_{11}$$

is the same as the sum $A + \frac{1}{n}B$ for A and B in Sect. 5.1.4.2, that is, for A given by (5.37) and B given by (5.73), with D given by (5.69).

To verify that for A in (5.37) and B given by (5.73) we have

$$A + \tfrac{1}{n}B = A_{11} = \tfrac{1}{n}X\big(I_n - \tfrac{1}{n}E_{nn}\big)X'$$

is not too hard but we will need to use some interesting features of the matrix D, shown in Appendix 4.

From (5.37) and (5.73) we may write

$$\begin{aligned} A + \tfrac{1}{n}B = &\underbrace{\tfrac{1}{n}X\left(I_n - M'(MM')^{-1}M\right)X'}_{A} \\ &\underbrace{+\tfrac{1}{n}XM'(MM')^{-1}\underbrace{DMM'D'}_{=DMM'}(MM')^{-1}MX'}_{B} \end{aligned} \tag{5.169}$$

for D given by (5.69).

Note that here we use M, and not M^*, since we are using the definitions of A and B in (5.37) and (5.73).

That $DMM'D' = DMM' = MM'D'$ is shown in Appendix 4. Using this fact we may write the matrix B in Sects. 5.1.4.2 and 5.1.1 and also in (5.169), as

$$B = XM'(MM')^{-1}DMM'(MM')^{-1}MX' = XM'(MM')^{-1}DMX'$$

where, for D in (5.69),

$$(MM')^{-1}D = (MM')^{-1} - \tfrac{1}{n}E_{qq}$$

so that

$$M'(MM')^{-1}DM = M'(MM')^{-1}M - \tfrac{1}{n}\underbrace{M'E_{qq}M}_{=E_{nn}}.$$

That $M'E_{qq}M = E_{nn}$ may be easily seen from the structure of M in (5.68), with $M'E_{qq} = E_{nq}$. But then we have

$$B = X\left(M'(MM')^{-1}M - \tfrac{1}{n}E_{nn}\right)X'$$

and hence

$$\begin{aligned} A + \tfrac{1}{n}B &= \tfrac{1}{n}X\left(I_n - M'(MM')^{-1}M + M(MM')^{-1}M - \tfrac{1}{n}E_{nn}\right)X' \\ &= \tfrac{1}{n}X\big(I_n - \tfrac{1}{n}E_{nn}\big)X' = A_{11}. \end{aligned}$$

However, to show now that $A_{11.2} = A$ for $A_{11.2}$ in (5.168) and A in (5.37) is a much harder task. Essentially, we need to show that

$$\begin{aligned} &I_n - M'(MM')^{-1}M = \\ &\quad \big(I_n - \tfrac{1}{n}E_{nn}\big) - \big(I_n - \tfrac{1}{n}E_{nn}\big)M^{*\prime}\left(M^*\big(I_n - \tfrac{1}{n}E_{nn}\big)M^{*\prime}\right)^{-1} M^*\big(I_n - \tfrac{1}{n}E_{nn}\big). \end{aligned} \tag{5.170}$$

Since $M'(MM')^{-1}M$ is the projector onto the space generated by the rows of M, which is the same as the space generated by the rows of M^*, we have

$$M'(MM')^{-1}MM^{*\prime} = M^{*\prime} \quad \text{and} \quad M'(MM')^{-1}ME_{nn}M^{*\prime} = E_{nn}M^{*\prime},$$

and as such we have

$$\big(I_n - \tfrac{1}{n}E_{nn}\big)M^{*\prime} = M'(MM')^{-1}M\big(I_n - \tfrac{1}{n}E_{nn}\big)M^{*\prime}$$

where, given the structure of M in (5.166), we may write

$$M\left(I_n-\tfrac{1}{n}E_{nn}\right)M^{*\prime}=\left[\begin{array}{c}\left[-\frac{n_1n_2}{n},\dots,-\frac{n_1n_q}{n}\right]\\ \hline M^*\left(I_n-\frac{1}{n}E_{nn}\right)M^{*\prime}\end{array}\right]\begin{array}{l}1\\ q-1\end{array}$$

so that

$$M\left(I_n-\tfrac{1}{n}E_{nn}\right)M^{*\prime}\left(M^*\left(I_n-\tfrac{1}{n}E_{nn}\right)M^{*\prime}\right)^{-1}=\left[\begin{array}{c}-E_{1,q-1}\\ \hline I_{q-1}\end{array}\right]$$

and thus

$$\begin{aligned}&M\left(I_n-\tfrac{1}{n}E_{nn}\right)M^{*\prime}\left(M^*\left(I_n-\tfrac{1}{n}E_{nn}\right)M^{*\prime}\right)^{-1}M^*\left(I_n-\tfrac{1}{n}E_{nn}\right)\\ &\qquad=\left[\begin{array}{c}-E_{1,q-1}\\ \hline I_{q-1}\end{array}\right]\left(M^*-\tfrac{1}{n}M^*E_{nn}\right)\\ &\qquad=\left[\begin{array}{c}M_1-E_{1n}\\ \hline M^*\end{array}\right]-\frac{1}{n}\left[\begin{array}{c}M_1E_{nn}-nE_{1n}\\ \hline M^*E_{nn}\end{array}\right]=M\left(I_n-\tfrac{1}{n}E_{nn}\right)\end{aligned}$$

so that we may finally write

$$\begin{aligned}&\left(I_n-\tfrac{1}{n}E_{nn}\right)-\left(I_n-\tfrac{1}{n}E_{nn}\right)M^{*\prime}\left(M^*\left(I_n-\tfrac{1}{n}E_{nn}\right)M^{*\prime}\right)^{-1}M^*\left(I_n-\tfrac{1}{n}E_{nn}\right)\\ &\quad=\left(I_n-\tfrac{1}{n}E_{nn}\right)\\ &\qquad-M'(MM')^{-1}M\left(I_n-\tfrac{1}{n}E_{nn}\right)M^{*\prime}\left(M^*\left(I_n-\tfrac{1}{n}E_{nn}\right)M^{*\prime}\right)^{-1}M^*\left(I_n-\tfrac{1}{n}E_{nn}\right)\\ &\quad=\left(I_n-\tfrac{1}{n}E_{nn}\right)-M'(MM')^{-1}M\left(I_n-\tfrac{1}{n}E_{nn}\right)\\ &\quad=I_n-\frac{1}{n}E_{nn}-M'(MM')^{-1}M'+\frac{1}{n}\underbrace{M'(MM')^{-1}ME_{nn}}_{=E_{nn}}=I_n-M'(MM')^{-1}M'\end{aligned}$$

which proves (5.170), showing that we indeed have $A_{11.2}=A$ for $A_{11.2}$ in (5.168) and A in (5.37).

In this way, the l.r.t. test statistic to test the independence of $\underline{X}$ and $\underline{M}^*$ is shown to be the same as the one in (5.72) in Sect. 5.1.4.2 and the one in (5.2) in Sect. 5.1.1, and as such, to test the independence between $\underline{X}$ and $\underline{M}^*$ is the same as to test the equality of the q mean vectors $\underline{\mu}_k$ $(k=1,\dots,q)$, that is, to test the null hypothesis in (5.1). This l.r.t. statistic is then given by (5.147) with $A_{11.2}$ given by (5.168) or alternatively defined as $X(I_n-M'(MM')^{-1}M)X'$ and A_{11} given by (5.164).

Alternatively we may take the l.r.t. statistic for this test to be the l.r.t. statistic in (5.148) with A_{22} and A_{12} given by (5.167) and A_{11} given by (5.164), or yet as the l.r.t. statistic in (5.145) with A given by (5.142), taking X_1 as X in the present subsection and X_2 as M^* and A_{11} and A_{22} given respectively by (5.164) and (5.167).

If we take p as the number of rows of X, then the exact distribution of Λ is given by (5.149) and (5.150), and, for even p or even $q-1$, that is, odd q, in terms of Corollary 4.2 with the set of parameters in (5.151) and (5.152), with p_1 replaced by p and p_2 replaced by $q-1$.

We should note here that there is indeed a slightly simpler approach for the implementation of the test of equality of mean vectors as a test of independence between $\underline{X}$ and $\underline{M}$. This is to simply take the whole original matrix M and to not center it, while centering the matrix X. That is, we will take A_{11}, A_{22} and A_{12} as

$$A_{11} = \tfrac{1}{n}X\big(I_n - \tfrac{1}{n}E_{nn}\big)X', \qquad A_{22} = \tfrac{1}{n}MM' \tag{5.171}$$

and

$$A_{12} = \tfrac{1}{n}X\big(I_n - \tfrac{1}{n}E_{nn}\big)M' = A_{21}'\,. \tag{5.172}$$

With these definitions of A_{11}, A_{22}, and A_{12} we will have, for example

$$\begin{aligned} A_{11.2} &= A_{11}A_{22}^{-1}A_{21} \\ &= \tfrac{1}{n}X\big(I_n - \tfrac{1}{n}E_{nn}\big)X' \\ &\qquad -\tfrac{1}{n}X\big(I_n - \tfrac{1}{n}E_{nn}\big)M'(MM')^{-1}M\big(I_n - \tfrac{1}{n}E_{nn}\big)X' \\ &= \tfrac{1}{n}X\Big(\underbrace{\big(I_n - \tfrac{1}{n}E_{nn}\big) - \big(I_n - \tfrac{1}{n}E_{nn}\big)M'(MM')^{-1}M\big(I_n - \tfrac{1}{n}E_{nn}\big)}_{=H}\Big)X'\,. \end{aligned} \tag{5.173}$$

We will now show that using these definitions for $A_{11.2}$, A_{11}, A_{12}, and A_{22}, the l.r.t. statistic in (5.147) is the same as the statistic Λ^{**} in (5.72), with D given by (5.69). We will do this by showing that the matrix $A_{11.2}$ in (5.173) is indeed the matrix A in (5.72), which is also matrix the A in (5.37). This may in turn be done by showing that the matrix H in (5.173) is equal to $I_n - M'(MM')^{-1}M$, which may be achieved through the use of the relation $E_{nq}M = E_{nn}$, easily derived from the structure of the matrix M. From this relation one may then write

$$\begin{aligned} E_{nq}M = E_{nn} \Rightarrow M'E_{qn} = E_{nn} &\Rightarrow ME_{nn} = MM'E_{qn} \\ &\Rightarrow (MM')^{-1}ME_{nn} = (MM')^{-1}MM'E_{qn} = E_{qn} \\ &\Rightarrow M'(MM')^{-1}ME_{nn} = M'E_{qn} = E_{nn} \\ &\Rightarrow E_{nn}M'(MM')^{-1}M = E_{nn}\,. \end{aligned}$$

Using these results it is then easy to show that

$$\begin{aligned} H &= (I_n - \tfrac{1}{n}E_{nn}) - (I_n - \tfrac{1}{n}E_{nn})M'(MM')^{-1}M(I_n - \tfrac{1}{n}E_{nn}) \\ &= I_n - \tfrac{1}{n}E_{nn} - M'(MM')^{-1}M + \tfrac{1}{n}\underbrace{M'(MM')^{-1}ME_{nn}}_{=E_{nn}} \\ &\qquad + \tfrac{1}{n}\underbrace{E_{nn}M'(MM')^{-1}M}_{=E_{nn}} - \tfrac{1}{n^2}E_{nn}\underbrace{M'(MM')^{-1}ME_{nn}}_{=E_{nn}} \\ &= I_n - \tfrac{1}{n}E_{nn} - M'(MM')^{-1}M + \tfrac{1}{n}E_{nn} + \tfrac{1}{n}E_{nn} - \tfrac{1}{n^2}\underbrace{E_{nn}E_{nn}}_{=nE_{nn}} \\ &= I_n - M'(MM')^{-1}M\,, \end{aligned} \tag{5.174}$$

which shows that indeed $A_{11.2}$ in (5.173) is the same matrix as matrix A in (5.37). We may note that this is enough to show that the statistic Λ in (5.147), and thus also the statistics in (5.145) and (5.148), are, for the definitions of the matrices A_{11}, A_{22}, and A_{12} in (5.171) and (5.172), and thus for the definition of the matrix A in (5.145) as

$$A = \begin{bmatrix} A_{11} & A_{12} \\ A_{21} & A_{22} \end{bmatrix} = \frac{1}{n}\left[\frac{X(I_n - \frac{1}{n}E_{nn})}{M}\right]\left[\frac{X(I_n - \frac{1}{n}E_{nn})}{M}\right]',$$

exactly the same statistic as that in (5.72), with D given by (5.69), given that we have already shown that A_{11} in (5.171) and (5.164) is the same matrix as the matrix that appears in the denominator of (5.72), for D in (5.69).

Module `PvalDataInd2R` is able to recognize matrices of indicator variables that correspond to the full set of indicator variables and will automatically not center them.

However, by now a question may already have come into the reader's mind. The derivation of the distribution of the l.r.t. in (5.145) was done under the assumption that both underlying subvectors, which in Sect. 5.1.9 were the subvectors $\underline{X}_1$ and $\underline{X}_2$, were random, when now only $\underline{X}$ is a random vector, while $\underline{M}$ or $\underline{M}^*$ may be seen as not random, or as having a multinomial distribution. Indeed for the l.r.t. statistic in (5.145) to have the distribution given by (5.149) and (5.150) we only need one of $\underline{X}_1$ or X_2 to have a multivariate Normal distribution. Then the other subvector may have any other distribution or even be nonrandom. This fact is supported for example by the derivations of the distributions of similar statistics in Sects. 8.4 and 9.1 of Kshirsagar (1972) where in order that the results obtained hold it is only necessary the multivariate normality of one of the subvectors. In fact, also in Sect. 5.1.4.2, where the l.r.t. statistic used and the distribution obtained were the same as the ones we are dealing with here, the derivation of the distribution of the l.r.t. statistic was done assuming the nonrandomness of the matrix M. This result is perfectly parallel with the one that shows that in the univariate case, the sample Pearson correlation coefficient between two random variables only needs that one of them has a normal distribution in order that its square has a Beta distribution, as well as the multiple correlation coefficient between one random variable and a vector of other variables only needs the normality of that single random variable for its square to have a Beta distribution. In both these cases the other variable or the vector of variables may have any other distribution or even be nonrandom.

Other Choices for the Matrix M Which Lead to the Same Results

Without going over too many details, it is worthwhile to remark here that there are other possible choices for the matrix M, which lead to the same results.

One other choice would be to use the matrix

$$M^{+} = \left[\begin{array}{c} E_{1,n} \\ \hline M^{*} \end{array}\right] \tag{5.175}$$

instead of the matrix M, once again not centering this matrix, while centering the matrix X, for the overall means. This would lead us to use the matrices A_{11} in (5.164) or (5.171) and A_{12} with a similar definition to that in (5.172) but now with M replaced by M^{+}, and A_{22} defined as

$$A_{22} = \tfrac{1}{n} M^{+} M^{+\prime} .$$

This implies that $A_{11.2}$ will be given by

$$A_{11.2} = \tfrac{1}{n} X \left(\left(I_n - \tfrac{1}{n} E_{nn} \right) - \left(I_n - \tfrac{1}{n} E_{nn} \right) M^{+\prime} \left(M^{+} M^{+\prime} \right)^{-1} M^{+} \left(I_n - \tfrac{1}{n} E_{nn} \right) \right) X' , \tag{5.176}$$

which, using (5.533) in Appendix 10, is shown to be equal to the matrix $A_{11.2}$ in (5.173), which was in turn shown to be equal to the matrix A in (5.72), for D in (5.69), in the sequel of (5.174). Since it was also shown before that the matrix in the denominator of (5.72), for D in (5.69), is equal to A_{11} in (5.171) or (5.164), this shows that using M^{+} in (5.175) instead of M and not centering this matrix indeed lead to a l.r.t. statistic identical to the one in (5.72).

We may note here that while the matrix M^{+} in (5.175) is obtained from M by replacing its first row by a row of 1's, we may indeed use instead any other matrix obtained from M by replacing any of its rows by a row of 1's.

The test being addressed in this subsection is indeed only a particular case of the test of the more general null hypothesis (5.49) in Sect. 5.1.4.1, for the reason that following the approach in the present subsection we are only able to carry out a test for $\beta^* = 0$, for β^* in (5.49). However, the present viewpoint over the test of equality of mean vectors has an important advantage: it allows this test to be placed into the framework of the general linear model.

Indeed, the model being addressed in this subsection is the MANOVA model, where the vector $\underline{X}$ is seen as the vector of response variables while the vector $\underline{M}$ or $\underline{M}^*$ of indicator variables for the subpopulations or the levels of the factor is seen as the vector of the set of explanatory variables. In the same way as the statistic Λ in (5.145) may be seen as the statistic to test the "fit" of a Multivariate Regression model when both subvectors $\underline{X}_1$ and $\underline{X}_2$ consist of continuous r.v.'s, this statistic may now be seen as the statistic for the test of the fit of a one-way MANOVA model. In case we reject the null hypothesis of independence of the vectors $\underline{X}$ and $\underline{M}$ or $\underline{M}^*$, we are reaching the conclusion that there are "significant" differences among the q mean vectors $\underline{\mu}_k$ $(k = 1, \ldots, q)$, or equivalently, that the "factor" whose levels are represented by the indicator variables in $\underline{M}$ or $\underline{M}^*$ has a "significant" effect on the mean values of the variables in $\underline{X}$, with changes in the levels of this factor affecting "significantly" the expected values of the variables in $\underline{X}$. The statistic Λ appears then as the statistic for the "test of fit" of the MANOVA model and as the "natural" generalization of the well-known F statistic used in the univariate ANOVA model.

An example of application of the test of independence of two sets of variables to the implementation of the test of equality of mean vectors, or the one-way MANOVA model, is shown in Fig. 5.39, where is displayed the command used with module `PvalDataInd2R` and the output obtained when using a data file with a data matrix with five columns, the first three of which form the matrix X, with the values for the three sales performance variables in Table 9.12 of Johnson and Wichern (2014) for the 30 (=5+15+10) observations in that table that have abstract reasoning scores of 8, 11, and 12 and that were used in Sect. 5.1.1.5 for the second real data example of the test of equality of mean vectors, with results reported in Fig. 5.17. The last two columns in the data file form the matrix M and are the indicator variables for each of the three sub-samples, that is, the ones with scores of 8, 11, and 12 for the abstract reasoning test, from which the first indicator variables was dropped. This data file was easily obtained from the data file that has the whole Table 9.12 in Johnson and Wichern (2014), using module `FileConv`. This module has only two arguments, the input and the output files and allows to read from the input file and choose one of its columns to setup a set of indicator variables. This column is then taken out of the data matrix and the set of indicator variables generated, with the one for the first level being taken out, is appended to

```
PvalDataInd2R[path <> "Tab9_12_JW_MANOVA.dat", , 3]
Computed value of Λ: 0.3859344456
p-value: 0.0003872679905
```

Fig. 5.39 Command used and output obtained with the Mathematica® module `PvalDataInd2R` to test the independence between the set of sales performance variables and the set of indicator variables for the set of observations with abstract reasoning scores of 8, 11 and 12—see Fig. 5.17

the output file as its last columns, allowing then for further selection of variables and sub-samples to be included in the output file.

The output obtained in Fig. 5.39, in terms of computed value of the statistic and p-value, as expected, exactly matches the output in Fig. 5.17, a consequence of the equivalence of the test of independence between the set $\underline{X}$ of three continuous r.v.'s and the set $\underline{M}^*$ of indicator variables and the test of equality of the three mean vectors for the sub-samples with scores of 8, 11, and 12 for the abstract reasoning test.

Then the extension to the two-way or the multi-way MANOVA model, with or without taking account of the possible interactions among factors will be quite straightforward. We only need to consider in $\underline{M}$ the set of all the indicator variables, considering for each factor the set of the indicator variables corresponding to all levels but the first one and for each interaction the indicator variables defined by the product of the indicator variables for the factors involved. Then the statistic to test the "fit" of the overall model is again the statistic Λ in (5.145), with a rejection of the null hypothesis of independence between $\underline{X}$ and $\underline{M}$ indicating that among the factors and interactions considered there is at least one of them that has a "significant" effect on the means of the variables in $\underline{X}$. Then, to test which of these factors or interactions are responsible for this "significant" effect, will be indeed a test between two nested multivariate models. This test will be addressed in the next subsection.

5.1.9.5 Testing Between Nested Multivariate Linear Models

In this subsection we will show how the test of independence of two sets of variables may be used to implement tests between two Multivariate Linear Models, one of them which will be considered the "original" or "complete" model, and which will be assigned to the alternative hypothesis, and a submodel of this original or complete model, which will be assigned to the null hypothesis.

We will address two main settings:

setting 1 – in which we test the nullity of a submatrix of parameters or regression coefficients (in the original or complete model)

setting 2 – in which we test the equality of model parameters or regression coefficients among several multivariate linear models.

In carrying out these tests we will use a similar setup to the one used in Sect. 5.1.4.4, in terms of hypotheses and l.r.t. statistics, with the following sequence of three hypotheses:

$$\begin{aligned} H_1 &: \text{ complete or original model}\\ H_0 &: \text{ submodel}\\ H_0^* &: \text{ null or empty model}\end{aligned}$$

and with

$$\Lambda_{H_a,H_b} = \frac{L_{H_b}}{L_{H_a}}$$

denoting the l.r.t. statistic to test the null hypothesis H_b versus the alternative hypothesis H_a, where L_{H_a} and L_{H_b} denote respectively the suprema of the likelihood function under H_a and H_b.

Then,

$$\Lambda_{H_1,H_0^*} = \frac{L_{H_0^*}}{L_{H_1}} \quad \text{and} \quad \Lambda_{H_0,H_0^*} = \frac{L_{H_0^*}}{L_{H_0}}$$

will be respectively the l.r.t. statistics to test the fit of the models in H_1 and H_0, and

$$\Lambda_{H_1,H_0} = \frac{\Lambda_{H_1,H_0^*}}{\Lambda_{H_0,H_0^*}} = \frac{L_{H_0}}{L_{H_1}} \tag{5.177}$$

will be the l.r.t. statistic to test between the original or complete model and the submodel, with a rejection of the null hypothesis, based on a computed value of Λ_{H_1,H_0} smaller than the α-quantile of Λ_{H_1,H_0}, leading us to stick with the original or complete model and a non-rejection leading us to use the submodel.

In (5.177), both statistics Λ_{H_1,H_0^*} and Λ_{H_0,H_0^*} will have expressions similar to those of Λ in (5.145), (5.147) or (5.148). In the computation of Λ_{H_1,H_0^*} we use A_{11}, which, if we take $\underline{X}_1$ as being the set of response variables, will be equal to $(n-1)/n$ times the covariance matrix for the sample of size n of the p_1 response variables, and as such will be the same for both the original model and the submodel, since the set of response variables is the same for both the H_1 and H_0 models, A_{22} which will in turn be equal to $(n-1)/n$ times the sample covariance matrix for the sample of size n of the set of p_2 explanatory variables in the original model, which will include any possible intercept, and A_{12} which will be equal to $(n-1)/n$ times the covariance matrix between the p_1 response variables and the p_2 explanatory variables, for the original model. In the computation of Λ_{H_0,H_0^*} we will use A_{11}, but only the part of A_{12} and A_{22} that corresponds to the explanatory variables left in the submodel.

The l.r.t. statistic Λ_{H_1,H_0} in (5.177) is indeed the multivariate generalization of the partial F test used in univariate Multiple Regression models and as such we will call it the partial Wilks' Lambda statistic, given that the statistic Λ used in Sects. 5.1.1 and 5.1.9 is usually called the Wilks' Lambda statistic, due to the fact that it was Wilks (1932, 1935) who first derived and studied this statistic.

Let us suppose that the set of response variables, which may be thought of as constituting $\underline{X}_1$, has p_1 variables and that the set of explanatory variables for the model in H_1 has p_2 variables, while the set of explanatory variables for the model in H_0 has $p_{21} (< p_2)$ variables. Then, according to (5.149) and (5.150), we will have

$$\Lambda_{H_1,H_0^*} \stackrel{d}{\equiv} \prod_{j=1}^{p_1} Y_j \stackrel{d}{\equiv} \prod_{k=1}^{p_2} Y_k^* \tag{5.178}$$

where, for $j = 1, \dots, p_1$ and $k = 1, \dots, p_2$,

$$Y_j \sim Beta\left(\frac{n-p_2-j}{2}, \frac{p_2}{2}\right) \quad \text{and} \quad Y_k^* \sim Beta\left(\frac{n-p_1-k}{2}, \frac{p_1}{2}\right) \tag{5.179}$$

are two sets of independent r.v.'s, and, also from (5.149) and (5.150),

$$\Lambda_{H_0,H_0^*} \stackrel{d}{\equiv} \prod_{j=1}^{p_1} Y_j^{**} \stackrel{d}{\equiv} \prod_{k=1}^{p_{21}} Y_k^{***} \tag{5.180}$$

where, for $j = 1, \dots, p_1$ and $k = 1, \dots, p_{21}$,

$$Y_j^{**} \sim Beta\left(\frac{n-p_{21}-j}{2}, \frac{p_{21}}{2}\right) \quad \text{and} \quad Y_k^{***} \sim Beta\left(\frac{n-p_1-k}{2}, \frac{p_1}{2}\right) \tag{5.181}$$

are two sets of independent r.v.'s. Then, for $p_{22} = p_2 - p_{21}$,

$$\Lambda_{H_1,H_0} = \frac{\Lambda_{H_1,H_0^*}}{\Lambda_{H_0,H_0^*}} \stackrel{d}{\equiv} \prod_{k=p_{21}+1}^{p_2} Y_k^{****} \stackrel{d}{\equiv} \prod_{j=1}^{p_1} Y_j^{*****} \stackrel{d}{\equiv} \prod_{\ell=1}^{p_{22}} Y_\ell^{(vi)} \tag{5.182}$$

where, for $j = 1, \dots, p_1$ and $k = p_{21}+1, \dots, p_2$ or $\ell = 1, \dots, p_{22}$,

$$Y_k^{****} \sim Beta\left(\frac{n-p_1-k}{2}, \frac{p_1}{2}\right), \quad Y_j^{*****} \sim Beta\left(\frac{n-p_2-j}{2}, \frac{p_{22}}{2}\right) \tag{5.183}$$

and

$$Y_\ell^{(vi)} \sim Beta\left(\frac{n-p_1-p_{21}-\ell}{2}, \frac{p_1}{2}\right) \tag{5.184}$$

are three sets of independent r.v.'s.

While the distribution of Λ_{H_1,H_0} in the form of the first product in (5.182) comes immediately from (5.178)–(5.181), the other two forms may then be obtained from this one. See Appendix 11 for further details on this.

The exact p.d.f. and c.d.f. of Λ_{H_1,H_0} are therefore given in terms of the Meijer G function by

$$f_{\Lambda_{H1,H0}}(z) = \left\{\prod_{j=1}^{p_1} \frac{\Gamma\left(\frac{n-p_{21}-j}{2}\right)}{\Gamma\left(\frac{n-p_2-j}{2}\right)}\right\} G_{p_1,p_1}^{p_1,0}\left(\begin{array}{c} \left\{\frac{n-p_{21}-j}{2}-1\right\}_{j=1:p_1} \\ \left\{\frac{n-p_2-j}{2}-1\right\}_{j=1:p_1} \end{array}\middle|\, z\right)$$

and

$$F_{\Lambda_{H1,H0}}(z)=\left\{\prod_{j=1}^{p_1}\frac{\Gamma\left(\frac{n-p_{21}-j}{2}\right)}{\Gamma\left(\frac{n-p_2-j}{2}\right)}\right\}G^{p_1,1}_{p_1+1,p_1+1}\left(\left.\begin{array}{c}\left\{1,\left\{\frac{n-p_{21}-j}{2}\right\}_{j=1:p_1}\right\}\\ \left\{\left\{\frac{n-p_2-j}{2}\right\}_{j=1:p_1},0\right\}\end{array}\right|\,z\right)$$

or by

$$f_{\Lambda_{H1,H0}}(z)=\left\{\prod_{\ell=1}^{p_{22}}\frac{\Gamma\left(\frac{n-p_{21}-\ell}{2}\right)}{\Gamma\left(\frac{n-p_1-p_{21}-\ell}{2}\right)}\right\}G^{p_{22},0}_{p_{22},p_{22}}\left(\left.\begin{array}{c}\left\{\frac{n-p_{21}-\ell}{2}-1\right\}_{\ell=1:p_{22}}\\ \left\{\frac{n-p_1-p_{21}-\ell}{2}-1\right\}_{\ell=1:p_{22}}\end{array}\right|\,z\right)$$

and

$$F_{\Lambda_{H1,H0}}(z)=\left\{\prod_{\ell=1}^{p_{22}}\frac{\Gamma\left(\frac{n-p_{21}-\ell}{2}\right)}{\Gamma\left(\frac{n-p_1-p_{21}-\ell}{2}\right)}\right\}G^{p_{22},1}_{p_{22}+1,p_{22}+1}\left(\left.\begin{array}{c}\left\{1,\left\{\frac{n-p_{21}-\ell}{2}\right\}_{\ell=1:p_{22}}\right\}\\ \left\{\left\{\frac{n-p_1-p_{21}-\ell}{2}\right\}_{\ell=1:p_{22}},0\right\}\end{array}\right|\,z\right).$$

As such, the exact distribution of Λ_{H_1,H_0} in (5.177) is for even p_1 or even p_{22} an EGIG distribution. More precisely, for even p_1 or even p_{22}, the exact distribution of Λ_{H_1,H_0} is of the type of the product in Theorem 3.2 and thus its exact p.d.f. and c.d.f. are given, for even p_1, by Corollary 4.2, with

$$m^*=1,\quad k_1=2,\quad a_1=\frac{n-p_2}{2},\quad n_1=\frac{p_1}{2}\quad\text{and}\quad m_1=p_{22}\,,\tag{5.185}$$

while for even p_{22}, the exact p.d.f. and c.d.f. of Λ_{H_1,H_0} will be given by Corollary 4.2, with

$$m^*=1,\quad k_1=2\quad,a_1=\frac{n-p_1-p_{21}}{2},\quad n_1=\frac{p_{22}}{2}\quad\text{and}\quad m_1=p_1\,.\tag{5.186}$$

The exact p.d.f. and c.d.f. of Λ_{H_1,H_0} in (5.177) are thus given in terms of the EGIG distribution representation, for even p_1 or even p_{22}, respectively by

$$f_{\Lambda_{H1,H0}}(z)=f^{EGIG}\left(z\,\left|\,\{r_j\}_{j=1:p_1+p_{22}-2};\left\{\frac{n-2-p_{21}-j}{2}\right\}_{j=1:p_1+p_{22}-2};p_1+p_{22}-2\right.\right)$$

and

$$F_{\Lambda_{H1,H0}}(z)=F^{EGIG}\left(z\,\left|\,\{r_j\}_{j=1:p_1+p_{22}-2};\left\{\frac{n-2-p_{21}-j}{2}\right\}_{j=1:p_1+p_{22}-1};p_1+p_{22}-2\right.\right),$$

where

$$r_j = \begin{cases} h_j, & j = 1, 2 \\ r_{j-2} + h_j, & j = 3, \ldots, p_1 + p_{22} - 2 \end{cases}$$

with

$$h_j = (\#\text{ of elements in } \{p_1, p_{22}\} \geq j) - 1, \quad j = 1, \ldots, p_1 + p_{22} - 2.$$

This shows that the distribution of Λ_{H_1,H_0} is the same as that of the statistic to test the independence between two sets of variables, one with p_1 and the other with p_{22} variables, but using a sample of size $n - p_{21}$ instead of size n.

Further details on the distribution of the three statistics, Λ_{H_1,H_0^*}, Λ_{H_0,H_0^*}, and Λ_{H_1,H_0}, mainly on obtaining the exact distribution of Λ_{H_1,H_0}, as well as in obtaining its expression in the form of (5.191) are considered in Appendix 11 and under the next heading.

A test that uses the l.r.t. statistic Λ_{H_1,H_0} bears a close resemblance to the test in Sect. 5.1.4.4, and we may see how the distributions of the r.v.'s Y_j^{*****} and $Y_\ell^{(vi)}$ in (5.183) and (5.184) agree with those of the r.v.'s Y_j^{**} and Y_k^{***} in (5.62) and how the distribution of Λ_{H_1,H_0} in (5.182) matches that of Λ^{**} in (5.61), for $p_1 = p^*$, $p_2 = q - 1$, and $p_{22} = q^*$.

Setting 1: The Test of Nullity of a Submatrix of Parameters or Regression Coefficients: A Test Between Two Multivariate Nested Models

Let us think of the subset $\underline{X}_1$ as the subset of response variables and let us split the subset $\underline{X}_2$ into two sub-sets, $\underline{X}_{21}$ with p_{21} variables, and $\underline{X}_{22}$ with p_{22} variables, with $p_{21} + p_{22} = p_2$, and let us suppose we want to test the null hypothesis $H_0 : \beta_2 = 0$ in the model

$$\underline{X}_1 = \beta_1 \underline{X}_{12} + \beta_2 \underline{X}_{22} + \mathcal{E}. \tag{5.187}$$

We should note that the test we want to carry out is different from testing $\beta_2^+ = 0$ in the model

$$\underline{X}_1 = \beta_2^+ \underline{X}_{22} + \mathcal{E},$$

since testing the nullity of β_2 in the model (5.187) is to test the nullity of β_2 in the presence of $\underline{X}_{21}$ in the model. As such, to be more precise, we will formulate our aim as being to test the null hypothesis

$$H_0 : \beta_2 = 0 \text{ in model (5.187)}. \tag{5.188}$$

This is the same as testing between the model in (5.187), which will be our H_1 model, and which may alternatively be written as

$$H_1 : \underline{X}_1 = \beta \underline{X}_2 + \mathcal{E}\,, \tag{5.189}$$

with $\beta = \left[\beta_1 | \beta_2\right]$, and the model

$$H_0 : \underline{X}_1 = \beta_1 \underline{X}_{21} + \mathcal{E}\,. \tag{5.190}$$

Following the exposition in the beginning of this subsection, and the details in Appendix 11, the l.r.t. statistic to be used to test H_0 in (5.188) is the partial Wilks' Lambda statistic

$$\Lambda_{H_1,H_0} = \frac{\Lambda_{H_1,H_0^*}}{\Lambda_{H_0,H_0^*}} = \frac{|A_{11.2}|}{|A_{11.2(1)}|}\,, \tag{5.191}$$

where

$$A_{11.2} = A_{11} - A_{12} A_{22}^{-1} A_{21} \quad \text{and} \quad A_{11.2(1)} = A_{11} - A_{12(1)} A_{22(11)}^{-1} A_{21(1)}\,, \tag{5.192}$$

for a split of the matrix A as in (5.549) in Appendix 11, corresponding to the split of the vector $\underline{X}_2$ into the two subvectors $\underline{X}_{21}$ and $\underline{X}_{22}$.

According to the details in Appendix 11, the exact distribution of Λ_{H_1,H_0} in (5.191) is given by (5.182)–(5.184) and is thus given for even p_1 by Corollary 4.2 with the parameters in (5.185), or, for even p_{22} with the parameters in (5.186).

In (5.191),

$$\Lambda_{H_1,H_0^*} = \frac{|A_{11.2}|}{|A_{11}|} \tag{5.193}$$

is the l.r.t. statistic to test the fit of model (5.187) or (5.189), where "to test the fit of model (5.187)" really means to test the null hypothesis $H_0^* : \beta = 0$ in model (5.189), or, equivalently, to test between the model in (5.187) or (5.189) and the empty model

$$H_0^* : \underline{X}_1 = \mathcal{E}\,. \tag{5.194}$$

Following the details in Appendix 11, Λ_{H_1,H_0^*} has its exact distribution given by (5.178) and (5.179).

Also in (5.191), the statistic

$$\Lambda_{H_0,H_0^*} = \frac{|A_{11.2(1)}|}{|A_{11}|} \tag{5.195}$$

is the l.r.t. statistic to test the fit of the model (5.190), where "to test the fit of model (5.190)" means to test the null hypothesis $\beta_1 = 0$ in model (5.190), or, equivalently, to test between the model in (5.190) and the empty model in (5.194). Following the details in Appendix 11, Λ_{H_0,H_0^*} has its exact distribution given by (5.180) and (5.181).

From (5.191) we may write

$$\Lambda_{H_1,H_0^*} = \Lambda_{H_1,H_0}\,\Lambda_{H_0,H_0^*}\,, \tag{5.196}$$

which gives us an interesting way to look at the statistic Λ_{H_1,H_0^*}. We should remember that, according to the exposition in Appendix 11, each statistic in (5.196) is indeed a l.r.t. statistic to test the nullity of a covariance matrix. As such, from Theorem 5 in Jensen (1988), and also from Lemma 10.4.1 in Anderson (2003), each of the statistics in (5.196) is independent of the m.l.e. of the covariance matrix being tested, under the corresponding null hypothesis. This means that Λ_{H_1,H_0} is independent of $A_{11.2(1)}$ and of

$$A^* = \begin{bmatrix} A_{11} & A_{12(1)} \\ A_{21(1)} & A_{22(11)} \end{bmatrix},$$

which is the upper left block of A in (5.549), delimited by the dotted line. As such, Λ_{H_1,H_0} is independent of Λ_{H_0,H_0^*}, which is only built on A^*.

This allows us to look at the test for which Λ_{H_1,H_0^*} is the l.r.t. statistic as the composition of two tests, one associated with Λ_{H_1,H_0}, which is the test of nullity of β_2 in model (5.187) and the other, associated with the statistic Λ_{H_0,H_0^*}, the test of nullity of β_1 in model (5.190), that is, the test of nullity of β_1 assuming β_2 null. We will say that these two tests are independent, given the independence of the two test statistics Λ_{H_1,H_0} and Λ_{H_0,H_0^*}.

We have shown that to test H_0^* vs H_1 is the same as to test H_0^* vs H_0 after testing H_0 vs H_1, or we may say that testing H_0^* is the same as testing $\beta_1 = 0$ after testing $\beta_2 = 0$ assuming $\beta_1 (\neq 0)$ in the model, i.e., we may write

$$\begin{aligned} H_0^* : \beta = 0 &\Leftrightarrow (\beta_1 = 0 \text{ (in model (5.190))} \circ (\beta_2 = 0 \text{ in model (5.187)}) \\ &\Leftrightarrow (\beta_1 = 0, \text{ assuming } \beta_2 = 0) \circ (\beta_2 = 0 \text{ (assuming } \beta_1 \neq 0))\,, \end{aligned}$$

where "$\circ$" is to be read as "after."

We may then say that we have split the null hypothesis $H_0^* : \beta = 0$ into a sequence of two conditionally independent null hypotheses.

We will use such decompositions of null hypotheses in a number of subsections ahead in this chapter. These decompositions are extremely useful in helping to establish l.r.t. statistics for elaborate hypothesis.

Just as a note, we may remark that from the independence relation of Λ_{H_1,H_0} and Λ_{H_0,H_0^*} and (5.196), we have the h-th moment of Λ_{H_1,H_0^*} given by

$$E\left(\Lambda_{H_1,H_0^*}^h\right) = E\left(\Lambda_{H_1,H_0}^h\right) E\left(\Lambda_{H_0,H_0^*}^h\right),$$

a relation that may be easily verified from the distributions of Λ_{H_1,H_0^*}, Λ_{H_1,H_0}, and Λ_{H_0,H_0^*} in (5.178)–(5.184).

Also for the partial Wilks Λ test, modules to compute the p.d.f., c.d.f., quantiles and p-values are made available. These are the modules `PDFInd2Rp`, `CDFInd2Rp`, `QuantInd2Rp`, `PvalInd2Rp`, and `PvalDataInd2Rp`. These modules have exactly the same arguments as the corresponding `Ind2R` modules, except that the first argument, which is nonoptional, for the first four of these modules is the value of p_{22}, that is, the number of variables from $\underline{X}_2$ for which we want to test the significance in the Canonical Analysis or Multivariate Regression model, or equivalently, the number of variables from $\underline{X}_2$ that are taken out from the original model to build the submodel. For module `PvalDataInd2Rp` the first, nonoptional, argument is a list with the order numbers of these p_{22} variables from $\underline{X}_2$ for which we want to test the significance in the model.

We will use the data in Table 9.12 in Johnson and Wichern (2014) to illustrate the test between a model and one of its submodels, testing for the significance in the original model of the variables that are taken out to build the submodel. We will test the significance of the scores for mechanical and abstract reasoning in the relation between the set $\underline{X}_1$ of the sales performance variables and the set $\underline{X}_2$ of the four test scores for creativity, mechanical and abstract reasoning, and mathematical ability. In Fig. 5.40 is illustrated how we can use module `PvalDataInd2Rp` to carry out this test by using a data file whose content is exactly the eight data columns in Table 9.12 of Johnson and Wichern (2014). We remember that according to the results in Fig. 5.38 in Sect. 5.1.9.2, the null hypothesis of independence between $\underline{X}_1$ and $\underline{X}_2$ was rejected.

In Fig. 5.40 the first optional argument `index1` of `PvalDataInd2Rp` is used since we have to select the last seven of the eight columns of the data file to enter the analysis, given that the first column in the file corresponds to the order number

```
PvalDataInd2Rp[{2, 3}, path <> "Tab9_12_JW.dat", , 3, 1]
 Original variables {2, 3, 4, 5, 6, 7, 8}
Computed value of Λ: 0.07609153879
p-value: 4.694028278 × 10^-22
```

Fig. 5.40 Command used and output obtained with the Mathematica® module `PvalDataInd2Rp` to test the significance of the test scores for mechanical and abstract reasoning in a model where the sales performance variables in Problem 9.19 and Table 9.12 in Johnson and Wichern (2014) are used as response variables and the four score tests as explanatory variables

of the observations in Table 9.12 of Johnson and Wichern (2014). The very low p-value obtained in Fig. 5.40 shows that the test scores for mechanical and abstract reasoning play a significant role in the association between $\underline{X}_1$ and $\underline{X}_2$ in a model where the scores for creativity and mathematical ability are kept in the model, or equivalently, that those scores are significant in explaining the values of the three sales performance variables in a model where these three variables are taken as the response variables and where the test scores for creativity and mathematical ability are already used as explanatory variables, or alternatively, that the coefficients for the scores of mechanical and abstract reasoning are not null in a model where the variables in $\underline{X}_1$ are taken as the response variables and the variables in $\underline{X}_2$ as explanatory variables.

The Multi-Way MANOVA Model

This setting of the test between an "original" model and a submodel is exactly the setting in which we may easily frame the multi-way MANOVA model as a particular case.

The "original" model will be a model of the type

$$\underline{X}_1 = \beta \underline{X}_2 + \mathcal{E} \tag{5.197}$$

where $\underline{X}_1$ is the set of p_1 continuous response variables and $\underline{X}_2$ will be formed by the indicator variables for all the factors and interactions in the model. Then any factor or interaction can be tested using the model in (5.197) as the original or complete model, associated with the alternative hypothesis, and a submodel which in $\underline{X}_2$ does not have the set of indicator variables corresponding to the factor or interaction whose significance we want to test. This can be done as long as in $\underline{X}_2$ in (5.197) there is no set of indicator variables corresponding to an interaction where such factor or interaction being tested is involved.

If, for example, one wants to test, in a model with two factors, the significance of the interaction between these two factors, onto a set of p_1 response variables, then all one has to do is use a l.r.t. statistic such as the one in (5.177), which will be the ratio of the l.r.t. statistic which tests the fit of the model (5.197), where $\underline{X}_1$ is the set of response variables, and $\underline{X}_2$ is formed by the indicator variables for factor 1, factor 2, and the interaction. More precisely, $\underline{X}_2$ is formed by the indicator variables corresponding to all the levels of factor 1, dropping one of them, which usually is the one corresponding to the first level, and a similar set of indicator variables for factor 2, and a third set corresponding to the interaction, which is formed by the product of all indicator variables for factor 1 by all indicator variables for factor 2. That is, we indeed have

$$\underset{p_1 \times p_2}{\beta} = \left[\underset{p_1 \times p_{21}}{\beta_1} \;\middle|\; \underset{p_1 \times p_{22}}{\beta_2} \;\middle|\; \underset{p_1 \times p_{23}}{\beta_3} \right] \tag{5.198}$$

where β_1 is the parameter matrix for factor 1 and $p_{21}+1$ is the number of levels of factor 1, β_2 is the parameter matrix for factor 2 and $p_{22}+1$ is the number of levels of factor 2, and β_3 is the parameter matrix for the interaction, with $p_{23}=p_{21}\times p_{22}$. The model in (5.197) may the be written as

$$\underline{X}_1=\beta_1\underline{X}_{21}+\beta_2\underline{X}_{22}+\beta_3\underline{X}_{23}+\mathcal{E} \tag{5.199}$$

where $\underline{X}_{21}$ is the set of indicator variables for factor 1, $\underline{X}_{22}$ the set of indicator variables for factor 2 and $\underline{X}_{23}$ the set of indicator variables for the interaction, and where we need to have a sample size of $n>p_1+p_2=p_1+p_{21}+p_{22}+p_{23}$.

Then, in testing for the significance of the interaction we want model (5.199) to be the model associated with H_1 and the model

$$\underline{X}_1=\beta_1\underline{X}_{21}+\beta_2\underline{X}_{22}+\mathcal{E} \tag{5.200}$$

to be the submodel associated with H_0.

Now we have to consider the matrix A given by (5.142), with

$$\underset{p\times n}{X}=\left[\begin{array}{c} X_1 \\ \hline X_2 \end{array}\right]\begin{array}{l} p_1 \\ p_2 \end{array}=\left[\begin{array}{c} X_1 \\ \hline X_{21} \\ X_{22} \\ X_{23} \end{array}\right]\begin{array}{l} p_1 \\ p_{21} \\ p_{22} \\ p_{23} \end{array}$$

so that the matrix A has to be taken split as

$$A=\left[\begin{array}{c|c} A_{11} & A_{12} \\ \hline A_{21} & A_{22} \end{array}\right]=\left[\begin{array}{c|ccc} A_{11} & A_{12(1)} & A_{12(2)} & A_{12(3)} \\ \hline A_{21(1)} & A_{22(11)} & A_{22(12)} & A_{22(13)} \\ A_{21(2)} & A_{22(21)} & A_{22(22)} & A_{22(23)} \\ A_{21(3)} & A_{22(31)} & A_{22(32)} & A_{22(33)} \end{array}\right]\begin{array}{l} p_1 \\ p_{21} \\ p_{22} \\ p_{23} \end{array},$$
$$\qquad\qquad\qquad\qquad p_1 \qquad p_{21} \qquad p_{22} \qquad p_{23}$$

so that the l.r. statistic to test between the two models will be the statistic Λ_{H_1,H_0} in (5.177) and (5.191) with $A_{12(1)}$ replaced by $[A_{12(1)}\ A_{12(2)}]$ and $A_{22(11)}$ replaced by

$$\begin{bmatrix} A_{22(11)} & A_{22(12)} \\ A_{22(21)} & A_{22(22)} \end{bmatrix}$$

in the definition of $A_{11.2(1)}$, and taking

$$A_{22}=\begin{bmatrix} A_{22(11)} & A_{22(12)} & A_{22(13)} \\ A_{22(21)} & A_{22(22)} & A_{22(23)} \\ A_{22(31)} & A_{22(32)} & A_{22(33)} \end{bmatrix}$$

and $A_{12}=\left[A_{12(1)}\;A_{12(2)}\;A_{12(3)}\right]$ in the definition of $A_{11.2}$.

The distribution of Λ_{H_1,H_0} will then be given by (5.182)–(5.184), with p_{22} replaced by p_{23} and p_{21} by $p_{21}+p_{22}$. Thus, the exact distribution of Λ_{H_1,H_0} will be given, for even p_1 by Corollary 4.2 with

$$m^*=1,\quad k_1=2,\quad a_1=\frac{n-p_2}{2},\quad n_1=\frac{p_1}{2}\quad \text{and}\quad m_1=p_{23}, \tag{5.201}$$

and for even p_{23} by Corollary 4.2 with

$$m^*=1,\quad k_1=2\quad ,a_1=\frac{n-p_1-p_{21}-p_{22}}{2},\quad n_1=\frac{p_{23}}{2}\quad \text{and}\quad m_1=p_1\,. \tag{5.202}$$

The exact p.d.f. and c.d.f. of Λ_{H_1,H_0} will be given in this case, in terms of the EGIG distribution representation, for any of the above cases, respectively by

$$f_{\Lambda_{H1,\,H0}}(z)=f^{EGIG}\left(z\,\middle|\,\{r_j\}_{j=1:p_1+p_{23}-2};\left\{\frac{n-2-p_{21}-p_{22}-j}{2}\right\}_{j=1:p_1+p_{23}-2};\; p_1+p_{23}-2\right)$$

and

$$F_{\Lambda_{H1,\,H0}}(z)=F^{EGIG}\left(z\,\middle|\,\{r_j\}_{j=1:p_1+p_{23}-2};\left\{\frac{n-2-p_{21}-p_{22}-j}{2}\right\}_{j=1:p_1+p_{23}-2};\; p_1+p_{23}-2\right)$$

where

$$r_j=\begin{cases} h_j\,, & j=1,2\\ r_{j-2}+h_j\,, & j=3,\dots,p_1+p_{23}-2\end{cases}$$

with

$$h_j=(\#\text{ of elements in }\{p_1,p_{23}\}\geq j)-1\,,\quad j=1,\dots,p_1+p_{23}-2\,.$$

Then, in case we do not reject the H_0 model in (5.200), which would mean that the interaction between the two factors is nonsignificant, we may think of testing the significance of any of the two factors. We may then for example test the significance of factor 1 by using now the model (5.200) as our H_1 model and the model

$$\underline{X}_1 = \beta_2 \underline{X}_{22} + \mathcal{E}$$

as our H_0 model.

In this case, the l.r. statistic to be used would be a similar statistic to Λ_{H_1,H_0} in (5.177), with $A_{11.2(1)}$ replaced by

$$A_{11.2(2)} = A_{11} - A_{12(2)} A_{22(22)}^{-1} A_{21(2)}\,,$$

and $A_{11.2}$ with a similar definition to the one in (5.192), but using $A_{12} = \left[A_{12(1)}\; A_{12(2)}\right]$ and

$$A_{22} = \begin{bmatrix} A_{22(11)} & A_{22(12)} \\ A_{22(21)} & A_{22(22)} \end{bmatrix}.$$

The distribution of Λ_{H_1,H_0} will thus be given by (5.182)–(5.184), with p_{22} replaced by p_{21} and p_{21} replaced by p_{22}. Thus, the exact distribution of Λ_{H_1,H_0} will be given, for even p_1 by Corollary 4.2 with

$$m^* = 1, \quad k_1 = 2, \quad a_1 = \frac{n-p_2}{2}, \quad n_1 = \frac{p_1}{2} \quad \text{and} \quad m_1 = p_{21}\,, \tag{5.203}$$

and for even p_{21} by Corollary 4.2 with

$$m^* = 1, \quad k_1 = 2 \quad , a_1 = \frac{n-p_1-p_{22}}{2}, \quad n_1 = \frac{p_{21}}{2} \quad \text{and} \quad m_1 = p_1\,. \tag{5.204}$$

The exact p.d.f. and c.d.f. of Λ_{H_1,H_0} will be given in this case, in terms of the EGIG distribution representation, for any of the above cases, by

$$f_{\Lambda_{H1,H0}}(z) = f^{EGIG}\left(z \,\middle|\, \{r_j\}_{j=1:p_1+p_{21}-2}; \left\{\frac{n-2-p_{22}-j}{2}\right\}_{j=1:p_1+p_{21}-2}; p_1+p_{21}-2 \right)$$

and

$$F_{\Lambda_{H1,H0}}(z) = F^{EGIG}\left(z \,\middle|\, \{r_j\}_{j=1:p_1+p_{21}-2}; \left\{\frac{n-2-p_{22}-j}{2}\right\}_{j=1:p_1+p_{21}-2}; p_1+p_{21}-2 \right),$$

where

$$r_j = \begin{cases} h_j\,, & j = 1, 2 \\ r_{j-2} + h_j\,, & j = 3, \ldots, p_1 + p_{21} - 2 \end{cases}$$

with

$$h_j = (\#\text{ of elements in } \{p_1, p_{21}\} \geq j) - 1\,, \quad j = 1, \ldots, p_1 + p_{21} - 2\,.$$

To illustrate the analysis of a multi-way MANOVA model we first consider a balanced two-way MANOVA model with interaction, based on the analysis of the steel bar data on a strain and torque study reported by Posten (1962) and analyzed by Kramer and Jensen (1970) and which data set is made available to us by Rencher (2002) and Rencher and Christensen (2012) in Table 6.6. The data concerns measurements of ultimate torque and strain on 32 steel bars in a two-way balanced experimental design where one factor is the type of lubricant, with 4 levels, and the other factor is the rotational velocity, with 2 levels. The data on the two response variables (ultimate torque and strain) for the 32 steel bars and the corresponding values for the three indicator variables for lubricant type and the indicator variable for rotational velocity, as well as the three indicator variables for the interaction levels are stored in this order in the file `SteelBar_MANOVA.dat`, which is used in the analysis reported in Fig. 5.41. In this figure the first command uses module `PvalDataInd2R` to assert that the set of seven (3+1+3) indicator variables for the two factors and the interaction explains significantly well the response variables. Then the second command uses module `PvalDataInd2Rp` to carry out a test for the interaction, using an original model with all seven indicator variables and a submodel without the three indicator variables for the interaction, which are the 5th, 6th, and 7th indicator variables in $\underline{X}_2$, the set of explanatory variables. The third and fourth commands in that figure test respectively the significance of rotational speed and lubricant type. We note once again that we are allowed to test for the individual factor effects since the interaction was considered nonsignificant, given the extremely high p-value obtained for its test. As such, the indicator variables for the interaction were taken out of the H_1 model when the test for the factors is carried out. Only the factor rotational velocity is considered to have a significant effect on the set of two response variables, or equivalently, its levels are considered to explain sufficiently well the changes in the values of the two response variables, given the rather low p-value obtained.

The file `SteelBar_MANOVA.dat` may be easily obtained from the data file `SteelBar.dat` which is easily obtained directly from Table 6.6 in Rencher (2002) or Rencher and Christensen (2012) and which has a first column with 1's and 2's indicating the level of rotational speed, a second column with 1's, 2's, 3's or 4's to indicate the lubricant type and then two last columns with the values of the two response variables. This file is then run through module `FileConv` to setup the indicator variables for the factors. On a first run we may setup the indicator variables

```
PvalDataInd2R[path <> "SteelBar_MANOVA.dat", , 2]
Computed value of Λ: 0.3197135833
p-value: 0.009225995202
PvalDataInd2Rp[{5, 6, 7}, path <> "SteelBar_MANOVA.dat", , 2]
Computed value of Λ: 0.9319287786
p-value: 0.9458303520
PvalDataInd2Rp[{4}, path <> "SteelBar_MANOVA.dat", , 2, 1]
 Original variables {1, 2, 3, 4, 5, 6}
Computed value of Λ: 0.4894503052
p-value: 0.00009251493492
PvalDataInd2Rp[{1, 2, 3}, path <> "SteelBar_MANOVA.dat", , 2, 1]
 Original variables {1, 2, 3, 4, 5, 6}
Computed value of Λ: 0.6942565836
p-value: 0.1314381471
```

Fig. 5.41 Commands used and output obtained with the Mathematica® modules `PvalDataInd2R` and `PvalDataInd2Rp` to analyze a two-way balanced MANOVA model based on the steel bars data set Table 6.6 in Rencher (2002) and Rencher and Christensen (2012)

for the lubricant type and on a second run the indicator variable for rotational speed, as is done in Fig. 5.42. Then, finally, with a run through the `FileInter` module we setup the indicator variables for the interaction, based on the indicator variables for the two factors, as indicated in the last command in Fig. 5.42.

We will now use the same data set from the torque and strain study on steel bars and the same file `SteelBar_MANOVA.dat` to illustrate how we would proceed to analyze a MANOVA hierarchical two-way model where the lubricant type is supposed to be nested into the rotational speed levels. The corresponding command and output is in Fig. 5.43. In this case the "effect" of, or rather, the sum of squares associated with, the lubricant type factor is the sum of what in the simple two-way model were the "effects," or rather, the sums of squares, associated with the factor "lubricant type" and the interaction, which means that now we have to test first the lubricant type significance using as "original" model the full model and as submodel the model without the indicator variables for lubricant type and for the interaction. Since from the result in Fig. 5.43 this factor in the hierarchical or nested model is taken as nonsignificant, given the rather high p-value obtained, we can proceed to test for the effect of rotational speed, which will be just the test of fit of a model with only this factor.

```
FileConv("<path>/SteelBar.dat","<path>/SteelBar_MANOVA.dat")$
 Is there a column in the data file with the sample assignments? (1-Yes)"1;
  Which is the column in the data file with the sample assignments ?2;
 Do you want to select samples ? (1=Yes) "0;
 Do you want to select variables ? (1=Yes) "0;

     There are 4 samples, with sizes: [8,8,8,8] and 3 variables
FileConv("<path>/SteelBar_MANOVA.dat",
                                   "<path>/SteelBar_MANOVA.dat")$
 Is there a column in the data file with the sample assignments? (1-Yes)"1;
  Which is the column in the data file with the sample assignments ?1;
 Do you want to select samples ? (1=Yes) "0;
 Do you want to select variables ? (1=Yes) "0;

     There are 2 samples, with sizes: [16,16] and 5 variables
FileInter("<path>/SteelBar_MANOVA.dat",
                         "<path>/SteelBar_MANOVA.dat",[3,4,5],[6])$
```

Fig. 5.42 Using modules `FileConv` and `FileInter` to setup a file for the implementation of a MANOVA on the steel bar data

```
PvalDataInd2Rp[{1, 2, 3, 5, 6, 7}, path <> "SteelBar_MANOVA.dat", , 2]
Computed value of Λ: 0.6469976900
p-value: 0.5239280799
```

Fig. 5.43 Command used and output obtained with the Mathematica® module `PvalDataInd2Rp` to analyze a two-way hierarchical MANOVA model based on the steel bars data set Table 6.6 in Rencher (2002) and Rencher and Christensen (2012)

In order to show that the analysis of unbalanced designs poses absolutely no problem when the MANOVA model is addressed under the umbrella of the Canonical Analysis or Multivariate Regression model and that under this framework an unbalanced and a balanced MANOVA design are analyzed in exactly the same way we will proceed to the analysis of a two-way MANOVA unbalanced model with interaction, based on the analysis of the three sales performance variables in Problem 9.19 and Table 9.12 in Johnson and Wichern (2014) for the set of 30 (=5+15+10) observations which correspond to scores of 8, 11 and 12 for the abstract reasoning test, which is used as a first factor with three levels, represented by the two indicator variables for its two last levels (obtained with a first use of module `FileConv`) and to which we add a second factor based on the scores of the mechanical reasoning test, setting up a first level for scores 8–12, a second level for scores 13–16 and a third level for scores 17–20. To these three levels also correspond two indicator variables, which together with the other two from the abstract reasoning factor are used to define the four indicator variables for the interaction of these two factors (through the use of module `FileInter`, after two utilizations of module `FileConv`, in exactly the same manner as illustrated in Fig. 5.42 for the steel bar example). We thus have a data file with 11 columns, the first three of which correspond to the three sales performance variables, taken as our continuous response variables, and for which we want to test the possible effect of the two factors based on the scores for abstract and mechanical reasoning and their possible interaction. The eight indicator variables for these two factors and their interaction make the set of explanatory variables $\underline{X}_2$, with the two first indicator variables pertaining to the first factor (abstract reasoning), the second two to the second factor (mechanical reasoning), and the last four to the interaction of these two factors.

The commands used and the output obtained with the Mathematica® version of module `PvalDataInd2Rp` to perform this analysis are shown in Fig. 5.44, where a first command uses module `PvalDataInd2R` to assert that the multivariate model that uses the three sales performance variables as response variables, and the set of 8 (=2+2+4) indicator variables for the two factors and their interaction as explanatory variables fits, that is, that these two groups of variables should be considered as non-independent, based on the rather low p-value of 0.030129 obtained.

We may remark here that the p-value obtained for the fit of this overall model is much larger than the one obtained in Fig. 5.39 when only the factor based on the abstract reasoning is used (although the computed value of the l.r.t. statistic, as it

```
PvalDataInd2R[path <> "Tab9_12_JW_MANOVA_2.dat", , 3]
Computed value of Λ: 0.1825439767
p-value: 0.03012885791

PvalDataInd2Rp[{5, 6, 7, 8}, path <> "Tab9_12_JW_MANOVA_2.dat", , 3]
Computed value of Λ: 0.6653345268
p-value: 0.7424086360

PvalDataInd2Rp[{3, 4}, path <> "Tab9_12_JW_MANOVA_2.dat", , 3, 1]
 Original variables {1, 2, 3, 4, 5, 6, 7}
Computed value of Λ: 0.7109088226
p-value: 0.2252471440

PvalDataInd2Rp[{1, 2}, path <> "Tab9_12_JW_MANOVA_2.dat", , 3, 1]
 Original variables {1, 2, 3, 4, 5, 6, 7}
Computed value of Λ: 0.4167156034
p-value: 0.001861815902
```

Fig. 5.44 Commands used and output obtained with the Mathematica® modules `PvalDataInd2R` and `PvalDataInd2Rp` to analyze a two-way MANOVA model based on the sales performance variables in Problem 9.19 and Table 9.12 in Johnson and Wichern (2014) used as response variables

should be, is now smaller, since the number of explanatory variables involved is larger now), showing that the second factor that was added and the interaction are either nonsignificant or much "less significant" than the factor based on the abstract reasoning scores.

The appropriate test for the interaction of the two factors is carried out with the first use of module `PvalDataInd2Rp` in Fig. 5.44. Given the quite large p-value obtained (0.742409) we should conclude that the interaction is nonsignificant and pursue with the analysis of the individual factors, noting that the test of the interaction has to be the first to be carried out and that in case it gives a significant result this would lead us to not test for the significance of individual factors. With the second command that uses module `PvalDataInd2Rp` in Fig. 5.44 we test the significance of the second factor, the one based on the mechanical reasoning scores, in the presence of the other factor, and dropping from the analysis the four indicator variables for the interaction, which is the reason why the first optional argument `index1` is given the value 1 in order to allow us to drop these variables from the analysis. Given the quite large p-value obtained (0.225247) we also conclude the nonsignificance of this factor, in the presence of the first factor in the model. Then we may want to test the significance of the first factor, the one based on the abstract reasoning scores, in the presence of the second factor, although this last one was nonsignificant. This is what is done with the third use of the module `PvalDataInd2Rp` in Fig. 5.44, and the very low p-value comes as no surprise, given the extremely low p-value obtained in Fig. 5.40 when this first factor was the only one considered, and, as it should be, showing in that case an even lower p-value than the one obtained when the second factor is included in the model. We thus conclude that the interaction of the two factors was nonsignificant and that from the two factors only the one based on the abstract reasoning had a significant effect (when only the individuals with abstract reasoning scores of 8, 11, and 12 are considered), so that in a step-down type strategy we would be left with only the first factor in the model.

In the analysis conducted, the mechanical reasoning scores were reduced to only three classes defined with some arbitrariness and the results obtained do not necessarily mean that the original mechanical reasoning scores would not be significant in a model with the set of explanatory variables being taken as the first factor and this variable considered as a continuous variable. This would be a MANCOVA (Multivariate Analysis of Covariance) model which will be considered in the next paragraph, where we will see that if we had used the original mechanical reasoning scores as a continuous variable the results would have been mostly the same.

In the next paragraph we use a similar approach to the one used in the preceding paragraphs to analyze MANCOVA (Multivariate Analysis of Covariance) models.

The approach followed of using the Canonical Analysis or Multivariate Regression model as an umbrella model for other models such as the MANOVA and MANCOVA models has several advantages and virtues, among which are those of (1) enabling for a simple and direct application of the MANOVA and MANCOVA models even in situations of unbalanced designs, which is not so easy a task when the classical Analysis of Variance framework is used, and (2) giving us a better understanding of the reason why in such models one cannot test the significance of factors or interactions involved in higher order interactions which are still present in the model. This last issue is related to the fact that the presence in the model of the indicator variables associated with such interactions does not allow for a "general" or "uniform" test for the underlying factors or interactions, given the fact that these interaction indicator variables are built from the indicator variables for these underlying factors or interactions of lower order. The need for the explicit use of module `FileInter` to build the indicator variables associated with the interactions also has the virtue of giving the user the explicit idea that the indicator variables for the interactions are indeed built from the product of the indicator variables for the underlying factors or interactions.

The MANCOVA (Multivariate Analysis of Covariance) Model

As its name clearly indicates, this model is closely related to the MANOVA model. The Analysis of Covariance model may indeed be seen as a hybrid model between the Analysis of Variance and Regression models. For a Covariance Analysis model with two factors, interaction and p_{24} covariates, we may think of a model with a similar structure to that of the model (5.199), where now $\underline{X}_2$ has a fourth subvector, say $\underline{X}_{24}$, formed by p_{24} continuous explanatory variables, in this setting usually called covariates, giving rise to a β matrix similar to the one in (5.198), but now with a fourth component of dimensions $p_1 \times p_{24}$. Then we may want to test the significance in the model of all or some of the p_{24} covariates in $\underline{X}_{24}$, in the presence of the factors and the interaction. In case we want to test the significance in the model of all the p_{24} covariates, in the presence of the two factors and their interaction, this would lead us to use the H_1 model

$$\underline{X}_1 = \beta_1 \underline{X}_{21} + \beta_2 \underline{X}_{22} + \beta_3 \underline{X}_{23} + \beta_4 \underline{X}_{24} + \mathcal{E} \tag{5.205}$$

and the H_0 model

$$\underline{X}_1 = \beta_1 \underline{X}_{21} + \beta_2 \underline{X}_{22} + \beta_3 \underline{X}_{23} + \mathcal{E}\,.$$

In this case,

$$\underset{p\times n}{X} = \begin{bmatrix} X_1 \\ \hline X_2 \end{bmatrix} \begin{matrix} p_1 \\ p_2 \end{matrix} = \begin{bmatrix} X_1 \\ \hline X_{21} \\ X_{22} \\ X_{23} \\ X_{24} \end{bmatrix} \begin{matrix} p_1 \\ p_{21} \\ p_{22} \\ p_{23} \\ p_{24} \end{matrix}$$

so that the matrix A has to be taken as

$$A=\left[\begin{array}{c|c} A_{11} & A_{12} \\ \hline A_{21} & A_{22} \end{array}\right]=\left[\begin{array}{c|cccc} A_{11} & A_{12(1)} & A_{12(2)} & A_{12(3)} & A_{12(4)} \\ \hline A_{21(1)} & A_{22(11)} & A_{22(12)} & A_{22(13)} & A_{22(14)} \\ A_{21(2)} & A_{22(21)} & A_{22(22)} & A_{22(23)} & A_{22(24)} \\ A_{21(3)} & A_{22(31)} & A_{22(32)} & A_{22(33)} & A_{22(34)} \\ A_{21(4)} & A_{22(41)} & A_{22(42)} & A_{22(43)} & A_{22(44)} \end{array}\right]\begin{array}{c} p_1 \\ p_{21} \\ p_{22} \\ p_{23} \\ p_{24} \end{array},$$
$$\qquad\qquad\qquad\qquad p_1 \quad p_{21} \quad p_{22} \quad p_{23} \quad p_{24}$$

and the l.r. statistic to be used to choose between the two models will be a l.r. statistic similar to the one in (5.191), with $A_{11.2}$ still given by (5.192), with

$$A_{22}=\left[\begin{array}{cccc} A_{22(11)} & A_{22(12)} & A_{22(13)} & A_{22(14)} \\ A_{22(21)} & A_{22(22)} & A_{22(23)} & A_{22(24)} \\ A_{22(31)} & A_{22(32)} & A_{22(33)} & A_{22(34)} \\ A_{22(41)} & A_{22(42)} & A_{22(43)} & A_{22(44)} \end{array}\right] \tag{5.206}$$

and $A_{11.2(1)}$ replaced by

$$A_{11.2(1\text{-}3)}=A_{11}-\left[\,A_{12(1)}\ A_{12(2)}\ A_{12(3)}\,\right]\left[\begin{array}{ccc} A_{22(11)} & A_{22(12)} & A_{22(13)} \\ A_{22(21)} & A_{22(22)} & A_{22(23)} \\ A_{22(31)} & A_{22(32)} & A_{22(33)} \end{array}\right]^{-1}\left[\begin{array}{c} A_{21(1)} \\ A_{21(2)} \\ A_{21(3)} \end{array}\right].$$

Thus, in this case

$$\Lambda_{H_1,H_0}\stackrel{\text{d}}{\equiv}\prod_{j=1}^{p_1} Y_j \stackrel{\text{d}}{\equiv}\prod_{k=1}^{p_{24}} Y_k^* \tag{5.207}$$

where, for $n>p_1+p_2$,

$$Y_j\sim Beta\left(\frac{n-p_2-j}{2},\frac{p_{24}}{2}\right)\quad\text{and}\quad Y_k^*\sim Beta\left(\frac{n-p_1-p_{21}-p_{22}-p_{23}-k}{2},\frac{p_1}{2}\right) \tag{5.208}$$

are two sets of independent r.v.'s. Therefore, the exact p.d.f. and c.d.f. of Λ_{H_1,H_0} is, for even p_1, given by Corollary 4.2 with

$$m^*=1,\quad k_1=2,\quad a_1=\frac{n-p_2}{2},\quad n_1=\frac{p_1}{2}\quad\text{and}\quad m_1=p_{24}\,, \tag{5.209}$$

and for even p_{24}, with

$$m^* = 1, \quad k_1 = 2, \quad a_1 = \frac{n - p_1 - p_{21} - p_{22} - p_{23}}{2}, \quad n_1 = \frac{p_{24}}{2} \quad \text{and} \quad m_1 = p_1 . \tag{5.210}$$

The exact p.d.f. and c.d.f. of Λ_{H_1,H_0} will be thus given in terms of the EGIG distribution representation, for any of the above cases, respectively by

$$f_{\Lambda_{H1,\,H0}}(z) = f^{EGIG}\left(z \,\middle|\, \{r_j\}_{j=1:p_1+p_{24}-2};\ \left\{\frac{n-2-p_{21}-p_{22}-p_{23}-j}{2}\right\}_{j=1:p_1+p_{24}-2};\ p_1+p_{24}-2\right)$$

and

$$F_{\Lambda_{H1,\,H0}}(z) = F^{EGIG}\left(z \,\middle|\, \{r_j\}_{j=1:p_1+p_{24}-2};\ \left\{\frac{n-2-p_{21}-p_{22}-p_{23}-j}{2}\right\}_{j=1:p_1+p_{24}-2};\ p_1+p_{24}-2\right)$$

where

$$r_j = \begin{cases} h_j\,, & j = 1, 2 \\ r_{j-2} + h_j\,, & j = 3, \ldots, p_1 + p_{24} - 2 \end{cases}$$

with

$$h_j = (\#\text{ of elements in } \{p_1, p_{24}\} \geq j) - 1\,, \quad j = 1, \ldots, p_1 + p_{24} - 2\,.$$

In case one wants to test the significance of only some of the covariates, then the H_0 model would be a model where the other covariates, to which we do not want to test the significance, would remain. The expression and distribution of Λ_{H_1,H_0} may then be easily deducted from the above results, which is left as an exercise.

Otherwise, we may want to test the significance of the interaction between the two factors in the presence of the covariates, using the H_1 model in (5.205) and the H_0 model

$$\underline{X}_1 = \beta_1 \underline{X}_{21} + \beta_2 \underline{X}_{22} + \beta_4 \underline{X}_{24} + \mathcal{E}\,.$$

In this case the statistic Λ_{H_1,H_0} would still have a similar formulation to the one in (5.191), with $A_{11.2}$ given by (5.192), with A_{22} given by (5.206) and $A_{11.2(1)}$ replaced by

$$A_{11.2(1,2,4)} = A_{11} - \begin{bmatrix} A_{12(1)} & A_{12(2)} & A_{12(4)} \end{bmatrix} \begin{bmatrix} A_{22(11)} & A_{22(12)} & A_{22(14)} \\ A_{22(21)} & A_{22(22)} & A_{22(24)} \\ A_{22(41)} & A_{22(42)} & A_{22(44)} \end{bmatrix}^{-1} \begin{bmatrix} A_{21(1)} \\ A_{21(2)} \\ A_{21(4)} \end{bmatrix},$$

so that now the exact distribution of Λ_{H_1,H_0} is given by (5.207) and (5.208) with p_{24} replaced by p_{23} and vice versa, p_{23} replaced by p_{24}. The exact p.d.f. and c.d.f. of Λ_{H_1,H_0} are thus now given, for even p_1, by Corollary 4.2 with the parameters in (5.209), with p_{24} replaced by p_{23} and for even p_{23} by Corollary 4.2 with the parameters in (5.210), with p_{23} replaced by p_{24}, and vice versa, p_{24} replaced by p_{23}.

To illustrate the analysis of a MANCOVA model we use the data matrix in file `Tab9_12_JW_MANCOVA.dat` which has 30 observations for a total of 21 variables, the first three of which are the sales performance variables in Problem 9.19 and Table 9.12 in Johnson and Wichern (2014), which will be our response variables. The 30 observations correspond to the observations in Table 9.12 for which the abstract reasoning scores have values of 8, 11, and 12. Since the idea is to illustrate the several types of explanatory variables that one can use in a MANCOVA model, we use a plethora of explanatory variables, including variables to test the existence of possible interactions between factors and continuous covariates. The first two explanatory variables are the scores for creativity and mathematical ability, used as continuous covariates. The third and fourth explanatory variables are then the two indicator variables for the three levels of abstract reasoning (scores 8, 11, and 12), with the indicator variable for the first level dropped. The fifth and sixth explanatory variables are the indicator variables for the three classes of mechanical reasoning scores already used in the MANOVA model studied in the previous paragraph, that is, for classes 8–12, 13–16, and 17–20, with the indicator for the first level (scores 8–12 of mechanical reasoning) dropped. The seventh through tenth explanatory variables are the four indicator variables for the interaction between these two factors, and the remaining eight explanatory variables represent the interactions between the two continuous covariates and the two factors. These interactions are used precisely to illustrate the possibility of using and analyzing such interactions to evaluate the possible existence of different slopes for the covariates across the different levels of the factors. The variables that represent these interactions are obtained, as all other interaction variables, as the product of the corresponding covariate by each of the indicator variables for the corresponding factors. This is exactly what is done when we test the equality of Multivariate Regression models as illustrated in the next paragraph. As such the 11th and 12th explanatory variables represent the interaction between the first covariate, the creativity score, and the first factor, the 13th and 14th the interaction of the first covariate and the second factor, the 15th and 16th the interaction of the second

x_1 x_2 F_{12} F_{13} F_{22} F_{23} I_1 I_2 I_3 I_4 x_1F_{12} x_1F_{13} x_2F_{12} x_2F_{13} x_1F_{22} x_1F_{23} x_2F_{22} x_2F_{23}

Fig. 5.45 List of the explanatory variables used in the MANCOVA model: x_1, x_2—continuous covariates; F_{ab}—indicator variable for level b of factor a; $I_1, \ldots, I_4$—indicator variables for the interaction of the two factors; x_cF_{ab}—variable associated with the interaction of covariate x_c with level b of factor a

```
PvalDataInd2Rp[{17, 18}, path <> "Tab9_12_JW_MANCOVA.dat", , 3]
Computed value of Λ: 0.4715503658
p-value: 0.2796675763
PvalDataInd2Rp[{15, 16}, path <> "Tab9_12_JW_MANCOVA.dat", , 3]
Computed value of Λ: 0.6095845396
p-value: 0.5537222550
PvalDataInd2Rp[{13, 14}, path <> "Tab9_12_JW_MANCOVA.dat", , 3]
Computed value of Λ: 0.5289637285
p-value: 0.3872352334
PvalDataInd2Rp[{11, 12}, path <> "Tab9_12_JW_MANCOVA.dat", , 3]
Computed value of Λ: 0.2060109388
p-value: 0.01580781150
PvalDataInd2Rp[{7, 8, 9, 10}, path <> "Tab9_12_JW_MANCOVA.dat", , 3, 1]
 Original variables {1, 2, 3, 4, 5, 6, 7, 8, 9, 10, 11, 12, 13, 14, 15}
Computed value of Λ: 0.5516538959
p-value: 0.6105381549
```

Fig. 5.46 Commands used and output obtained with the Mathematica® module `PvalDataInd2Rp` to analyze the MANCOVA model based on the sales performance variables in Problem 9.19 and Table 9.12 in Johnson and Wichern (2014) used as response variables—tests for interactions

covariate, the mathematical ability score, and the first factor, and finally the 17th and 18th the interaction of the second covariate and the second factor. In Fig. 5.45 is shown a scheme of the explanatory variables considered and their ordering in the data file `Tab9_12_JW_MANCOVA.dat` in order to help the reader to follow the tests carried out in Fig. 5.46 for the analysis of the MANCOVA model being considered.

In Figs. 5.46 and 5.47 we have the commands used to analyze the proposed MANCOVA model, using the Mathematica® version of `PvalDataInd2Rp`, where, adopting a general step-down type approach, we try to trim down the original model to a simpler model by dropping from the model all interactions that are

```
PvalDataInd2Rp[{5, 6}, path <> "Tab9_12_JW_MANCOVA.dat", , 3, 1]
 Original variables {1, 2, 3, 4, 5, 6, 7, 8, 9, 14, 15}
Computed value of Λ: 0.4315796767
p-value: 0.01019608954
PvalDataInd2Rp[{2}, path <> "Tab9_12_JW_MANCOVA.dat", , 3, 1]
 Original variables {1, 2, 3, 4, 5, 6, 7, 8, 9, 14, 15}
Computed value of Λ: 0.02711116247
p-value: 4.649386456 × 10^-15
PvalDataInd2Rp[{9, 10}, path <> "Tab9_12_JW_MANCOVA.dat", , 3, 1]
 Original variables {1, 2, 3, 4, 5, 6, 7, 8, 9, 14, 15, 16, 17}
Computed value of Λ: 0.4979699551
p-value: 0.05138527937
PvalDataInd2Rp[{9, 10}, path <> "Tab9_12_JW_MANCOVA.dat", , 3, 1]
 Original variables {1, 2, 3, 4, 5, 6, 7, 8, 9, 14, 15, 18, 19}
Computed value of Λ: 0.4025270229
p-value: 0.01209477827
PvalDataInd2Rp[{9, 10}, path <> "Tab9_12_JW_MANCOVA.dat", , 3, 1]
 Original variables {1, 2, 3, 4, 5, 6, 7, 8, 9, 14, 15, 20, 21}
Computed value of Λ: 0.6618254519
p-value: 0.2841813826
```

Fig. 5.47 Commands used and output obtained with the Mathematica® module PvalDataInd2Rp to analyze the MANCOVA model based on the sales performance variables in Problem 9.19 and Table 9.12 in Johnson and Wichern (2014) used as response variables—tests for factors and covariates, and again for interactions of covariates with the first factor

nonsignificant and leaving only the covariates and factors that are significant or those for which the test of significance cannot be carried out because of being involved in some significant interaction.

We will have to start by testing the existence of possible interactions, that is, the significance of the interactions considered in the model in order to be able to know whether or not we will be allowed to test the individual significance of the two factors and the two covariates.

The first four commands in Fig. 5.46 test the significance of the interactions of the two continuous covariates with the two factors, in the presence of all other covariate-factor interactions in the model, even if these are nonsignificant, in order to test each of these interactions in interchangeable situations. As such, to carry out each of these tests we drop from the full model the corresponding two explanatory variables. We may see that for the first three of these tests very high p-values are obtained, leading us to conclude the nonsignificance of these interactions, which means that we will have for the two covariates the same slope for all levels of the second factor (mechanical reasoning) as well as the same slope for mathematical ability for the three levels of abstract reasoning. This is indeed one of the steps of testing the equality of the Multivariate Regression models for the levels of the factors being considered—see the next heading 'Setting 2'. Concerning the output of the fourth command in Fig. 5.46, it shows a rather low p-value of 0.0158 for the interaction between the first covariate (creativity) and the first factor (abstract reasoning). As such we decide to leave the corresponding explanatory variables (the 11th and 12th in Fig. 5.45) in the model. Therefore, we will from here on drop only the last six explanatory variables, and we then pursue to test the significance of the interaction between the two factors with the fifth command in Fig. 5.46, where the last six original explanatory variables are not used as shown. This is the reason why we have to use the first optional argument of module `PvalDataInd2Rp` with a value of 1, in order to allow us to choose from the variables in the file the ones that are supposed to enter the original model. Given the very large p-value obtained we also conclude the nonsignificance of this interaction, which indeed comes at no surprise considering the results in the previous paragraph. Consequently we will also drop from further analyzes the corresponding four indicator variables.

However, since the first factor is involved in a significant interaction with the first covariate, we are only free to test the significance of the second factor (mechanical reasoning) and of the second covariate (mathematical ability). This is what is done with the first two commands in Fig. 5.47. The first command in this figure tests the significance of the second factor and leads us to conclude its nonsignificance only in case we use an α value smaller than around 0.0102, that is, with a rather low p-value, while the second command tests the significance of the second covariate, yielding a result that leaves very little doubt about concluding its significance in the estimation of the three sales performance response variables.

As such, and adopting a step-down type approach, we would say that we are left with a model with only the two covariates, the two factors and the interaction between the first covariate and the first factor.

We may now think about testing if in a 'forward' type of approach we should add to the model the interactions between covariates and factors that were dropped from the original model, now without the inclusion of the interaction between the two factors in the model. We will carry out these tests knowing that some authors may disagree with this second run of such tests, once similar tests were previously carried out. Anyway, we will do this also in order to illustrate how we have to proceed when selecting the variables to enter the models in these cases. The tests are carried out with the three last commands in Fig. 5.47. The first one implements the test to the interaction between the second covariate and the first factor and the last two the interactions between the two covariates and the second factor. We may see that the interaction between the first covariate and the second factor might indeed be considered for inclusion in the model, given the rather low p-value obtained. One may want to pay attention to the way the variables entering the original model and the ones dropped from this model to build the submodel have to be designated, with for example the first argument of module `PvalDataInd2Rp` in the last command in Fig. 5.47 indicating that are the ninth and tenth explanatory variables that are to be dropped from the original model to build the submodel (see Fig. 5.45) and with the list provided as output from the reading of the data file indicating the variables selected from the complete data file to enter the "original" or H_1 model, where the first three variables are the three sales performance response variables.

Setting 2: Test of Equality of Model Parameters or Test of Equality of Models

In this subsection we will show how the adequate use of indicator variables is a powerful tool in modeling and in comparing Multivariate Linear models.

To start with, let us first illustrate the problem we propose to solve, with a simple situation in a Univariate Linear model. Let us suppose we have a Univariate Linear model with one only response variable Y, an intercept, and four predictor or explanatory variables X_1, X_2, X_3, and X_4. We may write this model as

$$Y_i = \beta_0 + \beta_1 X_{1i} + \beta_2 X_{2i} + \beta_3 X_{3i} + \beta_4 X_{4i} + \epsilon_i\,, \quad i = 1, \ldots, n\,, \tag{5.211}$$

where ϵ_i are independent Normal r.v.'s with null expected value and common variance σ^2. Let us suppose that we want to test the null hypothesis $H_0 : \beta_2 = \beta_3$ in model (5.211).

Then we may use the statistic

$$T = \frac{\hat{\beta}_2 - \hat{\beta}_3}{\sqrt{\widehat{Var}(\hat{\beta}_2 - \hat{\beta}_3)}}$$

which will have a T distribution with $n-5$ degrees of freedom, or otherwise we may use a partial F test to test between the model in (5.211), which will be used as the H_1 model, and the model

$$Y_i = \beta_0^* + \beta_1^* X_{1i} + \beta_2^* (X_{2i} + X_{3i}) + \beta_4^* X_{4i} + \epsilon_i^*\,, \quad i = 1, \ldots, n\,, \tag{5.212}$$

which will be used as the H_0 model.

But, if we would like to test in the model (5.211) the null hypothesis $H_0 : \beta_2 = \beta_3 = \beta_4$, then the only way to do the test would be with a partial F test between the model (5.211) and the model

$$Y_i = \beta_0^{**} + \beta_1^{**} X_{1i} + \beta_2^{**} (X_{2i} + X_{3i} + X_{4i}) + \epsilon_i^{**}\,, \quad i = 1, \ldots, n\,. \tag{5.213}$$

A version of this test, which makes use of indicator variables in an adequate way, may be used to test the equality of linear models and its generalization to the multivariate setting, where we have not only one but several response variables, will be done by using the partial Wilks' Lambda statistic. Let us suppose that we have a set of p_1 response variables that we want to model in terms of a set of p_2 explanatory variables, and that we have L of these models that we want to compare. We may think of each model as relating to a different hospital, state, city, location, etc., and we want to test for each of the p_2 explanatory variables whether we may or may not use a common slope for all L "locations," or whether otherwise we should use different slopes for the different "locations" or even whether we may use the same slope for some of the "locations," only for some of the explanatory variables. Hereafter we will refer to the different settings simply as "locations."

Assume that we have n_ℓ observations for the ℓ-th location ($\ell = 1, \ldots, L$) and let us take $n = n_1 + \cdots + n_\ell + \cdots + n_L$. Then we will have to use a matrix X_1 where the observations from each location are stacked side by side, forming a matrix with dimensions $p_1 \times n$, and similarly for our base matrix X_2 with the values for the explanatory variables. But then, whichever model we plan to test we also have to define a set of L indicator variables for the L locations, somehow similar to the indicator variables for the samples used in the matrix M in Sect. 5.1.4.2, the ℓ-th of which will have a value of 1 for the observations that belong to the ℓ-th location and a value of 0 (zero) for all other observations.

Let us suppose we want to test the fit of a model in which the p_1 response variables are modeled using different intercepts and different slopes for each of the p_2 explanatory variables and for each of the L locations. Our X_2 matrix for this model will have dimensions $(L + Lp_2) \times n$, being formed by stacking side by side, for the ℓ-th location a $(L + Lp_2) \times n_\ell$ matrix whose first L rows are formed by the values of the L indicator variables for this location, which will be indeed all formed by zeros, except the values for the ℓ-th indicator variable which will be all 1's, and the result of multiplying the values of each of the p_2 explanatory variables by each of these indicator variables. In Fig. 5.48, where all matrices appear in the transposed position, because of space limitations, the matrix in the file `matX.dat` is the one that would correspond to this setting, with the two first columns corresponding to the matrix X_1, that is, the response variables, and the remaining columns corresponding to the matrix X_2. In this very simple example we have three locations and two explanatory variables. Although this may indeed constitute a very simple example, we will use it to illustrate several of the possible tests that we may carry out among such models.

Columns 3–5 in `matX.dat` in Fig. 5.48 have the values of the three indicator variables for the three locations and correspond to different intercepts for each location. Adding these three columns will give us the column of 1's in `matXA.dat` also in Fig. 5.48. Columns 6–8 in `matX.dat` result from multiplying the values of the first explanatory variable by the three indicator variables. Adding these three columns we get the fourth column in `matXA.dat`, which corresponds to the values of the first explanatory variable. Similarly, columns 9–11 in `matX.dat` result from multiplying the values of the second explanatory variable by the indicator variables for the three locations and adding these three columns gives the fifth column in `matXA.dat`, which corresponds to the values of the second explanatory variable. The matrix `matX` will correspond to our H_1 model.

file matX.dat
```
23.4 56.7  1 0 0 21.2    0    0 36.5    0    0
34.5 23.4  1 0 0 28.9    0    0 25.4    0    0
21.2 33.3  1 0 0 56.7    0    0 25.6    0    0
56.8 56.4  0 1 0    0 25.6    0    0 43.3    0
23.4 56.8  0 1 0    0 23.4    0    0 52.2    0
33.5 45.8  0 1 0    0 23.6    0    0 42.2    0
33.7 38.9  0 1 0    0 56.8    0    0 53.2    0
43.2 45.6  0 0 1    0    0 34.5    0    0 32.2
56.4 24.7  0 0 1    0    0 78.3    0    0 34.3
74.5 83.5  0 0 1    0    0 84.3    0    0 52.3
24.5 25.7  0 0 1    0    0 37.4    0    0 52.3
35.7 74.5  0 0 1    0    0 84.6    0    0 83.3
```

file matXA.dat
```
23.4 56.7  1 21.2 36.5
34.5 23.4  1 28.9 25.4
21.2 33.3  1 56.7 25.6
56.8 56.4  1 25.6 43.3
23.4 56.8  1 23.4 52.2
33.5 45.8  1 23.6 42.2
33.7 38.9  1 56.8 53.2
43.2 45.6  1 34.5 32.2
56.4 24.7  1 78.3 34.3
74.5 83.5  1 84.3 52.3
24.5 25.7  1 37.4 52.3
35.7 74.5  1 84.6 83.3
```

file matXB.dat
```
23.4 56.7  1 0 0 21.2    0 36.5    0    0
34.5 23.4  1 0 0 28.9    0 25.4    0    0
21.2 33.3  1 0 0 56.7    0 25.6    0    0
56.8 56.4  0 1 0 25.6    0    0 43.3    0
23.4 56.8  0 1 0 23.4    0    0 52.2    0
33.5 45.8  0 1 0 23.6    0    0 42.2    0
33.7 38.9  0 1 0 56.8    0    0 53.2    0
43.2 45.6  0 0 1    0 34.5    0    0 32.2
56.4 24.7  0 0 1    0 78.3    0    0 34.3
74.5 83.5  0 0 1    0 84.3    0    0 52.3
24.5 25.7  0 0 1    0 37.4    0    0 52.3
35.7 74.5  0 0 1    0 84.6    0    0 83.3
```

file matXC.dat
```
23.4 56.7  1 0 21.2    0 36.5    0
34.5 23.4  1 0 28.9    0 25.4    0
21.2 33.3  1 0 56.7    0 25.6    0
56.8 56.4  0 1 25.6    0    0 43.3
23.4 56.8  0 1 23.4    0    0 52.2
33.5 45.8  0 1 23.6    0    0 42.2
33.7 38.9  0 1 56.8    0    0 53.2
43.2 45.6  0 1    0 34.5    0 32.2
56.4 24.7  0 1    0 78.3    0 34.3
74.5 83.5  0 1    0 84.3    0 52.3
24.5 25.7  0 1    0 37.4    0 52.3
35.7 74.5  0 1    0 84.6    0 83.3
```

file matXD.dat
```
23.4 56.7  1 0 21.2 36.5    0
34.5 23.4  1 0 28.9 25.4    0
21.2 33.3  1 0 56.7 25.6    0
56.8 56.4  0 1 25.6    0 43.3
23.4 56.8  0 1 23.4    0 52.2
33.5 45.8  0 1 23.6    0 42.2
33.7 38.9  0 1 56.8    0 53.2
43.2 45.6  0 1 34.5    0 32.2
56.4 24.7  0 1 78.3    0 34.3
74.5 83.5  0 1 84.3    0 52.3
24.5 25.7  0 1 37.4    0 52.3
35.7 74.5  0 1 84.6    0 83.3
```

Fig. 5.48 Examples of data files to be used for tests between different multivariate linear models in Sect. 5.1.9.4

Then if we want to test between this model and a model that has a common intercept and a common slope for both explanatory variables for all three locations, we will use for our H_0 model the very same X_1 matrix that we used for the H_1 model, and which will in fact be always the same for any model, and an X_2 matrix as the one in file `matXA.dat` in Fig. 5.48. We may note that the column of 1's corresponds to the common intercept in the model and that it may be seen as the sum of the columns that in `matX.dat` correspond to the indicator variables for the three locations, while each of the two columns that now have the values for each one of the two explanatory variables in the model may be seen respectively as the sum of the second and third set of three columns of X_2, that is, columns 4–6 and 7–9, in `matX.dat`. Within this setting we may see the test between these two models as a particular case where we may apply the partial Wilks Lambda test, which now appears as a powerful tool to test the equality of Multivariate Linear models. A rejection of the null hypothesis would show that there is a significant difference between the fit of the H_1 and H_0 models, leading us to stick with the model associated with H_1. On the other hand, a non-rejection of the null hypothesis would show that we may use a model with a common intercept and common slope for both explanatory variables, for the $L(=3)$ locations. The statistic Λ_{H_1,H_0} has in this case its exact p.d.f. and c.d.f. given by Corollary 4.2, with the parameters given by (5.185) or (5.186) with $p_1 = 2$, $p_{21} = 3$ and $p_{22} = 6$.

In tests of this type we will have to use twice module `PvalDataInd2R` to obtain the computed value of the statistic Λ for the H_1 and the H_0 models and then the computed value for the partial Wilks' Λ statistic from the ratio of these two statistics, obtaining then the corresponding p-value by using module `PvalInd2R` with the due set of parameters.

In case we reject H_0 in the above test we may then plan on testing between the model in H_1 and another model that lies "in between" the model in H_1 and the above model associated with H_0, as any of the models associated with any of the other data matrices in Fig. 5.48.

In designing a hierarchy for the models corresponding to the data matrices in Fig. 5.48, where the model in a given item is a submodel of any of the models in previous items we have:

(i) the model corresponding to `matX`, with different intercepts for the three locations and different slopes for the three locations for each of the two explanatory variables,
(ii) the model corresponding to `matXB`, with different intercepts for the three locations, different slopes for the three locations for the second explanatory variable, and the same slope for the first and second locations for the first explanatory variable,
(iii) the model corresponding to `matXC`, with the same intercept for the second and third locations, the same slope for the first and second locations for the first explanatory variable, and the same slope for the second and third locations for the second explanatory variable,
(iv) the model corresponding to `matXD`, with the same intercept for the second and third locations, the same slope for all three locations for the first explanatory variable, and the same slope for the second and third locations for the second explanatory variable,
(v) the model corresponding to `matxA`, with the same intercept for all three locations and the same slope for all three locations for the first and the second explanatory variables.

Let us use the model that would be associated with the data matrix `matXB`, that is, the model in (ii) above, as our H_0 model. This is a model with different intercepts for the three locations, the same slope for the first explanatory variable for locations 1 and 2 and a different slope for location 3, and different slopes for the second explanatory variable for the three locations. Keeping as our H_1 model the model in (i) above, with different intercepts and different slopes for all three locations, the statistic Λ_{H_1,H_0} has now its exact p.d.f. and c.d.f. given by Corollary 4.2, with the parameters given by (5.185) with $p_1 = 2$, $p_{21} = 8$, and $p_{22} = 1$.

Let us suppose we do not reject H_0 so that we stick with the model in (ii) above. Then we may want to use this model as the H_1 model and the model associated with the data matrix `matXC.dat`, the model in (iii) above, as our H_0 model, since this latter model has the same intercept for locations 2 and 3 and also the same slope for the second explanatory variable for locations 2 and 3. In this case, the statistic Λ_{H_1,H_0} has its exact p.d.f. and c.d.f. given by Corollary 4.2, with the parameters given by (5.185) or (5.186) with $p_1 = 2$, $p_{21} = 6$, and $p_{22} = 2$.

In case we do not reject H_0 we may then plan to use the model in (iii) above as the H_1 model and the model associated with the data matrix `matXD`, the model in (iv) above, as the H_0 model, which has now the same slope for all 3 locations for the first explanatory variable. The statistic Λ_{H_1,H_0} has now its exact p.d.f. and c.d.f. given by Corollary 4.2, with the parameters given by (5.185) with $p_1 = 2$, $p_{21} = 5$, and $p_{22} = 1$.

Assume that we do not reject H_0 again. Then, we might want to test the model associated with `matX.dat` as the H_1 model versus the model associated with `matXD.dat` as the H_0 model. In this case the statistic Λ_{H_1,H_0} has its exact p.d.f. and c.d.f. given by Corollary 4.2, with the parameters given by (5.185) or (5.186) with $p_1 = 2$, $p_{21} = 5$, and $p_{22} = 4$.

A cautionary note is appropriate here. If each of the above tests is carried out at a given α level, then, because the tests may be not independent, the overall level of the full set of tests may be lower than α. This means that, for example, supposing that when doing the test between the model associated with the data matrix `MatXD`, that is, the model in (iv) above, as the H_0 model, and the model associated with the data matrix `matXC`, that is, the model in (iii) above, as the H_1 model, we did not reject the H_0 model, this would not mean that when carrying out the test between the model associated with the data matrix `matXD`, the model in (iv) above, as the H_1 model and the model associated with the data matrix `MatX`, the model in (i) above, as the H_0 model, we would not reject this H_0 model.

In the distribution of the statistic Λ_{H_1,H_0} for each of the above tests p_1 represents the number of response variables in the model, while p_{21} represents the number of explanatory variables in the H_0 model and p_{22} the difference between the number of explanatory variables in the H_1 and H_0 models.

5.1.9.6 Paired Samples, Repeated Measures, Longitudinal Studies, Random-Block Designs, and Growth Curve Models

All we have to do to be able to address a paired sample, repeated measures, or random-block design or a longitudinal study with any number of paired samples or repeated measures in the realm of the partial Wilks' Lambda test in the previous subsection is to setup a data matrix as follows:

$$\underset{(p\times nq)}{X} = \left[\begin{array}{c} \underset{(p_1\times nq)}{X_1} \\ \hline \underset{(p_2\times nq)}{X_2} \end{array}\right] = \left[\begin{array}{c} \underset{(p_1\times nq)}{X_1} \\ \hline \underset{((n-1)\times nq)}{I^*} \\ \hline M \end{array}\right] = \left[\begin{array}{c|c|c|c} X_{11}^* & X_{12}^* & \cdots & X_{1q}^* \\ \hline I_1^* & I_2^* & \cdots & I_q^* \\ \hline \multicolumn{4}{c}{M_{((q-1)\times nq)}} \end{array}\right] \qquad (5.214)$$

where $p = p_1 + p_2$, with $p_2 = n - 1 + q - 1$, where q is the number of paired samples or repeated measures, X_{1k}^* $(k = 1, \ldots, q)$ is the $p_1\times n$ data matrix with the k-th paired sample or repeated measure, taken on the n individuals and where I_k^* $(k = 1, \ldots, q)$ are $n\times n$ identity matrices with their first row removed and M is the $(q - 1)\times nq$ matrix of the $q - 1$ indicator variables for the q samples, that is, the indicator variables for the q levels of the factor in the study, with the one corresponding to the first level removed. We should note that with this setup we are assuming that each X_{1k}^* matrix yields the measurements of the p_1 continuous variables measured on the same n individuals, it being the case that the order of these n individuals is the same along all q samples, that is, along all X_{1k}^* matrices. The set of q matrices I_k^* $(k = 1, \ldots, q)$ form the set of $n-1$ indicator variables for the pairing of observations, or, in a random-block design, form the set of indicator variables for the blocks, with the indicator for the first level, that is, for the first pairing or the first block, removed.

We should note that in a more general setup we do not even need to have all samples ordered in the same way concerning the individuals observed or measured, neither need the samples to be complete, it being only the case that the matrix I^* has to be formed by the indicator variables for the pairings appropriately defined.

Then we will be interested in testing the significance of the set of indicator variables in M, in the presence in the model of all the indicator variables in the I^* matrix, in a model where the p_1 continuous variables, whose sample is found in X_1, are taken as the response variables, and the $n - 1 + q - 1$ indicator variables whose sample is found in X_2 are taken as the explanatory variables.

In a more elaborate setup M may be the matrix of indicator variables for a set of factors and interactions, since we may be interested in testing the significance of a given interaction or factor or set of interactions or factors, always in the presence in the model of the indicator variables for the pairings, that is, in the presence in the model of the I^* matrix of pairing indicators, and always abiding by the rule that a given factor or interaction can only be tested if it is not involved in a significant interaction of higher order, that is, if such interaction is no longer present in the model.

As a first example of the application of this approach we will use once again, as in Sect. 5.1.4.6 the bone mineral data in Tables 1.8 and 6.16 of Johnson and Wichern (2014) to test if there are differences between the mean vectors of bone mineral content of the 6 bones (dominant and nondominant radius, humerus, and ulna) for the two dates of measurement. We will use only the first 24 observations in Table 1.8 of Johnson and Wichern (2014) since these are the ones that correspond to the individuals appearing in Table 6.16, on which data was also collected 1 year later. Since the data file to be used with the module `PvalDataInd2Rp` has the transposed form of the matrix X in (5.214) we set the data matrix in Table 1.8 of Johnson and Wichern (2014) to be the transpose of X_{11}^* and the data matrix in Table 6.16 to be the transpose of X_{12}^*, while I_1^* and I_2^* are identity matrices of order 24 with their first row removed. The data matrix in file `Bones_2am.dat` has the aspect of the data matrix

$$\underset{(48\times 25)}{X'} = \left[\begin{array}{cc|c} X_{11}^{*\prime} & I_1^{*\prime} & \begin{matrix}0\\ \vdots\\ 0\end{matrix} \\ \hline X_{12}^{*\prime} & I_2^{*\prime} & \begin{matrix}1\\ \vdots\\ 1\end{matrix} \end{array}\right]$$

where the last column of zeros and ones is the transpose of the matrix M in (5.214), with the indicator for the second paired sample, or say, for the second level of the factor we are interested in analyzing, which here is "time," or rather, the effect of "time," with two levels (level 1—the starting date of the study, level 2—1 year after the beginning of the study).

We are interested in testing the significance of this indicator variable, that is, of the factor "time," in the presence of the indicator variables for the pairings. The command used and the output obtained with the Mathematica® version of module `PvalDataInd2Rp` are displayed in Fig. 5.49. The command used has the syntax displayed, with its first argument indicating that the variable whose significance in the model we want to test is the 24-th variable in the set of explanatory variables and its last argument indicates that the set of response variables has six variables. This last argument is necessary since the two data matrices, X_1 and X_2 are both included in a single file. We may see that the result is exactly the same as the one obtained in Sect. 5.1.4.6 when the test in Sect. 5.1.4 was used.

Growth Curves

Another possible choice for the study and modelization of Repeated measures or Longitudinal designs are the Growth Curve models (Wishart, 1938; Box, 1950; Rao,

```
PvalDataInd2Rp[{24}, path <> "bones_2am.dat", , 6]
Computed value of Λ: 0.7182600770
p-value: 0.3616271845
```

Fig. 5.49 Command used and output obtained with the Mathematica® module `PvalDataInd2Rp` to implement the test of equality of mean vectors, based on two paired samples, using a data matrix with the differences of the bone mineral values in Tables 1.8 (first 24 observations) and 6.16 of Johnson and Wichern (2014)

1958, 1959; Leech and Healy, 1959; Healy, 1961; Bock, 1963; Potthoff and Roy, 1964; Khatri, 1964, 1966, 1973; Kshirsagar and Williams, 1995).

These models may be used as a successful replacement for the MANOVA random-block or split-plot models when there is some trend, usually along time, in the measurements. In the more general form of these models the mean values for the q repeated measures are most commonly modeled along time with a polynomial in time of degree equal or less than $q - 1$ (Rao, 1959; Potthoff and Roy, 1964; Kshirsagar and Williams, 1995). As Kshirsagar and Williams (1995) state in the Introduction of their book, "Growth curve models are more general than repeated measures models. The basic premise in growth curve models is that there is a functional relationship between a treatment effect and its time of application, and that this relationship may be modeled. The actual functional relationship may be general, but in the absence of a functional form being specified by subject-matter area theory, the function may be approximated by a polynomial." Furthermore, the same authors even say that "A distinction between repeated measures and growth curve models is then analogous to the perceived differences between analysis of variance and regression models. The former is stated in terms of class effects and interactions, while the latter in linear or polynomial functions with unknown coefficients, but with known values of the regressor (input) variables." Under this setup the aim is to find an adequate polynomial that models the q repeated measures with $q - 1$ or fewer powers of time and which exhibits for the set of these $q - 1$ or less time powers a p-value that is equal or smaller than the p-value exhibited by the $q - 1$ indicator variables for the levels of the factor time. This is so since the p-value is able to precisely weight or balance the number of degrees of freedom used by the model, or equivalently, the number of explanatory variables it uses, with the amount of overall variance in the response variables that is explained by the model, making a model with fewer explanatory variables which displays a lower p-value for the overall set of these explanatory variables a preferable model. Although such a model will always have a larger sum of squares for the error than the model with the $q-1$ indicator variables or the $q-1$ powers of time, we will consider that model as a preferable model since it has a lower p-value for the overall set of time powers than the full model, that is, the model with the $q - 1$ powers of time, and also has a larger number of degrees of freedom for the error term.

We will first illustrate this process with the example used by Potthoff and Roy (1964), where the distance from the center of the pituitary to the pteryomaxillary fissure was measured on 11 girls and 16 boys at ages of 8, 10, 12, and 14 years. As such, there are $q = 4$ repeated measures and as Potthoff and Roy (1964) state, we want here to answer the two questions: "(a) Should the growth curves be represented by second degree equations in time, or are linear equations adequate? (b) Should two separate curves be used for boys and girls, or do both have the same growth curve?". The contents of the data file `P&G.dat` used is the transpose of the 31×108 matrix

$$\underset{(31\times108)}{X} = \begin{bmatrix} \underset{(1\times108)}{X_1} \\ \hline \underset{(30\times108)}{X_2} \end{bmatrix} = \left[\begin{array}{cccccccc} G_1 & G_2 & G_3 & G_4 & B_1 & B_2 & B_3 & B_4 \\ \hline I^*_{11} & I^*_{11} & I^*_{11} & I^*_{11} & 0_{10\times16} & 0_{10\times16} & 0_{10\times16} & 0_{10\times16} \\ 0_{15\times11} & 0_{15\times11} & 0_{15\times11} & 0_{15\times11} & I^*_{16} & I^*_{16} & I^*_{16} & I^*_{16} \\ 0_{1\times11} & 0_{1\times11} & 0_{1\times11} & 0_{1\times11} & E_{1,16} & E_{1,16} & E_{1,16} & E_{1,16} \\ E_{1,11} & 2\times E_{1,11} & 3\times E_{1,11} & 4\times E_{1,11} & E_{1,16} & 2\times E_{1,16} & 3\times E_{1,16} & 4\times E_{1,16} \\ E_{1,11} & 4\times E_{1,11} & 9\times E_{1,11} & 16\times E_{1,11} & E_{1,16} & 4\times E_{1,16} & 9\times E_{1,16} & 16\times E_{1,16} \\ 0_{1\times11} & 0_{1\times11} & 0_{1\times11} & 0_{1\times11} & E_{1,16} & 2\times E_{1,16} & 3\times E_{1,16} & 4\times E_{1,16} \\ 0_{1\times11} & 0_{1\times11} & 0_{1\times11} & 0_{1\times11} & E_{1,16} & 4\times E_{1,16} & 9\times E_{1,16} & 16\times E_{1,16} \end{array}\right] \begin{array}{c} 1 \\ 10 \\ 15 \\ 1 \\ 1 \\ 1 \\ 1 \\ 1 \end{array}$$
$$\qquad\qquad\qquad\qquad \begin{array}{cccccccc} 11 & 11 & 11 & 11 & 16 & 16 & 16 & 16 \end{array}$$

where the integer numbers placed on the right-hand side and at the bottom of the last form of the data matrix show the numbers of rows and columns of each submatrix, with G_j ($j = 1, \ldots, 4$) representing the 11×1 row vectors with the values for the j-th measurement on the 11 girls and B_j ($j = 1, \ldots, 4$) the 16×1 row vectors with the values for the j-th measurement on the 16 boys. The 10+15 first rows of X_2 form the set of 25 indicator variables for the pairings of the observations (where I^*_n stands for an identity matrix of order n with the first row removed). The next row is the indicator variable for sex (where $E_{1,n}$ stand for a row vector of n 1's), and the following two rows yield the first and second degrees of the time polynomial, while the last two rows of X_2 yield the interaction between sex and time, being exactly the product of the indicator variable for sex by the two explanatory variables with the time polynomials of degrees one and two.

The distance between the pituitary and the pteryomaxillary fissure will be our response variable and the variables corresponding to the rows of X_2 our explanatory variables. We will first answer question (a) above by testing the joint significance in the model of the variables corresponding to the 28-th and 30-th rows of X_2, which are the second power of time and its interaction with sex. This is done with the first command in Fig. 5.50, where the Mathematica® version of module `PvalDataInd2Rp` is used. The p-value obtained shows that those two variables should be considered as nonsignificant in the model and that as such a model with only the first power of time would be good enough. That the first power of time and its interaction with sex are important in our model is confirmed by the result obtained from the test carried out with the second command in Fig. 5.50, where the variables corresponding to the second power of time and its interaction with time are not anymore considered in the model. We may note that since we are using a single data file with both data matrices X_1 and X_2 we will need to leave the second argument of the module empty and provide as third argument the number of variables in X_1, which is just one. Then the use of the value 1 for the first optional argument will enable us to list the variables that will enter the model. The first variable, to which is given the number 1 refers to the response variable, and the variables whose significance in the model we want to test are variables 27 and 28 in the set of explanatory variables, once the variables for the second power of time and its interaction with sex are not considered in the model. Then the third command in Fig. 5.50 tries to help us answering question (b), by testing the significance of the interaction of the first power of time and sex in a model where the first power of time is already considered as an explanatory variable. The p-value obtained, being rather low, shows that there is some evidence that different growth curves should be considered for girls and boys.

```
(* testing simultaneously the 2nd power of time and
                              its interaction with sex *)
PvalDataInd2Rp[{28, 30}, path <> "P&R.dat", , 1]
Computed value of Λ: 0.9825720086
p-value: 0.5081932958
(* testing simultaneously the 1st power of time and
                              its interaction with sex *)
PvalDataInd2Rp[{27, 28}, path <> "P&R.dat", , 1, 1]
 Original variables {1, 2, 3, 4, 5, 6, 7, 8, 9, 10, 11,
   12, 13, 14, 15, 16, 17, 18, 19, 20, 21, 22, 23, 24, 25, 26, 27, 28, 30}
Computed value of Λ: 0.3802593947
p-value: 2.589155449×10^-17
(* testing significance of the interaction of
                         the 1st power of time and sex *)
PvalDataInd2Rp[{28}, path <> "P&R.dat", , 1, 1]
 Original variables {1, 2, 3, 4, 5, 6, 7, 8, 9, 10, 11,
   12, 13, 14, 15, 16, 17, 18, 19, 20, 21, 22, 23, 24, 25, 26, 27, 28, 30}
Computed value of Λ: 0.9261136136
p-value: 0.014097451
```

Fig. 5.50 Commands used and output obtained with the Mathematica® version of module `PvalDataInd2Rp` to analyze the example for a growth curve model used by Potthoff and Roy (1964)

For the analysis of the above problem we might have considered just a MANOVA model where the four measurement times would be considered as the four levels of a factor time, and as such consider the data matrix

$$
\underset{(33\times108)}{X} = \left[\begin{array}{c} \underset{(1\times108)}{X_1} \\ \hline \\ \underset{(32\times108)}{X_2} \\ \\ \end{array}\right] = \left[\begin{array}{cccccccc} G_1 & G_2 & G_3 & G_4 & B_1 & B_2 & B_3 & B_4 \\ \hline I_{11}^* & I_{11}^* & I_{11}^* & I_{11}^* & 0_{10\times16} & 0_{10\times16} & 0_{10\times16} & 0_{10\times16} \\ 0_{15\times11} & 0_{15\times11} & 0_{15\times11} & 0_{15\times11} & I_{16}^* & I_{16}^* & I_{16}^* & I_{16}^* \\ 0_{1\times11} & 0_{1\times11} & 0_{1\times11} & 0_{1\times11} & E_{1,16} & E_{1,16} & E_{1,16} & E_{1,16} \\ 0_{1\times11} & E_{1,11} & 0_{1\times11} & 0_{1\times11} & 0_{1\times16} & E_{1,16} & 0_{1\times16} & 0_{1\times16} \\ 0_{1\times11} & 0_{1\times11} & E_{1,11} & 0_{1\times11} & 0_{1\times16} & 0_{1\times16} & E_{1,16} & 0_{1\times16} \\ 0_{1\times11} & 0_{1\times11} & 0_{1\times11} & E_{1,11} & 0_{1\times16} & 0_{1\times16} & 0_{1\times16} & E_{1,16} \\ 0_{1\times11} & 0_{1\times11} & 0_{1\times11} & 0_{1\times11} & 0_{1\times16} & E_{1,16} & 0_{1\times16} & 0_{1\times16} \\ 0_{1\times11} & 0_{1\times11} & 0_{1\times11} & 0_{1\times11} & 0_{1\times16} & 0_{1\times16} & E_{1,16} & 0_{1\times16} \\ 0_{1\times11} & 0_{1\times11} & 0_{1\times11} & 0_{1\times11} & 0_{1\times16} & 0_{1\times16} & 0_{1\times16} & E_{1,16} \end{array}\right] \begin{array}{c} 1 \\ 10 \\ 15 \\ 1 \\ 1 \\ 1 \\ 1 \\ 1 \\ 1 \\ 1 \end{array}
$$
$$
\begin{array}{cccccccc} 11 & 11 & 11 & 11 & 16 & 16 & 16 & 16 \end{array}
$$

where we would now be interested in testing the significance in the model of the variables corresponding to the last six rows of X_2, the first three of which are the indicator variables for the last three levels of the factor time and the last three correspond to the interaction of this factor with sex, being just the product of the previous three indicator variables by the indicator variable for sex. This test is done with the command in Fig. 5.51 and we may see that the p-value obtained, 2.194460×10^{-14}, is larger than the one obtained with just the first power of time and its interaction with sex, which was 2.589155×10^{-17}, indicating a clear linear trend in the mean values for the response variable along time, and showing the value of the Growth Curve models in modeling these situations with fewer variables.

The model considered above is indeed a univariate model, which has a single response variable, and as such a univariate approach, with tests based on F statistics might have been used.

In order to illustrate the building process of truly multivariate Growth Curve models as well as the sharpness and sensitivity of these models based on time

```
PvalDataInd2Rp[{27, 28, 29, 30, 31, 32}, path <> "P&R_MANOVA.dat", , 1]
Computed value of Λ: 0.3709571849
p-value: 2.194460062 × 10^-14
```

Fig. 5.51 Commands used and output obtained with the Mathematica® version of module `PvalDataInd2Rp` to analyze the example for a growth curve model used by Potthoff and Roy (1964), now using a MANOVA model

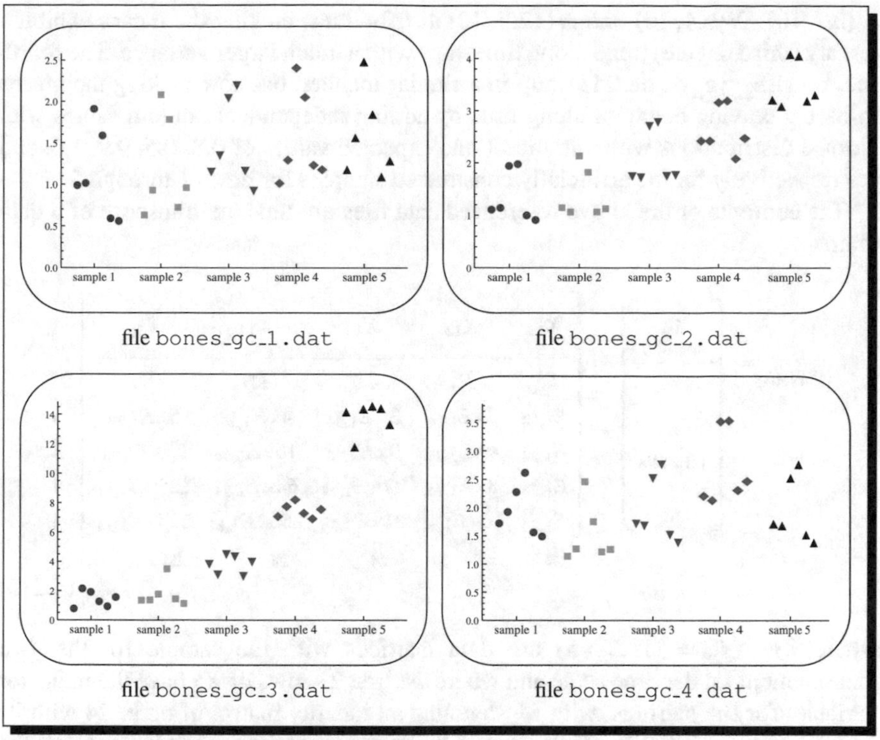

Fig. 5.52 Plots of means for the files used to illustrate the multivariate growth curve model (plots have different vertical scales)

polynomials we will use a set of four artificially generated data files, each with $q = 5$ samples supposedly taken along time. For the first two of these files the sample means exhibit a quite sharp evolution as time progresses, according to polynomials of degrees 1 and 2. In the third file the evolution of the sample means is less sharp and follows a time polynomial of degree 3, while in the fourth file they follow a wavelike pattern (see Fig. 5.52).

The origin of the samples in each of these files is the contents of Table 6.16 in Johnson and Wichern (2014) with the bone mineral content for the six bones (dominant and nondominant radius, humerus, and ulna) measured on 24 women to which a random quantity is added. For the first file, named `bones_gc_1.dat`, the first, second, third, fourth and fifth samples are generated by adding to the data in that table independent random quantities respectively with $N(0.1, 1)$, $N(0.2, 1)$, $N(0.3, 1)$, $N(0.4, 1)$ and $N(0.5, 1)$ distributions, thus making the means exhibit a linear trend along time, while for the file `bones_gc_2.dat` the means are made to exhibit a quadratic (second degree) trend by generating the first, second, third, fourth and fifth samples by adding independent random quantities respectively with $N(0.1, 1)$, $N(0.4, 1)$, $N(0.9, 1)$, $N(1.6, 1)$ and $N(2.5, 1)$ distributions. For file `bones_gc_3.dat` we added quantities respectively with $N(0.1, 10)$, $N(0.8, 10)$,

$N(2.7, 10)$, $N(6.4, 10)$ and $N(12.5, 10)$ distributions, making the means exhibit a tertiary (third degree) trend along time, now with a much larger variance. The fourth file, `bones_gc_4.dat` is setup in a similar manner, but now making the means to have a waving behavior along time by adding independent random values with Normal distributions with variance 1 and expected values of 0.8, 0.5, 0.9, 1.6, and 0.9 respectively for the artificially constructed samples for times 1 through 5.

The contents of the above referenced data files are thus the transpose of a data matrix

$$\underset{(33\times 120)}{X} = \begin{bmatrix} \underset{(6\times 120)}{X_1} \\ \hline \\ \underset{(27\times 120)}{X_2} \\ \\ \end{bmatrix} = \left[\begin{array}{ccccc} X_{11} & X_{12} & X_{13} & X_{14} & X_{15} \\ \hline I_{24}^* & I_{24}^* & I_{24}^* & I_{24}^* & I_{24}^* \\ E_{1,24} & 2\times E_{1,24} & 3\times E_{1,24} & 4\times E_{1,24} & 5\times E_{1,24} \\ E_{1,24} & 4\times E_{1,24} & 9\times E_{1,24} & 16\times E_{1,24} & 25\times E_{1,24} \\ E_{1,24} & 8\times E_{1,24} & 27\times E_{1,24} & 64\times E_{1,24} & 125\times E_{1,24} \\ E_{1,24} & 16\times E_{1,24} & 81\times E_{1,24} & 256\times E_{1,24} & 625\times E_{1,24} \end{array}\right]\begin{array}{c} 6 \\ 23 \\ 1 \\ 1 \\ 1 \\ 1 \end{array}$$
$$\begin{array}{ccccc} 24 & 24 & 24 & 24 & 24 \end{array} \tag{5.215}$$

where X_{1j} $(j = 1, \ldots, 5)$ are data matrices with the sample for the j-th measurement on the time scale and where the first 23 rows of X_2 have the indicator variables for the pairings, with I_{24}^* denoting an identity matrix of order 24 with its first row removed. The last four rows of X_2 correspond to the variables that bear the values for the $q-1$ powers of time to be used in the model.

We should note here that, as it will be shown, in these models the p-value for the whole set of these $q-1$ powers of time and the p-value for the whole set of $q-1$ indicator variables for the q levels of the factor time, that is, for the q dates of observation, where the indicator for one of the levels, usually the first one, is dropped, is exactly the same. To show this feature, also files `bones_ind_1.dat`, `bones_ind_2.dat`, `bones_ind_3.dat`, and `bones_ind_4.dat` whose content is the transpose of a data matrix

$$\underset{(33\times 120)}{X} = \begin{bmatrix} \underset{(6\times 120)}{X_1} \\ \hline \\ \underset{(27\times 120)}{X_2} \\ \\ \end{bmatrix} = \left[\begin{array}{ccccc} X_{11} & X_{12} & X_{13} & X_{14} & X_{15} \\ \hline I_{24}^* & I_{24}^* & I_{24}^* & I_{24}^* & I_{24}^* \\ 0_{1\times 24} & E_{1,24} & 0_{1\times 24} & 0_{1\times 24} & 0_{1\times 24} \\ 0_{1\times 24} & 0_{1\times 24} & E_{1,24} & 0_{1\times 24} & 0_{1\times 24} \\ 0_{1\times 24} & 0_{1\times 24} & 0_{1\times 24} & E_{1,24} & 0_{1\times 24} \\ 0_{1\times 24} & 0_{1\times 24} & 0_{1\times 24} & 0_{1\times 24} & E_{1,24} \end{array}\right]\begin{array}{c} 6 \\ 23 \\ 1 \\ 1 \\ 1 \\ 1 \end{array} \tag{5.216}$$
$$\begin{array}{ccccc} 24 & 24 & 24 & 24 & 24 \end{array}$$

where the last four rows of X_2 are exactly the indicator variables for the last four of the five measurements along time, will be used.

Our objective will be to find a polynomial in time of degree equal or less than $q-1$ which displays for the set of powers of time in the model a p-value equal or lower than that of the set of $q-1$ indicator variables for the repeated measures taken along time, and to show that the powers of time in the polynomial chosen make complete sense when reported to the contents of each file.

To illustrate the trend of the sample means along time, plots of these means for each of the four data files considered are displayed in Fig. 5.52.

The empirical method most commonly used to obtain an adequate polynomial in time is to start with the first power and then, in case this power "belongs in the model," check if in a model where this first power is included we should or not add the second power. In the affirmative case we would then check if we should add the third power, continuing till we reach power $q-1$ or otherwise stop the procedure as soon as we find any given power of time that should not be included in the model. We will call this method the "empirical forward" method. We will then carry out a backward procedure to check if after adding further powers of time, the lower powers in the model are still relevant or not, leaving only the significant powers. Furthermore we will also use a truly "forward" method and a "backward" method.

In Fig. 5.53 we have the commands used to implement the "empirical forward" method with the data in file bones_gc_1.dat, where a linear trend in the means

```
(* computing p-value for the set of 4 indicator variables
      corresponding to the five measurement times *)
PvalDataInd2Rp [{24, 25, 26, 27}, path <> "bones_ind_1.dat", , 6]
Computed value of Λ: 0.7345352070
p-value: 0.2639833687
(* computing p-value for the set of 4 powers of time *)
PvalDataInd2Rp [{24, 25, 26, 27}, path <> "bones_gc_1.dat", , 6]
Computed value of Λ: 0.7345352070
p-value: 0.2639833687
(* checking if in a model with power 1 of time we should
                              or not include power 2 of time *)
PvalDataInd2Rp [{25}, path <> "bones_gc_1.dat", , 6, 1]
 Original variables {1, 2, 3, 4, 5, 6, 7, 8, 9, 10, 11, 12, 13,
  14, 15, 16, 17, 18, 19, 20, 21, 22, 23, 24, 25, 26, 27, 28, 29, 30, 31}
Computed value of Λ: 0.9548911990
p-value: 0.6497211411
```

Fig. 5.53 Commands used and output obtained with the Mathematica® version of module PvalDataInd2Rp to obtain the p-values for the whole set of indicator variables and the whole set of four powers of time for the data in file bones_gc_1.dat and to implement the "empirical forward" method

is present, and the corresponding output obtained. All three methods, the "empirical forward," the forward and the backward methods all give the same final model, which is the model with just the first power of time. It is interesting to note that while the model with the full set of indicator variables for the samples has what we would consider a rather high p-value of 0.263983, indicating that there are no significant differences among the mean vectors of the six response variables for the five repeated measures, the model with only the first power of time has a much lower p-value of 0.006587 (see the output of the first command in Fig. 5.54), showing that there is a significant linear trend in the means.

From the output obtained from the two first commands in Fig. 5.53 we may see that both the p-value for the set of four indicator variables for the levels of time and the p-value for the set of the first four powers of time are exactly the same.

```
(* Forward method - checking which power of time should
                                first enter the model *)

PvalDataInd2Rp[{24}, path <> "bones_gc_1.dat", , 6, 1]

 Original variables {1, 2, 3, 4, 5, 6, 7, 8, 9, 10, 11, 12,
  13, 14, 15, 16, 17, 18, 19, 20, 21, 22, 23, 24, 25, 26, 27, 28, 29, 30}

Computed value of Λ: 0.8234544565

p value: 0.006587463852


PvalDataInd2Rp[{24}, path <> "bones_gc_1.dat", , 6, 1]

 Original variables {1, 2, 3, 4, 5, 6, 7, 8, 9, 10, 11, 12,
  13, 14, 15, 16, 17, 18, 19, 20, 21, 22, 23, 24, 25, 26, 27, 28, 29, 31}

Computed value of Λ: 0.8395607987

p-value: 0.01332576052


PvalDataInd2Rp[{24}, path <> "bones_gc_1.dat", , 6, 1]

 Original variables {1, 2, 3, 4, 5, 6, 7, 8, 9, 10, 11, 12,
  13, 14, 15, 16, 17, 18, 19, 20, 21, 22, 23, 24, 25, 26, 27, 28, 29, 32}

Computed value of Λ: 0.8579712156

p-value: 0.02868074194


PvalDataInd2Rp[{24}, path <> "bones_gc_1.dat", , 6, 1]

 Original variables {1, 2, 3, 4, 5, 6, 7, 8, 9, 10, 11, 12,
  13, 14, 15, 16, 17, 18, 19, 20, 21, 22, 23, 24, 25, 26, 27, 28, 29, 33}

Computed value of Λ: 0.8734279061

p-value: 0.05274047715
```

Fig. 5.54 Commands used and output obtained with the Mathematica® version of module `PvalDataInd2Rp` to implement the forward method for the data in file `bones_gc_1.dat`

The output from the third command in Fig. 5.53 shows that once the first power of time is in the model, the second power of time is nonsignificant in that model. This fact together with the quite low p-value obtained for the model that only has the first power of time, which is the output obtained for the first command in Fig. 5.54, shows that there is indeed a significant linear trend along time among the means. Also from the results obtained from the first four commands in this figure we see that the first power of time is the first one to enter the model for the forward method. Then we know from the last command in Fig. 5.53 that the second power of time should not enter the model, as would happen with the third and fourth powers.

The "backward" method, which starts with a model with the full set of $q-1=4$ powers of time, is usually not the best method to obtain a growth curve model with fewer powers of time than the full set of $q-1$ powers. Given the fact that we are not using orthogonal polynomials, the introduction in the model of higher powers than those that indeed model the basic trend in the means, may sometimes introduce some distortion in the model, which may result in models that although fitting the data well and having all powers significant may be not the best submodel. This fact occurs more frequently in situations such as those of the files `bones_gc_3.dat` or `bones_gc_4.dat` where the values have a larger variance or a waving trend along time, but as we will see it also generally happens in all situations where the trend in the means is not so sharp.

To spare printing space in the book all further figures with commands and outputs concerning the building of Growth Curve models for the "backward" method with regard to the file `bones_gc_1.dat` and concerning the other data files have been placed in the "supplementary material" of the book where they are set as Fig. 5.54a to s. Also the data files `bones_gc_*.dat` and `bones_ind_*.dat` are available in this supplementary volume in Fig. 5.54t1 to w4, and may be easily copied in ASCII format from these figures, or downloaded directly from the supplementary material web site.

With regard to the results of the "backward" method applied to the data in file `bones_gc_1.dat`, we may see in Fig. 5.54a that the first power of time to leave the model should be the fourth power, the one that presents a higher p-value in the presence of all other three powers and then as Fig. 5.54b shows, powers 1 and 3 should also be sequentially dropped from the model, leaving only power 2, which we know to be not the best submodel with only one power of time. In fact it is not uncommon, even in situations as the present one where the "noise around the data" is rather slim, the "backward" method to give as a final model a model with a higher power of time than the other methods, although as we will see next, this is not always the case, mainly in situations where the trend in the means is quite clear.

Concerning the data in file `bones_gc_2.dat` we may see in Fig. 5.54c that once again the full set of four indicator variables and the set of four powers of time have exactly the same overall p-value, and that the "empirical forward" method gives a final model with only power 2, after leading us to add this power to the first power (third command in Fig. 5.54c) and to not add power 3 to a model where powers 1 and 2 are already present (fourth command in Fig. 5.54c). Then the fifth command in Fig. 5.54c shows that the first power should be dropped from the model.

The pure "forward" method also gives as final model a model with only the second power of time in it, with the results in Fig. 5.54d showing that this should be the first power to enter the model and the results in Fig. 5.54e showing that no other power of time should then be added to the model. The results in Fig. 5.54f and g show that also the "backward" method will give the same final model, with powers 4, 3 and 1 being dropped in this sequence from a model that started with the first four powers of time, leaving only power 2. We may see from the result of the second command in Fig. 5.54d that the model with only the second power of time has indeed a much lower p-value than the model with the four indicator variables or the four powers of time, capturing in a much more efficient way the second degree trend in the means.

For the data in file `bones_gc_3.dat` we have a slightly different scenario, given the larger variances that affect the data in this file. The three methods of finding a submodel will give different results, but will still be able to find submodels that fit the data with fewer than the four powers of time and that have a lower p-value than this model, still grasping very well the trend present in the means and showing the usefulness and resilience of the Growth Curve models.

Since now we will have to deal with some p-values that will be on the borderline of the most common significance levels used, we will find ourselves forced to set an α-level for our decisions. We will use the more than common values of $\alpha = 0.10$, $\alpha = 0.05$, and $\alpha = 0.01$ and argue about possible differences in our decisions for different values of α, being aware of all the risks that such a rather strict approach may entail, but also intending to show that this overcriticized approach in recent times gives most of the time quite sensible results.

From Fig. 5.54h we see that once again the set of four indicator variables and the set of the first four powers of time exhibit the same overall p-value and how after power 1 is considered in the model we should sequentially add powers 2 and 3 but not power 4, with power 3 showing an almost borderline p-value, but still to be included in the model if we use and α-level of 0.05. Then from the results in Fig. 5.54i we see that if we use an α-value of 0.05, powers 1 and 2 of time should not be individually dropped from the model, leaving us with a model with powers 1, 2, and 3 of time, with a p-value for the set of these three powers which is quite lower than the p-value for the set of all four powers of time. Would we use an α-value of 0.10, which would not make much sense once we used an α-value of 0.05 to include power 3 in the model, powers 1 and 2 should be dropped from the model, leaving us with a model with only power 3 in it and which, as we may see from the output of the third command in Fig. 5.54p, would have an even lower p-value than the model with powers 1, 2, and 3. The inclusion of lower order powers of time in a model where the trend in the means is governed mainly by a given higher power is indeed quite common in practical applications. But actually the p-value for the set of powers 1 and 2 in the model with powers 1, 2, and 3 is 0.298545, which indicates that powers 1 and 2 could indeed be dropped from this model. If we had decided to use an α-value of 0.01, we would have been left with a model with only powers 1 and 2, which, as shown in Fig. 5.54i, would have an even lower p-value than the model with powers 1, 2, and 3, displaying a very good fit to the data.

From Fig. 5.54j and k we may see that the final model obtained by the true "forward" method is a model with only power 3 since this is the first power to enter the model and after it is included in the model no other power of time should be added to that model.

From Fig. 5.54l–n we see that the "backward" method gives us a final model with powers 2 and 4, with power 3 being the first power of time to leave the initial model where the four powers of time were present, followed by power 1, leaving a model with powers 2 and 4, which indeed presents once again a lower p-value than the model with all four powers of time or all four indicator variables, indicating that although power 3 is not present in the model, still the trend in the means is very well modeled.

However, in situations in which the means show a pattern with some oscillation with a not so clear trend, the "backward" method may then show clear advantages over the other methods of finding submodels which having fewer than the $q-1$ first powers of time may exhibit for these set of powers a lower p-value than the one that is obtained for the overall set of $q-1$ powers of time or indicator variables.

We use the data in file `bones2_gc_4.dat` to illustrate this fact in a situation where the means exhibit a not so clear wavelike pattern as time evolves, as we may see in Fig. 5.52. The results are shown in Fig. 5.54o–s in the supplementary volume, where we may see how using the "empirical forward" method we would be left with a model with the first power of time as the only explanatory variable, which, from the output of the first command in Fig. 5.54p we know to have a much larger p-value than the model with the four powers of time, being thus a not so good choice, while from Fig. 5.54r and s we may see that the "backward" method leads us to drop exactly the first power of time from the original model with all four powers of time and remain with the model with powers 2, 3, and 4 as the final model, since any of these three powers of time exhibit a very low p-value in this model, in the presence of the other two powers of time. The "forward" method would lead us to include as first explanatory variable the first power of time and then no other power of time would be included in the model. We may note from Fig. 5.54s that the model with powers 2, 3, and 4 of time exhibits, for the overall set of these three powers, a quite lower p-value than the model with all four powers of time, being thus a good submodel to be used to model the evolution of the means with time.

These examples are intended to show the reliability and usefulness of the Growth Curve models. They are able to adequately model trends in the means, when these really exist, quite often with a lower number of powers of time than the number of variables in the full set of indicator variables. These models display a lower p-value than the model with the full set of indicator variables, while having a larger number of degrees of freedom for the error term.

5.1.9.7 Discriminant Analysis (and a Test for Nullity of Canonical Correlations)

In simple terms, the Discriminant Analysis model may be seen as a model or test procedure where one wants to evaluate whether a set of p_1 continuous r.v.'s is or is not "good enough" to discriminate among p_2 classes of a discrete r.v., or, rather, among p_2 groups or subpopulations, represented by the set of their p_2 indicator variables. Under this point of view, the Discriminant Analysis may be seen as closely related to the one-way MANOVA model and the test of equality of mean vectors, and as such also with the test of independence of two sets of variables.

Actually, while in the MANOVA model we see the set of continuous variables as the response variables set, in Discriminant Analysis we tend to reverse the roles of the sets, looking at the set of continuous variables rather as the set of explanatory variables and at the set of indicator variables as the set of response variables.

Using a similar lineup to the one we have been using so far, $\underline{X}_1$ will be the set of continuous random variables and $\underline{X}_2$ the set of indicator variables for the subpopulations. We may switch these two sets but, for the sake of simplicity in the exposition, we will hereafter stick with this lineup of sets $\underline{X}_1$ and $\underline{X}_2$.

Indeed Kshirsagar (1971) clearly states in the Introduction of his paper that "The problem of discrimination among the groups then reduces to the study of the relationship between the two vectors **x** and **y**," and Bartlett (1951) also states in the Preamble section of his paper that "It should be remembered that discriminant analysis (as introduced by Fisher, 1936) is a special case of the analysis of the association between two sets of variables (see, for example, Bartlett, 1938)."

In case we reject, for a given α level, the null hypothesis of independence between the two sets of variables, that is, the set of p_1 continuous r.v.'s and the set of indicator variables, we will assume that, for that given α level, the set of p_1 continuous r.v.'s is "good enough" to discriminate among the p_2 classes of the discrete r.v., or p_2 groups or subpopulations. If we assume that the p_1 continuous r.v.'s have a joint multivariate Normal or elliptically contoured distribution, then the l.r.t. statistic to be used is, according to the exposition in Sect. 5.1.9.4, the statistic studied in the main Sect. 5.1.9, with $\underline{X}_2$ formed by the set of $p_2 - 1$ indicator variables corresponding to all but one, usually the first one, of the subpopulations or classes of the discrete r.v., and $\underline{X}_1$ formed by the p_1 continuous r.v.'s.

In case one wants to test if a given subset of p_{12} of these p_1 continuous r.v.'s is or not significant in this discrimination process, the test to be used is the partial Wilks Lambda test described in Sect. 5.1.9.5.

To assume that the p_1 continuous r.v.'s discriminate "well enough" among the p_2 subpopulations or classes of the discrete r.v., is equivalent to assume that the p_2 population mean vectors for the p_1 continuous r.v.'s differ for the p_2 subpopulations defined by the classes of the discrete r.v., or more precisely, that at least two of them differ, or, more precisely yet, that there is at least one contrast of them that is not null, which corresponds to rejecting the null hypothesis in (5.1), for $q = p_2$ and $p = p_1$.

But then, as Fisher (1936) says "When two or more populations have been measured in several characters, $x_1, \ldots, x_s$, special interest attaches to certain linear functions of the measurements by which the populations are best discriminated." These "linear functions" are the so-called "discriminant functions," which are now known to be the canonical variables for the set of continuous variables.

In case there are only two groups or subpopulations, in which case just one single indicator variable would be enough to assign the individuals or observations to the two groups or subpopulations (the number of indicator variables being always equal to the number of groups or subpopulations minus one), there will be just one single discriminant function, and in this case as Bartlett (1938) remarks, the test of significance for this single discriminant function will be carried out by using the statistic Λ in (5.145), (5.147), or (5.148). This yields indeed an equivalent approach to the one delineated by Fisher (1936), given the relation that in this case the statistic Λ has with an F statistic, as remarked in Sects. 5.1.1 and 5.1.3.

But what happens when there are more than two groups or subpopulations? While the overall test for the discriminant analysis model is provided by the statistic Λ in (5.145), (5.147), or (5.148), with a rejection of the null hypothesis of independence of the two sets of variables leading to the conclusion that the set of p_1 continuous r.v.'s discriminates well enough among the p_2 groups or subpopulations, there will, in this case, remain the question of how many discriminant functions we should consider in order to adequately discriminate among the p_2 groups or subpopulations. The answer to this question is: "as many as the number of non-null canonical correlations." Indeed Kshirsagar (1971) says "The adequate number of discriminant functions is then the number of true nonzero canonical correlations between $\mathbf{x}$ and $\mathbf{y}$ and the discriminant functions are the canonical variables corresponding to these canonical correlations." But then the following question will arise: how can one test for the significance of single canonical correlations?

Although a test for the significance of a single canonical correlation is not that easy to obtain, mainly due to the complicated structure of the distribution of what would be the associated test statistic, Bartlett (1939), Lawley (1959) and Kshirsagar (1972, Sect. 8.7) give us a simple method to test the significance of the residual canonical correlations, that is, to test the null hypothesis

$$H_{0(s)} : \rho_1 \neq 0, \ldots, \rho_s \neq 0, \rho_{s+1} = \cdots = \rho_k = 0 \tag{5.217}$$

for $s \in \{0, \ldots, k-1\}$, where $\rho_1 \geq \cdots \geq \rho_k$ with $k = \min(p_1, p_2)$.

The statistic for the test of this null hypothesis is the statistic

$$\Lambda^* = \prod_{\alpha=1}^{s} (1 - \widehat{\rho}_\alpha^2) \quad (s \leq k-1,\ k = \min(p_1, p_2)), \tag{5.218}$$

where $\widehat{\rho}_1^2 \geq \widehat{\rho}_2^2 \geq \cdots \geq \widehat{\rho}_k^2$ are the squares of the sample canonical correlations in Sect. 5.1.9.1, i.e., eigenvalues of the matrices $A_{11}^{-1}A_{12}A_{22}^{-1}A_{21}$ or $A_{22}^{-1}A_{21}A_{11}^{-1}A_{12}$,

with

$$\Lambda^* \stackrel{d}{\equiv} \prod_{j=1}^{p_1-s} Y_j \stackrel{d}{\equiv} \prod_{\ell=1}^{p_2-s} Y_\ell^*, \qquad (5.219)$$

where for $n > p_1 + p_2 - s$, $j = 1, \ldots, p_1 - s$, and $\ell = 1, \ldots, p_2 - s$

$$Y_j \sim Beta\left(\frac{n-p_2-j}{2}, \frac{p_2-s}{2}\right) \quad \text{and} \quad Y_\ell^* \sim Beta\left(\frac{n-p_1-\ell}{2}, \frac{p_1-s}{2}\right)$$

form two sets of independent r.v.'s.

The exact p.d.f. and c.d.f. of Λ^* are thus given, in terms of the Meijer G function, by

$$f_{\Lambda^*}(z) = \left\{\prod_{j=1}^{p_1} \frac{\Gamma\left(\frac{n-s-j}{2}\right)}{\Gamma\left(\frac{n-p_2-j}{2}\right)}\right\} G^{p_1-s,0}_{p_1-s,p_1-s}\left(\left.\begin{array}{c}\left\{\frac{n-s-j}{2}-1\right\}_{j=1:p_1-s}\\ \left\{\frac{n-p_2-j}{2}-1\right\}_{j=1:p_1-s}\end{array}\right| z\right)$$

and

$$F_{\Lambda^*}(z) = \left\{\prod_{j=1}^{p_1} \frac{\Gamma\left(\frac{n-s-j}{2}\right)}{\Gamma\left(\frac{n-p_2-j}{2}\right)}\right\} G^{p_1-s,1}_{p_1-s+1,p_1-s+1}\left(\left.\begin{array}{c}\left\{1, \left\{\frac{n-s-j}{2}\right\}_{j=1:p_1-s}\right\}\\ \left\{\left\{\frac{n-p_2-j}{2}\right\}_{j=1:p_1-s}, 0\right\}\end{array}\right| z\right)$$

or

$$f_{\Lambda^*}(z) = \left\{\prod_{k=1}^{p_2} \frac{\Gamma\left(\frac{n-s-\ell}{2}\right)}{\Gamma\left(\frac{n-p_1-\ell}{2}\right)}\right\} G^{p_2-s,0}_{p_2-s,p_2-s}\left(\left.\begin{array}{c}\left\{\frac{n-s-\ell}{2}-1\right\}_{\ell=1:p_2-s}\\ \left\{\frac{n-p_1-\ell}{2}-1\right\}_{\ell=1:p_2-s}\end{array}\right| z\right)$$

and

$$F_{\Lambda^*}(z) = \left\{\prod_{k=1}^{p_2} \frac{\Gamma\left(\frac{n-s-\ell}{2}\right)}{\Gamma\left(\frac{n-p_1-\ell}{2}\right)}\right\} G^{p_2-s,1}_{p_2-s+1,p_2-s+1}\left(\left.\begin{array}{c}\left\{1, \left\{\frac{n-s-\ell}{2}\right\}_{\ell=1:p_2-s}\right\}\\ \left\{\left\{\frac{n-p_1-\ell}{2}\right\}_{\ell=1:p_2-s}, 0\right\}\end{array}\right| z\right).$$

and for even $p_1 - s$ its distribution is given by Theorem 3.2 and its p.d.f. and c.d.f. by Corollary 4.2, with

$$m^* = 1\,, \quad k_1 = 2\,, \quad a_1 = \frac{n-p_2}{2}\,, \quad n_1 = \frac{p_1-s}{2}\,, \quad m_1 = p_2 - s \qquad (5.220)$$

while for even $p_2 - s$ the exact distribution and the exact p.d.f. and c.d.f. of Λ are given by the same Theorem and Corollary, now with

$$m^* = 1, \quad k_1 = 2, \quad a_1 = \frac{n - p_1}{2}, \quad n_1 = \frac{p_2 - s}{2}, \quad m_1 = p_1 - s, \tag{5.221}$$

that is, switching in (5.220) p_1 and p_2 in the parameters that define the distribution.

The exact p.d.f. and c.d.f. of Λ^* in (5.218) or (5.148) are thus given, in terms of the EGIG distribution representation (see Appendix in Chap. 2), for any of the above cases, respectively by

$$f_{\Lambda^*}(z) = f^{EGIG}\left(z \,\middle|\, \{r_j\}_{j=1:p^*}; \left\{ \frac{n-s-2-j}{2} \right\}_{j=1:p^*}; p^* \right)$$

and

$$F_{\Lambda^*}(z) = F^{EGIG}\left(z \,\middle|\, \{r_j\}_{j=1:p^*}; \left\{ \frac{n-s-2-j}{2} \right\}_{j=1:p^*}; p^* \right),$$

where $p^* = p_1 + p_2 - 2s - 2$ and

$$r_j = \begin{cases} h_j, & j = 1, 2 \\ r_{j-2} + h_j, & j = 3, \ldots, p^* \end{cases} \tag{5.222}$$

with

$$h_j = (\text{\# of elements in } \{p_1 - s, p_2 - s\} \geq j) - 1, \quad j = 1, \ldots, p^*. \tag{5.223}$$

We should note here that in this whole formulation concerning the test of nullity of the remaining $k - s$ canonical correlations, in expression (5.218) and thereafter, in the case of the Discriminant Analysis, p_2 should really be taken as the number of groups or subpopulations minus 1.

Then a test to the null hypothesis $H_{0(s)}$ for $s = 0$ will clearly be equivalent to the overall test or the test of fit for the Discriminant Analysis or Canonical Analysis model, that is, equivalent to the test of independence of the two sets of variables $\underline{X}_1$ and $\underline{X}_2$. In case of rejection of $H_{0(s)}$ for a given $s \in \{0, \ldots, \min(p_1, p_2)\}$ then one may think about keeping on testing further hypotheses of this type for $s + 1, s + 2, \ldots$, till one of these hypotheses is not rejected. Let us suppose that the first one for which that happens is for some $s^* \in \{0, \ldots, \min(p_1, p_2)\}$. Then the first s^* population canonical correlations would be judged to be different from zero and s^* discriminant functions should be used to properly discriminate among the p_2 groups or subpopulations. Otherwise one may start by testing $H_{0(s)}$ for $s = k - 1$ and then in case of a non-rejection of this null hypothesis consider testing $H_{0(s)}$ for $s = k-2, k-3$ and so on, till one finds a value of s for which $H_{0(s)}$ is rejected. This would be the number of canonical correlations that are different from zero and the number of discriminant functions to be used. However, these procedures would have the drawback that the overall level of significance for the set of all tests carried out would not be anymore equal to α. If one wants to keep a given α level for the test, then one should choose a value of $s \in \{0, \ldots, \min(p_1, p_2)\}$ and carry out the test for the corresponding null hypothesis $H_{0(s)}$. Then, in case of rejection of this null hypothesis, this would tell us that at least s canonical correlations are different from zero and that as such one should use at least s discriminant functions to properly discriminate among the p_2 groups or subpopulations, and vice versa, in case of a non-rejection would tell us that at most s canonical correlations are different from zero and that we need at most s discriminant functions to properly discriminate among the p_2 groups or subpopulations.

Modules `PDFInd2Rs`, `CDFInd2Rs`, `QuantInd2Rs`, `PvalInd2Rs`, and `PvalDataInd2Rs` are provided for the test to the null hypotheses $H_{0(s)}$ in (5.217). Their usage and arguments are exactly the same as those of the corresponding `Ind2R` modules, except that their first mandatory argument is the value of $s \in \{0, \ldots, \min(p_1, p_2)\}$, where as already stated before, in Discriminant Analysis or in the case of the test in Sect. 5.1.1, p_2 should be taken as the number of groups or subpopulations minus 1.

To illustrate the test of the hypotheses $H_{0(s)}$ in (5.217) and the use of module `PvalDataInd2Rs` we use the data file used for the one-way MANOVA test in Sect. 5.1.9.4. The commands used and the output obtained are shown in Fig. 5.55 where we can see how the use of a value of $s = 0$ yields exactly the same computed value for the statistic and the same p-value as those obtained for the test of equality of mean vectors, or equivalently the test for the fit of the one-way MANOVA model in Fig. 5.39, with the low p-value obtained showing that the mean vectors for the three groups or subpopulations defined by the scores of 8, 11, and 12 for the abstract reasoning test scores are different, or equivalently that the set of three sales performance variables is "good enough" to discriminate among the three groups or subpopulations.

```
PvalDataInd2Rs[0, path <> "Tab9_12_JW_MANOVA.dat", , 3]
Computed value of Λ: 0.3859344456
p-value: 0.0003872679905
PvalDataInd2Rs[1, path <> "Tab9_12_JW_MANOVA.dat", , 3]
Computed value of Λ: 0.8990211065
p-value: 0.2506158548
```

Fig. 5.55 Commands used and output obtained with the Mathematica® module `PvalDataInd2Rs` to analyze a Discriminant Analysis model in what concerns the number of significant discriminant functions

The quite large p-value obtained with the second command in Fig. 5.55 where a value of $s = 1$ was used shows that just one single discriminant function will be enough to discriminate among the three groups or subpopulations, since only the first population canonical correlation should be taken as non-null. This result is actually in complete agreement with our perception when we look at the plot in Fig. 5.16 where the profiles for the three groups or subpopulations seem quite parallel, suggesting that one single linear discriminant function should be enough to separate these groups or subpopulations and also in complete agreement with the rejection of the hypothesis of equality of the mean vectors in Sect. 5.1.1.5 and the non-rejection of the hypothesis of profile parallelism in Sect. 5.1.3.4.

Another illustration of a Discriminant Analysis uses the data in Table 6.2 of Rencher (2002) or Rencher and Christensen (2012) on six different rootstocks onto which four variables were measured: trunk girth at 4 years (in mm$\times$100), extension growth at 4 years (in m), trunk girth at 15 years (in mm$\times$100), and weight of tree above ground at 15 years (in lb$\times$1000). Measurements were taken on eight plants, for each rootstock. We want to test if the four variables measured discriminate well enough among the six different rootstocks and if so, how many linear discriminant functions we need to consider in order to accomplish an appropriate discrimination. The data file used, `Rootstock_disc.dat`, has nine columns, the first four of which are the measurements on the 48 (=6 $\times$ 8) plants and the last five the set of indicator variables for the six rootstocks. A plot of the six rootstock profiles may be seen in Fig. 5.56, while the commands used to carry out the discriminant analysis and the output obtained are displayed in Fig. 5.57. In this figure the Mathematica® version of module `PvalDataInd2R` is first used to test if the four variables

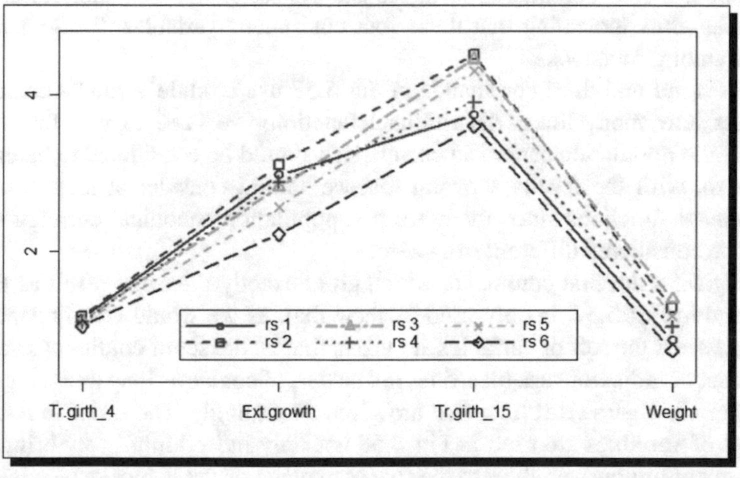

Fig. 5.56 Plot of the profiles of the six rootstocks for the four variables measured (trunk girth at 4 years (mm$\times$100), extension growth at 4 years (m), trunk girth at 15 years (mm$\times$100), weight of tree above ground at 15 years (lb$\times$1000)) for the rootstock data in Table 6.2 of Rencher (2002) and Rencher and Christensen (2012)

```
PvalDataInd2R[path <> "Rootstock_disc.dat", , 4]
Computed value of Λ: 0.1540076673
p-value: 7.717445955×10^-9
PvalDataInd2Rs[0, path <> "Rootstock_disc.dat", , 4]
Computed value of Λ: 0.1540076673
p-value: 7.717445955×10^-9
PvalDataInd2Rs[1, path <> "Rootstock_disc.dat", , 4]
Computed value of Λ: 0.4428754012
p-value: 0.0006383063599
PvalDataInd2Rs[2, path <> "Rootstock_disc.dat", , 4]
Computed value of Λ: 0.7930545610
p-value: 0.1363020403
```

Fig. 5.57 Commands used and output obtained with the Mathematica® modules `PvalDataInd2R` and `PvalDataInd2Rs` to analyze a Discriminant Analysis model concerning a study of six different rootstocks

measured discriminate well enough among the six rootstocks, with a very low p-value indicating that we should reject the null hypothesis of non-association between the set of this four continuous variables and the set of five indicator variables for rootstocks, thus indicating that these four continuous variables discriminate well enough among rootstocks.

The second and third commands in Fig. 5.57 use module `PvalDataInd2Rs` to assess how many linear discriminant functions we need to use, that is, how many of the population canonical correlations should be considered to be different from zero, with the results showing that we should consider at least two linear discriminant functions since there are two population canonical correlations that should be considered different from zero.

In Fig. 5.58 the first command, which gives exactly the same result as the first command in Fig. 5.57, is only used to show that, as we would clearly expect, the order in which the sets of variables are taken, that is, the set of continuous variables and the set of indicator variables, does not matter, of course as long as the arguments for the module `PvalDataInd2R` are given accordingly. The order in which the two sets of variables are used in Fig. 5.58 not only indeed intuitively brings more the set of continuous variables to the role of explanatory variables, which is exactly what happens in Discriminant Analysis, as well as it has to be the one to be used in the other tests carried out in that same figure as it is explained next.

The second command in Fig. 5.58 uses module `PvalDataInd2Rp` to test if the variables "trunk girth at 4 years" and "trunk girth at 15 years," taken

```
PvalDataInd2R[path <> "Rootstock_disc.dat", , 5, 1]
 Original variables {5, 6, 7, 8, 9, 1, 2, 3, 4}
Computed value of Λ: 0.1540076673
p-value: 7.717445955 × 10^-9
PvalDataInd2Rp[{1, 3}, path <> "Rootstock_disc.dat", , 5, 1]
 Original variables {5, 6, 7, 8, 9, 1, 2, 3, 4}
Computed value of Λ: 0.5131115853
p-value: 0.002330512280
PvalDataInd2Rp[{1}, path <> "Rootstock_disc.dat", , 5, 1]
 Original variables {5, 6, 7, 8, 9, 1, 2, 3}
Computed value of Λ: 0.9292550174
p-value: 0.693399504
PvalDataInd2Rp[{1}, path <> "Rootstock_disc.dat", , 5, 1]
 Original variables {5, 6, 7, 8, 9, 1, 2, 4}
Computed value of Λ: 0.9133743652
p-value: 0.584852334
PvalDataInd2Rp[{1, 2, 3}, path <> "Rootstock_disc.dat", , 5, 1]
 Original variables {5, 6, 7, 8, 9, 1, 2, 3, 4}
Computed value of Λ: 0.3769102224
p1 and p22 are both odd, and none of them is equal to 1,
to allow the use of the F distribution
```

Fig. 5.58 Commands used and output obtained with the Mathematica® modules PvalDataInd2R and PvalDataInd2Rs to analyze a Discriminant Analysis model concerning a study of six different rootstocks

together, are important in a linear Discriminant Analysis model where the other two variables are included, that is, to test whether these two variables add significantly to the discriminant power of the other two variables ("extension growth" and "weight above ground at 15 years"). The rather low p-value obtained shows that, considered together, those two variables should be considered even in a model where "extension growth" and "weight above ground at 15 years" are already included in order to significantly improve the discriminant power. Note that the way module PvalDataInd2Rp is programmed forces the variables that are removed from the "original" model in order to build the submodel to be part of the second set of variables, which forces us to give the first optional argument of the module a value of 1 to allow us to rearrange the order of the variables in order to allow for the test to be carried out. On the other hand the third and fourth commands show that "trunk girth at 4 years" just by itself does not add significant discriminant power to a model where "extension growth" and "trunk girth at 15 years" or "extension

```
PvalDataEqMeanVecR[path <> "Rootstock.dat"]
 There are 6 samples, with sizes {8, 8, 8, 8, 8, 8} and 4 variables
Computed value of Λ: 0.1540076673
p-value: 7.717445955×10^-9
```

Fig. 5.59 Command used and output obtained with the Mathematica® module `PvalDataEqMeanVecR` to carry out a test of equality of mean vectors for the six different rootstocks

growth" and "weight above ground at 15 years" are already in the model. The last command in Fig. 5.58 shows the common behavior of all `PvalData` modules that compute the value of the statistic but do not give a p-value in situations where the exact distribution does not have a finite form representation through the EGIG distribution. As noted at the end of Sect. 5.1.9, these situations may be handled through the use of a near-exact distribution or alternatively by taking the approach outlined in Chap. 7.

As happens with the first example in this paragraph, also the overall test for the Discriminant Analysis model for the rootstock data, which is carried out by the first commands in Figs. 5.57 and 5.58, is equivalent to a test of equality of mean vectors for the six rootstocks, which may be carried out with module `PvalDataEqMeanVecR` on the "original" data file `Rootstock.dat`, which has 48 rows, one for each plant, with a first column with integer values ranging from 1 to 6, indexing the rootstock, followed by four columns with the values for each of the four variables measured on each plant. The command used and corresponding output may be examined in Fig. 5.59.

5.1.10 The Likelihood Ratio Statistic to Test the Independence of Two Groups of Complex Variables [Ind2C]

If $\underline{X} = [\underline{X}_1', \underline{X}_2']' \sim CN_p(\underline{\mu}, \Sigma)$, that is, if the random vector $\underline{X}$ has a complex p-variate Normal distribution with mean vector $\underline{\mu}$ and Hermitian covariance matrix Σ (see Sect. 5.1.5 for further details on the complex multivariate Normal distribution), and we are interested in testing a similar hypothesis to that in (5.140) or (5.141), that is, the hypothesis of independence of $\underline{X}_1$ and $\underline{X}_2$, then, for a sample of size n, the $(1/n)$-th power of the l.r.t. statistic will have a similar formulation to the one in (5.145), (5.147), or (5.148), with A, given by a similar expression to that in (5.142), where the transpose of the matrix X has to be replaced by the transpose of the complex conjugate of X, as the m.l.e. of Σ. See Wooding (1956), Goodman (1963a), and Anderson (2003, Problem 3.11) for references concerning the m.l.e. of Σ in the complex case.

In this case, for a sample of size n, and taking p_1 and p_2 respectively as the dimensions of $\underline{X}_1$ and $\underline{X}_2$,

$$\Lambda \equiv \prod_{j=1}^{p_1} Y_j \stackrel{d}{\equiv} \prod_{j=1}^{p_2} Y_j^* \tag{5.224}$$

where, for $n > p_1 + p_2$,

$$Y_j \sim Beta\,(n - p_2 - j, p_2) \quad \text{and} \quad Y_j^* \sim Beta\,(n - p_1 - j, p_1) \tag{5.225}$$

form two sets of respectively p_1 and p_2 independent r.v.'s (see Appendix 8 for further details).

As such, the exact p.d.f. and c.d.f. of Λ is given in terms of the Meijer G function by

$$f_\Lambda(z) = \left\{\prod_{j=1}^{p_1} \frac{\Gamma\,(n-j)}{\Gamma\,(n-p_2-j)}\right\} G^{p_1,0}_{p_1,p_1}\left(\begin{matrix}\left\{n-j-1\right\}_{j=1:p_1} \\ \left\{n-p_2-j-1\right\}_{j=1:p_1}\end{matrix}\,\middle|\, z\right)$$

and

$$F_\Lambda(z) = \left\{\prod_{j=1}^{p_1} \frac{\Gamma\,(n-j)}{\Gamma\,(n-p_2-j)}\right\} G^{p_1,1}_{p_1+1,p_1+1}\left(\begin{matrix}\left\{1, \left\{n-j\right\}_{j=1:p_1}\right\} \\ \left\{\left\{n-p_2-j\right\}_{j=1:p_1}, 0\right\}\end{matrix}\,\middle|\, z\right)$$

or by similar expressions with p_1 and p_2 exchanged.

Therefore, the distribution of Λ is given by Theorem 3.2, and its p.d.f. and c.d.f. by Corollary 4.2, with

$$m^* = 1\,, \quad k_1 = 1\,, \quad a_1 = n - p_2\,, \quad n_1 = p_1\,, \quad m_1 = p_2 \tag{5.226}$$

or

$$m^* = 1\,, \quad k_1 = 1\,, \quad a_1 = n - p_1\,, \quad n_1 = p_2\,, \quad m_1 = p_1\,. \tag{5.227}$$

The exact p.d.f. and c.d.f. of Λ are thus expressible, in terms of the EGIG distribution representation, for any p_1 and p_2, respectively by

$$f_\Lambda(z) = f^{EGIG}\left(z \,\middle|\, \{r_j\}_{j=1:p_1+p_2-1}; \{n-1-j\}_{j=1:p_1+p_2-1}\,; p_1+p_2-1\right)$$

and

$$F_\Lambda(z) = F^{EGIG}\left(z \,\middle|\, \{r_j\}_{j=1:p_1+p_2-1}; \{n-1-j\}_{j=1:p_1+p_2-1}\,; p_1+p_2-1\right),$$

where

$$r_j = \begin{cases} h_j\,, & j = 1 \\ r_{j-1} + h_j\,, & j = 2, \ldots, p_1 + p_2 - 1 \end{cases}$$

with

$$h_j = (\#\text{ of elements in } \{p_1, p_2\} \geq j) - 1\,, \quad j = 1, \ldots, p_1 + p_2 - 1\,,$$

which is an equivalent result to that in Coelho et al. (2015).

In Mathai (1973) the reader may find an alternative finite form expression for the p.d.f. of Λ in (5.224), from which expressions for the c.d.f. are left to be obtained by term by term integration.

As for all other tests in this chapter, Mathematica®, MAXIMA, and R modules `PDFInd2C`, `CDFInd2C`, `PvalInd2C`, `QuantInd2C`, and `PvalDataInd2C` are provided. They have usages analogous to those of their real counterparts in Sect. 5.1.9.

5.1.10.1 The Use of Non-centered Data Matrices, the Test of Fit for a Multivariate Regression Model, and the Test in Sect. 5.1.8

As happens in the real case, also in the complex case the test of independence of two sets of variates may be carried out using one of the data sub-matrices, either X_1 or X_2, not centered for sample means, it being then necessary to add to this data sub-matrix a row of ones. The proof of the fact that the l.r. statistic obtained in this way is exactly the same as the one obtained by centering both data matrices X_1 and X_2 and using A_{11}, A_{12}, and A_{22} given by (5.143) and (5.144), with the transpose of the matrices X_1^* and X_2^* replaced by the transpose of their complex conjugate, would follow exactly the same steps as the ones described in Sect. 5.1.9.1.

Also, as in the real case, the test in Sect. 5.1.8, more precisely, the test of the null hypothesis in (5.113) may be carried out as a test of independence of two sets of variables, by taking X_1 as the matrix X in Sect. 5.1.8 and the matrix X_2 as the matrix M, also in Sect. 5.1.8, and not centering them. Then the statistic Λ, with a similar formulation to that in (5.145), (5.147), or (5.148), will be the same as the statistic Λ in (5.114).

For further details see the remarks in the last paragraph of Sect. 5.1.9.1 entitled “Implementing the test in Sect. 5.1.4 through a test of independence,” and replace all references to Sect. 5.1.4 by 5.1.8 and the references to (5.33) and (5.34) by references to (5.113) and (5.114), respectively.

The only difference from the real case is that now in the case where $\underline{X} = [\underline{X}_1', \underline{X}_2']'$ has a complex multivariate Normal distribution (see Sect. 5.1.5 for more details on this distribution), we always have, for any p_1 and p_2, the exact p.d.f. and c.d.f. of Λ given by Corollary 4.2, as described above.

5.1.10.2 The Test of Equality of Mean Vectors as a Test of Independence: The MANOVA Model

Parallel to what is explained in Sect. 5.1.9.4, and also according to the results in the previous subsection, it is possible in the complex case to address the test of equality of mean vectors as a test of independence, since it is indeed a test of independence between the variables in, say, $\underline{X}_1$, and the population indicator variables represented in the rows of M, the matrix that will then take the place of the matrix X_2, similar to what happens in the real case. See Sect. 5.1.9.4 for further details. Also, since the test of equality of mean vectors in Sect. 5.1.5 may be seen as a particular case of an extension of the test to the null hypothesis in (5.113), that is, as a particular case of the extended version of the test in Sect. 5.1.8, by using a similar setup to the one in Sect. 5.1.9.4 for the real case, it is possible to implement a one-way MANOVA for complex r.v.'s based on the test of independence of two sets of variables presented in this subsection.

All the details and options in Sect. 5.1.9.4 apply, with the only difference being that the exact distribution of Λ is now given by (5.224)–(5.227) for any values of p_1 and p_2.

5.1.10.3 Testing Between Multivariate Linear Models

All derivations and explanations in Sect. 5.1.9.5 are applicable in the present complex case where Λ_{H_1,H_0^*} has now a distribution with a similar structure to the one in (5.178) but with

$$Y_j \sim Beta(n-p_2-j,\, p_2) \quad \text{and} \quad Y_k^* \sim Beta(n-p_1-k,\, p_1)$$

and Λ_{H_0,H_0^*} has a distribution with a similar structure to the one in (5.180) but now with

$$Y_j^{**} \sim Beta(n-p_{21}-j,\, p_{21}) \quad \text{and} \quad Y_k^{***} \sim Beta(n-p_1-k,\, p_1)$$

and Λ_{H_1,H_0} has a distribution with a similar structure to the one in (5.182) but with

$$Y_k^{****} \sim Beta(n-p_2-j,\, p_{22})\,, \quad Y_j^{*****} \sim Beta(n-p_1-k,\, p_1)$$

and

$$Y_\ell^{(vi)} \sim Beta(n-p_1-p_{21}-\ell,\, p_1)\,,$$

so that, for $n > p_1 + p_2$, the exact p.d.f. and c.d.f. of Λ_{H_1,H_0} are given by Corollary 4.2 with

$$m^* = 1, \quad k_1 = 1, \quad a_1 = n - p_2, \quad n_1 = p_1 \quad \text{and} \quad m_1 = p_{22}\,, \tag{5.228}$$

or

$$m^* = 1, \quad k_1 = 1, \quad a_1 = n - p_1 - p_{21}, \quad n_1 = p_{22} \quad \text{and} \quad m_1 = p_1 , \tag{5.229}$$

for any p_1 and any p_{22}.

The exact p.d.f. and c.d.f. of Λ_{H_1,H_0} are thus given, in terms of the EGIG distribution representation, for any p_1 and any p_{22}, respectively by

$$f_{\Lambda_{H1,\,H0}}(z) = f^{EGIG}\left(z \,\middle|\, \{r_j\}_{j=1:p_1+p_{22}-1};\, \{n-1-p_{21}-j\}_{j=1:p_1+p_{22}-1}\,;\, p_1+p_{22}-1\right)$$

and

$$F_{\Lambda_{H1,\,H0}}(z) = F^{EGIG}\left(z \,\middle|\, \{r_j\}_{j=1:p_1+p_{22}-1};\, \{n-1-p_{21}-j\}_{j=1:p_1+p_{22}-1}\,;\, p_1+p_{22}-1\right),$$

where

$$r_j = \begin{cases} h_j\,, & j=1 \\ r_{j-1}+h_j\,, & j=2,\dots,p_1+p_{22}-1 \end{cases}$$

with

$$h_j = (\#\text{ of elements in } \{p_1, p_{22}\} \geq j) - 1\,, \qquad j = 1,\dots,p_1+p_{22}-1\,.$$

Alternatively the exact p.d.f. of Λ_{H_1,H_0} may be given in terms of the Meijer G function by

$$f_{\Lambda_{H1,\,H0}}(z) = \left\{\prod_{j=1}^{p_1} \frac{\Gamma(n-p_{21}-j)}{\Gamma(n-p_2-j)}\right\} G^{p_1,0}_{p_1,p_1}\left(\begin{matrix} \left\{n-p_{21}-j-1\right\}_{j=1:p_1} \\ \left\{n-p_2-j-1\right\}_{j=1:p_1} \end{matrix} \,\middle|\, z \right)$$

or

$$f_{\Lambda_{H1,\,H0}}(z) = \left\{\prod_{\ell=1}^{p_{22}} \frac{\Gamma(n-p_{21}-\ell)}{\Gamma(n-p_1-p_{21}-\ell)}\right\} G^{p_{22},0}_{p_{22},p_{22}}\left(\begin{matrix} \left\{n-p_{21}-\ell-1\right\}_{\ell=1:p_{22}} \\ \left\{n-p_1-p_{21}-\ell-1\right\}_{\ell=1:p_{22}} \end{matrix} \,\middle|\, z \right)$$

and its c.d.f. by

$$F_{\Lambda_{H1,\,H0}}(z) = \left\{\prod_{j=1}^{p_1} \frac{\Gamma(n-p_{21}-j)}{\Gamma(n-p_2-j)}\right\} G^{p_1,1}_{p_1+1,p_1+1}\left(\begin{matrix} \left\{1, \left\{n-p_{21}-j\right\}_{j=1:p_1}\right\} \\ \left\{\left\{n-p_2-j\right\}_{j=1:p_1}, 0\right\} \end{matrix} \,\middle|\, z \right)$$

or

$$F_{\Lambda_{H1,H0}}(z) = \left\{ \prod_{\ell=1}^{p_{22}} \frac{\Gamma(n-p_{21}-\ell)}{\Gamma(n-p_1-p_{21}-\ell)} \right\} G^{p_{22},1}_{p_{22}+1,p_{22}+1}\left(\left. \begin{array}{c} \left\{1, \left\{n-p_{21}-\ell\right\}_{\ell=1:p_{22}}\right\} \\ \left\{\left\{n-p_1-p_{21}-\ell\right\}_{\ell=1:p_{22}}, 0\right\} \end{array} \right| z \right).$$

As for the real case, Mathematica®, MAXIMA, and R modules `PDFInd2Cp`, `CDFInd2Cp`, `PvalInd2Cp`, `QuantInd2Cp`, and `PvalDataInd2Cp` are provided, with similar arguments and similar usage as their real counterparts in Sect. 5.1.9.

Setting 1: The Test of Nullity of a Submatrix of Parameters or Regression Coefficients: A Test Between Two Multivariate Nested Models (Complex r.v.'s)

The statistic Λ_{H_1,H_0} is the statistic to be used to test for the nullity of a submatrix of parameters or regression coefficients in a Multivariate Linear model where the variables in $\underline{X}_1$ are taken as the response variables and the variables in $\underline{X}_2$ as the explanatory variables, p_{22} of which are taken away when specifying the H_0 model. It is exactly relative to these p_{22} variables that we want to test the nullity of the associated regression parameters.

The Multi-Way MANOVA Model

In what concerns the multi-way MANOVA model, the whole exposition in Sect. 5.1.9.4 applies also here in the complex case, now with the exact p.d.f. and c.d.f. of the statistic Λ_{H_1,H_0} used to test the significance of the interaction between the two factors being given by Corollary 4.2 for any p_1 and any p_{23}, with

$$m^* = 1, \quad k_1 = 1, \quad a_1 = n - p_2, \quad n_1 = p_1 \quad \text{and} \quad m_1 = p_{23}\,,$$

or

$$m^* = 1, \quad k_1 = 1 \quad , a_1 = n - p_1 - p_{21} - p_{22}, \quad n_1 = p_{23} \quad \text{and} \quad m_1 = p_1\,,$$

instead of the parameters in (5.201) and (5.202).

The exact p.d.f. and c.d.f. of Λ_{H_1,H_0} are therefore given, in terms of the EGIG distribution representation, for any p_1 and any p_{23}, by

$$f_\Lambda(z) = f^{EGIG}\Big(z \,\Big|\, \{r_j\}_{j=1:p_1+p_{23}-1};\, \{n-1-p_{21}-p_{22}-j\}_{j=1:p_1+p_{23}-1}; \\ p_1+p_{23}-1\Big)$$

and

$$F_\Lambda(z) = F^{EGIG}\left(z \,\middle|\, \{r_j\}_{j=1:p_1+p_{23}-1}; \{n-1-p_{21}-p_{22}-j\}_{j=1:p_1+p_{23}-1}; \\ p_1+p_{23}-1\right),$$

where

$$r_j = \begin{cases} h_j, & j=1 \\ r_{j-1}+h_j, & j=2,\ldots,p_1+p_{23}-1 \end{cases}$$

with

$$h_j = (\#\text{ of elements in }\{p_1, p_{23}\} \geq j) - 1, \quad j=1,\ldots,p_1+p_{23}-1.$$

Then, supposing that the interaction between the two factors is not significant, and that, as in Sect. 5.1.9.3, one wants to test the significance of factor 1, the whole explanation in that subsection still applies, now with the exact p.d.f. and c.d.f. of the l.r. statistic being given by Corollary 4.2, for any p_1 and any p_{21}, with

$$m^* = 1, \quad k_1 = 1, \quad a_1 = n - p_2, \quad n_1 = p_1 \quad \text{and} \quad m_1 = p_{21},$$

or

$$m^* = 1, \quad k_1 = 1 \quad, a_1 = n - p_1 - p_{22}, \quad n_1 = p_{21} \quad \text{and} \quad m_1 = p_1,$$

instead of the parameters in (5.203) and (5.204), with the exact p.d.f. and c.d.f. of Λ_{H_1,H_0} given, in terms of the EGIG distribution representation, for any p_1 and any p_{21}, by

$$f_\Lambda(z) = f^{EGIG}\left(z \,\middle|\, \{r_j\}_{j=1:p_1+p_{21}-1}; \{n-1-p_{22}-j\}_{j=1:p_1+p_{21}-1}; p_1+p_{21}-1\right)$$

and

$$F_\Lambda(z) = F^{EGIG}\left(z \,\middle|\, \{r_j\}_{j=1:p_1+p_{21}-1}; \{n-1-p_{22}-j\}_{j=1:p_1+p_{21}-1}; p_1+p_{21}-1\right),$$

where

$$r_j = \begin{cases} h_j, & j=1 \\ r_{j-1}+h_j, & j=2,\ldots,p_1+p_{21}-1 \end{cases}$$

with

$$h_j = (\#\text{ of elements in }\{p_1, p_{21}\} \geq j) - 1, \quad j=1,\ldots,p_1+p_{21}-1.$$

The MANCOVA (Multivariate Analysis of Covariance) Model

Concerning the MANCOVA model, the whole exposition in Sect. 5.1.9.5 still applies to the complex case, with the exact p.d.f. and c.d.f. of the l.r. statistic to test the significance of the covariates in presence of the two factors and their interaction given by Corollary 4.2 for any p_1 and p_{24}, with

$$m^* = 1, \quad k_1 = 1, \quad a_1 = n - p_2, \quad n_1 = p_1 \quad \text{and} \quad m_1 = p_{24}\,, \tag{5.230}$$

or

$$m^* = 1, \quad k_1 = 1 \quad , a_1 = n - p_1 - p_{21} - p_{22} - p_{23}, \quad n_1 = p_{24} \quad \text{and} \quad m_1 = p_1\,, \tag{5.231}$$

instead of the parameters in (5.209) and (5.210), with

$$f_\Lambda(z) = f^{EGIG}\Big(z\,\Big|\,\{r_j\}_{j=1:p_1+p_{24}-1};\{n-1-p_{21}-p_{22}-p_{23}-j\}_{j=1:p_1+p_{24}-1};\; p_1+p_{24}-1\Big)$$

and

$$F_\Lambda(z) = F^{EGIG}\Big(z\,\Big|\,\{r_j\}_{j=1:p_1+p_{24}-1};\{n-1-p_{21}-p_{22}-p_{23}-j\}_{j=1:p_1+p_{24}-1};\; p_1+p_{24}-1\Big),$$

where

$$r_j = \begin{cases} h_j\,, & j=1\\ r_{j-1}+h_j\,, & j=2,\dots,p_1+p_{24}-1 \end{cases}$$

with

$$h_j = (\#\text{ of elements in }\{p_1, p_{24}\} \geq j) - 1\,, \quad j = 1,\dots,p_1+p_{24}-1\,,$$

and with the exact p.d.f. and c.d.f. of the l.r. statistic used to test the significance of the interaction between the two factors in presence of the covariates given by Corollary 4.2 for any p_1 and p_{23} with the parameters in (5.230) and (5.231) above, with p_{24} replaced by p_{23} and vice versa, p_{23} replaced by p_{24}.

Setting 2: Test of Equality of Model Parameters or Test of Equality of Models

The whole exposition and explanation in the paragraph with a similar heading in Sect. 5.1.9.5 are fully applicable to the present complex case. The only change to be made is that in the complex case we always have the distribution of the partial Wilks' Lambda statistic given by Corollary 4.2, with all references to (5.185) and/or (5.186) in Sect. 5.1.9.5 to be replaced by references to (5.228) or (5.229).

5.1.10.4 Discriminant Analysis

Also in the complex case the link between the test of equality of mean vectors and the Discriminant Analysis may be established in exactly the same way as it was done in Sect. 5.1.9.7 for the real case.

The only difference is that now the l.r.t. statistic to test $H_{0(s)}$ in (5.217), although it still has a formulation similar to that in (5.218), it has, for $n > p_1 + p_2 - s$, a distribution with the structure in (5.219) but with

$$Y_j \sim Beta\,(n - p_2 - j, p_2 - s) \quad \text{and} \quad Y_\ell^* \sim Beta\,(n - p_1 - \ell, p_1 - s)\,,$$

so that now the statistic Λ^* has, for any p_1, p_2, and s, its distribution given by Theorem 3.2 with

$$m^* = 1\,, \quad k_1 = 1\,, \quad a_1 = n - p_2\,, \quad n_1 = p_1 - s\,, \quad m_1 = p_2 - s$$

or

$$m^* = 1\,, \quad k_1 = 1\,, \quad a_1 = n - p_1\,, \quad n_1 = p_2 - s\,, \quad m_1 = p_1 - s\,,$$

and as such, its exact p.d.f. and c.d.f., under $H_{0(s)}$, given by Corollary 4.2 respectively as

$$f_{\Lambda^*}(z) = f^{EGIG}\left(z \,\middle|\, \{r_j\}_{j=1:p^*}; \{n-s-1-j\}_{j=1:p^*}\,; p^*\right)$$

and

$$F_{\Lambda^*}(z) = F^{EGIG}\left(z \,\middle|\, \{r_j\}_{j=1:p^*}; \{n-s-1-j\}_{j=1:p^*}\,; p^*\right),$$

where $p^* = p_1 + p_2 - 2s - 1$ and

$$r_j = \begin{cases} h_j\,, & j = 1 \\ r_{j-1} + h_j\,, & j = 2, \ldots, p_1 + p_2 - 2s - 1 \end{cases}$$

with

$$h_j = (\#\text{ of elements in } \{p_1 - s, p_2 - s\} \geq j) - 1\,, \quad j = 1, \ldots, p_1 + p_2 - 2s - 1\,.$$

Alternatively the p.d.f. and c.d.f. of Λ^* may be given, in terms of Meijer G functions, respectively by

$$f_{\Lambda^*}(z) = \left\{\prod_{j=1}^{p_1} \frac{\Gamma(n-s-j)}{\Gamma(n-p_2-j)}\right\} G^{\,p_1-s,0}_{\,p_1-s,p_1-s}\left(\begin{array}{c} \{n-s-j-1\}_{j=1:p_1-s} \\ \{n-p_2-j-1\}_{j=1:p_1-s} \end{array} \middle|\, z \right)$$

and

$$F_{\Lambda^*}(z) = \left\{\prod_{j=1}^{p_1} \frac{\Gamma(n-s-j)}{\Gamma(n-p_2-j)}\right\} G^{\,p_1-s,1}_{\,p_1-s+1,p_1-s+1}\left(\begin{array}{c} \left\{1, \{n-s-j\}_{j=1:p_1-s}\right\} \\ \left\{\{n-p_2-j\}_{j=1:p_1-s}, 0\right\} \end{array} \middle|\, z \right)$$

or

$$f_{\Lambda^*}(z) = \left\{\prod_{k=1}^{p_2} \frac{\Gamma(n-s-\ell)}{\Gamma(n-p_1-\ell)}\right\} G^{\,p_2-s,0}_{\,p_2-s,p_2-s}\left(\begin{array}{c} \{n-s-\ell-1\}_{\ell=1:p_2-s} \\ \{n-p_1-\ell-1\}_{\ell=1:p_2-s} \end{array} \middle|\, z \right)$$

and

$$F_{\Lambda^*}(z) = \left\{\prod_{k=1}^{p_2} \frac{\Gamma(n-s-\ell)}{\Gamma(n-p_1-\ell)}\right\} G^{\,p_2-s,1}_{\,p_2-s+1,p_2-s+1}\left(\begin{array}{c} \left\{1, \{n-s-\ell\}_{\ell=1:p_2-s}\right\} \\ \left\{\{n-p_1-\ell\}_{\ell=1:p_2-s}, 0\right\} \end{array} \middle|\, z \right).$$

As in the real case, modules `PDFInd2Cs`, `CDFInd2Cs`, `PvalInd2Cs`, `QuantInd2Cs`, and `PvalDataInd2Cs` are provided to implement this test, with similar arguments and similar usage as their real counterparts in Sect. 5.1.9.

5.1.11 *The Likelihood Ratio Statistic to Test the Independence of Several Groups of Real Variables [IndR]*

Let us consider a random vector $\underline{X} \sim N_p(\underline{\mu}, \Sigma)$, which is split into m subvectors $\underline{X}_k$ $(k = 1, \ldots, m)$, the k-th of which has dimension p_k, with $p = \sum_{k=1}^{m} p_k$, and that we want to test the null hypothesis of independence of the m subvectors $\underline{X}_k$.

The partition of the vector $\underline{X}$ into m subvectors $\underline{X}_k$ $(k = 1, \ldots, m)$ induces in Σ the corresponding partition

$$\Sigma = \begin{bmatrix} \Sigma_{11} & \cdots & \Sigma_{1k} & \cdots & \Sigma_{1m} \\ \vdots & \ddots & \vdots & & \vdots \\ \Sigma_{k1} & \cdots & \Sigma_{kk} & \cdots & \Sigma_{km} \\ \vdots & & \vdots & \ddots & \vdots \\ \Sigma_{m1} & \cdots & \Sigma_{mk} & \cdots & \Sigma_{mm} \end{bmatrix}, \qquad (5.232)$$

and thus the hypothesis of mutual independence of all the random subvectors $\underline{X}_k$ may be written as

$$\begin{aligned} H_0 : \Sigma &= bdiag(\Sigma_{11}, \ldots, \Sigma_{kk}, \ldots, \Sigma_{mm}) \\ &\iff \Sigma_{ij} = 0,\ i \neq j,\ i, j = 1, \ldots, m\,. \end{aligned} \qquad (5.233)$$

The $(2/n)$-th power of the l.r.t. statistic to test this null hypothesis is

$$\Lambda = \frac{|A|}{\prod_{k=1}^{m} |A_{kk}|} \qquad (5.234)$$

where A is the m.l.e. of Σ and A_{kk} is its k-th diagonal block of dimensions $p_k \times p_k$.

The h-th moment of Λ, under H_0 in (5.233), may be written, for $n > p$ and $h > -(n-p)/2$, as (Coelho, 2004, expr. (26))

$$E\left(\Lambda^h\right) = \prod_{k=1}^{m-1} \prod_{j=1}^{p_k} \frac{\Gamma\left(\frac{n-j}{2}\right) \Gamma\left(\frac{n-j-q_k}{2} + h\right)}{\Gamma\left(\frac{n-j-q_k}{2}\right) \Gamma\left(\frac{n-j}{2} + h\right)} \qquad (5.235)$$

where

$$q_k = p_{k+1} + \cdots + p_m\,, \qquad (5.236)$$

and where p_k and q_k are interchangeable. The expression for $E(\Lambda^h)$ was first derived, with a slightly different appearance, by Wilks (1935, expr. (28)) (see also Wilks (1932)) and it may also be found in Kshirsagar (1972, Sect. 10.3, expr. (3.23)) and Lee et al. (1977, expr. (2.2)), while other equivalent expressions, with a somewhat different look, may be found in Anderson (2003, Theorem 9.3.4) and Muirhead (2005, Theorem 11.2.3). See also Appendix 12 for a full derivation of the expression in (5.235) directly from the Wishart distribution of the m.l.e. of Σ.

Since the range of Λ is delimited, with $0 < \Lambda < 1$, the moments of Λ determine its distribution, and as such we may say that, for $n > p$,

$$\Lambda \stackrel{d}{\equiv} \prod_{k=1}^{m-1} \prod_{j=1}^{p_k} Y_{jk} \tag{5.237}$$

where

$$Y_{jk} \sim Beta\left(\frac{n-q_k-j}{2}, \frac{q_k}{2}\right), \tag{5.238}$$

is a set of $\sum_{k=1}^{m-1} p_k = p - p_m$ independent r.v.'s, with q_k given by (5.236).

The exact p.d.f. and c.d.f. of Λ in (5.234) are given in terms of the Meijer G function by

$$f_\Lambda(z) = \left\{ \prod_{k=1}^{m-1} \prod_{j=1}^{p_k} \frac{\Gamma\left(\frac{n-j}{2}\right)}{\Gamma\left(\frac{n-q_k-j}{2}\right)} \right\} G^{p^*,0}_{p^*,p^*} \left(\begin{array}{c} \left\{\frac{n-j}{2} - 1\right\}_{\substack{j=1:p_k \\ k=1:m-1}} \\ \left\{\frac{n-q_k-j}{2} - 1\right\}_{\substack{j=1:p_k \\ k=1:m-1}} \end{array} \middle| \, z \right)$$

and

$$F_\Lambda(z) = \left\{ \prod_{k=1}^{m-1} \prod_{j=1}^{p_k} \frac{\Gamma\left(\frac{n-j}{2}\right)}{\Gamma\left(\frac{n-q_k-j}{2}\right)} \right\} G^{p^*,1}_{p^*+1,p^*+1} \left(\begin{array}{c} \left\{1, \left\{\frac{n-j}{2}\right\}_{\substack{j=1:p_k \\ k=1:m-1}} \right\} \\ \left\{ \left\{\frac{n-q_k-j}{2}\right\}_{\substack{j=1:p_k \\ k=1:m-1}}, 0 \right\} \end{array} \middle| \, z \right)$$

for $p^* = \sum_{k=1}^{m-1} p_k = p - p_m$, and where p_k and q_k are interchangeable.

As such, when at most one of the p_k is odd, and if that happens, without any loss of generality, let it be p_m, the exact distribution of Λ in (5.234) is given by Theorem 3.2 and its p.d.f. and c.d.f. by Corollary 4.2 with

$$m^* = m - 1\,, \quad k_\nu = 2\,, \quad n_\nu = \tfrac{p_\nu}{2}\,,$$
$$m_\nu = q_\nu = p_{\nu+1} + \cdots + p_{m^*} + p_m\,, \quad a_\nu = \tfrac{n-q_\nu}{2}\,, \quad (\nu = 1, \ldots, m^*)\,.$$

The exact p.d.f. and c.d.f. of Λ in (5.234) are thus given, in terms of the EGIG distribution representation, when at most one of the p_k's is odd, respectively by

$$f_\Lambda(z) = f^{EGIG}\left(z \,\middle|\, \{r_j\}_{j=1:p-2}; \left\{\frac{n-2-j}{2}\right\}_{j=1:p-2}; p-2 \right) \tag{5.239}$$

and

$$F_\Lambda(z) = F^{EGIG}\left(z \,\middle|\, \{r_j\}_{j=1:p-2};\left\{\frac{n-2-j}{2}\right\}_{j=1:p-2};p-2\right), \tag{5.240}$$

where

$$r_j = \begin{cases} h_j\,, & j=1,2 \\ r_{j-2}+h_j\,, & j=3,\dots,p-2 \end{cases} \tag{5.241}$$

with

$$h_j = (\#\text{ of } p_k\text{'s } (k=1,\dots,m) \geq j) - 1\,, \quad j=1,\dots,p-2\,. \tag{5.242}$$

We may note that this result confirms the ones in Coelho (1998, 1999), Coelho et al. (2010), Marques et al. (2011), and Arnold et al. (2013) and even the results in Wald and Brookner (1941). These latter authors, although they did not obtain a general expression for either the p.d.f. or the c.d.f. of Λ in (5.234), devised a method to build case by case the expression for the c.d.f. of Λ based on the identification of the poles of the integrand function of the Meijer G function and their multiplicities, when there is at most one subset with an odd number of variables. These poles are the rate parameters in the EGIG distribution and their multiplicities the corresponding shape parameters, the r_j, given by (5.241) and (5.242).

We may also note the extremely compact and manageable form of the exact distribution of Λ given by (5.239) and (5.240), and also that the definition of the shape parameters r_j derived from Theorem 3.2, given by (3.13) and (3.14), matches the one in Coelho et al. (2010) and Marques et al. (2011), yielding a much more compact formulation than the original definition of these parameters in Coelho (1998).

As happens with the test in Sect. 5.1.9, also the distribution of Λ in (5.234), under the null hypothesis in (5.233), remains valid if the distribution of $\underline{X}$ is not multivariate Normal, but some elliptically contoured or left orthogonal-invariant distribution—see Theorem 5.3.3 in Fang and Zhang (1990, Chap. V) and Sect. 9.11 in Anderson (2003), and also Jensen and Good (1981), Kariya (1981), Anderson et al. (1986), and Anderson and Fang (1990).

When more than one p_k is odd, near-exact distributions for the statistic Λ in (5.234) are available in Coelho (2004), Coelho et al. (2010), and Marques et al. (2011) and the reader may also refer to the material in Chap. 7.

Similar to what happens with the statistic Λ in Sect. 5.1.9, also the statistic Λ in (5.234) is invariant under linear transformations within each subset $\underline{X}_k$, that is, under transformations of the type $\underline{X} \to C\underline{X} + \underline{b}$, where $C = bdiag(C_1,\dots,C_k,\dots,C_m)$ and $\underline{b} = [\underline{b}_1',\dots,\underline{b}_k',\dots,\underline{b}_m']'$, or in other words, under transformations $\underline{X}_k \rightarrow C_k\underline{X}_k + \underline{b}_k$, where C_k are full-rank nonrandom $p_k \times p_k$ matrices and $\underline{b}_k$ is a real nonrandom $p_k \times 1$ vector, for $k = 1, \ldots, m$.

It is not hard to show that Λ in (5.234) may be written as (see Anderson (2003, Sec. 9.3) and Coelho (1992, Sec. 4.7))

$$\Lambda = \prod_{k=1}^{m-1} \Lambda_{k,(k+1,\ldots,m)} \tag{5.243}$$

where $\Lambda_{k,(k+1,\ldots,m)}$ is the $(2/n)$-th power of the l.r.t. statistic to test the independence between the k-th set and the superset formed by joining sets $k+1$ through m, that is

$$\Lambda_{k,(k+1,\ldots,m)} = \frac{|A_{k,\ldots,m;k,\ldots,m}|}{|A_{kk}|\,|A_{k+1,\ldots,m;k+1,\ldots,m}|}, \tag{5.244}$$

where

$$A_{k,\ldots,m;k,\ldots,m} = \begin{bmatrix} A_{kk} & \ldots & A_{km} \\ \vdots & \ddots & \vdots \\ A_{mk} & \ldots & A_{mm} \end{bmatrix} \tag{5.245}$$

is the m.l.e. of

$$\Sigma_{k,\ldots,m;k,\ldots,m} = \begin{bmatrix} \Sigma_{kk} & \ldots & \Sigma_{km} \\ \vdots & \ddots & \vdots \\ \Sigma_{mk} & \ldots & \Sigma_{mm} \end{bmatrix}.$$

Under H_0 in (5.233) the $m-1$ statistics $\Lambda_{k,(k+1,\ldots,m)}$ in (5.243) are independent (see Anderson (2003, Sec. 9.3) and Coelho (1992, Sec. 4.7)). From this fact and the fact that from Sect. 5.1.9 we know that, for $n > p_k + q_k$,

$$\Lambda_{k(k+1,\ldots,m)} \stackrel{d}{\equiv} \prod_{j=1}^{p_k} Y_{jk} \stackrel{d}{\equiv} \prod_{j=1}^{q_k} Y_{jk}^*$$

where Y_{jk} are the set of independent r.v.'s in (5.237) and (5.238) and

$$Y_{jk}^* \sim Beta\left(\frac{n-p_k-j}{2}, \frac{p_k}{2}\right), \quad j = 1, \ldots, q_k$$

also form a set of independent r.v.'s, we may write the h-th moment of Λ in (5.234) as

$$E\left(\Lambda^h\right) = \prod_{k=1}^{m-1} E\left(\Lambda^h_{k,(k+1,\dots,m)}\right) = \prod_{k=1}^{m-1}\prod_{j=1}^{p_k} E\left(Y^h_{jk}\right) = \prod_{k=1}^{m-1}\prod_{j=1}^{q_k} E\left(Y^{*h}_{jk}\right), \tag{5.246}$$

which taking into account the expression for the h-th moment of a Beta r.v., yields (5.235) or that same expression with p_k and q_k switched.

To allow for an easy implementation and application of the test of independence of several sets of variables, the authors provide modules `PDFIndR`, `CDFIndR`, `QuantIndR`, `PvalIndR`, and `PvalDataIndR`. The first four of these modules have as mandatory arguments the sample size and a vector or list with the values for the numbers of variables in each set, that is, the p_k's ($k = 1, \dots, m$), and they have exactly the same optional arguments as the corresponding modules in Sect. 5.1.1. Module `PvalDataIndR` has as mandatory arguments the name of the data file with the $n \times p$ data matrix, and the vector or list with the p_k values ($k = 1, \dots, m$) and has the same optional arguments as the `PvalData` module in Sect. 5.1.1.

In Fig. 5.60 its use is illustrated. The first optional argument is given the value 1 in order to allow for the selection of the variables to enter the test and their reordering. With the first command in that figure, which uses the Mathematica® version of module `PvalDataIndR`, we test the independence of three sets of variables, the

```
PvalDataIndR[path <> "Tab9_12_JW.dat", {3, 2, 2}, 1]
 Original variables {2, 3, 4, 5, 8, 6, 7}
Computed value of Λ: 0.0008542783607
p-value: 5.211733396×10^-58
PvalDataIndR[path <> "Tab9_12_JW.dat", {3, 4}, 1]
 Original variables {2, 3, 4, 5, 8, 6, 7}
Computed value of Λ: 0.002148472296
p-value: 1.254086916×10^-51
PvalDataIndR[path <> "Tab9_12_JW.dat", {2, 2}, 1]
 Original variables {5, 8, 6, 7}
Computed value of Λ: 0.3976213062
p-value: 1.103879969×10^-8
```

Fig. 5.60 Commands used and output obtained with the Mathematica® module `PvalDataIndR` for a test of independence of three sets of variables with $p_1 = 2$, $p_2 = 2$, and $p_3 = 2$ variables

first formed by the three sales performance variables in problem 9.19 and Table 9.12 in Johnson and Wichern (2014), the second one formed by the test scores for creativity and mathematical ability and the third by the test scores for mechanical and abstract reasoning.

To illustrate the relation in (5.243), the last two commands in Fig. 5.60 use module `PvalDataIndR` to test first the independence between the set of three sales performance variables and the set of four test scores and then the independence between the set of the test scores for creativity and mathematical ability and the set of the test scores for mechanical and abstract reasoning. One may check that the product of the computed values for these last two l.r.t. statistics exactly matches the computed value for the l.r.t. statistic to test the independence of the three sets of variables. Of course one could alternatively have used module `PvalDataInd2R` to carry out these last two tests, obtaining exactly the same result.

In Fig. 5.61 we may take a look at the plots of some p.d.f.'s of Λ in (5.234) for various values of n for three different situations where $\underline{X}$ has a real 21-variate Normal distribution, being subdivided into three subvectors with different numbers of variables. The same sample sizes are used for the plots of the three different sets of values for p_1, p_2 and p_3, showing how more unbalanced values for the p_k's make the p.d.f.'s shift towards the right. The plots in this figure have the same scales as those in Fig. 5.64 for the case of complex random variables, in order to allow for a comparison of the shape of the distributions in the real and complex cases.

In Fig. 5.62 we have the plots of p.d.f.'s of Λ for different sample sizes, and for a situation where $\underline{X}$ has a 51-variate Normal distribution, which is split into five subvectors with $p_1 = 6, p_2 = 12, p_3 = 11, p_4 = 10, p_5 = 12$. We may see how we need extremely large sample sizes to make the p.d.f. shift towards the right, which is one of the reasons why common asymptotic distributions have so much trouble handling the distribution of Λ when the number of variables involved grows large. All plots in both figures were obtained with module `PDFIndR`.

The statistic Λ in (5.234), which is also called the generalized Wilks' Lambda (Wilks, 1935), given its expression in (5.234) and its relation in (5.243) with the "common" Wilks' Lambda statistics $\Lambda_{k,(k+1,\dots,m)}$, may be seen as the statistic for the test of fit of the Generalized Canonical Analysis model (Coelho, 1992; SenGupta, 2004), where the approach to be followed is that in Carroll (1968) and not quite that of Kettenring (1971). Once the statistic Λ in (5.234) is identified as the statistic used to test the fit of the Generalized Canonical Analysis model, that is, to test the null hypothesis of independence of the m subsets $\underline{X}_1, \dots, \underline{X}_m$, we will be able to see that all the statistics used in the tests of fit of all the common linear models are particular cases or variations of this statistic. Figure 5.63 provides a diagram with the interrelations among most of the well-known linear models that may be seen as particular cases of the Generalized Canonical Analysis model, some of which were addressed in Sect. 5.1.9, as particular cases of the Canonical Analysis model.

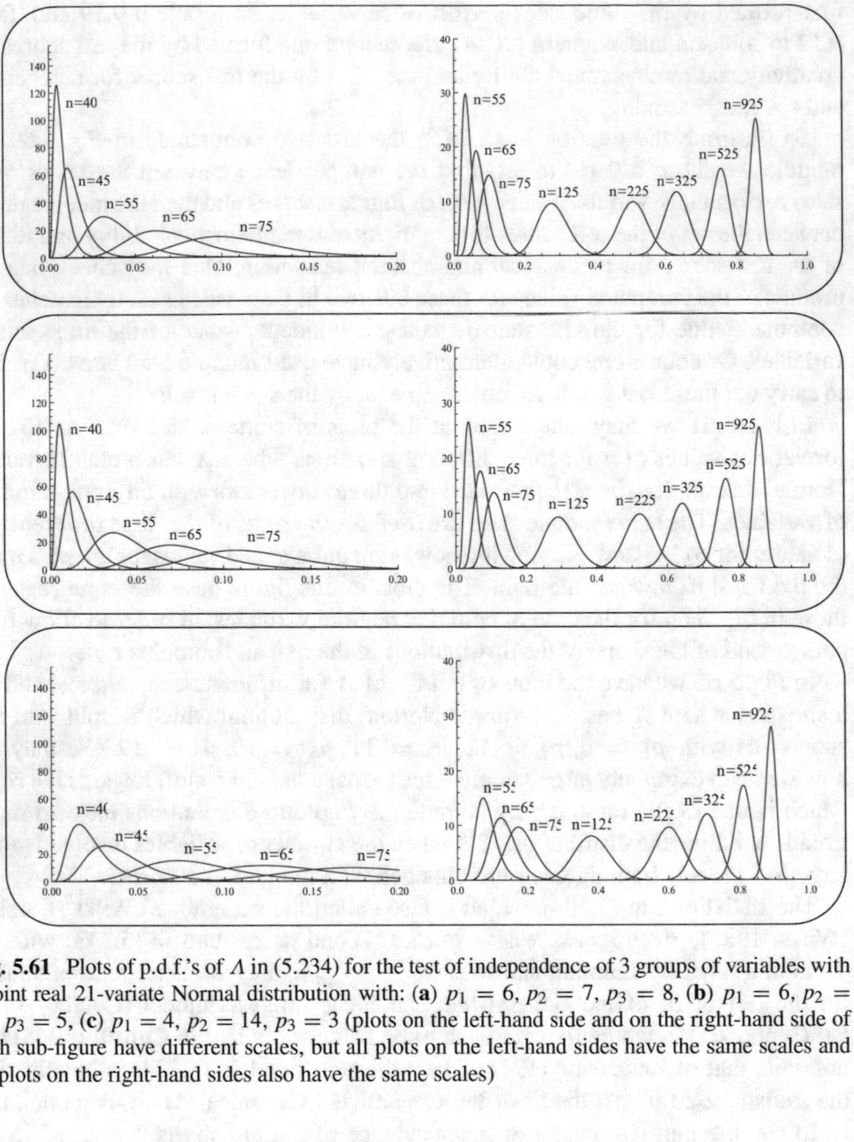

Fig. 5.61 Plots of p.d.f.'s of Λ in (5.234) for the test of independence of 3 groups of variables with a joint real 21-variate Normal distribution with: (**a**) $p_1 = 6$, $p_2 = 7$, $p_3 = 8$, (**b**) $p_1 = 6$, $p_2 = 10$, $p_3 = 5$, (**c**) $p_1 = 4$, $p_2 = 14$, $p_3 = 3$ (plots on the left-hand side and on the right-hand side of each sub-figure have different scales, but all plots on the left-hand sides have the same scales and all plots on the right-hand sides also have the same scales)

There are very limited tables of quantiles available for Λ in (5.234). Davis and Field (1971) list "correction" factors "c" for $\alpha = 0.05$ and $\alpha = 0.01$, from which by taking

$$e^{-c\chi_f^2(1-\alpha)/\rho^*}$$

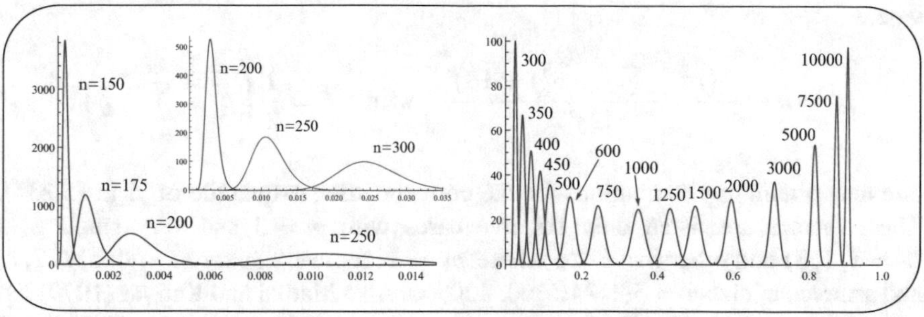

Fig. 5.62 Plots of p.d.f.'s of Λ in (5.234) for the test of independence of 5 groups of variables with a joint real 51-variate Normal distribution, with $p_1 = 6$, $p_2 = 12$, $p_3 = 11$, $p_4 = 10$, $p_5 = 12$

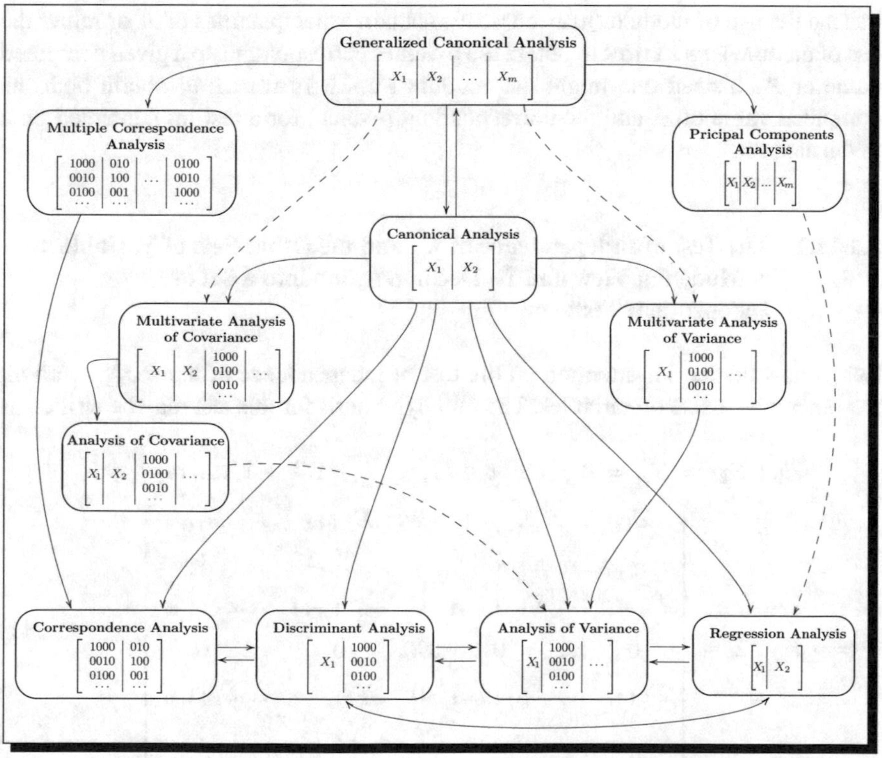

Fig. 5.63 Diagram with the main multivariate and univariate linear models that may be seen as particular cases of the Generalized Canonical Analysis model, and whose statistics of fit are particular cases of the generalized Wilks' Lambda

where

$$\rho^* = n - \frac{2\left(p^3 - \sum_{k=1}^m p_k^3\right) + 18f}{12f} \quad \text{with} \quad f = \frac{1}{2}\left(p^2 - \sum_{k=1}^m p_k^2\right),$$

one may obtain approximations for the corresponding α-quantile of Λ in (5.234). These factors are listed only for five cases with $m = 3$ and very small p_k's ($k = 1, 2, 3$) and other five cases where $p_k = 1$, for every k, for $m = 4, 5, 6, 7, 8$, and samples of size $n = 5(1)24(5)60, 120, \infty$. Also Mathai and Katiyar (1979) list exact 0.05 and 0.01 quantiles for $-\{n - 1 - (2p + 5)/6\} \log \Lambda$ for Λ in (5.234) and for $p = 3, \ldots, 10$, with $p_k = 1$ for all k, that is, only for the test of independence of individual variables, for samples of size $n = 3, \ldots, 20, \infty$.

Of course, given the nature of this test it is impossible to build quantile tables that cover a wide enough variety of situations and as such the authors strongly advise instead the use of module `QuantIndR` to obtain exact quantiles of Λ or rather the use of module `PvalIndR` to obtain the p-values corresponding to a given computed value of Λ. Instead one might use module `PvalDataIndR` to obtain both the computed value of Λ and the corresponding p-value, for a test implemented on a given data set.

5.1.11.1 The Test of Independence of $\underline{X}_k$ and the Other Sets of Variables: A Modeling View and Its Decomposition into a Set of Independent Tests

Let us now center our attention on the test of independence of the set $\underline{X}_k$ with all the other $m - 1$ sets of variables. The null hypothesis for this test may be written as

$$H_0 : \Sigma_{k\ell} = \Sigma_{\ell k}' = 0 \;\text{ for all } \ell \in \{1, \ldots, k-1, k+1, \ldots, m\}$$
$$\Longleftrightarrow H_0 : \Sigma = \begin{bmatrix} \Sigma_{11} & \ldots & \Sigma_{1,k-1} & 0 & \Sigma_{1,k+1} & \ldots & \Sigma_{1m} \\ \vdots & \ddots & \vdots & \vdots & \vdots & & \vdots \\ \Sigma_{k-1,1} & \ldots & \Sigma_{k-1,k-1} & 0 & \Sigma_{k-1,k+1} & \ldots & \Sigma_{k-1,m} \\ 0 & \ldots & 0 & \Sigma_{kk} & 0 & \ldots & 0 \\ \Sigma_{k+1,1} & \ldots & \Sigma_{k+1,k-1} & 0 & \Sigma_{k+1,k+1} & \ldots & \Sigma_{k+1,m} \\ \vdots & & \vdots & \vdots & \vdots & \ddots & \vdots \\ \Sigma_{m1} & \ldots & \Sigma_{m,k-1} & 0 & \Sigma_{m,k+1} & \ldots & \Sigma_{mm} \end{bmatrix}. \tag{5.247}$$

There are a number of different ways in which we can look at this test. One of them, possibly the simplest one and also the one most directly related to the writing of the null hypothesis in the form in (5.247), is to look at it as the test of independence between two sets of variables which are $\underline{X}_k$ and the superset formed by joining all the other $m-1$ sets. From this viewpoint, for a sample of size n, the $(2/n)$-th power of the l.r.t. statistic to test H_0 in (5.247) would be written as

$$\Lambda_{(k)} = \frac{|A|}{|A_{kk}|\,|A^*_{(k)}|} \tag{5.248}$$

where $A^*_{(k)}$ is the matrix that is obtained from A by deleting the set of rows and columns corresponding to $\underline{X}_k$.

From Sect. 5.1.9 we know that, for $n > p$,

$$\Lambda_{(k)} \stackrel{\text{d}}{\equiv} \prod_{j=1}^{p_k} Y_j \stackrel{\text{d}}{\equiv} \prod_{k=1}^{p-p_k} Y_k^*\,, \tag{5.249}$$

where, for $j = 1, \ldots, p_k$ and $k = 1, \ldots, p - p_k$,

$$Y_j \sim Beta\left(\frac{n-(p-p_k)-j}{2}, \frac{p-p_k}{2}\right), \quad \text{and} \quad Y_k^* \sim Beta\left(\frac{n-p_k-k}{2}, \frac{p_k}{2}\right) \tag{5.250}$$

form two sets of independent r.v.'s, so that for even p_k the distribution of $\Lambda_{(k)}$ is given by Theorem 3.2 and its p.d.f. and c.d.f. by Corollary 4.2, with

$$m^* = 1\,, \quad k_1 = 2\,, \quad a_1 = \frac{n-(p-p_k)}{2}\,, \quad n_1 = \frac{p_k}{2}\,, \quad m_1 = p - p_k$$

and for even $p - p_k$ the exact distribution and the exact p.d.f. and c.d.f. of $\Lambda_{(k)}$ are given by the same results, now with

$$m^* = 1\,, \quad k_1 = 2\,, \quad a_1 = \frac{n-p_k}{2}\,, \quad n_1 = \frac{p-p_k}{2}\,, \quad m_1 = p_k\,.$$

The exact p.d.f. and c.d.f. of $\Lambda_{(k)}$ are thus given, in terms of the EGIG distribution representation, for any of the above cases, respectively by

$$f_{\Lambda_{(k)}}(z) = f^{EGIG}\left(z \,\middle|\, \{r_j\}_{j=1:p-2}; \left\{\frac{n-2-j}{2}\right\}_{j=1:p-2}; p-2\right)$$

and

$$F_{\Lambda_{(k)}}(z) = F^{EGIG}\left(z \,\middle|\, \{r_j\}_{j=1:p-2}; \left\{\frac{n-2-j}{2}\right\}_{j=1:p-2}; p-2\right),$$

where

$$r_j = \begin{cases} h_j\,, & j=1,2 \\ r_{j-2}+h_j\,, & j=3,\dots,p-2 \end{cases}$$

with

$$h_j = (\#\text{ of elements in } \{p_k, p-p_k\} \geq j) - 1\,, \quad j=1,\dots,p-2\,.$$

or, alternatively for any p_k and any $p-p_k$, in terms of Meijer G functions, by

$$f_{\Lambda_{(k)}}(z) = \left\{\prod_{j=1}^{p_k} \frac{\Gamma\left(\frac{n-j}{2}\right)}{\Gamma\left(\frac{n-p+p_k-j}{2}\right)}\right\} G^{\,p_k,0}_{\,p_k,p_k}\left(\begin{array}{c} \left\{\frac{n-j}{2}-1\right\}_{j=1:p_k} \\ \left\{\frac{n-p+p_k-j}{2}-1\right\}_{j=1:p_k} \end{array}\middle|\, z\right)$$

and

$$F_{\Lambda_{(k)}}(z) = \left\{\prod_{j=1}^{p_k} \frac{\Gamma\left(\frac{n-j}{2}\right)}{\Gamma\left(\frac{n-p+p_k-j}{2}\right)}\right\} G^{\,p_k,1}_{\,p_k+1,p_k+1}\left(\begin{array}{c} \left\{1, \left\{\frac{n-j}{2}\right\}_{j=1:p_k}\right\} \\ \left\{\left\{\frac{n-p+p_k-j}{2}\right\}_{j=1:p_k}, 0\right\} \end{array}\middle|\, z\right).$$

But, one alternative and interesting way to obtain and represent the l.r.t. statistic in (5.248) is to consider the l.r.t. statistic to test the independence of the $m-1$ sets of variables $\underline{X}_1,\dots,\underline{X}_{k-1},\underline{X}_{k+1},\dots,\underline{X}_m$, which is

$$\Lambda^*_{(k)} = \frac{|A^*_{(k)}|}{\left\{\prod_{j=1}^{k-1}|A_{jj}|\right\}\left\{\prod_{j=k+1}^{m}|A_{jj}|\right\}}$$

and then, for Λ in (5.234), to take

$$\Lambda'_{(k)} = \frac{\Lambda}{\Lambda^*_{(k)}}\,. \tag{5.251}$$

It is almost immediate to recognize that $\Lambda'_{(k)} \equiv \Lambda_{(k)}$, since from (5.251) we may write

$$\Lambda'_{(k)} = \frac{\Lambda}{\Lambda^*_{(k)}} = \frac{|A|}{|A_{kk}|\left\{\prod\limits_{j=1}^{k-1}|A_{jj}|\right\}\left\{\prod\limits_{j=k+1}^{m}|A_{jj}|\right\}} \Bigg/ \frac{|A^*_{(k)}|}{\left\{\prod\limits_{j=1}^{k-1}|A_{jj}|\right\}\left\{\prod\limits_{j=k+1}^{m}|A_{jj}|\right\}} = \frac{|A|}{|A_{kk}|\,|A^*_{(k)}|}\,.$$

This way to express $\Lambda_{(k)}$ is extremely useful in obtaining a decomposition of this statistic into a number of independent factors, themselves also l.r.t. statistics, as we will see next. Directly from (5.251) and from the fact that we may obtain for Λ

in (5.234) one other factorization in line with the one in (5.243), but, say, done the "reverse" way (see Anderson (2003, Sec. 9.3.1)), which allows us to write

$$\Lambda = \prod_{k=2}^{m} \Lambda_{k,(1,\dots,k-1)}\,, \tag{5.252}$$

we may obtain a third expression for $\Lambda_{(k)}$, this one yielding a direct factorization of this statistic into a set of independent l.r.t. statistics. Indeed, by applying to $\Lambda^*_{(k)}$ a similar factorization to that of Λ in (5.252) we may write

$$\Lambda^*_{(k)} = \left\{ \prod_{j=2}^{k-1} \Lambda_{j,(1,\dots,j-1)} \right\} \left\{ \prod_{j=k+1}^{m} \Lambda_{j,(1,\dots,k-1,k+1,\dots,j-1)} \right\}$$

where for $j = k+1$ we have $\Lambda_{j,(1,\dots,k-1,k+1,\dots,j-1)} \equiv \Lambda_{k+1,(1,\dots,k-1)}$ and for $j = k+2$ we have $\Lambda_{k+2,(1,\dots,k-1,k+1)}$, and so on. But then, combining this factorization of $\Lambda^*_{(k)}$ with that of Λ in (5.252) and using (5.251), we may write

$$\begin{aligned} \Lambda_{(k)} = \frac{\Lambda}{\Lambda^*_{(k)}} &= \Lambda_{k,(1,\dots,k-1)} \left\{ \prod_{j=k+1}^{m} \frac{\Lambda_{j,(1,\dots,j-1)}}{\Lambda_{j,(1,\dots,k-1,k+1,\dots,j-1)}} \right\} \\ &= \Lambda_{k,(1,\dots,k-1)} \left\{ \prod_{j=k+1}^{m} \Lambda_{jk|(1,\dots,k-1,k+1,\dots,j-1)} \right\}. \end{aligned} \tag{5.253}$$

The $k-1$ statistics $\Lambda_{jk|(1,\dots,k-1,k+1,\dots,j-1)} = \Lambda_{j,(1,\dots,j-1)}/\Lambda_{j,(1,\dots,k-1,k+1,\dots,j-1)}$ in (5.253) are, for a sample of size n, the $(2/n)$-th powers of the l.r.t. statistics used to test the nullity of the covariance of $\underline{X}_j$ and $\underline{X}_k$, accounting for the covariances of $\underline{X}_j$, and also $\underline{X}_k$, with sets $\underline{X}_1, \dots, \underline{X}_{k-1}, \underline{X}_{k+1}, \dots, \underline{X}_{j-1}$, that is, each of these statistics is the l.r.t. statistic to test between the model

$$\begin{aligned} H_1 : \underline{X}_j = \beta_{j,1}\underline{X}_1 + \beta_{j,2}\underline{X}_2 + \dots + \beta_{j,k-1}\underline{X}_{k-1} + \beta_{j,k}\underline{X}_k \\ +\beta_{j,k+1}\underline{X}_{k+1} + \dots + \beta_{j,j-1}\underline{X}_{j-1} + \underline{\mathcal{E}} \end{aligned} \tag{5.254}$$

and the submodel

$$\begin{aligned} H_0 : \underline{X}_j = \beta_{j,1}\underline{X}_1 + \beta_{j,2}\underline{X}_2 + \dots + \beta_{j,k-1}\underline{X}_{k-1} \\ +\beta_{j,k+1}\underline{X}_{k+1} + \dots + \beta_{j,j-1}\underline{X}_{j-1} + \underline{\mathcal{E}}\,, \end{aligned} \tag{5.255}$$

where $\underline{\mathcal{E}}$ is a p_j-variate Normal distributed error vector with null expected value, and where for $\ell = 1, \dots, j-1$, $\beta_{j,\ell}$ is a $p_j \times p_\ell$ matrix of model parameters. The statistic $\Lambda_{jk|(1,\dots,k-1,k+1,\dots,j-1)}$ is thus the l.r.t. statistic to test the null hypothesis

$$H_{0jk} : \beta_{j,k} = 0 \ \text{ in model (5.254)}\,.$$

Let

$$A_j^* = \begin{bmatrix} A_{11} & \dots & A_{1j} \\ \vdots & \ddots & \vdots \\ A_{j1} & \dots & A_{jj} \end{bmatrix}.$$

Then,

$$\Lambda_{j,(1,\dots,j-1)} = \frac{|A_j^*|}{|A_{jj}|\,|A_{j-1}^*|}$$

is the l.r.t. statistic to test the fit of the H_1 model in (5.254), and

$$\Lambda_{j,(1,\dots,k-1,k+1,\dots,j-1)} = \frac{|A_{j(k)}^*|}{|A_{jj}|\,|A_{j-1(k)}^*|},$$

is the l.r.t. statistic to test the fit of the H_0 model in (5.255), where

$$A_{j(k)}^* = \begin{bmatrix} A_{11} & \dots & A_{1,k-1} & A_{1,k+1} & \dots & A_{1j} \\ \vdots & \ddots & \vdots & \vdots & & \vdots \\ A_{k-1,1} & \dots & A_{k-1,k-1} & A_{k-1,k+1} & \dots & A_{k-1,j} \\ A_{k+1,1} & \dots & A_{k+1,k-1} & A_{k+1,k+1} & \dots & A_{k+1,j} \\ \vdots & & \vdots & \vdots & \ddots & \vdots \\ A_{j,1} & \dots & A_{j,k-1} & A_{j,k+1} & \dots & A_{jj} \end{bmatrix},$$

is equal to A_j^* without the set of rows and columns corresponding to $\underline{X}_k$.

It is then possible to prove the following claim.

Claim 1 The $m-k+1$ l.r.t. statistics on the right-hand side of (5.253), that is the $m-k$ statistics $\Lambda_{jk|(1,\dots,k-1,k+1,\dots,j-1)}$ and $\Lambda_{k,(1,\dots,k-1)}$, are all independent.

Proof This may be shown by using the fact that while from (5.249) and (5.250) we may write the h-th moment of $\Lambda_{(k)}$ as

$$E\left(\Lambda_{(k)}^h\right) = \prod_{j=1}^{p_k} \frac{\Gamma\left(\frac{n-j}{2}\right)\Gamma\left(\frac{n-(p-p_k)-j}{2}+h\right)}{\Gamma\left(\frac{n-(p-p_k)-j}{2}\right)\Gamma\left(\frac{n-j}{2}+h\right)}, \tag{5.256}$$

we also know that for $n > p_k^*$, with $p_k^* = p_1 + \cdots + p_k$,

$$\Lambda_{k,(1,\dots,k-1)} \stackrel{d}{\equiv} \prod_{j=1}^{p_k} Y_j^*$$

where, for $j = 1, \ldots, p_k$,

$$Y_j^* \sim Beta\left(\frac{n - p_{k-1}^* - j}{2}, \frac{p_{k-1}^*}{2}\right)$$

form a set of independent r.v.'s, so that

$$E\left(\Lambda_{k,(1,\ldots,k-1)}^h\right) = \prod_{j=1}^{p_k} \frac{\Gamma\left(\frac{n-j}{2}\right)}{\Gamma\left(\frac{n-p_{k-1}^*-j}{2}\right)} \frac{\Gamma\left(\frac{n-p_{k-1}^*-j}{2}+h\right)}{\Gamma\left(\frac{n-j}{2}+h\right)}.$$

Then, from (5.178) and (5.179), we know that for each of the $m-k$ statistics $\Lambda_{jk|(1,\ldots,k-1,k+1,\ldots,j-1)}$ we have

$$\Lambda_{jk|(1,\ldots,k-1,k+1,\ldots,j-1)} \stackrel{d}{\equiv} \prod_{\ell=1}^{p_k} Y_{j\ell} \tag{5.257}$$

where, for $\ell = 1, \ldots, p_k$, $j = k+1, \ldots, m$, and $n > p_j^*$, with $p_j^* = \sum_{\ell'=1}^{j} p_{\ell'}$,

$$Y_{j\ell} \sim Beta\left(\frac{n - p_j - (p_{j-1}^* - p_k) - \ell}{2}, \frac{p_j}{2}\right)$$

form $m-k$ sets of p_k independent r.v.'s, so that we may write the h-th moment of each statistic $\Lambda_{jk|(1,\ldots,k-1,k+1,\ldots,j-1)}$ as

$$E\left(\Lambda_{jk|(1,\ldots,k-1,k+1,\ldots,j-1)}^h\right) = \prod_{\ell=1}^{p_k} \frac{\Gamma\left(\frac{n-(p_{j-1}^*-p_k)-\ell}{2}\right)\Gamma\left(\frac{n-p_j-(p_{j-1}^*-p_k)-\ell}{2}+h\right)}{\Gamma\left(\frac{n-p_j-(p_{j-1}^+-p_k)-\ell}{2}\right)\Gamma\left(\frac{n-(p_{j-1}^*-p_k)-\ell}{2}+h\right)}. \tag{5.258}$$

But then, if the $m-k$ statistics $\Lambda_{jk|(1,\ldots,k-1,k+1,\ldots,j-1)}$ were all independent and independent from $\Lambda_{k,(1,\ldots,k-1)}$, we could write

$$\begin{aligned} E\left(\Lambda_{(k)}^h\right) &= \left\{\prod_{j=k+1}^{m} E\left(\Lambda_{jk|(1,\ldots,k-1,k+1,\ldots,j-1)}^h\right)\right\} E\left(\Lambda_{k,(1,\ldots,k-1)}^h\right) \\ &= \left\{\prod_{j=k+1}^{m}\prod_{\ell=1}^{p_k} \frac{\Gamma\left(\frac{n-(p_{j-1}^*-p_k)-\ell}{2}\right)\Gamma\left(\frac{n-p_j-(p_{j-1}^*-p_k)-\ell}{2}+h\right)}{\Gamma\left(\frac{n-p_j-(p_{j-1}^+-p_k)-\ell}{2}\right)\Gamma\left(\frac{n-(p_{j-1}^*-p_k)-\ell}{2}+h\right)}\right\} \\ &\quad \times \left\{\prod_{j=1}^{p_k} \frac{\Gamma\left(\frac{n-j}{2}\right)}{\Gamma\left(\frac{n-p_{k-1}^*-j}{2}\right)} \frac{\Gamma\left(\frac{n-p_{k-1}^*-j}{2}+h\right)}{\Gamma\left(\frac{n-j}{2}+h\right)}\right\}, \end{aligned} \tag{5.259}$$

where we can reverse the order of the two products in the double product and, by noticing that $p_j + p^*_{j-1} = p^*_j$, write it as

$$\prod_{\ell=1}^{p_k}\prod_{j=k+1}^{m}\frac{\Gamma\left(\frac{n-p^*_{j-1}+p_k-\ell}{2}\right)\Gamma\left(\frac{n-p^*_j+p_k-\ell}{2}+h\right)}{\Gamma\left(\frac{n-p^*_j+p_k-\ell}{2}\right)\Gamma\left(\frac{n-p^*_{j-1}+p_k-\ell}{2}+h\right)}$$

where it is clear now that consecutive factors in j cancel out to yield

$$\prod_{j=k+1}^{m}\frac{\Gamma\left(\frac{n-p^*_{j-1}+p_k-\ell}{2}\right)\Gamma\left(\frac{n-p^*_j+p_k-\ell}{2}+h\right)}{\Gamma\left(\frac{n-p^*_j+p_k-\ell}{2}\right)\Gamma\left(\frac{n-p^*_{j-1}+p_k-\ell}{2}+h\right)}=\frac{\Gamma\left(\frac{n-p^*_k+p_k-\ell}{2}\right)\Gamma\left(\frac{n-p^*_m+p_k-\ell}{2}+h\right)}{\Gamma\left(\frac{n-p^*_m+p_k-\ell}{2}\right)\Gamma\left(\frac{n-p^*_k+p_k-\ell}{2}+h\right)} \tag{5.260}$$

where $p^*_m = p = \sum_{k=1}^m p_k$ and $p^*_k - p_k = p^*_{k-1}$ so that from (5.259) and (5.260) we finally have

$$\begin{aligned}E\left(\Lambda^h_{(k)}\right) &= \left\{\prod_{j=1}^{p_k}\frac{\Gamma\left(\frac{n-j}{2}\right)}{\Gamma\left(\frac{n-p^*_{k-1}-j}{2}\right)}\frac{\Gamma\left(\frac{n-p^*_{k-1}-j}{2}+h\right)}{\Gamma\left(\frac{n-j}{2}+h\right)}\right\}\\ &\qquad\times\left\{\prod_{j=1}^{p_k}\frac{\Gamma\left(\frac{n-p^*_k+p_k-\ell}{2}\right)\Gamma\left(\frac{n-p^*_m+p_k-\ell}{2}+h\right)}{\Gamma\left(\frac{n-p^*_m+p_k-\ell}{2}\right)\Gamma\left(\frac{n-p^*_k+p_k-\ell}{2}+h\right)}\right\}\\ &= \prod_{j=1}^{p_k}\frac{\Gamma\left(\frac{n-j}{2}\right)\Gamma\left(\frac{n-(p-p_k)-j}{2}+h\right)}{\Gamma\left(\frac{n-(p-p_k)-j}{2}\right)\Gamma\left(\frac{n-j}{2}+h\right)},\end{aligned}$$

which exactly matches (5.256), thus proving our claim.

This is so since all Λ statistics have their supports delimited to (0, 1) and as such the expressions for their h-th moments determine their distributions. Furthermore, all these moment expressions remain valid for any $h \in \mathbb{C}$ and as such we may take

$$W_{(k)} = -\log \Lambda_{(k)}\,, \quad W_{jk|(1,\dots,k-1,k+1,\dots,j-1)} = -\log \Lambda_{jk|(1,\dots,k-1,k+1,\dots,j-1)}$$

and

$$W_{k,(1,\dots,k-1)} = -\log \Lambda_{k,(1,\dots,k-1)}$$

and, from (5.259) write

$$
\begin{aligned}
\Phi_{W_{(k)}}(t) &= E\left(e^{\mathrm{i}t\,W_{(k)}}\right) = E\left(e^{-\mathrm{i}t\log\Lambda_{(k)}}\right)\\
&= E\left(\Lambda_{(k)}^{-\mathrm{i}t}\right) = \left\{\prod_{j=k+1}^{m} E\left(\Lambda_{jk|(1,\dots,k-1,k+1,\dots,j-1)}^{-\mathrm{i}t}\right)\right\} E\left(\Lambda_{k,(1,\dots,k-1)}^{-\mathrm{i}t}\right)\\
&= \left\{\prod_{j=k+1}^{m} \Phi_{W_{jk|(1,\dots,k-1,k+1,\dots,j-1)}}(t)\right\} \Phi_{W_{k,(1,\dots,k-1)}}(t)
\end{aligned}
$$

which clearly shows that the distribution of $W_{(k)}$ is the same as that of

$$
\left\{\sum_{j=k+1}^{m} W_{jk|(1,\dots,k-1,k+1,\dots,j-1)}\right\} + W_{k,(1,\dots,k-1)}
$$

where all $m-k+1$ statistics are independent. This independence entails the independence of the corresponding Λ statistics in (5.253). □

Since $\Lambda_{(k)} \equiv \Lambda_{k,(1,\dots,k-1,k+1,\dots,m)}$, we may write

$$
\Lambda_{k,(k+1,\dots,m)|(1,\dots,k-1)} = \frac{\Lambda_{k,(1,\dots,k-1,k+1,\dots,m)}}{\Lambda_{k,(1,\dots,k-1)}} = \frac{\Lambda_{(k)}}{\Lambda_{k,(1,\dots,k-1)}}\,,
$$

which yields

$$
\Lambda_{(k)} = \Lambda_{k,(1,\dots,k-1)}\,\Lambda_{k,(k+1,\dots,m)|(1,\dots,k-1)}\,, \tag{5.261}
$$

with (5.253) showing that we may write

$$
\begin{aligned}
\Lambda_{k,(k+1,\dots,m)|(1,\dots,k-1)} &= \prod_{j=k+1}^{m} \Lambda_{jk|(1,\dots,k-1,k+1,\dots,j-1)}\\
&= \prod_{j=k+1}^{m} \frac{\Lambda_{j,(1,\dots,j-1)}}{\Lambda_{j,(1,\dots,k-1,k+1,\dots,j-1)}}
\end{aligned} \tag{5.262}
$$

In Coelho and Marques (2014) a parallel factorization of the one in (5.253) was obtained for $\Lambda_{(k)}$ as

$$
\Lambda_{(k)} = \left\{\prod_{j=1}^{k-1} \Lambda_{jk|(j+1,\dots,k-1,k+1,\dots,m)}\right\} \Lambda_{k,(k+1,\dots,m)}\,. \tag{5.263}
$$

Since we may also write

$$\Lambda_{k,(1,\dots,k-1)|(k+1,\dots,m)} = \frac{\Lambda_{k,(1,\dots,k-1,k+1,\dots,m)}}{\Lambda_{k,(k+1,\dots,m)}} = \frac{\Lambda_{(k)}}{\Lambda_{k,(k+1,\dots,m)}}$$

thus yielding

$$\Lambda_{(k)} = \Lambda_{k,(k+1,\dots,m)}\Lambda_{k(1,\dots,k-1)|(k+1,\dots,m)}\,, \tag{5.264}$$

expression (5.263) shows that we may write, for $k \geq 2$,

$$\begin{aligned}\Lambda_{k,(1,\dots,k-1)|(k+1,\dots,m)} &= \prod_{j=1}^{k-1} \Lambda_{jk|(j+1,\dots,k-1,k+1,\dots,m)} \\ &= \prod_{j=1}^{k-1} \frac{\Lambda_{j,(j+1,\dots,m)}}{\Lambda_{j,(j+1,\dots,k-1,k+1,\dots,m)}}\,.\end{aligned} \tag{5.265}$$

Expressions (5.261) and (5.264) show two other conceptually interesting factorizations of $\Lambda_{(k)}$, while expressions (5.262) and (5.265) show interesting factorizations of $\Lambda_{k,(k+1,\dots,m)|(1,\dots,k-1)}$ and $\Lambda_{k,(1,\dots,k-1)|(k+1,\dots,m)}$. Expression (5.265) will show itself to be the key for the double factorization of the overall l.r.t. statistic Λ into a set of $m(m-1)/2$ independent statistics in the next subsection.

5.1.11.2 A Double Factorization of Λ into a Set of Independent Statistics

From the decomposition of Λ in (5.252) we may write the overall null hypothesis of independence of the m sets of variables in (5.233) as

$$H_0 : \bigwedge_{k=2}^{m} \left[\Sigma_{k1} \dots \Sigma_{k,k-1}\right] = 0 \tag{5.266}$$

where $\bigwedge$ denotes a conjunction, so that testing the independence of the m sets of variables ends up being equivalent to test simultaneously the fit of the $m-1$ models

$$\underline{X}_k = \beta_{k,1}\underline{X}_1 + \dots + \beta_{k,k-1}\underline{X}_{k-1} + \underline{\mathcal{E}}\,, \tag{5.267}$$

for $k = 2, \dots, m$, where $\beta_{k,\ell}$ ($\ell = 1, \dots, k-1$) are $p_k \times p_\ell$ parameter matrices.

Using a somewhat simpler and more direct approach than the one used in Coelho and Marques (2014), we may, from (5.265), by "ignoring" sets $\underline{X}_{k+1}, \ldots, \underline{X}_m$ on both sides of that expression, write, for $k \geq 2$,

$$\Lambda_{k,(1,\ldots,k-1)} = \prod_{j=1}^{k-1} \Lambda_{jk|(j+1,\ldots,k-1)} = \prod_{j=1}^{k-1} \frac{\Lambda_{j,(j+1,\ldots,k)}}{\Lambda_{j,(j+1,\ldots,k-1)}}, \tag{5.268}$$

where, for $j = k-1$, $\Lambda_{j,(j+1,\ldots,k-1)} = 1$, so that, for $j = k-1$, we have $\Lambda_{jk|(j+1,\ldots,k-1)} = \Lambda_{k-1,k}$.

In (5.268) the statistics $\Lambda_{jk|(j+1,\ldots,k-1)}$ $(j = 1, \ldots, k-1)$ form a set of $k-1$ independent statistics (see Appendix 13 for further details). But then, from (5.268) and from the factorization of Λ in (5.252) we have

$$\Lambda = \prod_{k=2}^{m} \prod_{j=1}^{k-1} \Lambda_{jk|(j+1,\ldots,k-1)} = \prod_{k=2}^{m} \prod_{j=1}^{k-1} \frac{\Lambda_{j,(j+1,\ldots,k)}}{\Lambda_{j,(j+1,\ldots,k-1)}},$$

where all $\sum_{k=2}^{m-1}(k-1) = (m-1)m/2$ statistics $\Lambda_{jk|(j+1,\ldots,k-1)}$ are independent.

Each statistic $\Lambda_{jk|(j+1,\ldots,k-1)}$ is the l.r.t. statistic to test the nullity of the parameter matrix $\beta_{j,k}$ in the model

$$\underline{X}_j = \beta_{j,j+1}\underline{X}_{j+1} + \cdots + \beta_{j,k}\underline{X}_k + \mathcal{E}\,, \tag{5.269}$$

or, equivalently, with $\Lambda_{j(j+1,\ldots,k)}$ being the l.r.t. statistic to test the fit of the model (5.269), and $\Lambda_{j(j+1,\ldots,k-1)}$ the l.r.t. statistic to test the fit of the model

$$\underline{X}_j = \beta_{j,j+1}\underline{X}_{j+1} + \cdots + \beta_{j,k-1}\underline{X}_{k-1} + E\,.$$

From Sect. 5.1.9.5 we know that

$$\Lambda_{jk|(j+1,\ldots,k-1)} \stackrel{d}{\equiv} \prod_{\nu=1}^{p_j} Y_{\nu k} \stackrel{d}{\equiv} \prod_{\ell=1}^{p_k} Y^*_{\ell k}\,,$$

where, for $\nu = 1, \ldots, p_j$ and $\ell = 1, \ldots, p_k$,

$$Y_{\nu k} \sim Beta\left(\frac{n-(p_{j+1}+\cdots+p_k)-j}{2}, \frac{p_k}{2}\right),$$

and

$$Y^*_{\ell k} \sim Beta\left(\frac{n-(p_j+\cdots+p_{k-1})-\ell}{2}, \frac{p_j}{2}\right)$$

form two sets of independent r.v.'s, so that for even p_j the distribution of the statistic $\Lambda_{jk|(j+1,\dots,k-1)}$, under the null hypothesis of independence of $\underline{X}_j$ and $\underline{X}_k$, when accounting for the covariances of $\underline{X}_j$, and also $\underline{X}_k$, with $\underline{X}_{j+1}, \dots, \underline{X}_{k-1}$, $\underline{X}_{k+1}, \dots, \underline{X}_m$, is given by Theorem 3.2 and its p.d.f. and c.d.f. by Corollary 4.2, with

$$m^* = 1\,, \quad k_1 = 2\,, \quad a_1 = \frac{n - (p_{j+1} + \cdots + p_k)}{2}\,, \quad n_1 = \frac{p_j}{2}\,, \quad m_1 = p_k$$

and for even p_k the exact distribution and the exact p.d.f. and c.d.f. of $\Lambda_{jk|(j+1,\dots,k-1)}$ are given by the same results, now with

$$m^* = 1\,, \quad k_1 = 2\,, \quad a_1 = \frac{n - (p_j + \cdots + p_{k-1})}{2}\,, \quad n_1 = \frac{p_k}{2}\,, \quad m_1 = p_j\,,$$

the exact p.d.f. and c.d.f. of $\Lambda_{jk|(j+1,\dots,k-1)}$ being thus given in terms of the EGIG distribution representation, for any of the above cases, respectively by

$$f_\Lambda(z) = f^{EGIG}\left(z \,\middle|\, \{r_j\}_{j=1:p_k+p_j-2};\, \left\{ \frac{n-2-(p_{j+1}+\cdots+p_{k-1})-j}{2} \right\}_{j=1:p_k+p_j-2};\; p_k + p_j - 2 \right)$$

and

$$F_\Lambda(z) = F^{EGIG}\left(z \,\middle|\, \{r_j\}_{j=1:p_k+p_j-2};\, \left\{ \frac{n-2-(p_{j+1}+\cdots+p_{k-1})-j}{2} \right\}_{j=1:p_k+p_j-2};\; p_k + p_j - 2 \right),$$

where

$$r_j = \begin{cases} h_j\,, & j = 1, 2 \\ r_{j-2} + h_j\,, & j = 3, \dots, p_j + p_k - 2 \end{cases}$$

with

$$h_j = (\#\text{ of elements in } \{p_j, p_k\} \geq j) - 1\,, \quad j = 1, \dots, p_j + p_k - 2\,.$$

This shows that to test the independence of the m sets of variables $\underline{X}_k$ $(k = 1, \dots, m)$ is the same as to test, for $k = 2, \dots, m$, the fit of the $m - 1$ models in (5.267) or yet to test simultaneously the nullity of the covariance between $\underline{X}_j$

and $\underline{X}_k$ for $j = 1, \ldots, k-1$ and $k = 2, \ldots, m$, accounting for the covariances of $\underline{X}_j$ and $\underline{X}_k$ with $\underline{X}_{j+1}, \ldots, \underline{X}_{k-1}$.

In other words, to test the independence of the m groups of variables is equivalent to test, for $j = 1, \ldots, k-1$ and $k = 2, \ldots, m$, the null hypotheses

$$H_{0jk} : \beta_{j,k} = 0 \;\text{ in model (5.269)}\,, \tag{5.270}$$

so that one other representation for the null hypothesis of independence of the m groups of variables in (5.233) is, for H_{0jk} in (5.270),

$$H_0 : \bigwedge_{k=2}^{m} \bigwedge_{j=1}^{k-1} H_{0jk} \tag{5.271}$$

which gives us another modeling view of this test, in that we may consider the $(m-1)m/2$ sub-hypotheses in (5.271) as "independent" in the sense that the associated test statistics are all independent. This representation is somewhat simpler than the one obtained in Coelho and Marques (2014).

Since the statistic Λ in (5.234) is the l.r.t. statistic to test the fit of the Generalized Canonical Analysis model (Coelho, 1992; SenGupta, 2004), expression (5.266) shows that to test the fit of this model is equivalent to testing the fit of the $m-1$ models in (5.267) for $k = 2, \ldots, m$, where these $m-1$ tests are independent in the sense that under H_0 in (5.233), the associated test statistics are independent. In turn, expression (5.271) shows that it is possible to further break down each one of these $m-1$ hypotheses into a set of $k-1$ independent hypotheses, which are the hypotheses of nullity of the covariance between $\underline{X}_k$ and $\underline{X}_j$, for $j = 1, \ldots, k-1$, in the presence in the model of the sets $\underline{X}_1, \ldots, \underline{X}_{j-1}, \underline{X}_{j+1}, \ldots, \underline{X}_{k-1}$, this way splitting the overall null hypothesis H_0 in (5.233) into a set of $m(m-1)/2$ null hypotheses which correspond to independent tests.

The usefulness of being able to obtain the factorization of Λ into the set of $m(m-1)/2$ independent statistics $\Lambda_{jk|(j+1,\ldots,k-1)}$ is that when the null hypothesis H_0 in (5.233) is rejected we may then search for the pair or pairs of sets of variables that were responsible for this rejection. Šidák's correction (Šidák, 1967; Abdi, 2007) may then be used to adjust the α-level of these tests by using for each one of them a level $\alpha^* = 1-(1-\alpha)^{1/(m(m-1)/2)}$, where α is the level intended to be used for the test of the overall hypothesis in (5.233). Moreover, Šidák's correction gives the exact adjustment for the α level of these "partial" tests when the associated test statistics are independent.

5.1.12 The Likelihood Ratio Statistic to Test the Independence of Several Groups of Complex Variables [IndC]

If the random vector in the previous subsection has a p-variate complex Normal distribution, that is, if $\underline{X} \sim CN_p(\mu, \Sigma)$, and we are interested in testing the null hypothesis in (5.233), then the $(1/n)$-th power of the l.r.t. statistic is still given by

$$\Lambda = \frac{|A|}{\prod_{k=1}^{m} |A_{kk}|}, \tag{5.272}$$

now, for $n > p$, with

$$E\left(\Lambda^h\right) = \prod_{k=1}^{m-1} \prod_{j=1}^{p_k} \frac{\Gamma(n-j)\,\Gamma(n-j-q_k+h)}{\Gamma(n-j-q_k)\,\Gamma(n-j+h)},$$

(see expression (2.6) in Krishnaiah et al. (1976) for an equivalent expression, and Appendix 14 for a full derivation of the result, directly from the complex Wishart distribution of the m.l.e. of Σ), which allows us to infer that, for $n > p$,

$$\Lambda \stackrel{d}{\equiv} \prod_{k=1}^{m-1} \prod_{j=1}^{p_k} Y_{jk}$$

where

$$Y_{jk} \sim Beta\,(n-q_k-j, q_k)\,,$$

is a set of $\sum_{k=1}^{m-1} p_k = p - p_m$ independent r.v.'s, with $q_k = p_{k+1} + \cdots + p_m$, and where p_k and q_k are interchangeable.

Thus, in this case, the exact distribution of Λ is given by Theorem 3.2 and its p.d.f. and c.d.f. by Corollary 4.2 with

$$m^* = m-1\,, \quad k_\nu = 1\,, \quad n_\nu = p_\nu\,,$$
$$m_\nu = q_\nu = n_{\nu+1} + \cdots + n_{m^*} + p_m = p_{\nu+1} + \cdots + p_{m^*} + p_m\,,$$
$$a_\nu = n - q_\nu = n - m_\nu\,, \quad (\nu = 1, \ldots, m^*)\,.$$

The exact p.d.f. and c.d.f. of Λ are therefore given, in terms of the EGIG distribution representation, for any combination of p_k's, respectively by

$$f_\Lambda(z) = f^{EGIG}\left(z \,\middle|\, \{r_j\}_{j=1:p-1};\, \{n-1-j\}_{j=1:p-1}\,;\, p-1\right)$$

and

$$F_\Lambda(z) = F^{EGIG}\left(z \,\middle|\, \{r_j\}_{j=1:p-1}; \{n-1-j\}_{j=1:p-1}; p-1\right),$$

where

$$r_j = \begin{cases} h_j\,, & j=1 \\ r_{j-1}+h_j\,, & j=2,\ldots,p-1 \end{cases}$$

with

$$h_j = (\#\text{ of } p_k\text{'s } (k=1,\ldots,m) \geq j) - 1\,, \quad j=1,\ldots,p-1\,,$$

a result that confirms the one obtained in Coelho et al. (2015).

Alternatively, the exact p.d.f. and c.d.f. of Λ may be given in terms of the Meijer G function by

$$f_\Lambda(z) = \left\{\prod_{k=1}^{m-1}\prod_{j=1}^{p_k}\frac{\Gamma(n-j)}{\Gamma(n-q_k-j)}\right\} G^{p^*,0}_{p^*,p^*}\left(\begin{array}{c} \{n-j-1\}_{\substack{j=1:p_k\\k=1:m-1}} \\ \{n-q_k-j-1\}_{\substack{j=1:p_k\\k=1:m-1}} \end{array} \middle|\, z \right)$$

and

$$F_\Lambda(z) = \left\{\prod_{k=1}^{m-1}\prod_{j=1}^{p_k}\frac{\Gamma(n-j)}{\Gamma(n-q_k-j)}\right\} G^{p^*,1}_{p^*+1,p^*+1}\left(\begin{array}{c} \left\{1, \{n-j\}_{\substack{j=1:p_k\\k=1:m-1}}\right\} \\ \left\{\{n-q_k-j\}_{\substack{j=1:p_k\\k=1:m-1}}, 0\right\} \end{array} \middle|\, z \right)$$

for $p^* = \sum_{k=1}^{m-1} p_k = p - p_m$, and where p_k and q_k are interchangeable.

Mathai (1983) gives explicit finite equivalent expressions for the p.d.f. of Λ in (5.272), but he leaves the expression for the c.d.f. to be obtained by term by term integration.

In Fig. 5.64 we may take a look at the plots of some p.d.f.'s of Λ in (5.272), for various values of n for a situation in which $\underline{X}$ has a complex 21-variate Normal distribution, which is subdivided into three subvectors, respectively with six, ten, and five variables. Plots in Fig. 5.64 have the same vertical and horizontal scales as plots in Fig. 5.61, where plots of p.d.f.'s of Λ are shown for cases where $\underline{X}$ has a real 21-variate Normal distribution with similar p_k's, in order to make the plots comparable.

The plots in Fig. 5.64 were obtained with the Mathematica® version of module `PDFIndC`. As for all the other tests, modules `PDFIndC`, `CDFIndC`, `PvalIndC`,

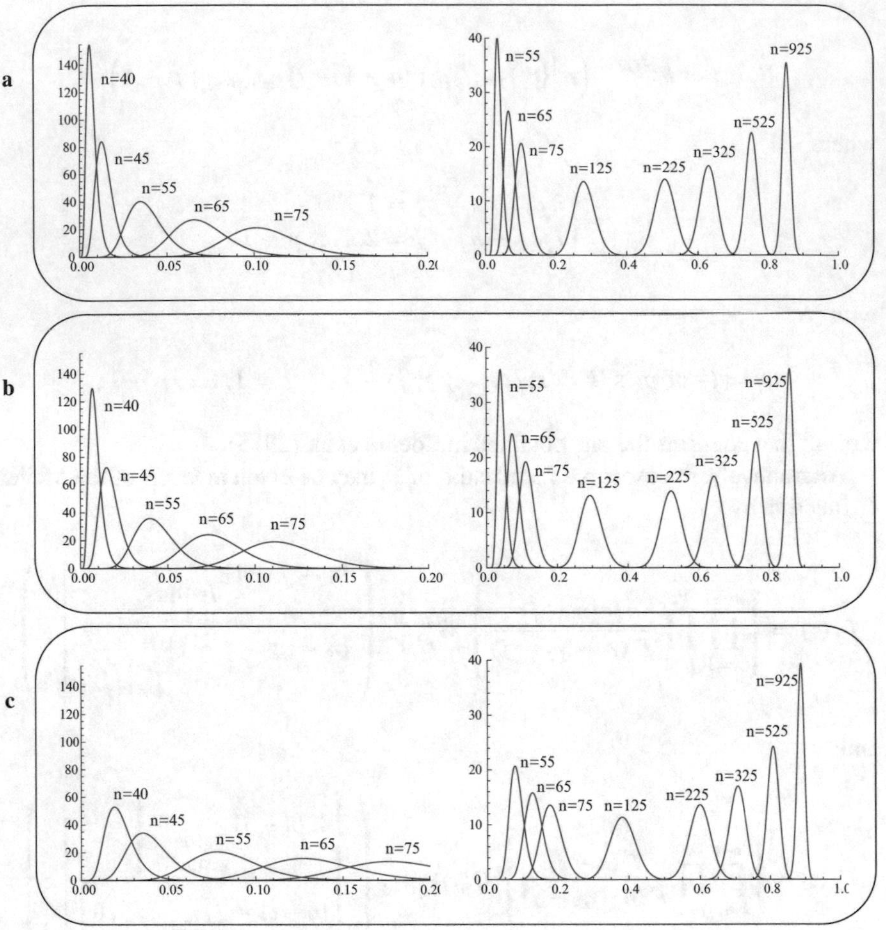

Fig. 5.64 Plots of p.d.f.'s of Λ in (5.234) for the test of independence of 3 groups of variables with a joint complex 21-variate Normal distribution with: (**a**) $p_1 = 6, p_2 = 7, p_3 = 8$, (**b**) $p_1 = 6, p_2 = 10, p_3 = 5$, (**c**) $p_1 = 4, p_2 = 14, p_3 = 3$ (plots on the left-hand side and on the right-hand side of each sub-figure have different scales, but all plots on the left-hand sides have the same scales and all plots on the right-hand sides also have the same scales)

QuantIndC, and PvalDataIndC are made available, with a usage similar to that of their real counterparts in Sect. 5.1.11.

Krishnaiah et al. (1976) provide some small tables of approximate quantiles for $-2\log\Lambda$, only for $\alpha = 0.05$ and only for $m = 3, 4, 5$, with all p_k equal to 1, 2, or 3. However, many of the quantiles lack precision on the third decimal place and for $m = 5$ and $p_k = 3$, for $n = 22$ the value listed even lacks precision on the second decimal place. But, we can use module QuantIndR to obtain exact quantiles for Λ in (5.272) for any combination of p_k values and for any desired α-value, as well as, virtually with any number of decimal places desired. Alternatively, module `PvalIndC` may be used to obtain the p-value for a computed value of the statistic Λ or module `PvalDataIndC` to obtain the computed value of the statistic Λ and the corresponding p-value, for a given sample stored in a data file.

5.1.13 A Test for Outliers (Real r.v.'s) [OutR]

Let us suppose we have a random sample of size n from $\underline{X} \sim N_p(\underline{\mu}, \Sigma)$ and that we are interested in testing whether the $k(< n)$ observations numbered $\eta_1, \ldots, \eta_k$ (with $\eta_1, \ldots, \eta_k \in \{1, \ldots, n\}$) should or should not be considered as outliers. More precisely, we are interested in testing, for $\eta_1, \ldots, \eta_k \in \{1, \ldots, n\}$, with $k < n$, the null hypothesis

$$H_0 : \text{observations } \eta_1, \ldots, \eta_k \text{ are not outliers}\,. \tag{5.273}$$

Wilks (1963) suggests the use of a very simple but intuitively appealing statistic to test H_0, which is

$$\Lambda = \frac{|A^*|}{|A|} \tag{5.274}$$

where A is equal to n times the usual m.l.e. of Σ, based on the whole sample, that is, on the n observations, and A^* is equal to $n - k$ times the m.l.e. of Σ based on the $n - k$ observations that remain after removing observations $\eta_1, \ldots, \eta_k$. More precisely, let

$$\underset{p\times n}{X} = \Big[\ \underset{p\times(n-k)}{X^{\approx}}\ \Big|\ \underset{p\times k}{X^{\sim}}\ \Big]$$

where X is the complete $p{\times}n$ sample data matrix, $X^{\sim}$ is the $p{\times}k$ data matrix formed by the k observations $\eta_1, \ldots, \eta_k$ for which we want to test if they should be considered as outliers, and $X^{\approx}$ is the $p{\times}(n-k)$ data matrix formed by the remaining $n-k$ observations. Without any loss of generality we may take $X^{\approx}$ as being formed by the first $n-k$ columns of X and $X^{\sim}$ as being formed by the last k columns of X. Then we have A and A^* respectively defined as

$$A = \left(X - \overline{\underline{X}} E_{1n}\right)\left(X - \overline{\underline{X}} E_{1n}\right)' = X\Big(\underbrace{I_n - \tfrac{1}{n}E_{nn}}_{Q_1}\Big)X'\,, \tag{5.275}$$

where $\underline{\overline{X}} = \frac{1}{n} X E_{n1}$ is the vector of sample means for the n observations, and

$$A^* = \left(X^{\approx} - \underline{\overline{X}}^{\approx} E_{1,n-k}\right)\left(X^{\approx} - \underline{\overline{X}}^{\approx} E_{1,n-k}\right)' = X^{\approx}\Big(\underbrace{I_{n-k} - \tfrac{1}{n-k}E_{n-k,n-k}}_{Q_2}\Big)X^{\approx\prime} \tag{5.276}$$

where $\underline{\overline{X}}^{\approx} = \frac{1}{n-k} X^{\approx} E_{n-k,1}$ is the vector of sample means for the $n-k$ observations obtained from the original set of n observations after removing observations $\eta_1, \ldots, \eta_k$.

It is clear from (5.275) and (5.276) that $A^* \sim W_p(n-k-1, \Sigma)$ and that, under H_0 in (5.273), $A \sim W_p(n-1, \Sigma)$, since they may be respectively written as $A^* = X Q_2 X'$ and $A = X Q_1 X'$ where X is the matrix of a random sample from a p-variate Normal distribution and Q_2 and Q_1 are idempotent matrices, respectively with

$$rank(Q_2) = tr(Q_2) = n-k-1 \quad \text{and} \quad rank(Q_1) = tr(Q_1) = n-1\,.$$

Although Λ in (5.274) is not the l.r.t. statistic to test H_0 in (5.273), not even a power of that l.r.t. statistic, it is a statistic that has the great virtue of being possible to be included in the domain of the Wilks Lambda statistics, since we are indeed able to write A as

$$A = A^* + B$$

where B is independent of A^* and, under H_0 in (5.273), also has a Wishart distribution.

In order to do that we need first to write A^* as a function of X, the overall sample matrix, instead of just a function of $X^{\approx}$. But this is an easy task since we have

$$X^{\approx} = X I_{n,n-k}$$

where

$$I_{n,n-k} = \left[\frac{I_{n-k}}{0_{k\times(n-k)}}\right]$$

with I_{n-k} an identity matrix of order $n-k$ and $0_{k\times(n-k)}$ a matrix of zeros with dimensions $k\times(n-k)$. As such, we may write

$$A^* = X \underbrace{I_{n,n-k} Q_2 I'_{n,n-k}}_{Q_2^*} X'$$

where Q_2^* is clearly idempotent, given the idempotency of Q_2 and the fact that

$$Q_2^* Q_2^* = I_{n,n-k} Q_2 \underbrace{I'_{n,n-k} I_{n,n-k}}_{=I_{n-k}} Q_2 I'_{n,n-k} = I_{n,n-k} Q_2 I'_{n,n-k} = Q_2^*$$

with

$$\begin{aligned} rank(Q_2^*) = tr(Q_2^*) &= tr(I_{n,n-k} Q_2 I'_{n,n-k}) \\ &= tr(Q_2 I'_{n,n-k} I_{n,n-k}) = tr(Q_2) = n-k-1 \,. \end{aligned}$$

But then we may write

$$B = A - A^* = X Q_1 X' - X Q_2^* X' = X(Q_1 - Q_2^*)X'$$

where it remains to show that $Q_1 - Q_2^*$ is an idempotent matrix for B to have a Wishart distribution and to show that $(Q_1 - Q_2^*)Q_2^* = 0$ to show that A^* and B are independent.

Indeed

$$\begin{aligned} (Q_1 - Q_2^*)(Q_1 - Q_2^*) &= Q_1 Q_1 - Q_1 Q_2^* - Q_2^* Q_1 + Q_2^* Q_2^* \\ &= Q_1 - Q_1 Q_2^* - Q_2^* Q_1 + Q_2^* \end{aligned}$$

where it is not hard to show that $Q_1 Q_2^* = Q_2^* Q_1 = Q_2^*$ since

$$Q_1 Q_2^* = (I_n - \tfrac{1}{n} E_{nn}) Q_2^* = Q_2^* - \tfrac{1}{n} E_{nn} Q_2^*$$

where $E_{nn} Q_2^* = 0_{n\times n}$, given that

$$\begin{aligned} E_{nn} Q_2^* &= E_{nn} I_{n,n-k} Q_2 I'_{n,n-k} \\ &= E_{nn} I_{n,n-k} \left(I_{n-k} - \tfrac{1}{n-k} E_{n-k,n-k} \right) I'_{n,n-k} \\ &= E_{nn} I_{n,n-k} I'_{n,n-k} - \tfrac{1}{n-k} E_{nn} I_{n,n-k} E_{n-k,n-k} I'_{n,n-k} \\ &= E_{nn} \left[\begin{array}{c|c} I_{n-k} & 0_{(n-k)\times k} \\ \hline 0_{k\times(n-k)} & 0_{k\times k} \end{array} \right] - \tfrac{1}{n-k} E_{nn} \left[\begin{array}{c|c} E_{n-k,n-k} & 0_{(n-k)\times k} \\ \hline 0_{k\times(n-k)} & 0_{k\times k} \end{array} \right] \\ &= \left[E_{n,n-k} \,\middle|\, 0_{n\times k} \right] - \tfrac{1}{n-k} \left[(n-k) E_{n,n-k} \,\middle|\, 0_{n\times k} \right] = 0 \,. \end{aligned}$$

As such we have in fact

$$(Q_1 - Q_2^*)(Q_1 - Q_2^*) = Q_1 - Q_1 Q_2^* - Q_2^* Q_1 + Q_2^* = Q_1 - Q_2^* - Q_2^* + Q_2^* = Q_1 - Q_2^*$$

and

$$(Q_1 - Q_2^*)Q_2^* = Q_1Q_2^* - Q_2^*Q_2^* = Q_2^* - Q_2^* = 0\,,$$

so that A^* and B are independent, and, under H_0 in (5.273), $B \sim W_p(k, \Sigma)$ since, given that $Q_1 - Q_2^*$ is idempotent

$$rank(Q_1 - Q_2^*) = tr(Q_1 - Q_2^*) = tr(Q_1) - tr(Q_2^*) = n - 1 - (n - k - 1) = k\,.$$

But then, it is possible to write Λ in (5.274) as

$$\Lambda = \frac{|A^*|}{|A^* + B|} \tag{5.277}$$

with A^* and B exhibiting the distributions stated above, and as such, from Sects. 5.1.1 and 5.1.9, and also Appendix 1, we know that, for $n > p + k$, under H_0 in (5.273),

$$\Lambda \stackrel{d}{\equiv} \prod_{j=1}^{p} Y_j \stackrel{d}{\equiv} \prod_{\ell=1}^{k} Y_\ell^* \tag{5.278}$$

where, for $j = 1, \ldots, p$ and $\ell = 1, \ldots, k$,

$$Y_j \sim Beta\left(\frac{n-k-j}{2}, \frac{k}{2}\right) \quad \text{and} \quad Y_\ell^* \sim Beta\left(\frac{n-p-\ell}{2}, \frac{p}{2}\right) \tag{5.279}$$

are two sets of independent r.v.'s.

In fact the statistic Λ in (5.274) or (5.277) is precisely the $(2/n)$-th power of the l.r.t. statistic to test the null hypothesis in (5.273).

But so, from (5.279) and Sect. 2.1, we know that the exact p.d.f. of Λ in (5.274) or (5.277) is given, in terms of the Meijer G function, by

$$f_\Lambda(z) = \left\{\prod_{j=1}^{p} \frac{\Gamma\left(\frac{n-j}{2}\right)}{\Gamma\left(\frac{n-k-j}{2}\right)}\right\} G_{p,p}^{p,0}\left(\begin{array}{c} \left\{\frac{n-j}{2} - 1\right\}_{j=1:p} \\ \left\{\frac{n-k-j}{2} - 1\right\}_{j=1:p} \end{array} \middle|\, z\right)$$

or

$$f_\Lambda(z) = \left\{\prod_{j=1}^{k} \frac{\Gamma\left(\frac{n-j}{2}\right)}{\Gamma\left(\frac{n-p-j}{2}\right)}\right\} G_{k,k}^{k,0}\left(\begin{array}{c} \left\{\frac{n-j}{2} - 1\right\}_{j=1:k} \\ \left\{\frac{n-p-j}{2} - 1\right\}_{j=1:k} \end{array} \middle|\, z\right)$$

while the c.d.f. is given by

$$F_\Lambda(z) = \left\{ \prod_{j=1}^{p} \frac{\Gamma\left(\frac{n-j}{2}\right)}{\Gamma\left(\frac{n-k-j}{2}\right)} \right\} G_{p+1,p+1}^{p,1}\left(\begin{array}{c} \left\{1, \left\{\frac{n-j}{2}\right\}_{j=1:p}\right\} \\ \left\{\left\{\frac{n-k-j}{2}\right\}_{j=1:p}, 0\right\} \end{array} \middle| \, z \right)$$

or

$$F_\Lambda(z) = \left\{ \prod_{j=1}^{k} \frac{\Gamma\left(\frac{n-j}{2}\right)}{\Gamma\left(\frac{n-p-j}{2}\right)} \right\} G_{k+1,k+1}^{k,1}\left(\begin{array}{c} \left\{1, \left\{\frac{n-j}{2}\right\}_{j=1:k}\right\} \\ \left\{\left\{\frac{n-p-j}{2}\right\}_{j=1:k}, 0\right\} \end{array} \middle| \, z \right),$$

and therefore, for even p the exact distribution of Λ is given by Theorem 3.2 and its exact p.d.f. and c.d.f. given by Corollary 4.2 with

$$m^* = 1\,, \quad k_1 = 2\,, \quad a_1 = \frac{n-k}{2}\,, \quad n_1 = \frac{p}{2}\,, \quad m_1 = k$$

while, for even k the exact distribution and the exact p.d.f. and c.d.f. of Λ are given by the same results, now with

$$m^* = 1\,, \quad k_1 = 2\,, \quad a_1 = \frac{n-p}{2}\,, \quad n_1 = \frac{k}{2}\,, \quad m_1 = p\,,$$

with the exact p.d.f. and c.d.f. of Λ in (5.274), being given, in any of these cases, in terms of the EGIG distribution representation, respectively by

$$f_\Lambda(z) = f^{EGIG}\left(z \,\middle|\, \{r_j\}_{j=1:p+k-2}; \left\{\frac{n-2-j}{2}\right\}_{j=1:p+k-2}; p+k-2 \right)$$

and

$$F_\Lambda(z) = F^{EGIG}\left(z \,\middle|\, \{r_j\}_{j=1:p+k-2}; \left\{\frac{n-2-j}{2}\right\}_{j=1:p+k-2}; p+k-2 \right),$$

where

$$r_j = \begin{cases} h_j\,, & j = 1,2 \\ r_{j-2} + h_j\,, & j = 3, \ldots, p+k-2 \end{cases}$$

with

$$h_j = (\text{\# of elements in } \{p,k\} \geq j) - 1\,, \quad j = 1, \ldots, p+k-2\,.$$

In case the k observations $\eta_1, \ldots, \eta_k$ are randomly picked from among the n observations, then an α level test will be the one that rejects H_0 in (5.273) if the computed value of Λ in (5.274) or (5.277) is smaller than the α-quantile of Λ. Wilks (1963) states that in case the k observations chosen are those that, taken one by one, produce the k largest computed values for the statistic Λ, among all possible n computed values for each one of the n observations taken one by one, then we should use an α-quantile with a value of α somewhere between the intended α value for the test and $\alpha/\binom{n}{k}$.

Modules `PDFOutR`, `CDFOutR`, `PvalOutR`, `QuantOutR`, and `PvalData OutR` are available for the implementation of the present test, and have a usage parallel to that of the corresponding modules in Sect. 5.1.9, with the only difference that the first three mandatory arguments of the first four of these modules are now replaced by the sample size n, the number of variables p, and the value of k, the number of observations for which we want to test if they should be considered outliers. Module `PvalDataOutR` has as mandatory arguments the name of the data file and a list with the ordering-numbers of the observations for which we want to test if they should be considered as outliers. All these modules have exactly the same optional arguments as their counterparts in Sect. 5.1.1.

To illustrate the application of the present test we use the data set on costs of fuel, repair, and capital per mile associated with milk transportation from farm to dairy plants, concerning 36 gasoline trucks in Table 6.10 of Johnson and Wichern (2014). These were also the data used by Caroni and Prescott (1992) who developed a sequential strategy using Wilks' statistic in (5.274) and by Bacon-Stone and Fung (1987) who used a graphical method to detect outliers. First we randomly picked observations 4 and 18 to test the set of these two observations for outliers, where by "randomly" we mean that we did not look at any scatter plot of the data or took into account any other considerations related to any aspects of the data set when choosing these two observations. The large p-value obtained as output for the first command in Fig. 5.65 leads us to not reject the null hypothesis that these two observations are not outliers.

We may take a look at a three-dimensional scatter plot of the data in Fig. 5.66. By looking at this plot one would not be shocked if observations 9 and 21 would be considered outliers. Actually, the second command in Fig. 5.65 which tests this hypothesis gives a very low p-value. Anyway we should be careful since we picked up these two observations by looking at the scatter plot and choosing the pair of observations which seemed to lie furthest away from the center of the cloud of points. As such, we actually did implicitly a number of tests and by choosing the two most isolated observations we should follow the approach delineated in Wilks (1963) by using a lower-bound for our α-value equal to $\alpha/\binom{n}{k} = \alpha/\binom{36}{2} = \alpha/630$. This means that if, for example, we intended to use an α-value of 0.01, we should rather use a value of α somewhere between 0.01 and $0.01/630 = 0.000016$, still much larger than the p-value obtained, leading us to reject the null hypothesis that the set of these two points are not outliers. Equivalently, since we got a p-value of around 3.024914×10^{-7}, we should only not reject the null hypothesis that the observations 9 and 21 are not outliers if our α-value would be smaller than

```
PvalDataOutR[path <> "Outliers.dat", {4, 18}]
Computed value of Λ: 0.8224746463
p-value: 0.3957352957
PvalDataOutR[path <> "Outliers.dat", {9, 21}]
Computed value of Λ: 0.2778191737
p-value: 3.024914033×10^-7
PvalDataOutR[path <> "Outliers.dat", {23, 36}]
Computed value of Λ: 0.7420093158
p-value: 0.1453228752
PvalDataOutR[path <> "Outliers.dat", {23, 36}, , 1]
 Original observations {1, 2, 3, 4, 5, 6, 7, 8,
   10, 11, 12, 13, 14, 15, 16, 17, 18, 19, 20, 22, 23,
   24, 25, 26, 27, 28, 29, 30, 31, 32, 33, 34, 35, 36}
Computed value of Λ: 0.5517151076
p-value: 0.006697122054
```

Fig. 5.65 Commands used and output obtained for outliers tests carried out with the milk distribution costs

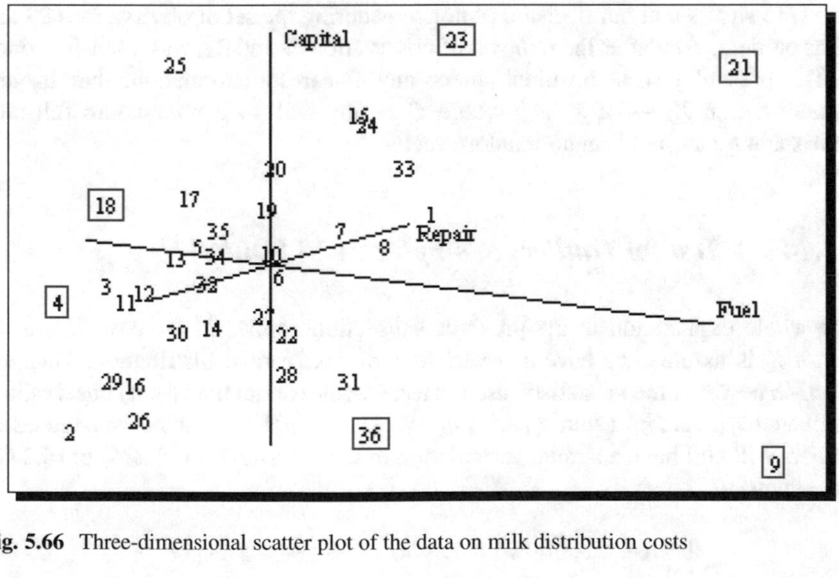

Fig. 5.66 Three-dimensional scatter plot of the data on milk distribution costs

$630 \times 3.024914 \times 10^{-7} = 0.000191$, which would be an unusually low value. As such, we should either way reject the null hypothesis and consider observations 9 and 21 as outliers. Another set of observations that seems to be somewhat sticking out of the main cloud of points are observations 23 and 36, but according to the p-value obtained with the third command in Fig. 5.65 we would not consider these two points as outliers. However, one may question what would happen if the points 9 and 21 were removed from the analysis, bringing then the two points 23 and 36 closer to the edge of the convex hull of the cloud of points. With the last command in Fig. 5.65 we test if the set of observations 23 and 36 should or should not be considered outliers when the observations 9 and 21 are removed. This is also a good chance to use the second optional argument of `PvalDataOutR` in order to be able to select the observations to be included in the analysis. The p-value obtained, although quite low, has to be once again analyzed with care. Since the two observations 23 and 36 were chosen among the 34 remaining observations by picking them from the scatter plot it is actually as if we made by eye $\binom{34}{2} = 561$ tests among the 34 points in order to choose the two observations that lie further away from the center of the cloud. But, even if we would use the absurdly high initial value of 0.5 for α, we should be looking for an α-value of $0.5/561 = 0.000891$, still much lower than the p-value obtained, indicating that we should not reject the null hypothesis that the set of two observations 23 and 36 are not outliers, even after removing observations 9 and 21. Although we think it is a much better strategy to use p-values rather than some fixed quantile to carry out the tests, we may check that indeed while for example for $\alpha = 0.05$ the α-quantile for Λ (for $n = 34$, $p = 3$, and $k = 2$) is 0.656968, the $\alpha/\binom{34}{2} = 0.000089$ quantile is 0.391033, which compared with the computed value for Λ of 0.551715 shows that our decision of not considering the set of observations 23 and 36 as outliers, even after the removal of observations 9 and 21, was a sensible one.

The present test is invariant under any linear transformation, that is, any transformation $\underline{X} \to C\underline{X} + \underline{b}$ where C is any real $p \times p$ nonrandom full-rank matrix and $\underline{b}$ any $p \times 1$ real nonrandom vector.

5.1.14 A Test for Outliers (Complex r.v.'s) [OutC]

The whole explanation in the previous subsection applies in the case where the vector $\underline{X}$ is assumed to have a p-variate complex Normal distribution. Then the $(1/n)$-th power of the l.r. statistic used to test whether or not the $k(< n)$ observations numbered $\eta_1, \ldots, \eta_k$ $\big($with $\eta_1, \ldots, \eta_k \in \{1, \ldots, n\}\big)$ should be considered as outliers will still have a similar formulation to the statistic Λ in (5.274) or (5.277), where now

$$A = (X - \overline{\underline{X}}E_{1n})(X - \overline{\underline{X}}E_{1n})^{\#} = X\left(I_n - \tfrac{1}{n}E_{nn}\right)X^{\#},$$

$$A^* = \left(X^{\approx} - \overline{\underline{X}}^{\approx}E_{1,n-k}\right)\left(X^{\approx} - \overline{\underline{X}}^{\approx}E_{1,n-k}\right)^{\#} = X^{\approx}\left(I_{n-k} - \tfrac{1}{n-k}E_{n-k,n-k}\right)X^{\approx\#}$$

with $(\cdot)^{\#}$ denoting the transpose of the complex conjugate of $(\cdot)$. Then, for $n > p + k$, the statistic Λ will have in this case a somewhat similar distribution to the one in (5.278), now with

$$Y_j \sim Beta\,(n-k-j, k) \quad \text{and} \quad Y_\ell^* \sim Beta\,(n-p-\ell, p)$$

forming two sets of independent r.v.'s, for $j = 1, \ldots, p$ and $\ell = 1, \ldots, k$, so that for any p and k the exact distribution of Λ is given by Theorem 3.2 and its exact p.d.f. and c.d.f. given by Corollary 4.2 with

$$m^* = 1\,, \quad k_1 = 1\,, \quad a_1 = n-k\,, \quad n_1 = p\,, \quad m_1 = k$$

or

$$m^* = 1\,, \quad k_1 = 1\,, \quad a_1 = n-p\,, \quad n_1 = k\,, \quad m_1 = p\,.$$

The exact p.d.f. and c.d.f. of the statistic Λ are, in this case, given, for any p and k, in terms of the EGIG distribution representation, respectively by

$$f_\Lambda(z) = f^{EGIG}\left(z \,\middle|\, \{r_j\}_{j=1:p+k-1}; \{n-1-j\}_{j=1:p+k-1}\,; p+k-1\right)$$

and

$$F_\Lambda(z) = F^{EGIG}\left(z \,\middle|\, \{r_j\}_{j=1:p+k-1}; \{n-1-j\}_{j=1:p+k-1}\,; p+k-1\right),$$

where

$$r_j = \begin{cases} h_j\,, & j=1 \\ r_{j-1}+h_j\,, & j=2,\ldots,p+k-1 \end{cases}$$

with

$$h_j = (\#\text{ of elements in } \{p,k\} \geq j) - 1\,, \quad j = 1,\ldots,p+k-1\,,$$

or, alternatively, in terms of Meijer G functions by

$$f_\Lambda(z) = \left\{\prod_{j=1}^{p} \frac{\Gamma\,(n-j)}{\Gamma\,(n-k-j)}\right\} G^{p,0}_{p,p}\left(\begin{matrix}\{n-j-1\}_{j=1:p} \\ \{n-k-j-1\}_{j=1:p}\end{matrix} \,\middle|\, z\right)$$

and

$$F_\Lambda(z) = \left\{\prod_{j=1}^{p} \frac{\Gamma(n-j)}{\Gamma(n-k-j)}\right\} G^{p,1}_{p+1,p+1}\left(\begin{array}{c} \left\{1, \{n-j\}_{j=1:p}\right\} \\ \left\{\{n-k-j\}_{j=1:p}, 0\right\} \end{array} \middle| \, z \right)$$

or by similar expressions with p and k switched.

Modules `PDFOutC`, `CDFOutC`, `PvalOutC`, `QuantOutC`, and `PvalDataOutC` are made available for the implementation of the present test, and have a usage parallel to that of the corresponding modules in the previous subsection.

5.1.15 Testing the "Symmetrical Equivalence" of Two Sets of Real Random Variables [SymEqR]

Let us suppose that $\underline{X} = \left[\underline{X}_1', \underline{X}_2'\right]' \sim N_{2p}(\underline{\mu}, \Sigma)$, where each subvector $\underline{X}_1$ and $\underline{X}_2$ is p-dimensional. Let us then suppose that we want to test the hypothesis

$$H_0 : \Sigma = Var(\underline{X}) = \begin{bmatrix} \Sigma_1 & \Sigma_2 \\ \Sigma_2 & \Sigma_1 \end{bmatrix}, \tag{5.280}$$

that is, the hypothesis that

$$Var(\underline{X}_1) = Var(\underline{X}_2) \quad \text{and} \quad Cov(\underline{X}_1, \underline{X}_2) = Cov(\underline{X}_2, \underline{X}_1) \tag{5.281}$$

or, equivalently stated, that, for any $j, j' \in \{1, \dots, p\}$,

$$Cov(X_{1j}, X_{1j'}) = Cov(X_{2j}, X_{2j'}) \quad \text{and} \quad Cov(X_{1j}, X_{2j'}) = Cov(X_{2j}, X_{1j'}) \tag{5.282}$$

where X_{kj} represents the j-th variable in the k-th set ($k = 1, 2$; $j = 1, \dots, p$).

If (5.280), (5.281), or (5.282) holds, then we will say that $\underline{X}_1$ and $\underline{X}_2$ are "symmetrically equivalent."

The vector $\underline{X}_1$ may be for example the set of the values of some mineral bone content for p bones of the left side of the body, or the set of p mineral bone contents for a given bone on the left side of the body, and $\underline{X}_2$ the set of similar values for the right side of the body. Another example would be to have $\underline{X}_1$ as the set of the breaking point strengths of p components of the left side of a structure which is symmetrical about some plane, and $\underline{X}_2$ as the set of similar strengths for the components on the right side of the structure. One may then be interested in testing a null hypothesis such as the one in (5.280), (5.281), or (5.282).

In order for Σ in (5.280) to be positive-definite, we need Σ_1 and Σ_2 to be such that $\Sigma_1 - \Sigma_2$ and $\Sigma_1 + \Sigma_2$ are also positive-definite.

The present test is not to be confused with the usual test of equality of covariance matrices based on independent samples (see Muirhead (2005, Sec. 8.2), Anderson (2003, Chap. 10), Kshirsagar (1972, Chap. 10), Coelho et al. (2010), Marques et al. (2011), Coelho and Marques (2012)). In the present case, not only are the samples obtained for $\underline{X}_1$ and $\underline{X}_2$ not independent but instead they are paired, as well as, besides testing the equality of the covariance matrices of $\underline{X}_1$ and $\underline{X}_2$, we are also simultaneously testing the symmetry of the covariance matrix of $\underline{X}_1$ with $\underline{X}_2$.

We will show that it is possible to reduce the test of the hypothesis in (5.280) to the test of independence of two sets of variables in Sect. 5.1.9.

Let us take the matrix

$$\Gamma = \begin{bmatrix} \frac{1}{\sqrt{2}} & \frac{1}{\sqrt{2}} \\ \frac{1}{\sqrt{2}} & -\frac{1}{\sqrt{2}} \end{bmatrix}. \tag{5.283}$$

The matrix Γ is a Helmert matrix of order 2, that is, a 2×2 orthogonal matrix whose first row is proportional to a constant vector of dimension 2. Note also that Γ is a symmetric matrix.

Then let us consider the matrix $\Gamma \otimes I_p$. Not only is this matrix a symmetric matrix, since

$$\left(\Gamma \otimes I_p\right)' = \Gamma' \otimes I_p' = \Gamma \otimes I_p = \left[\begin{array}{c|c} \frac{1}{\sqrt{2}} I_p & \frac{1}{\sqrt{2}} I_p \\ \hline \frac{1}{\sqrt{2}} I_p & -\frac{1}{\sqrt{2}} I_p \end{array}\right]$$

but also, for Σ with the structure in (5.280), we have

$$\begin{aligned}\left(\Gamma \otimes I_p\right) \Sigma \left(\Gamma \otimes I_p\right)' &= \left[\begin{array}{c|c} \frac{1}{\sqrt{2}} I_p & \frac{1}{\sqrt{2}} I_p \\ \hline \frac{1}{\sqrt{2}} I_p & -\frac{1}{\sqrt{2}} I_p \end{array}\right] \begin{bmatrix} \Sigma_1 & \Sigma_2 \\ \Sigma_2 & \Sigma_1 \end{bmatrix} \left[\begin{array}{c|c} \frac{1}{\sqrt{2}} I_p & \frac{1}{\sqrt{2}} I_p \\ \hline \frac{1}{\sqrt{2}} I_p & -\frac{1}{\sqrt{2}} I_p \end{array}\right] \\ &= \left[\begin{array}{c|c} \frac{1}{\sqrt{2}} (\Sigma_1 + \Sigma_2) & \frac{1}{\sqrt{2}} (\Sigma_2 + \Sigma_1) \\ \hline \frac{1}{\sqrt{2}} (\Sigma_1 - \Sigma_2) & \frac{1}{\sqrt{2}} (\Sigma_2 - \Sigma_1) \end{array}\right] \left[\begin{array}{c|c} \frac{1}{\sqrt{2}} I_p & \frac{1}{\sqrt{2}} I_p \\ \hline \frac{1}{\sqrt{2}} I_p & -\frac{1}{\sqrt{2}} I_p \end{array}\right] \\ &= \begin{bmatrix} \Sigma_1 + \Sigma_2 & 0 \\ 0 & \Sigma_1 - \Sigma_2 \end{bmatrix}. \end{aligned} \tag{5.284}$$

That the matrix $\left(\Gamma \otimes I_p\right) \Sigma \left(\Gamma \otimes I_p\right)'$ is a block-diagonal matrix if and only if the matrix Σ has the structure in (5.280) is shown in Appendix 15.

But then, since the elements in $\Gamma \otimes I_p$ and the structure of this matrix are not functions of the elements in Σ, but only of its dimension, to test the null hypothesis in (5.280) will be equivalent to testing the null hypothesis

$$H_0 : (\Gamma \otimes I_p)\, \Sigma\, (\Gamma \otimes I_p)' = \begin{bmatrix} \Sigma_1^* & 0 \\ 0 & \Sigma_2^* \end{bmatrix} \tag{5.285}$$

where $\Sigma_1^* = \Sigma_1 + \Sigma_2$ and $\Sigma_2^* = \Sigma_1 - \Sigma_2$.

We may note that the matrix in (5.284) is indeed the covariance matrix of

$$(\Gamma \otimes I_p)\, \underline{X} = \left[\begin{array}{c|c} \frac{1}{\sqrt{2}} I_p & \frac{1}{\sqrt{2}} I_p \\ \hline \frac{1}{\sqrt{2}} I_p & -\frac{1}{\sqrt{2}} I_p \end{array}\right] \begin{bmatrix} \underline{X}_1 \\ \underline{X}_2 \end{bmatrix} = \begin{bmatrix} \frac{1}{\sqrt{2}}(\underline{X}_1 + \underline{X}_2) \\ \frac{1}{\sqrt{2}}(\underline{X}_1 - \underline{X}_2) \end{bmatrix}$$

with

$$\begin{aligned} Var\left(\tfrac{1}{\sqrt{2}}(\underline{X}_1 + \underline{X}_2)\right) &= \tfrac{1}{2}\big(Var(\underline{X}_1) + Var(\underline{X}_2) + Cov(\underline{X}_1, \underline{X}_2) + Cov(\underline{X}_2, \underline{X}_1)\big) \\ &= \Sigma_1 + \Sigma_2 = \Sigma_1^* \\ Var\left(\tfrac{1}{\sqrt{2}}(\underline{X}_1 - \underline{X}_2)\right) &= \tfrac{1}{2}\big(Var(\underline{X}_1) + Var(\underline{X}_2) - Cov(\underline{X}_1, \underline{X}_2) - Cov(\underline{X}_2, \underline{X}_1)\big) \\ &= \Sigma_1 - \Sigma_2 = \Sigma_2^*\,, \end{aligned}$$

$$\begin{aligned} Cov\left(\tfrac{1}{\sqrt{2}}(\underline{X}_1 + \underline{X}_2), \tfrac{1}{\sqrt{2}}(\underline{X}_1 - \underline{X}_2)\right) &= \tfrac{1}{2}\big(Var(\underline{X}_1) - Cov(\underline{X}_1, \underline{X}_2) \\ &\qquad + Cov(\underline{X}_2, \underline{X}_1) - Var(\underline{X}_2)\big) = 0 \\ &= Cov\left(\tfrac{1}{\sqrt{2}}(\underline{X}_1 - \underline{X}_2), \tfrac{1}{\sqrt{2}}(\underline{X}_1 + \underline{X}_2)\right) \end{aligned}$$

and

$$Var\big((\Gamma \otimes I_p)\, \underline{X}\big) = (\Gamma \otimes I_p)\, \Sigma\, (\Gamma \otimes I_p)' = \Sigma^*\,.$$

Since the m.l.e. of Σ is

$$A = \tfrac{1}{n} X (I_n - \tfrac{1}{n} E_{nn}) X' = \begin{bmatrix} A_{11} & A_{12} \\ A_{21} & A_{22} \end{bmatrix},$$

where X is the $2p{\times}n$ sample data matrix of a sample of size n from $\underline{X}$, the m.l.e. of $\Sigma^* = (\Gamma \otimes I_p)\, \Sigma\, (\Gamma \otimes I_p)'$ is

$$\begin{aligned} A^* = (\Gamma \otimes I_p)\, A\, (\Gamma \otimes I_p)' &= \tfrac{1}{2} \left[\begin{array}{c|c} A_{11} + A_{22} + A_{12} + A_{21} & A_{11} - A_{22} + A_{21} - A_{12} \\ \hline A_{11} - A_{22} - A_{21} + A_{12} & A_{11} + A_{22} - A_{12} - A_{21} \end{array}\right] \\ &= \begin{bmatrix} A_{11}^* & A_{12}^* \\ A_{21}^* & A_{22}^* \end{bmatrix} \sim W_{2p}\left(n-1, \tfrac{1}{n}\Sigma^*\right), \end{aligned} \tag{5.286}$$

where Σ^*, under H_0 in (5.285), is the block-diagonal matrix in (5.285).

But then, for a sample of size n, the $(2/n)$-th power of the l.r.t. statistic to test H_0 in (5.285) is, from Sect. 5.1.9,

$$\Lambda = \frac{|A^*|}{|A_{11}^*|\,|A_{22}^*|} = \frac{|A_{11}^* - A_{12}^*(A_{22}^*)^{-1}A_{21}^*|}{|A_{11}^*|} = \frac{|A_{22}^* - A_{21}^*(A_{11}^*)^{-1}A_{12}^*|}{|A_{22}^*|} \tag{5.287}$$

with

$$\Lambda \stackrel{d}{\equiv} \prod_{j=1}^{p} Y_j\,,$$

where, for $n > 2p$, under H_0 in (5.285),

$$Y_j \sim Beta\left(\frac{n-p-j}{2}, \frac{p}{2}\right), \qquad (j = 1, \ldots, p)$$

form a set of p independent r.v.'s.

The exact p.d.f. and c.d.f. of Λ in (5.287) are thus given in terms of the Meijer G function by

$$f_\Lambda(z) = \left\{\prod_{j=1}^{p} \frac{\Gamma\left(\frac{n-j}{2}\right)}{\Gamma\left(\frac{n-p-j}{2}\right)}\right\} G_{p,p}^{p,0}\left(\begin{array}{c} \left\{\frac{n-j}{2}-1\right\}_{j=1:p} \\ \left\{\frac{n-p-j}{2}-1\right\}_{j=1:p} \end{array} \middle| \, z\right)$$

and

$$F_\Lambda(z) = \left\{\prod_{j=1}^{p} \frac{\Gamma\left(\frac{n-j}{2}\right)}{\Gamma\left(\frac{n-p-j}{2}\right)}\right\} G_{p+1,p+1}^{p,1}\left(\begin{array}{c} \left\{1, \left\{\frac{n-j}{2}\right\}_{j=1:p}\right\} \\ \left\{\left\{\frac{n-p-j}{2}\right\}_{j=1:p}, 0\right\} \end{array} \middle| \, z\right),$$

and for even p, through the results in Theorem 3.2, by Corollary 4.2 with

$$m^* = 1\,, \quad k_1 = 2\,, \quad a_1 = \frac{n-p}{2}\,, \quad n_1 = \frac{p}{2}\,, \quad m_1 = p\,.$$

The exact p.d.f. and c.d.f. of Λ in (5.287) are thus given, for even p, in terms of the EGIG distribution representation respectively by

$$f_\Lambda(z) = f^{EGIG}\left(z \,\middle|\, \{r_j\}_{j=1:2p-2}; \left\{\frac{n-2-j}{2}\right\}_{j=1:2p-2}; 2p-2\right)$$

and

$$F_\Lambda(z) = F^{EGIG}\left(z \,\middle|\, \{r_j\}_{j=1:2p-2};\left\{\frac{n-2-j}{2}\right\}_{j=1:2p-2};2p-2\right),$$

where

$$r_j = \begin{cases} \left\lfloor \frac{j+1}{2} \right\rfloor, & j=1,\dots,p \\ p-\left\lfloor \frac{j+1}{2} \right\rfloor, & j=p+1,\dots,2p-2, \end{cases}$$

or, equivalently,

$$r_j = \frac{p}{2} - \left|\frac{p}{2} - \left\lfloor \frac{j+1}{2} \right\rfloor\right|, \qquad j=1,\dots,2p-2.$$

The present test is indeed a particular case of the l.r. test for the block compound symmetric covariance structure in Coelho and Roy (2017) for $u = 2$, and consequently the near-exact distribution developed in this reference may be used for the case of odd p.

The statistic Λ in (5.287) is invariant under any linear transformations within each one of the two subsets $\underline{X}_1$ and $\underline{X}_2$ that use the same full-rank matrix, that is, it is invariant under the class of transformations $\underline{X} \to C^*\underline{X} + \underline{b}^*$ for $\underline{b}^*$ any real nonrandom $2p\times 1$ vector and $C^* = bdiag(C, C)$, where C is any $p\times p$ nonrandom full-rank real matrix.

For the test of symmetrical equivalence modules `PDFSymEqR`, `CDFSymEqR`, `PvalSymEqR`, `QuantSymEqR`, and `PvalDataSymEqR` are made available. The first four of these modules have as first two mandatory arguments the sample size n, and the value of p. The third mandatory argument for the first three of these modules is the computed value of the statistic, while for the fourth module this argument is the value of α for which the quantile is to be computed. Module `PvalDataSymEqR` has as its only mandatory arguments the names of the files with the sample data matrices for $\underline{X}_1$ and $\underline{X}_2$, or optionally the name of a single data file with the data matrix of dimensions $n\times 2p$ of the sample of $\underline{X}$, that is, the combined sample of $\underline{X}_1$ and $\underline{X}_2$. The first four of these modules have exactly the same optional arguments as the corresponding modules in Sect. 5.1.1, and module `PvalDataSymEqR` has the same optional arguments as the corresponding module in Sect. 5.1.9.

For an example of use of module `PvalDataSymEqR` and of the implementation of the present test we use the bone mineral content data available in Tables 1.18 and 6.16 in Johnson and Wichern (2014). These data refer to measurements of mineral content taken on the dominant and nondominant radius, humerus, and ulna of 24 older women taken at the beginning of the study (Table 1.8) and 1 year after the kickoff of the study (Table 6.16). We consider only the first 24 observations in

Table 1.18, since these are the ones that correspond to the observations in Table 6.16. We thus have a sample of size $n = 24$ from a set of $2p = 12$ variables, split into two groups of six variables each: the first set corresponding to the measurements taken at the start of the study and the second group to the measurements taken 1 year after. The six variables in each group are the mineral content on the dominant and non-dominant radius, humerus, and ulna. Given the nature of the data involved, we have good reasons to believe that the overall covariance matrix, that is, the covariance matrix for the whole set of 12 variables may have a 2-block compound symmetric or symmetrical equivalent structure as the one in (5.280) and we are going to test this hypothesis using the data set in the file `Tabl_8_6_16_JW.dat`. The contents of this data file are an exact copy of the data in Tables 1.8 and 6.16 of Johnson and Wichern (2014), placed side by side, except for the first columns in these tables, which list the subject numbers. The sample covariance matrix for these data with values rounded to three decimal places (to make it fit on the page) is the matrix

$$A = \left[\begin{array}{cccccc|cccccc}
0.013 & 0.010 & 0.023 & 0.021 & 0.009 & 0.008 & 0.013 & 0.011 & 0.026 & 0.020 & 0.009 & 0.009 \\
0.010 & 0.011 & 0.018 & 0.021 & 0.008 & 0.008 & 0.011 & 0.011 & 0.022 & 0.022 & 0.008 & 0.009 \\
0.023 & 0.018 & 0.082 & 0.069 & 0.016 & 0.012 & 0.026 & 0.019 & 0.088 & 0.072 & 0.019 & 0.015 \\
0.021 & 0.021 & 0.069 & 0.072 & 0.018 & 0.016 & 0.023 & 0.023 & 0.077 & 0.076 & 0.021 & 0.020 \\
0.009 & 0.008 & 0.016 & 0.018 & 0.011 & 0.007 & 0.009 & 0.009 & 0.021 & 0.018 & 0.011 & 0.007 \\
0.008 & 0.008 & 0.012 & 0.016 & 0.007 & 0.010 & 0.009 & 0.009 & 0.016 & 0.017 & 0.007 & 0.010 \\
\hline
0.013 & 0.011 & 0.026 & 0.023 & 0.009 & 0.009 & 0.016 & 0.011 & 0.029 & 0.024 & 0.009 & 0.009 \\
0.011 & 0.011 & 0.019 & 0.023 & 0.009 & 0.009 & 0.011 & 0.012 & 0.023 & 0.024 & 0.009 & 0.010 \\
0.026 & 0.022 & 0.088 & 0.077 & 0.021 & 0.016 & 0.029 & 0.023 & 0.105 & 0.084 & 0.024 & 0.018 \\
0.020 & 0.022 & 0.072 & 0.076 & 0.018 & 0.017 & 0.024 & 0.024 & 0.084 & 0.084 & 0.021 & 0.020 \\
0.009 & 0.008 & 0.019 & 0.021 & 0.011 & 0.007 & 0.009 & 0.009 & 0.024 & 0.021 & 0.012 & 0.008 \\
0.009 & 0.009 & 0.015 & 0.020 & 0.007 & 0.010 & 0.009 & 0.010 & 0.018 & 0.020 & 0.008 & 0.013
\end{array}\right],$$

which indeed displays a structure that may lead us to believe that it may make sense to not reject the hypothesis that the population covariance matrix has a symmetrical equivalent structure.

The command used and the output obtained when using the Mathematica® version of module `PvalDataSymEqR`, with the option for the implementation of the multivariate normality tests are in Fig. 5.67. We may see that all multivariate normality tests, except the Shapiro-Wilk test (and the Mardia kurtosis test for the combined sample), give quite large p-values both for the tests based on the combined sample and for the tests based on the samples from each subvector. The test of symmetrical equivalence then gives a large p-value, which leads us to not reject the null hypothesis of symmetrical equivalence of the subvectors $\underline{X}_1$ and $\underline{X}_2$, that is, of the mineral content measurements taken at the beginning of the study and 1 year later.

We may note that, given the definition of the matrix A^* in (5.286) and given that $\Gamma \otimes I_p$ is an orthogonal matrix, we may replace $|A^*|$ by $|A|$ in (5.287).

```
PvalDataSymEqR[path <> "Tabl_8_6_16_JW.dat", , , , , , , , norm → 1]
```

Combined sample of X1 and X2

	Statistic	P-Value
Anderson-Darling	10.6039	1.
Cramér-von Mises	10.6359	1.
Jarque-Bera ALM	5.2713	0.235428
Kolmogorov-Smirnov	10.4953	1.
Kuiper	9.14954	0.99944
Mardia Combined	365.21	0.487059
Mardia Kurtosis	−2.56723	0.0102515
Mardia Skewness	312.888	0.569724
Pearson χ^2	7.99746	0.977584
Shapiro-Wilk	0.569061	2.8403×10^{-7}
Watson U^2	9.55246	0.999906

Sample of X1

	Statistic	P-Value
Anderson-Darling	5.31769	0.99986
Cramér-von Mises	5.34341	0.999889
Jarque-Bera ALM	2.31221	0.169716
Kolmogorov-Smirnov	5.31519	0.999857
Kuiper	4.52058	0.98554
Mardia Combined	59.1951	0.395387
Mardia Kurtosis	−1.01535	0.30994
Mardia Skewness	49.9289	0.395594
Pearson χ^2	3.7371	0.847312
Shapiro-Wilk	0.815233	0.000528256
Watson U^2	4.76723	0.995127

Sample of X2

	Statistic	P-Value
Anderson-Darling	5.28622	0.999816
Cramér-von Mises	5.29248	0.999826
Jarque-Bera ALM	2.9591	0.477514
Kolmogorov-Smirnov	5.18015	0.999578
Kuiper	4.62896	0.990797
Mardia Combined	68.7251	0.13736
Mardia Kurtosis	−0.602944	0.546546
Mardia Skewness	58.6824	0.124315
Pearson χ^2	4.26036	0.962868
Shapiro-Wilk	0.774806	0.000117315
Watson U^2	4.78523	0.995538

```
Computed value of Λ: 0.1041955654

p-value: 0.4434096135
```

Fig. 5.67 Commands used and output obtained with the Mathematica® module `PvalDataSymEqR` to implement the test of symmetrical equivalence, using the bone mineral data in Tables 1.8 and 6.16 of Johnson and Wichern (2014)

5.1.16 Testing the "Complete Symmetrical Equivalence" of Two Sets of Real Random Variables [CompSymEqR]

Let us suppose that, as in the previous subsection, $\underline{X} = [\underline{X}_1', \underline{X}_2']' \sim N_{2p}(\underline{\mu}, \Sigma)$, where both subvectors $\underline{X}_1$ and $\underline{X}_2$ are p-dimensional and

$$\underline{\mu} = \begin{bmatrix} \underline{\mu}_1 \\ \underline{\mu}_2 \end{bmatrix}, \quad \text{with} \quad \underline{\mu}_1 = E(\underline{X}_1) \quad \text{and} \quad \underline{\mu}_2 = E(\underline{X}_2)\,,$$

and let us suppose that now we want to test the null hypothesis

$$H_0 : \Sigma = \begin{bmatrix} \Sigma_1 & \Sigma_2 \\ \Sigma_2 & \Sigma_1 \end{bmatrix} \text{ and } \underline{\mu}_1 = \underline{\mu}_2\,, \tag{5.288}$$

where we continue to have Σ_1 and Σ_2 non-specified but with Σ_1, $\Sigma_1 + \Sigma_2$ and $\Sigma_1 - \Sigma_2$ being positive-definite matrices.

We will call the hypothesis in (5.288) the "complete symmetrical equivalence" hypothesis.

Then, all we have to do is to use a similar technique to the one used in the previous subsection, by considering a matrix Γ identical to the one in (5.283) and then defining the transformed random vector

$$\underline{X}^* = (\Gamma \otimes I_p)\,\underline{X}$$

where, as in the previous subsection, we have

$$\underline{X}^* \sim N_{2p}\left(\underline{\mu}^*, \Sigma^*\right)$$

with

$$\underline{\mu}^* = \begin{bmatrix} \underline{\mu}_1^* \\ \underline{\mu}_2^* \end{bmatrix} = \begin{bmatrix} \frac{1}{\sqrt{2}}\left(\underline{\mu}_1 + \underline{\mu}_2\right) \\ \frac{1}{\sqrt{2}}\left(\underline{\mu}_2 - \underline{\mu}_1\right) \end{bmatrix} \quad \text{and} \quad \Sigma^* = \begin{bmatrix} \Sigma_1^* & 0 \\ 0 & \Sigma_2^* \end{bmatrix} = \begin{bmatrix} \Sigma_1 + \Sigma_2 & 0 \\ 0 & \Sigma_1 - \Sigma_2 \end{bmatrix},$$

so that to test H_0 in (5.288) is indeed equivalent to testing

$$H_0 : \Sigma^* = bdiag(\Sigma_1^*, \Sigma_2^*) \quad \text{and} \quad \underline{\mu}_2^* = \underline{0}_{(p\times 1)}\,. \tag{5.289}$$

But then we may split the null hypothesis in (5.289) as

$$H_0 \equiv H_{0b} \circ H_{0a}$$

where

$$H_{0a} : \Sigma^* = bdiag(\Sigma_1^*, \Sigma_2^*) \tag{5.290}$$

and

$$H_{0b} : \underline{\mu}_2^* = \underline{0}_{(p\times 1)}\,. \tag{5.291}$$

While from Sect. 5.1.9 the $(2/n)$-th power of the l.r.t. statistic to test H_{0a} in (5.290) is

$$\Lambda_a = \frac{|A^*|}{|A_{11}^*|\,|A_{22}^*|} \tag{5.292}$$

where A^* is the m.l.e. of Σ^*, that is, the matrix in (5.286), in the previous subsection, and, as in that same subsection, A_{11}^* and A_{22}^* are its two diagonal blocks of dimensions $p\times p$, the m.l.e.'s respectively of Σ_1^* and Σ_2^*, from

Sects. 5.1.4.1, 5.1.4.4, and 5.1.4.5, the $(2/n)$-th power of the l.r.t. statistic to test H_{0b} in (5.291), is, for

$$\underset{(p\times 2p)}{C} = \left[\underset{(p\times p)}{0} \;\middle|\; I_p\right], \quad D = I_1, \quad \beta^* = \underline{0}_{(p\times 1)}, \quad \text{and} \quad M = E_{1n},$$

the statistic

$$\Lambda_b = \frac{|CA^*C'|}{\left|CA^*C' + \frac{1}{n}C\overline{\underline{X}}^*\overline{\underline{X}}^{*\prime}C'\right|} = \frac{|A_{22}^*|}{|A_{22}^* + \frac{1}{n}\overline{\underline{X}}_2^*\overline{\underline{X}}_2^{*\prime}|}, \tag{5.293}$$

where

$$\overline{\underline{X}}^* = \frac{1}{n}X^*E_{n1} = \begin{bmatrix} \overline{\underline{X}}_1^* \\ \hline \overline{\underline{X}}_2^* \end{bmatrix}, \quad \text{with} \quad \overline{\underline{X}}_2^* = \frac{1}{n}X_2^*E_{n1}$$

and with

$$X^* = \left(\Gamma \otimes I_p\right)X = \begin{bmatrix} X_1^* \\ X_2^* \end{bmatrix} \begin{matrix} p \\ p \end{matrix},$$

where X is the $2p\times n$ matrix of a sample of size n from $\underline{X}$ and where

$$CX^* = X_2^* \quad \text{and} \quad C\underline{X}^* = \underline{X}_2^* \sim N_p\left(\underline{\mu}_2^*, \Sigma_2^*\right).$$

Therefore, for a sample of size n $(> 2p)$, the $(2/n)$-th power of the l.r.t. statistic to test H_0 in (5.289) is

$$\Lambda = \Lambda_a\,\Lambda_b = \frac{|A^*|}{|A_{11}^*|\,|A_{22}^* + \frac{1}{n}\overline{\underline{X}}_2^*\overline{\underline{X}}_2^{*\prime}|}, \tag{5.294}$$

where, similar to what happens with Λ in (5.287), and for similar reasons, $|A^*|$ may be replaced by $|A|$.

Since, as shown in Sect. 5.1.9, Λ_a in (5.292) is independent of A_{11}^* and A_{22}^*, and Λ_b in (5.293) is only a function of A_{22}^* and $\overline{\underline{X}}_2^*$, the two statistics Λ_a and Λ_b are independent, and since from the results in Sects. 5.1.9 and 5.1.4 we have, for $n > 2p$,

$$\Lambda_a \stackrel{d}{\equiv} \prod_{j=1}^{p} Y_j$$

where

$$Y_j \sim Beta\left(\frac{n-p-j}{2}, \frac{p}{2}\right)$$

are p independent r.v.'s, and

$$\Lambda_b \stackrel{d}{\equiv} Y$$

where

$$Y \sim Beta\left(\frac{n-p}{2}, \frac{p}{2}\right),$$

we have

$$\Lambda \stackrel{d}{\equiv} \prod_{j=1}^{p+1} Y_j\,,$$

where

$$Y_j \sim Beta\left(\frac{n-p-j+1}{2}, \frac{p}{2}\right)$$

form a set of $p+1$ independent r.v.'s. But, in much the same way we may switch p_1 and p_2 in (5.149) and (5.150), we may also write

$$\Lambda \stackrel{d}{\equiv} \prod_{j=1}^{p} Y_j^*\,,$$

where

$$Y_j^* \sim Beta\left(\frac{n-p-j}{2}, \frac{p+1}{2}\right)$$

form a set of p independent r.v.'s.

As such, for odd p, the exact distribution of Λ in (5.294) is given by Theorem 3.2 with

$$m^*=1\,,\quad k_1=2\,,\quad a_1=\frac{n-p+1}{2}\,,\quad n_1=\frac{p+1}{2}\,,\quad m_1=p \tag{5.295}$$

while for even p the exact distribution of Λ in (5.294) is given by Theorem 3.2 for

$$m^* = 1\,, \quad k_1 = 2\,, \quad a_1 = \frac{n-p}{2}\,, \quad n_1 = \frac{p}{2}\,, \quad m_1 = p+1\,, \tag{5.296}$$

but with the W_{1j} Gamma distributed r.v.'s in (3.12) clearly having the same distribution either for even or odd p, with rate parameters, from (3.12), given by

$$a_1 - n_1 + \frac{j-1}{k_1} = \frac{n-2p+j-1}{2}\,, \quad j = 1, \ldots, 2p-1\,, \tag{5.297}$$

and shape parameters given by (3.13) and (3.14) as

$$r_j = \begin{cases} h_j & j = 1, 2 \\ h_j + r_{j-2} & j = 3, \ldots, 2p-1\,, \end{cases} \tag{5.298}$$

with

$$h_j = \begin{cases} 1 & j = 1, \ldots, p \\ 0 & j = p+1 \\ -1 & j = p+2, \ldots, 2p-1\,, \end{cases} \tag{5.299}$$

or, alternatively, as

$$r_j = \begin{cases} \left\lceil \frac{j}{2} \right\rceil & j = 1, \ldots, p \\ \left\lceil \frac{2p-j}{2} \right\rceil & j = p+1, \ldots, 2p-1 \end{cases} \quad \text{or} \quad r_j = \begin{cases} \left\lfloor \frac{j+1}{2} \right\rfloor & j = 1, \ldots, p \\ p - \left\lfloor \frac{j}{2} \right\rfloor & j = p+1, \ldots, 2p-1\,. \end{cases} \tag{5.300}$$

Given the symmetry of the r_j we may indeed reverse the indexing of the rate parameters in (5.297) and write the exact p.d.f. and c.d.f. of Λ in (5.294) as given by Corollary 4.2 in terms of the EGIG distribution representation, for any p, as

$$f_\Lambda(z) = f^{EGIG}\left(z \,\middle|\, \{r_j\}_{j=1:2p-1}; \left\{ \frac{n-1-j}{2} \right\}_{j=1:2p-1}; 2p-1 \right)$$

and

$$F_\Lambda(z) = F^{EGIG}\left(z \,\middle|\, \{r_j\}_{j=1:2p-1}; \left\{ \frac{n-1-j}{2} \right\}_{j=1:2p-1}; 2p-1 \right),$$

for r_j given by (5.298)–(5.299) or (5.300) above.

Alternatively, the p.d.f. and c.d.f. of Λ may be written, in terms of the Meijer G function as

$$f_\Lambda(z) = \left\{\prod_{j=1}^{p} \frac{\Gamma\left(\frac{n-j+1}{2}\right)}{\Gamma\left(\frac{n-p-j}{2}\right)}\right\} G^{p,0}_{p,p}\left(\begin{array}{c}\left\{\frac{n-j+1}{2}-1\right\}_{j=1:p} \\ \left\{\frac{n-p-j}{2}-1\right\}_{j=1:p}\end{array}\middle|\, z\right)$$

and

$$F_\Lambda(z) = \left\{\prod_{j=1}^{p} \frac{\Gamma\left(\frac{n-j+1}{2}\right)}{\Gamma\left(\frac{n-p-j}{2}\right)}\right\} G^{p,1}_{p+1,p+1}\left(\begin{array}{c}\left\{1,\left\{\frac{n-j+1}{2}\right\}_{j=1:p}\right\} \\ \left\{\left\{\frac{n-p-j}{2}\right\}_{j=1:p},0\right\}\end{array}\middle|\, z\right).$$

The statistic in (5.294) remains invariant under any identical linear transformations within each of the two subsets $\underline{X}_1$ and $\underline{X}_2$, that is, under linear transformations $\underline{X} \to C^*\underline{X} + \underline{b}^*$, for $C^* = bdiag(C, C)$ and $\underline{b}^* = [\underline{b}', \underline{b}']'$, where C is any real nonrandom full-rank $p\times p$ matrix and $\underline{b}$ any real $p\times 1$ vector.

We may actually want to take our testing procedure, one step further, by willing to test the null hypothesis

$$H_0 : \Sigma = \begin{bmatrix} \Sigma_1 & \Sigma_2 \\ \Sigma_2 & \Sigma_1 \end{bmatrix}, \quad \underline{\mu}_1 = \underline{\mu}_2 = \underline{0}_{(p\times 1)} . \qquad (5.301)$$

Taking a similar approach to the one taken in testing the null hypothesis in (5.288) we may see that this would be equivalent to testing the null hypothesis

$$H_0 : \Sigma^* = bdiag(\Sigma_1^*, \Sigma_2^*), \quad \underline{\mu}^* = \underline{0}_{(2p\times 1)} . \qquad (5.302)$$

We may split this null hypothesis as

$$H_{0b|a} \circ H_{0a}$$

where H_{0a} is the same as H_{0a} in (5.290) and

$$\begin{array}{l} H_{0b|a} : \; \underline{\mu}^* = \underline{0}_{(2p\times 1)} \\ \qquad\quad \text{assuming } H_{0a} \end{array} \iff \left\{\begin{array}{l}\underline{\mu}_1^* = \underline{0}_{(p\times 1)} \\ \underline{\mu}_2^* = \underline{0}_{(p\times 1)}\end{array}\right. \text{(two independent tests).} \qquad (5.303)$$

For a sample of size n, the $(2/n)$-th power of the l.r.t. statistic to test $H_{0b|a}$ in (5.303) is

$$\Lambda_{b|a} = \frac{|A_{11}^*|}{|A_{11}^* + \frac{1}{n}\overline{\underline{X}}_1^*\overline{\underline{X}}_1^{*\prime}|} \, \frac{|A_{22}^*|}{|A_{22}^* + \frac{1}{n}\overline{\underline{X}}_2^*\overline{\underline{X}}_2^{*\prime}|}$$

so that the $(2/n)$-th power of the l.r.t. statistic to test H_0 in (5.301) or (5.302) is

$$\begin{aligned}\Lambda = \Lambda_a\, \Lambda_{b|a} &= \frac{|A^*|}{|A_{11}^*|\,|A_{22}^*|}\frac{|A_{11}^*|}{|A_{11}^*+\frac{1}{n}\underline{\overline{X}}_1^*\underline{\overline{X}}_1^{*\prime}|}\frac{|A_{22}^*|}{|A_{22}^*+\frac{1}{n}\underline{\overline{X}}_2^*\underline{\overline{X}}_2^{*\prime}|}\\ &= \frac{|A^*|}{|A_{11}^*+\frac{1}{n}\underline{\overline{X}}_1^*\underline{\overline{X}}_1^{*\prime}|\,|A_{22}^*+\frac{1}{n}\underline{\overline{X}}_2^*\underline{\overline{X}}_2^{*\prime}|}\end{aligned} \tag{5.304}$$

with

$$\Lambda_{b|a} \stackrel{\mathrm{d}}{\equiv} Y_1\, Y_2$$

where Y_1 and Y_2 are two independent r.v.'s, both with a $Beta\left(\frac{n-p}{2}, \frac{p}{2}\right)$ distribution.

Given the independence of Λ_a and both A_{11}^* and A_{22}^*, and since $\Lambda_{b|a}$ is only a function of these two matrices, $\underline{\overline{X}}_1^*$ and $\underline{\overline{X}}_2^*$, then

$$\Lambda \stackrel{\mathrm{d}}{\equiv} \underbrace{\left\{\prod_{j=1}^{p+1} Y_j\right\}}_{Y^*} Y \tag{5.305}$$

where

$$Y_j \sim Beta\left(\frac{n-p-j+1}{2}, \frac{p}{2}\right) \quad \text{and} \quad Y \sim Beta\left(\frac{n-p}{2}, \frac{p}{2}\right)$$

form a set of $p+2$ independent r.v.'s. We may note that Y^* in (5.305) has the same distribution as the statistic Λ in (5.294).

As such, the exact p.d.f. and c.d.f. of Λ in (5.304) are given in terms of the Meijer G function respectively by

$$f_\Lambda(z) = \left\{\frac{\Gamma\left(\frac{n}{2}\right)}{\Gamma\left(\frac{n-p}{2}\right)}\prod_{j=1}^{p+1}\frac{\Gamma\left(\frac{n-j+1}{2}\right)}{\Gamma\left(\frac{n-p-j+1}{2}\right)}\right\} G^{p+1,0}_{p+1,p+1}\left(\left.\begin{matrix}\left\{\frac{n-2}{2}, \left\{\frac{n-j-1}{2}\right\}_{j=1:p+1}\right\}\\ \left\{\frac{n-p}{2}, \left\{\frac{n-p-j-1}{2}\right\}_{j=1:p+1}\right\}\end{matrix}\right|\, z\right)$$

and

$$F_\Lambda(z) = \left\{\frac{\Gamma\left(\frac{n}{2}\right)}{\Gamma\left(\frac{n-p}{2}\right)}\prod_{j=1}^{p+1}\frac{\Gamma\left(\frac{n-j+1}{2}\right)}{\Gamma\left(\frac{n-p-j+1}{2}\right)}\right\} G^{p+1,1}_{p+2,p+2}\left(\left.\begin{matrix}\left\{1, \frac{n}{2}, \left\{\frac{n-j+1}{2}\right\}_{j=1:p+1}\right\}\\ \left\{\frac{n-p}{2}, \left\{\frac{n-p-j+1}{2}\right\}_{j=1:p+1}, 0\right\}\end{matrix}\right|\, z\right),$$

while for even p the distribution of Λ in (5.304) will be given by Theorem 3.2 with

$$m^* = 2\,,\ \ k_\nu = \{2,1\}\,,\ \ a_\nu = \left\{\frac{n-p}{2}, \frac{n-p+2}{2}\right\},\ \ n_\nu = \left\{\frac{p}{2}, 1\right\},$$
$$m_\nu = \left\{p+1, \frac{p}{2}\right\}, \quad (\nu = 1,2)\,,$$

and its exact p.d.f. and c.d.f. by Corollary 4.2, in terms of the EGIG p.d.f. and c.d.f., as

$$f_\Lambda(z) = f^{EGIG}\left(z \,\middle|\, \{r_j\}_{j=1:2p-1}; \left\{\frac{n-1-j}{2}\right\}_{j=1:2p-1}; 2p-1\right)$$

and

$$F_\Lambda(z) = F^{EGIG}\left(z \,\middle|\, \{r_j\}_{j=1:2p-1}; \left\{\frac{n-1-j}{2}\right\}_{j=1:2p-1}; 2p-1\right),$$

with

$$r_j = \begin{cases} \left\lceil \frac{j}{2} \right\rceil + \mod(j,2) & j = 1,\dots,p \\ \left\lceil \frac{2p-j}{2} \right\rceil & j = p+1,\dots,2p-1\,. \end{cases}$$

The statistic in (5.304) is invariant under linear transformations of the type $\underline{X} \to C^*\underline{X}$, for $C^* = bdiag(C,C)$ where C is any real nonrandom full-rank $p\times p$ matrix.

Modules `PDFCompSymEqR`, `CDFCompSymEqR`, `PvalCompSymEqR`, `QuantCompSymEqR`, and `PvalDataCompSymEqR`, with exactly the same arguments as the corresponding modules in the previous subsection, are available to help in carrying out the first test in the present subsection, and modules `PDFCompSymEqR0`, `CDFCompSymEqR0`, `PvalCompSymEqR0`, `QuantCompSymEqR0`, and `PvalDataCompSymEqR0` are available to help carrying out this second test.

As an example of the implementation of the first test we use exactly the same bone mineral content data set used in the previous subsection and carry out the test of complete symmetrical equivalence between the set of six bone mineral content measured at baseline and the set of similar variables measured after 1 year of follow-up. The sample means for these two sets of variables, rounded to the fifth decimal place, are

$$\overline{\underline{X}}_1' = [\,0.84083,\ 0.81342,\ 1.78525,\ 1.72925,\ 0.69754,\ 0.68658\,]$$

and

$$\overline{\underline{X}}_2' = [\,0.84096,\ 0.81017,\ 1.77808,\ 1.71692,\ 0.71267,\ 0.68675\,]$$

```
PvalDataCompSymEqR[path <> "Tabl_8_6_16_JW.dat"]
Computed value of Λ: 0.1025199900
p-value: 0.6529753696
```

Fig. 5.68 Commands used and output obtained with the Mathematica® module `PvalDataSymEqR` to implement the test of symmetrical equivalence, using the bone mineral data in Tables 1.8 and 6.16 of Johnson and Wichern (2014)

so that it comes at no surprise that the null hypothesis in (5.288) is not rejected, as a consequence of the high p-value obtained in Fig. 5.68, with module `PvalDataCompSymEqR`, there being no enough evidence of a change in the mean values of the six variables that measure bone mineral content after 1 year of follow-up. This fact, together with the covariance structure of the two sets of variables will lead us to say that the two sets of bone mineral content variables are "completely symmetrically equivalent," in the sense that using one of them or the other in any statistical analysis will generally lead to similar conclusions.

5.1.17 Testing for "Symmetrical Spherical Equivalence" or Independent Two-Block Compound Symmetry [SymSphEq]

Let us assume a similar setup to the one in the two previous subsections, now with $p = 2$ and where we want to test the null hypothesis

$$H_0 : \Sigma = \begin{bmatrix} a & b & 0 & 0 \\ b & a & 0 & 0 \\ 0 & 0 & a & b \\ 0 & 0 & b & a \end{bmatrix}, \tag{5.306}$$

where a and b are not specified but have to verify the relation $-a < b < a$, with $a > 0$.

The test of such hypothesis may be relevant in spatial or temporal statistical analyses, as for example the analysis of seismological data, where a given variable is measured at four different points, scattered, in space or time, in such a way that points 1 and 2 are close enough so that we may assume that their data is correlated, and that they have between them a similar distance, in space or time, to that of points 3 and 4, so that the correlation between the data in these two points will be similar to that between the data in points 1 and 2, while the first set of points, 1 and 2, lies so far apart from the second set of points, 3 and 4, that we may assume that the values of this variable at these two sets of points are uncorrelated. Moreover, since the variable being measured is the "same" in all four points, we may assume that the variance of the variable measured is the same in all four points. Other applications may be for example in analyzing the amplitude of vibrations at the four end-points of a bridge or even in the analysis of sociological, learning, or perceptional studies—see the two examples at the end of this subsection.

We may split the null hypothesis in (5.306) into a sequence of "conditionally independent" sub-hypotheses, in the sense that when testing each one of these sub-hypotheses one will assume that the "previous" one holds, at the same time that the associated test statistics are independent, if we write

$$H_0 \equiv H_{0c|b,a} \circ H_{0b|a} \circ H_{0a} \tag{5.307}$$

where

$$\begin{aligned}
&H_{0a} : \Sigma = bdiag(\Sigma_1, \Sigma_2) \;\; \text{(where } \Sigma_1 \text{ and } \Sigma_2 \text{ are } 2{\times}2 \text{ unspecified matrices)}\\
&H_{0b|a} : \Sigma_1 = \Sigma_2\,(= \Xi, \text{ unspecified}), \text{ assuming } \Sigma = bdiag(\Sigma_1, \Sigma_2)\\
&H_{0c|b,a} : \Xi = \begin{bmatrix} a & b \\ b & a \end{bmatrix} \text{(compound symmetry or circularity test for a } 2{\times}2 \text{ matrix)},\\
&\qquad \text{assuming } \Sigma = bdiag(\Sigma_1, \Sigma_2) \text{ and } \Sigma_1 = \Sigma_2\\
&\qquad \text{(and where } a \text{ and } b \text{ are not specified but satisfy } -a < b < a).
\end{aligned} \tag{5.308}$$

The $(2/n)$-th power of the l.r.t. statistic to test the null hypothesis H_{0a} in (5.308) is

$$\Lambda_a = \frac{|A|}{|A_{11}|\,|A_{22}|} \tag{5.309}$$

where A is the m.l.e. of Σ and A_{11} and A_{22} its diagonal blocks of dimensions $2{\times}2$, with

$$\Lambda_a \stackrel{d}{\equiv} \prod_{j=1}^{2} Y_j$$

where, provided $n > 4$,

$$Y_j \sim Beta\left(\frac{n-2-j}{2}, 1\right) \quad (j = 1, 2)$$

are two independent r.v.'s.

The $(2/n)$-th power of the l.r.t. statistic to test $H_{0b|a}$ in (5.308) is (see for example Anderson (2003, Sec. 10.2) and Marques et al. (2011, expr. (2.13))),

$$\Lambda_{b|a} = 2^4 \frac{|A_{11}||A_{22}|}{|A_{11} + A_{22}|^2}, \tag{5.310}$$

which is the $(2/n)$-th power of the l.r.t. statistic to test the equality of two 2×2 covariance matrices, with

$$\Lambda_{b|a} \stackrel{d}{\equiv} Y_1\, Y_2\, Y_3\,,$$

where, for $n > 2$ (see Marques et al. (2011, expr. (2.14))),

$$Y_1 \sim Beta\left(\frac{n-1}{2}, \frac{1}{2}\right), \quad Y_2 \sim Beta\left(\frac{n-2}{2}, \frac{1}{4}\right) \text{ and } Y_3 \sim Beta\left(\frac{n-2}{2}, \frac{3}{4}\right)$$

are three independent r.v.'s.

The $(2/n)$-th power of the l.r.t. statistic to test $H_{0c|b,a}$ in (5.308) is the $(2/n)$-th power of the l.r.t. statistic to test the compound symmetry or the circularity of a 2×2 covariance matrix, based on a sample of size $2n$, given the fact that once assumed $\Sigma_1 = \Sigma_2$ and their independence, we will have to use a pooled estimator of Ξ, which ends-up being based on a sample of size $2n$ (see Wilks (1946) and Olkin and Press (1969) respectively for the definition of the compound symmetric or equivariance-equicorrelation structure and corresponding l.r. test and for the definition of the circular or circulant structure and corresponding l.r. tests; see also Sect. 5.1.24 for the definition of the compound symmetric or equivariance-equicorrelation structure and Sect. 5.1.22 for the definition of the circular or circulant structure and Sect. 5.2.1 for the corresponding l.r. test, which for $p = 2$ is also the l.r. test for compound symmetry). This statistic is, from (5.418)–(5.419) in Sect. 5.2.1,

$$\Lambda_{c|b,a} = \left(\frac{|A_{11} + A_{22}|}{v_{11}\, v_{22}}\right)^2, \tag{5.311}$$

where v_{11} and v_{22} are the two diagonal elements of $\Gamma(A_{11} + A_{22})\Gamma'$, where Γ is the matrix in (5.283), which is also equal to the matrix U in Sect. 5.2.1 for $p = 2$. Then, since the pooled estimator of Ξ in (5.311), which is $A_{11} + A_{22}$, has, under H_{0a}, a Wishart distribution with $2(n-1)$ degrees of freedom, we have

$$\Lambda_{c|b,a} \stackrel{d}{\equiv} Y^2$$

where

$$Y \sim Beta\left(\frac{2(n-1)-1}{2}, \frac{1}{2}\right).$$

But, at least at first sight, we seem to be unable to recognize the distribution of the statistic

$$\Lambda = \Lambda_a \Lambda_{b|a} \Lambda_{c|ba} = 2^4 \frac{|A|}{v_{11}^2 \, v_{22}^2}, \tag{5.312}$$

which is the $(2/n)$-th power of the l.r. statistic to test H_0 in (5.306) or (5.307), in the scope of Theorems 3.1–3.3.

However, we may try a slightly different approach. Let us consider the orthogonal matrix

$$\Gamma^* = (\Gamma \otimes I_2)\, I_{23}\,, \tag{5.313}$$

where Γ is the matrix in (5.283), I_2 is the identity matrix of order 2 and

$$I_{23} = \begin{bmatrix} 1 & 0 & 0 & 0 \\ 0 & 0 & 1 & 0 \\ 0 & 1 & 0 & 0 \\ 0 & 0 & 0 & 1 \end{bmatrix}$$

is also a symmetric and orthogonal matrix. Then (5.306) holds if and only if

$$H_0 : \Sigma^* = \Gamma^* \Sigma \Gamma^{*\prime} = \begin{bmatrix} a+b & 0 & 0 & 0 \\ 0 & a+b & 0 & 0 \\ 0 & 0 & a-b & 0 \\ 0 & 0 & 0 & a-b \end{bmatrix} \tag{5.314}$$

also holds, and given that the elements in Γ^* are not functions of the elements in Σ, to test H_0 in (5.306) is equivalent to test H_0 in (5.314). This latter null hypothesis may be decomposed as

$$H_0 \equiv H_{0b|a} \circ H_{0a} \tag{5.315}$$

where

$$\begin{array}{l} H_{0a} : \Sigma^* = bdiag(\Sigma_1^*, \Sigma_2^*) \text{ (where } \Sigma_1^* \text{ and } \Sigma_2^* \text{ are } 2{\times}2 \text{ unspecified matrices)} \\ H_{0b|a} : \Sigma_1^* = \sigma_1^2 I_2,\ \Sigma_2^* = \sigma_2^2 I_2, \text{ assuming } \Sigma^* = bdiag(\Sigma_1^*, \Sigma_2^*) \\ \qquad\qquad (\sigma_1^2 \text{ and } \sigma_2^2 \text{ non-specified}). \end{array} \tag{5.316}$$

The test associated with $H_{0b|a}$ is the combination of two independent sphericity tests for two 2×2 matrices (note that we call a matrix "spherical" if it is proportional to the identity matrix of the same order and also note that since we have $-a < b < a$, and $a > 0$, both 2×2 diagonal blocks of Σ^* above are positive-definite).

But then, for

$$A^* = \Gamma^* A \Gamma^{*\prime}, \tag{5.317}$$

where A is the m.l.e. of Σ, while the $(2/n)$-th power of the l.r.t. statistic to test H_{0a} in (5.316) is the statistic

$$\Lambda_a^* = \frac{|A^*|}{|A_{11}^*|\,|A_{22}^*|},$$

the $(2/n)$-th power of the l.r.t. statistic to test $H_{0b|a}$ will be (see for example Anderson (2003, Sec. 10.7) and Marques et al. (2011, expr. (2.10)))

$$\Lambda_{b|a}^* = 2^4 \frac{|A_{11}^*|}{\left(tr\,(A_{11}^*)\right)^2} \frac{|A_{22}^*|}{\left(tr\,(A_{22}^*)\right)^2}, \tag{5.318}$$

where A_{11}^* and A_{22}^* are the two 2×2 diagonal blocks of $A^* = \Gamma^* A \Gamma^{*\prime}$, with A being the m.l.e. of Σ. Then, while Λ_a^* has a similar distribution to that of Λ_a in (5.309),

$$\Lambda_{b|a}^* \stackrel{d}{\equiv} Y_1\, Y_2$$

where Y_1 and Y_2 are independent r.v.'s with

$$Y_j \sim Beta\left(\frac{n-2}{2}, 1\right), \quad j = 1, 2\,.$$

For a sample of size n the $(2/n)$-th power of the l.r.t. statistic to test H_0 in (5.314)–(5.316) is thus

$$\Lambda = \Lambda_a^*\, \Lambda_{b|a}^* = 2^4 \frac{|A^*|}{\left(tr(A_{11}^*)\, tr(A_{22}^*)\right)^2} \tag{5.319}$$

where Λ_a^* and $\Lambda_{b|a}^*$ are independent, given the independence of Λ_a^* and A_{11}^* and A_{22}^* and the fact that $\Lambda_{b|a}^*$ is a function of only A_{11}^* and A_{22}^*, so that

$$\Lambda \stackrel{d}{\equiv} \left\{ \prod_{j=1}^{3} Y_j \right\} Y$$

where, for $n > 4$,

$$Y_j \sim Beta\left(\frac{n-j-1}{2}, 1\right) \quad (j = 1, 2, 3) \quad \text{and} \quad Y \sim Beta\left(\frac{n-2}{2}, 1\right) \tag{5.320}$$

form a set of four independent r.v.'s.

As such, the exact distribution of Λ is given by Theorem 3.2 with

$$m^* = 3, \quad k_\nu = \{2, 1, 1\}, \; a_\nu = \left\{\frac{n-2}{2}, \frac{n}{2}, \frac{n}{2}\right\}, \; n_\nu = \{1, 1, 1\}, \; m_\nu = \{2, 1, 1\},$$

for $\nu = 1, 2, 3$, and hence the exact p.d.f. and c.d.f. of Λ are given by Corollary 4.2 in terms of the EGIG p.d.f. and c.d.f. as

$$f_\Lambda(z) = f^{EGIG}\left(z \,\middle|\, \{r_j\}_{j=1:3}; \left\{\frac{n-1-j}{2}\right\}_{j=1:3}; 3\right)$$

and

$$F_\Lambda(z) = F^{EGIG}\left(z \,\middle|\, \{r_j\}_{j=1:3}; \left\{\frac{n-1-j}{2}\right\}_{j=1:3}; 3\right),$$

with

$$r_j = \begin{cases} 2 & j = 1 \\ 1 & j = 2, 3, \end{cases}$$

or in terms of the Meijer G function by

$$f_\Lambda(z) = \left\{\frac{\Gamma\left(\frac{n}{2}\right)}{\Gamma\left(\frac{n-2}{2}\right)} \prod_{j=1}^{3} \frac{\Gamma\left(\frac{n-j+1}{2}\right)}{\Gamma\left(\frac{n-j-1}{2}\right)}\right\} G^{4,0}_{4,4}\left(\begin{array}{c} \left\{\frac{n}{2}, \left\{\frac{n-j+1}{2}\right\}_{j=1:3}\right\} \\ \left\{\frac{n-2}{2}, \left\{\frac{n-j-1}{2}\right\}_{j=1:3}\right\} \end{array} \middle|\, z\right)$$

and

$$F_\Lambda(z) = \left\{\frac{\Gamma\left(\frac{n}{2}\right)}{\Gamma\left(\frac{n-2}{2}\right)} \prod_{j=1}^{3} \frac{\Gamma\left(\frac{n-j+1}{2}\right)}{\Gamma\left(\frac{n-j-1}{2}\right)}\right\} G^{4,1}_{5,5}\left(\begin{array}{c} \left\{1, \frac{n}{2}, \left\{\frac{n-j+1}{2}\right\}_{j=1:3}\right\} \\ \left\{\frac{n-2}{2}, \left\{\frac{n-j-1}{2}\right\}_{j=1:3}, 0\right\} \end{array} \middle|\, z\right).$$

Although the above distribution for the statistic Λ was derived under the assumption of a multivariate Normal distribution for the random vector $\underline{X}$, the results in Chmielewski (1980), Anderson (2003, Sect. 9.11), and Fang and Zhang (1990, Sects. 5.3.1, 5.3.2) allow us to extend this distribution to the cases where $\underline{X}$ has any elliptically contoured distribution.

Since the tests of the hypotheses in (5.306) and (5.314) are indeed the same test, the statistics Λ in (5.312) and (5.319) are indeed the same and have the same distribution. In order to show that they have the same distribution it is enough to show that the distribution of $\Lambda_{b|a}\Lambda_{c|b,a}$ for $\Lambda_{b|a}$ in (5.310) and $\Lambda_{c|b,a}$ in (5.311) is the same as that of $\Lambda^*_{b|a}$ in (5.318). In order to do this, let

$$W_1 = -\log\left(\Lambda_{b|a}\,\Lambda_{c|b,a}\right) \quad \text{and} \quad W_2 = -\log\,\Lambda^*_{b|a}\,.$$

Then we have the c.f. (characteristic function) of W_2 given by

$$\begin{aligned}
\Phi_{W_2}(t) &= E\left(e^{\mathrm{i}t\,W_2}\right) = E\left(e^{-\mathrm{i}t\,\log\,\Lambda^*_{b|a}}\right) = E\left(\left(\Lambda^*_{b|a}\right)^{-\mathrm{i}t}\right)\\
&= \left(\frac{\Gamma\left(\frac{n}{2}\right)}{\Gamma\left(\frac{n-2}{2}\right)}\,\frac{\Gamma\left(\frac{n-2}{2}-\mathrm{i}t\right)}{\Gamma\left(\frac{n}{2}-\mathrm{i}t\right)}\right)^2
\end{aligned}$$

while the c.f. of W_1, applying the Gamma duplication formula in (5.583) and using the fact that $\Lambda_{b|a}$ and $\Lambda_{c|b,a}$ are independent under H_{0a} in (5.307), given that $\Lambda_{b|a}$ is independent of $A_{11}+A_{22}$ (see Lemma 10.4.1 in Anderson (2003) and Theorem 5 in Jensen (1988)), may be written as

$$\begin{aligned}
\Phi_{W_1}(t) &= E\left(e^{\mathrm{i}t\,W_1}\right) = E\left(e^{-\mathrm{i}t\,\log(\Lambda_{b|a}\Lambda_{c|b,a})}\right) = E\left(\Lambda_{b|a}^{-\mathrm{i}t}\,\Lambda_{c|b,a}^{-\mathrm{i}t}\right)\\
&= \frac{\Gamma\left(\frac{n}{2}\right)\Gamma\left(\frac{n-1}{2}-\mathrm{i}t\right)}{\Gamma\left(\frac{n-1}{2}\right)\Gamma\left(\frac{n}{2}-\mathrm{i}t\right)}\,\frac{\Gamma\left(\frac{n-2}{2}+\frac{1}{4}\right)\Gamma\left(\frac{n-2}{2}-\mathrm{i}t\right)}{\Gamma\left(\frac{n-2}{2}\right)\Gamma\left(\frac{n-2}{2}+\frac{1}{4}-\mathrm{i}t\right)}\\
&\qquad \frac{\Gamma\left(\frac{n-2}{2}+\frac{3}{4}\right)\Gamma\left(\frac{n-2}{2}-\mathrm{i}t\right)}{\Gamma\left(\frac{n-2}{2}\right)\Gamma\left(\frac{n-2}{2}+\frac{3}{4}-\mathrm{i}t\right)}\,\frac{\Gamma(n-1)\,\Gamma\left(n-1-\frac{1}{2}-2\mathrm{i}t\right)}{\Gamma\left(n-1-\frac{1}{2}\right)\Gamma\left(n-1-2\mathrm{i}t\right)}\\
&= \frac{\Gamma\left(\frac{n}{2}\right)}{\Gamma\left(\frac{n-2}{2}\right)}\,\frac{\Gamma\left(\frac{n-2}{2}-\mathrm{i}t\right)}{\Gamma\left(\frac{n}{2}-\mathrm{i}t\right)}\,\frac{\Gamma\left(\frac{n-2}{2}-\mathrm{i}t\right)}{\Gamma\left(\frac{n-2}{2}\right)}\,\frac{\cancel{\Gamma\left(\frac{n-1}{2}-\mathrm{i}t\right)}}{\cancel{\Gamma\left(\frac{n-1}{2}\right)}}\,\frac{\Gamma\left(\frac{n-2}{2}+\frac{1}{4}\right)}{\Gamma\left(\frac{n-2}{2}+\frac{1}{4}-\mathrm{i}t\right)}\\
&\qquad \frac{\Gamma\left(\frac{n-2}{2}+\frac{3}{4}\right)}{\Gamma\left(\frac{n-2}{2}+\frac{3}{4}-\mathrm{i}t\right)}\,\frac{\cancel{\Gamma\left(\frac{n-1}{2}\right)}\,\Gamma\left(\frac{n}{2}\right)\Gamma\left(\frac{n-1}{2}-\frac{1}{4}-\mathrm{i}t\right)\Gamma\left(\frac{n-1}{2}+\frac{1}{4}-\mathrm{i}t\right)}{\Gamma\left(\frac{n-1}{2}-\frac{1}{4}\right)\Gamma\left(\frac{n-1}{2}+\frac{1}{4}\right)\cancel{\Gamma\left(\frac{n-1}{2}-\mathrm{i}t\right)}\,\Gamma\left(\frac{n}{2}-\mathrm{i}t\right)}\\
&= \left(\frac{\Gamma\left(\frac{n}{2}\right)}{\Gamma\left(\frac{n-2}{2}\right)}\,\frac{\Gamma\left(\frac{n-2}{2}-\mathrm{i}t\right)}{\Gamma\left(\frac{n}{2}-\mathrm{i}t\right)}\right)^2\frac{\Gamma\left(\frac{n}{2}-\frac{3}{4}\right)}{\Gamma\left(\frac{n}{2}-\frac{3}{4}-\mathrm{i}t\right)}\,\frac{\Gamma\left(\frac{n}{2}-\frac{1}{4}\right)}{\Gamma\left(\frac{n}{2}-\frac{1}{4}-\mathrm{i}t\right)}\\
&\qquad \frac{\Gamma\left(\frac{n}{2}-\frac{1}{4}-\mathrm{i}t\right)\Gamma\left(\frac{n}{2}-\frac{3}{4}-\mathrm{i}t\right)}{\Gamma\left(\frac{n}{2}-\frac{1}{4}\right)\Gamma\left(\frac{n}{2}-\frac{3}{4}\right)}\\
&= \left(\frac{\Gamma\left(\frac{n}{2}\right)}{\Gamma\left(\frac{n-2}{2}\right)}\,\frac{\Gamma\left(\frac{n-2}{2}-\mathrm{i}t\right)}{\Gamma\left(\frac{n}{2}-\mathrm{i}t\right)}\right)^2.
\end{aligned}$$

Since the c.f.'s of W_1 and W_2 match, then the distribution of $\Lambda_{b|a}\Lambda_{c|b,a}$ and that of $\Lambda^*_{b|a}$ are the same and as such, also the distributions of the l.r.t. statistics in (5.312) and (5.319) are the same. In fact, it is possible to verify that the two likelihood ratio test statistics are algebraically equivalent functions of the original data, i.e., they are the same statistic.

There are a number of other equivalent forms for the distribution of Λ in (5.312) and (5.319) in terms of products of independent Beta r.v.'s, among which we have

$$\Lambda \stackrel{d}{\equiv} \left\{\prod_{j=1}^{2} Y_j\right\} Y^* \tag{5.321}$$

where Y_j $(j = 1, 2)$ are the same as in (5.320) and where

$$Y^* \sim Beta\left(\frac{n-4}{2}, 2\right), \tag{5.322}$$

or

$$\Lambda \stackrel{d}{\equiv} Y^* \left(Y^{**}\right)^2 \tag{5.323}$$

where

$$Y^{**} \sim Beta(n-3, 2)\,. \tag{5.324}$$

These representations may give rise to other equivalent representations for the p.d.f. and c.d.f. of Λ in terms of Meijer G and Fox H functions, but they will still yield the very same representations in terms of the EGIG p.d.f. and c.d.f.. From (5.321) and (5.322) we may write the p.d.f. and c.d.f. of Λ in terms of the Meijer G function as

$$f_\Lambda(z) = \left\{\frac{\Gamma\left(\frac{n}{2}\right)}{\Gamma\left(\frac{n-4}{2}\right)} \prod_{j=1}^{2} \frac{\Gamma\left(\frac{n-j+1}{2}\right)}{\Gamma\left(\frac{n-j-1}{2}\right)}\right\} G^{3,0}_{3,3}\left(\begin{array}{c} \left\{\frac{n-2}{2}, \left\{\frac{n-j-1}{2}\right\}_{j=1:2}\right\} \\ \left\{\frac{n-6}{2}, \left\{\frac{n-j-3}{2}\right\}_{j=1:2}\right\} \end{array} \middle|\, z \right)$$

and

$$F_\Lambda(z) = \left\{\frac{\Gamma\left(\frac{n}{2}\right)}{\Gamma\left(\frac{n-4}{2}\right)} \prod_{j=1}^{2} \frac{\Gamma\left(\frac{n-j+1}{2}\right)}{\Gamma\left(\frac{n-j-1}{2}\right)}\right\} G^{3,1}_{4,4}\left(\begin{array}{c} \left\{1, \frac{n}{2}, \left\{\frac{n-j+1}{2}\right\}_{j=1:2}\right\} \\ \left\{\frac{n-4}{2}, \left\{\frac{n-j-1}{2}\right\}_{j=1:2}, 0\right\} \end{array} \middle|\, z \right).$$

and from (5.323) and (5.324), in terms of the Fox H function, as

$$f_\Lambda(z) = \left\{\frac{\Gamma\left(\frac{n}{2}\right)}{\Gamma\left(\frac{n-4}{2}\right)}\frac{\Gamma(n-1)}{\Gamma(n-3)}\right\} H^{2,0}_{2,2}\left(\begin{array}{c}\left\{\left(\frac{n-2}{2},1\right),(n-3,2)\right\} \\ \left\{\left(\frac{n-6}{2},1\right),(n-5,2)\right\}\end{array}\middle|\, z\right)$$

and

$$F_\Lambda(z) = \left\{\frac{\Gamma\left(\frac{n}{2}\right)}{\Gamma\left(\frac{n-4}{2}\right)}\frac{\Gamma(n-1)}{\Gamma(n-3)}\right\} H^{2,1}_{3,3}\left(\begin{array}{c}\left\{(1,1),\left(\frac{n}{2},1\right),(n-1,2)\right\} \\ \left\{\left(\frac{n-4}{2},1\right),(n-3,2),(0,1)\right\}\end{array}\middle|\, z\right).$$

The present test is invariant under any linear transformations within each one of the two subsets $\underline{X}_1$ and $\underline{X}_2$ that use the same nonrandom full-rank 2×2 compound symmetric or circular matrix, that is, for any linear transformation $\underline{X} \to C^*\underline{X}+\underline{b}^*$, as long as $\underline{b}^*$ is any 4×1 real nonrandom vector and $C^* = bdiag(C, C)$, where the matrix C is any nonrandom 2×2 full-rank compound symmetric or circular matrix.

Also for this test we make available modules `PDFSymSphEq`, `CDFSymSphEq`, `PvalSymSphEq`, `QuantSymSphEq`, and `PvalDataSymSphEq`, with exactly the same arguments as the modules in the previous subsection, except for the second non-optional argument for the first four of these modules, which for the modules in the previous subsection would indicate the number of variables and which now is not present.

To illustrate the application of this test we use the data in Table 5.8 of Johnson and Wichern (2014) on five types of overtime hours for 12 pay periods of about half a year for the Madison, Wisconsin, police department. The data file `police.dat` mimics exactly that table, and we are interested in using the "legal appearance," "extraordinary event," "holdover," and "compensatory overtime allowed" hours to test if we may consider that all these four overtime hours have the same variance and at the same time we may also consider that the first two have an independent distribution from the last two and yet that the covariance between "legal appearance" and "extraordinary event" hours is equal to that of "holdover" and "compensatory overtime allowed" hours. The output in Fig. 5.69, where a p-value of 0.053767 was obtained, indicates that there is some evidence against such an hypothesis, but that evidence may be not strong enough to lead us to clearly reject it.

Interestingly enough the multivariate normality tests indicate that the multivariate normality assumption should not be rejected for the set of "holdover" and "compensatory overtime allowed" hours, but these tests give somewhat low p-values for the set of the other two overtime hours, thus indicating that we should also be somewhat careful with our decisions. In fact, since our initial assumption is that the whole set of four variables has a multivariate Normal distribution, one should indeed pay more attention to the results of the multivariate normality tests for the whole set of four variables. These show rather low p-values for five of the tests.

A second example uses the data in Table 3.9.7 of Timm (2002) on reaction times recorded for 11 subjects for the response to a question which asked for the word

```
PvalDataSymSphEq["<Path>/Police.dat", , 1, , , , , norm → 1]
Original variables {1, 2, 3, 4}
```

Combined sample of X1 and X2

	Statistic	P-Value
Anderson-Darling	2.43257	0.765775
Cramér-von Mises	2.54047	0.818355
Jarque-Bera ALM	0.863248	0.0231383
Kolmogorov-Smirnov	2.62282	0.853491
Kuiper	1.75337	0.340118
Mardia Combined	36.2013	0.0364971
Mardia Kurtosis	−0.388219	0.0464402
Mardia Skewness	28.2154	0.0346
Pearson χ^2	2.30011	0.692079
Shapiro-Wilk	0.667516	0.0000756633
Watson U^2	1.89747	0.431991

Sample of X1

	Statistic	P-Value
Anderson-Darling	0.724626	0.262541
Cramér-von Mises	0.785835	0.308768
Jarque-Bera ALM	0.0211286	0.000223208
Kolmogorov-Smirnov	0.95439	0.45543
Kuiper	0.310007	0.0480522
Mardia Combined	25.0918	0.0264474
Mardia Kurtosis	2.06075	0.0177079
Mardia Skewness	15.4887	0.0298756
Pearson χ^2	0.641049	0.205472
Shapiro-Wilk	0.743257	0.000527535
Watson U^2	0.34425	0.0592541

Sample of X2

	Statistic	P-Value
Anderson-Darling	1.70794	0.957351
Cramér-von Mises	1.75464	0.969899
Jarque-Bera ALM	0.84212	0.354583
Kolmogorov-Smirnov	1.66843	0.945031
Kuiper	1.44336	0.845078
Mardia Combined	2.64516	0.34602
Mardia Kurtosis	−1.23032	0.569208
Mardia Skewness	0.840713	0.344158
Pearson χ^2	1.65906	0.941882
Shapiro-Wilk	0.966388	0.777252
Watson U^2	1.55322	0.900193

```
Computed value of Λ: 0.3135370383
p-value: 0.05376721716
```

Fig. 5.69 Command used and output obtained with the Mathematica® module `PvalDataSymSphEq` to implement the test of symmetrical spherical equivalence, using the overtime hours data in Table 5.8 of Johnson and Wichern (2014)

that in a tape-recorded sentence followed a given "token word" taken from one of five positions in the sentence. One may be interested in using the data referring to positions 1, 2, 4, and 5, that is, referring to the two first and the last two positions in the sentence to test the hypothesis that all four response times have the same variance, with the response times for positions 1 and 2 and those for positions 4 and 5 being independent and the covariance between response times for positions 1 and 2 being the same as that for response times for positions 4 and 5. The data is stored in file `response_times.dat`, from which we extract the time for positions 1, 2, 4, and 5. The result obtained is in Fig. 5.70, showing that there is no enough evidence to reject this null hypothesis, at the same time that there seems also to be no enough evidence against the multivariate normality of the set of four variables.

We may note here that A^* in (5.317) is indeed the m.l.e. of Σ^* in (5.314), but, similar to what happens with the l.r.t. statistics in the two previous subsections, also in the definition of Λ in (5.319) we may replace $|A^*|$ by $|A|$, given the definition of the matrix A^* in (5.317) and the fact that the matrix Γ^* in (5.313) is an orthogonal matrix.

```
PvalDataSymSphEq["<Path>/Response_time.dat", , 1, , , , , , norm → 1]
Original variables {1, 2, 4, 5}
```

Combined sample of X1 and X2

	Statistic	P-Value
Anderson-Darling	2.94029	0.947457
Cramér-von Mises	2.7283	0.891933
Jarque-Bera ALM	2.37918	0.737198
Kolmogorov-Smirnov	2.63231	0.857254
Kuiper	2.13333	0.588133
Mardia Combined	17.4378	0.864792
Mardia Kurtosis	−1.73234	0.759199
Mardia Skewness	10.2089	0.853276
Pearson χ^2	1.88564	0.424238
Shapiro-Wilk	0.840347	0.030378
Watson U^2	1.92684	0.451356

Sample of X1

	Statistic	P-Value
Anderson-Darling	1.25151	0.71988
Cramér-von Mises	1.17526	0.659905
Jarque-Bera ALM	0.991415	0.491452
Kolmogorov-Smirnov	1.04222	0.541328
Kuiper	0.734602	0.26982
Mardia Combined	4.15427	0.840711
Mardia Kurtosis	−1.03761	0.877571
Mardia Skewness	2.01511	0.774821
Pearson χ^2	0.857861	0.367963
Shapiro-Wilk	0.91844	0.303761
Watson U^2	0.760591	0.289249

Sample of X2

	Statistic	P-Value
Anderson-Darling	1.68878	0.951572
Cramér-von Mises	1.55304	0.900112
Jarque-Bera ALM	1.38776	0.812584
Kolmogorov-Smirnov	1.59009	0.915988
Kuiper	1.39872	0.819233
Mardia Combined	2.67124	0.304002
Mardia Kurtosis	−1.47957	0.190446
Mardia Skewness	0.315659	0.0969527
Pearson χ^2	1.02778	0.527395
Shapiro-Wilk	0.946063	0.595244
Watson U^2	1.16625	0.652433

```
Computed value of Λ: 0.3579363777
p-value: 0.3960206770
```

Fig. 5.70 Command used and output obtained with the Mathematica® module `PvalDataSymSphEq` to implement the test of symmetrical spherical equivalence, using the response time data in Table 3.9.7 of Timm (2002)

5.1.18 Testing for "Complete Symmetrical Spherical Equivalence" [CompSymSphEq]

Let us suppose a similar setup to the one in the previous subsection, where now besides testing the null hypothesis (5.306) for the covariance structure one wants to test

$$\begin{aligned} &\text{i) } \mu_1 = \mu_2 \text{ and } \mu_3 = \mu_4 \\ &\text{ii) } \mu_1 = \mu_2 = \mu_3 = \mu_4 = 0 \end{aligned} \qquad (5.325)$$

where the subscripts index the original variables.

Then, for Γ^* in (5.313), we will have the expected value of $\underline{X}^* = \Gamma^* \underline{X}$ equal to

$$\underline{\mu}^* = \begin{bmatrix} \underline{\mu}^*_{1\,(2\times1)} \\ \hline \underline{\mu}^*_{2\,(2\times1)} \end{bmatrix} = \frac{1}{\sqrt{2}} \begin{bmatrix} \mu_1 + \mu_2 \\ \mu_3 + \mu_4 \\ \hline \mu_1 - \mu_2 \\ \mu_3 - \mu_4 \end{bmatrix}. \qquad (5.326)$$

As such, testing for

$$H_0 : \Sigma \text{ has the structure in (5.306) and } \mu_1 = \mu_2 \wedge \mu_3 = \mu_4 \tag{5.327}$$

will be the same as testing

$$H_{0c|b,a} \circ H_{0b|a} \circ H_{0a}$$

where $H_{0b|a}$ and H_{0a} are the same as in (5.315) and (5.316), and

$$\begin{aligned} H_{0c|b,a} : \underline{\mu}_2^* = \underline{0}_{(2\times 1)} & \text{ (for } \underline{\mu}_2^* \text{in (5.326))} \\ & \text{assuming } \Sigma_1^* \text{ and } \Sigma_2^* \text{ are spherical} \\ & \text{and } \Sigma^* = bdiag(\Sigma_1^*, \Sigma_2^*). \end{aligned} \tag{5.328}$$

The $(2/n)$-th power of the l.r.t. statistic to test the null hypothesis in (5.328) is (see Coelho et al. (2016, Sec. 7)),

$$\Lambda_{c|b,a} = \left(\frac{tr(A_{22}^*)}{tr(X_2^* X_2^{*\prime})} \right)^2$$

where A_{22}^* is the second 2×2 diagonal block of the matrix A^* in (5.317), for Γ^* given by (5.313), and where X_2^* is the $2\times n$ lower submatrix of

$$\underset{(4\times n)}{X^*} = \Gamma^* X = \left[\begin{array}{c} X_1^* \\ \hline X_2^* \end{array} \right]$$

where X is the $4\times n$ matrix of the original sample, with

$$\Lambda_{c|b,a} \stackrel{\text{d}}{\equiv} (Y^*)^2$$

where

$$Y^* \sim Beta(n-1, 1)\,. \tag{5.329}$$

The $(2/n)$-th power of the l.r.t. statistic to test H_0 in (5.327) is thus

$$\begin{aligned} \Lambda = \Lambda_a \, \Lambda_{b|a} \, \Lambda_{c|b,a} &= 2^4 \, \frac{|A^*|}{|A_{11}^*| \, |A_{22}^*|} \, \frac{|A_{11}^*|}{\left(tr(A_{11}^*)\right)^2} \, \frac{|A_{22}^*|}{\left(tr(A_{22}^*)\right)^2} \left(\frac{tr(A_{22}^*)}{tr(X_2^* X_2^{*\prime})} \right)^2 \\ &= 2^4 \, \frac{|A^*|}{\left(tr(A_{11}^*) \, tr(X_2^* X_2^{*\prime})\right)^2} \end{aligned} \tag{5.330}$$

with

$$\Lambda \stackrel{\text{d}}{\equiv} \left\{\prod_{j=1}^{3} Y_j\right\} Y\,(Y^*)^2 \tag{5.331}$$

where Y_j and Y are the same r.v.'s as those in (5.320) and Y^* is the r.v. in (5.329), and where $|A^*|$ may be replaced by $|A|$.

As such, the exact distribution of Λ in (5.330) is given by Theorem 3.2 with

$$m^* = 4\,,\quad k_\nu = \{2,1,1,2\}\,,\ a_\nu = \left\{\frac{n-2}{2},\frac{n}{2},\frac{n}{2},\frac{n+1}{2}\right\},\ n_\nu = \{1,1,1,1\}\,,$$
$$m_\nu = \{2,1,1,1\}\,,\quad (\nu = 1,\dots,4)\,,$$

and as such the exact p.d.f. and c.d.f. of Λ are given by Corollary 4.2 in terms of the EGIG p.d.f. and c.d.f. as

$$f_\Lambda(z) = f^{EGIG}\left(z\,\middle|\,\{r_j\}_{j=1:4};\left\{\frac{n-j}{2}\right\}_{j=1:4};4\right)$$

and

$$F_\Lambda(z) = F^{EGIG}\left(z\,\middle|\,\{r_j\}_{j=1:4};\left\{\frac{n-j}{2}\right\}_{j=1:4};4\right),$$

with

$$r_j = \begin{cases} 1 & j=1,3,4 \\ 2 & j=2\,. \end{cases}$$

The exact p.d.f. and c.d.f. of Λ may also, of course, be represented through the use of Meijer G or Fox H functions, but to this end the form of the distribution given by (5.331) is not the most convenient one. Besides this representation of the distribution of Λ there are indeed a number of other equivalent representations through products of independent Beta r.v.'s, some of which yield directly more convenient representations of the p.d.f. and c.d.f. of Λ through Meijer G or Fox H functions, as it is the case of the representation

$$\Lambda \stackrel{\text{d}}{\equiv} \left\{\prod_{j=1}^{2} Y_j^*\right\} Y$$

where Y still is the same r.v. in (5.320) and where

$$Y_j^* \sim Beta\left(\frac{n-2-j}{2}, 2\right), \quad j = 1, 2, \tag{5.332}$$

which directly yields the p.d.f. and c.d.f. given by

$$f_\Lambda(z) = \left\{ \frac{\Gamma\left(\frac{n}{2}\right)}{\Gamma\left(\frac{n-2}{2}\right)} \prod_{j=1}^{2} \frac{\Gamma\left(\frac{n-j+2}{2}\right)}{\Gamma\left(\frac{n-j-2}{2}\right)} \right\} G_{3,3}^{3,0}\left(\left. \begin{array}{l} \left\{\frac{n}{2}, \left\{\frac{n-j+2}{2}\right\}_{j=1:2}\right\} \\ \left\{\frac{n-2}{2}, \left\{\frac{n-j-2}{2}\right\}_{j=1:2}\right\} \end{array} \right| z \right)$$

and

$$F_\Lambda(z) = \left\{ \frac{\Gamma\left(\frac{n}{2}\right)}{\Gamma\left(\frac{n-2}{2}\right)} \prod_{j=1}^{2} \frac{\Gamma\left(\frac{n-j+2}{2}\right)}{\Gamma\left(\frac{n-j-2}{2}\right)} \right\} G_{4,4}^{3,1}\left(\left. \begin{array}{l} \left\{1, \frac{n}{2}, \left\{\frac{n-j+2}{2}\right\}_{j=1:2}\right\} \\ \left\{\frac{n-2}{2}, \left\{\frac{n-j-2}{2}\right\}_{j=1:2}, 0\right\} \end{array} \right| z \right).$$

Instead one might consider the following alternative representation:

$$\Lambda \stackrel{d}{\equiv} \left\{ \prod_{j=1}^{2} \left(Y_j^{**}\right)^2 \right\} Y_3^{**}$$

where

$$Y_j^{**} \sim Beta\,(n-2j+1, j)\,,\ j = 1, 2 \quad \text{and} \quad Y_3^{**} \sim Beta\left(\frac{n-4}{2}, 2\right), \tag{5.333}$$

which directly yields the p.d.f. and c.d.f. of Λ given in terms of Fox H function as

$$f_\Lambda(z) = \left\{ \frac{\Gamma\left(\frac{n}{2}\right)}{\Gamma\left(\frac{n-4}{2}\right)} \prod_{j=1}^{2} \frac{\Gamma\,(n-j+1)}{\Gamma\,(n-2j+1)} \right\} H_{3,3}^{3,0}\left(\left. \begin{array}{l} \left\{\left(\frac{n-2}{2}, 1\right), \{(n-j-1, 2)\}_{j=1:2}\right\} \\ \left\{\left(\frac{n-6}{2}, 1\right), \{(n-2j-1, 2)\}_{j=1:2}\right\} \end{array} \right| z \right)$$

and

$$F_\Lambda(z) = \left\{ \frac{\Gamma\left(\frac{n}{2}\right)}{\Gamma\left(\frac{n-4}{2}\right)} \prod_{j=1}^{2} \frac{\Gamma\,(n-j+1)}{\Gamma\,(n-2j+1)} \right\}$$
$$\times H_{4,4}^{3,1}\left(\left. \begin{array}{l} \left\{(1,1), \left(\frac{n}{2}, 1\right), \{(n-j+1, 2)\}_{j=1:2}\right\} \\ \left\{\left(\frac{n-4}{2}, 1\right), \{(n-2j+1, 2)\}_{j=1:2}, (0,1)\right\} \end{array} \right| z \right).$$

The present test is invariant under any linear transformations within each one of the two subsets $\underline{X}_1$ and $\underline{X}_2$ that use the same full-rank 2×2 compound symmetric matrix and similar vectors in each subset, that is, for any linear transformation $\underline{X} \to C^*\underline{X} + \underline{b}^*$, as long as $\underline{b}^* = [\underline{b}_1', \underline{b}_2']'$ and $C^* = bdiag(C, C)$, where $\underline{b}_1 = aE_{21}$ and $\underline{b}_2 = bE_{21}$, with a and b any reals, and where the matrix C is any nonrandom 2×2 full-rank compound symmetric or circular matrix.

To test

$$H_0 : \Sigma \text{ has the structure in (5.306) and } \mu_1 = \mu_2 = \mu_3 = \mu_4 = 0 \tag{5.334}$$

is the same as to test

$$H_{0c|b,a} \circ H_{0b|a} \circ H_{0a}$$

where $H_{0b|a}$ and H_{0a} are the same as in (5.315) and (5.316), and

$$\begin{aligned} H_{0c|b,a} : \underline{\mu}_2^* = \underline{0}_{(2\times 1)} \wedge \underline{\mu}_1^* = \underline{0}_{(2\times 1)} & \text{ (for } \underline{\mu}_1^* \text{ and } \underline{\mu}_2^* \text{ in (5.326))} \\ & \text{assuming } \Sigma_1^* \text{ and } \Sigma_2^* \text{ are spherical} \\ & \text{and } \Sigma^* = bdiag(\Sigma_1^*, \Sigma_2^*). \end{aligned} \tag{5.335}$$

The $(2/n)$-th power of the l.r.t. statistic to test the null hypothesis in (5.335) is (see Coelho et al. (2016, Sec. 7)),

$$\Lambda_{c|b,a} = \left(\frac{tr(A_{11}^*)}{tr(X_1^* X_1^{*\prime})} \frac{tr(A_{22}^*)}{tr(X_2^* X_2^{*\prime})} \right)^2$$

with

$$\Lambda_{c|b,a} \stackrel{d}{\equiv} (Y_1^* \, Y_2^*)^2 \tag{5.336}$$

where, given the independence of A_{11}^* and A_{22}^*, Y_1^* and Y_2^* are two independent r.v.'s, each one with the same distribution as Y^* in (5.329).

The $(2/n)$-th power of the l.r.t. statistic to test H_0 in (5.334) is thus

$$\begin{aligned} \Lambda &= \Lambda_a \, \Lambda_{b|a} \, \Lambda_{c|b,a} \\ &= 2^4 \frac{|A^*|}{|A_1^*|, \| A_2^*|} \frac{|A_1^*|}{\left(tr(A_1^*)\right)^2} \frac{|A_2^*|}{\left(tr(A_2^*)\right)^2} \left(\frac{tr(A_1^*)}{tr(X_1^* X_1^{*\prime})} \frac{tr(A_2^*)}{tr(X_2^* X_2^{*\prime})} \right)^2 \\ &= 2^4 \frac{|A^*|}{\left(tr(X_1^* X_1^{*\prime})\, tr(X_2^* X_2^{*\prime})\right)^2} \end{aligned} \tag{5.337}$$

with

$$\Lambda \stackrel{d}{\equiv} \left\{ \prod_{j=1}^{3} Y_j \right\} Y \, (Y_1^* \, Y_2^*)^2$$

where Y_j and Y are the same r.v.'s as those in (5.320) and Y_1^* and Y_2^* are the r.v.'s in (5.336), and where, as it happened with the statistic Λ in (5.330), $|A^*|$ may be replaced by $|A|$.

As such, the exact distribution of Λ in (5.337) is given by Theorem 3.2 with

$$m^* = 5\,, \;\; k_\nu = \{2,1,1,2,2\}\;, \;\; a_\nu = \left\{ \frac{n-2}{2}, \frac{n}{2}, \frac{n}{2}, \frac{n+1}{2}, \frac{n+1}{2} \right\},$$
$$n_\nu = \{1,1,1,1,1\}\;, \quad m_\nu = \{2,1,1,1,1\}\;, \quad (\nu = 1, \ldots, 5)\,,$$

and the exact p.d.f. and c.d.f. of Λ are given by Corollary 4.2 in terms of the EGIG p.d.f. and c.d.f. as

$$f_\Lambda(z) = f^{EGIG}\left(z \,\middle|\, \{r_j\}_{j=1:4}; \left\{ \frac{n-j}{2} \right\}_{j=1:4}; 4 \right)$$

and

$$F_\Lambda(z) = F^{EGIG}\left(z \,\middle|\, \{r_j\}_{j=1:4}; \left\{ \frac{n-j}{2} \right\}_{j=1:4}; 4 \right),$$

with

$$r_j = \begin{cases} 2 & j = 1,2 \\ 1 & j = 3,4\,. \end{cases}$$

If one wants to write the p.d.f. and c.d.f. of Λ in (5.337) in terms of the Meijer G function, then it is better to use the representation

$$\Lambda \stackrel{d}{\equiv} \left\{ \prod_{j=1}^{2} Y_j^* \right\} Y \left\{ \prod_{j=1}^{2} Y_j^{***} \right\}$$

where all r.v.'s are independent, with the distribution of Y_j^* $(j = 1,2)$ given by (5.332), that of Y given by (5.320) and with

$$Y_j^{***} \sim Beta\left(\frac{n+1-j}{2}, \frac{1}{2} \right), \quad j = 1,2\,,$$

which yields the p.d.f. and c.d.f. of Λ given by

$$f_\Lambda(z) = \left\{ \frac{\Gamma\left(\frac{n}{2}\right)}{\Gamma\left(\frac{n-2}{2}\right)} \left\{ \prod_{j=1}^{2} \frac{\Gamma\left(\frac{n+2-j}{2}\right)}{\Gamma\left(\frac{n-2-j}{2}\right)} \right\} \left\{ \prod_{j=1}^{2} \frac{\Gamma\left(\frac{n+2-j}{2}\right)}{\Gamma\left(\frac{n+1-j}{2}\right)} \right\} \right\} \times G^{5,0}_{5,5}\left(\begin{array}{c} \left\{\frac{n-2}{2}, \left\{\frac{n-j}{2}\right\}_{j=1:2}, \left\{\frac{n-j}{2}\right\}_{j=1:2}\right\} \\ \left\{\frac{n-4}{2}, \left\{\frac{n-4-j}{2}\right\}_{j=1:2}, \left\{\frac{n-1-j}{2}\right\}_{j=1:2}\right\} \end{array} \middle| \, z \right)$$

and

$$F_\Lambda(z) = \left\{ \frac{\Gamma\left(\frac{n}{2}\right)}{\Gamma\left(\frac{n-2}{2}\right)} \left\{ \prod_{j=1}^{2} \frac{\Gamma\left(\frac{n+2-j}{2}\right)}{\Gamma\left(\frac{n-2-j}{2}\right)} \right\} \left\{ \prod_{j=1}^{2} \frac{\Gamma\left(\frac{n+2-j}{2}\right)}{\Gamma\left(\frac{n+1-j}{2}\right)} \right\} \right\} \times G^{5,1}_{6,6}\left(\begin{array}{c} \left\{1, \frac{n}{2}, \left\{\frac{n+2-j}{2}\right\}_{j=1:2}, \left\{\frac{n+2-j}{2}\right\}_{j=1:2}\right\} \\ \left\{\frac{n-2}{2}, \left\{\frac{n-2-j}{2}\right\}_{j=1:2}, \left\{\frac{n+1-j}{2}\right\}_{j=1:2}, 0\right\} \end{array} \middle| \, z \right).$$

Alternatively one may write the distribution of Λ in (5.337) as

$$\Lambda \stackrel{d}{\equiv} \left\{ \prod_{j=1}^{2} \left(Y_j^{**}\right)^2 \right\} Y_3^{**} \left(Y_4^{**}\right)^2$$

where the distributions of Y_j^{**} $(j = 1, 2)$ and Y_3^{**} are given in (5.333) and

$$Y_4^{**} \sim Beta(n-1, 1)\,,$$

yielding the p.d.f. and c.d.f. of Λ given by

$$f_\Lambda(z) = \left\{ \frac{\Gamma\left(\frac{n}{2}\right)}{\Gamma\left(\frac{n-4}{2}\right)} \frac{\Gamma(n)}{\Gamma(n-1)} \prod_{j=1}^{2} \frac{\Gamma(n-j+1)}{\Gamma(n-2j+1)} \right\} \times H^{4,0}_{4,4}\left(\begin{array}{c} \left\{\left(\frac{n-2}{2}, 1\right), (n-2, 2), \{(n-j-1, 2)\}_{j=1:2}\right\} \\ \left\{\left(\frac{n-6}{2}, 1\right), (n-3, 2), \{(n-2j-1, 2)\}_{j=1:2}\right\} \end{array} \middle| \, z \right)$$

and

$$F_\Lambda(z) = \left\{ \frac{\Gamma\left(\frac{n}{2}\right)}{\Gamma\left(\frac{n-4}{2}\right)} \frac{\Gamma(n)}{\Gamma(n-1)} \prod_{j=1}^{2} \frac{\Gamma(n-j+1)}{\Gamma(n-2j+1)} \right\} \times H_{5,5}^{4,1}\left(\begin{array}{c} \{(1,1), \left(\frac{n}{2},1\right), (n,2), \{(n-j+1,2)\}_{j=1:2}\} \\ \left\{\left(\frac{n-4}{2},1\right), (n-1,2), \{(n-2j+1,2)\}_{j=1:2}, (0,1)\right\} \end{array} \middle| z \right).$$

This test is invariant under any linear transformations $\underline{X} \to bdiag(C, C)\underline{X}$, with $C = aI_2$, where a is any non-null real.

For the implementation of the first test in this subsection the authors make available modules `PDFCompSymSphEq`, `CDFCompSymSphEq`, `PvalCompSymSphEq`, `QuantCompSymSphEq` and `PvalDataCompSymSphEq` and, for the implementation of the second test, modules `PDFCompSymSphEq0`, `CDFCompSymSphEq0`, `PvalCompSymSphEq0`, `QuantCompSymSphEq0`, and `PvalDataCompSymSphEq0`. These modules have exactly the same arguments as the corresponding modules in the previous subsection.

As an example of application of the first test in this subsection we will use the data on the response times described and used in the previous subsection to test the structure on the covariance matrix for the response times for the first, second, fourth and fifth positions in the sentence at the same time that we want to test if we can consider that the mean times for the first and second positions are the same and also the response times for the fourth and fifth positions are also the same. The result is in Fig. 5.71 and it shows that there is quite some evidence that we should reject this null hypothesis, which given the result obtained in the previous subsection shows that its rejection is certainly due to the fact that the mean values for the response times for the first and second and for the fourth and fifth positions are not equal. Indeed, the sample means for these four variables, rounded to four decimal places, are

$$[36.0909, 25.5455, 27.2727, 30.7273].$$

```
PvalDataCompSymSphEq[path <> "Response_time.dat", , 1]
 Original variables {1, 2, 4, 5}
Computed value of Λ: 0.06570979656
p-value: 0.01109938277
```

Fig. 5.71 Command used and output obtained with the Mathematica® module `PvalDataCompSymSphEq` to implement the test of complete symmetrical spherical equivalence, using the response time data in Table 3.9.7 of Timm (2002)

5.1.19 Testing the "Symmetrical Equivalence" of Two Sets of Complex Random Variables [SymEqC]

All deductions and explanations in Sect. 5.1.15 are applicable in the case where the vector $\underline{X} = \left[\underline{X}_1', \underline{X}_2'\right]'$ is assumed to have a p-variate complex Normal distribution (see Sect. 5.1.5 for the definition of the complex multivariate Normal distribution being used). Then, for a sample of size n, the $(1/n)$-th power of the l.r. statistic used to test the "symmetrical equivalence" between $\underline{X}_1$ and $\underline{X}_2$ will have a similar formulation to the one in (5.287), with the only difference being that in the computation of the matrix A one will have now to use the transpose of the complex conjugate of the data matrix where in the real case one uses the transpose. In this setting we have

$$\Lambda \stackrel{d}{\equiv} \prod_{j=1}^{p} Y_j\,,$$

where, for a sample of size n $(> 2p)$, for $j = 1, \ldots, p$,

$$Y_j \sim Beta\,(n - p - j,\, p)$$

form a set of p independent r.v.'s, so that for any p the exact distribution of Λ is given by Theorem 3.2 and its exact p.d.f. and c.d.f. given by Corollary 4.2 with

$$m^* = 1\,, \quad k_1 = 1\,, \quad a_1 = n - p\,, \quad n_1 = p\,, \quad m_1 = p\,.$$

The exact p.d.f. and c.d.f. of Λ are in this case given, for any p, in terms of the EGIG distribution representation, respectively by

$$f_\Lambda(z) = f^{EGIG}\left(z \,\middle|\, \{r_j\}_{j=1:2p-1};\, \{n-1-j\}_{j=1:2p-1}\,;\, 2p-1\right)$$

and

$$F_\Lambda(z) = F^{EGIG}\left(z \,\middle|\, \{r_j\}_{j=1:2p-1};\, \{n-1-j\}_{j=1:2p-1}\,;\, 2p-1\right),$$

where

$$r_j = p - |p - j|\,, \quad j = 1, \ldots, 2p-1\,,$$

and in terms of the Meijer G function by

$$f_\Lambda(z) = \left\{\prod_{j=1}^{p} \frac{\Gamma\,(n-j)}{\Gamma\,(n-p-j)}\right\} G^{p,0}_{p,p}\left(\begin{array}{c} \{n-1-j\}_{j=1:p} \\ \{n-p-j-1\}_{j=1:p} \end{array} \,\middle|\, z \right)$$

and

$$F_\Lambda(z) = \left\{\prod_{j=1}^{p} \frac{\Gamma(n-j)}{\Gamma(n-p-j)}\right\} G^{p,1}_{p+1,p+1}\left(\begin{array}{c} \left\{1, \{n-j\}_{j=1:p}\right\} \\ \left\{\{n-p-j\}_{j=1:p}, 0\right\} \end{array} \middle| \, z \right).$$

Modules PDFSymEqC, CDFSymEqC, PvalSymEqC, QuantSymEqC, and PvalDataSymEqC with exactly the same arguments as the corresponding modules in Sect. 5.1.15 are available for the implementation of this test.

5.1.20 Testing the "Complete Symmetrical Equivalence" of Two Sets of Complex Random Variables [CompSymEqC]

Let us suppose a similar setup to the one in Sects. 5.1.15 and 5.1.16, where the hypothesis to be tested is similar to that in (5.288), but where now the r.v.'s have a joint $2p$-multivariate complex Normal distribution (see Sect. 5.1.5 for the definition of the complex multivariate Normal distribution being used).

Then all deductions and derivations in Sect. 5.1.16 still hold, with our objective of testing the null hypothesis in (5.288) still being equivalent to test the null hypothesis in (5.289), and with, for a sample of size n $(> 2p)$, the $(1/n)$-th power of the l.r.t. statistic having the formulation in (5.294), with the only change being, similar to what happens in the previous subsection, that in the computation of the matrix A one has to use the transpose of the complex conjugate of the data matrix where in the real case one uses just the transpose. Then, from Sects. 5.1.10 and 5.1.19, we have, for $n > 2p$,

$$\Lambda_a \stackrel{d}{\equiv} \prod_{j=1}^{p} Y_j \quad \text{with} \quad Y_j \sim Beta(n-p-j, p) \tag{5.338}$$

forming a set of p independent r.v.'s, and, from Sect. 5.1.8, or more precisely, Sect. 5.1.8.1,

$$\Lambda_b \stackrel{d}{\equiv} Y \quad \text{with} \quad Y \sim Beta(n-p, p),$$

so that

$$\Lambda \stackrel{d}{\equiv} \prod_{j=1}^{p+1} Y_j \quad \text{with} \quad Y_j \sim Beta(n-p-j+1, p),$$

forming a set of $p+1$ independent r.v.'s, or, equivalently,

$$\Lambda \stackrel{d}{\equiv} \prod_{j=1}^{p} Y_j^* \quad \text{with} \quad Y_j^* \sim Beta(n-p-j, p+1)\,,$$

forming a set of p independent r.v.'s.

As such, for any p, the exact distribution of Λ is given by Theorem 3.2, with

$$m^* = 1\,, \quad k_1 = 1\,, \quad a_1 = n-p+1\,, \quad n_1 = p+1\,, \quad m_1 = p\,,$$

or

$$m^* = 1\,, \quad k_1 = 1\,, \quad a_1 = n-p\,, \quad n_1 = p\,, \quad m_1 = p+1\,.$$

The exact p.d.f. and c.d.f. of Λ are thus given by Corollary 4.2, in terms of the EGIG p.d.f. and c.d.f. as

$$f_\Lambda(z) = f^{EGIG}\left(z \,\Big|\, \{r_j\}_{j=1:2p}; \{n-j\}_{j=1:2p}; 2p\right)$$

and

$$F_\Lambda(z) = F^{EGIG}\left(z \,\Big|\, \{r_j\}_{j=1:2p}; \{n-j\}_{j=1:2p}; 2p\right),$$

where

$$r_j = \begin{cases} j\,, & j = 1, \ldots, p \\ p - |p-j+1|\,, & j = p+1, \ldots, 2p\,, \end{cases}$$

or, in terms of the Meijer G function, by

$$f_\Lambda(z) = \left\{ \prod_{j=1}^{p} \frac{\Gamma\,(n-j+1)}{\Gamma\,(n-p-j)} \right\} G_{p,p}^{p,0}\left(\begin{matrix} \{n-j\}_{j=1:p} \\ \{n-p-j-1\}_{j=1:p} \end{matrix} \,\middle|\, z \right)$$

and

$$F_\Lambda(z) = \left\{ \prod_{j=1}^{p} \frac{\Gamma\,(n-j+1)}{\Gamma\,(n-p-j)} \right\} G_{p+1,p+1}^{p,1}\left(\begin{matrix} \left\{1, \{n-j+1\}_{j=1:p}\right\} \\ \left\{\{n-p-j\}_{j=1:p}, 0\right\} \end{matrix} \,\middle|\, z \right).$$

Modules `PDFCompSymEqC`, `CDFCompSymEqC`, `PvalCompSymEqC`, `QuantCompSymEqC`, and `PvalDataCompSymEqC` are available for the implementation of the present test and they have exactly the same arguments as the corresponding modules in Sects. 5.1.15 and 5.1.16.

If our objective is to test a null hypothesis similar to that in (5.301), which is, as it was in Sect. 5.1.16, equivalent to testing the null hypothesis in (5.302), then the $(1/n)$-th power of the l.r.t. statistic to test these null hypotheses is a statistic with a similar formulation to that of the statistic Λ in (5.304), now with Λ_a distributed as in (5.338) above and

$$\Lambda_{b|a} \stackrel{d}{\equiv} Y_1^* Y_2^* \quad \text{with} \quad Y_j^* \sim Beta(n-p, p)$$

being two independent r.v.'s.

The exact distribution of Λ is thus in this case the same as that of

$$\left\{ \prod_{j=1}^{p+1} Y_j \right\} Y$$

with

$$Y_j \sim Beta(n-p-j+1, p) \ (j = 1, \dots, p+1) \quad \text{and} \quad Y \sim Beta(n-p, p)\,,$$

so that the exact distribution of Λ is given by Theorem 3.2 with

$$m^* = 2\,, \quad k_\nu = \{1, 1\}\,, \quad a_\nu = \{n-p+1, n-p+1\}\,, \quad n_\nu = \{p+1, 1\}\,, \\ m_\nu = \{p, p\}\,, \quad (\nu = 1, 2)\,,$$

or

$$m^* = 2\,, \quad k_\nu = \{1, 1\}\,, \quad a_\nu = \{n-p, n-p+1\}\,, \quad n_\nu = \{p, 1\}\,, \\ m_\nu = \{p+1, p\}\,, \quad (\nu = 1, 2)\,,$$

and thus its exact p.d.f. and c.d.f. are given by Corollary 4.2 as

$$f_\Lambda(z) = f^{EGIG}\left(z \,\middle|\, \{r_j\}_{j=1:2p}; \{n-j\}_{j=1:2p}; 2p\right)$$

and

$$F_\Lambda(z) = F^{EGIG}\left(z \,\middle|\, \{r_j\}_{j=1:2p}; \{n-j\}_{j=1:2p}; 2p\right),$$

where

$$r_j = \begin{cases} j+1\,, & j = 1, \dots, p \\ p - |p-j+1|\,, & j = p+1, \dots, 2p\,, \end{cases}$$

or, in terms of the Meijer G function by

$$f_\Lambda(z) = \left\{ \frac{\Gamma(n)}{\Gamma(n-p)} \prod_{j=1}^{p} \frac{\Gamma(n-j+1)}{\Gamma(n-p-j)} \right\} \times G^{\,p+1,0}_{\,p+1,p+1}\left(\begin{array}{c} \left\{n-1, \left\{n-j\right\}_{j=1:p+1}\right\} \\ \left\{n-p-1, \left\{n-p-j\right\}_{j=1:p+1}\right\} \end{array} \middle| \, z \right)$$

and

$$F_\Lambda(z) = \left\{ \frac{\Gamma(n)}{\Gamma(n-p)} \prod_{j=1}^{p} \frac{\Gamma(n-j+1)}{\Gamma(n-p-j)} \right\} \times G^{\,p+1,1}_{\,p+2,p+2}\left(\begin{array}{c} \left\{1, n, \left\{n-j+1\right\}_{j=1:p+1}\right\} \\ \left\{n-p, \left\{n-p-j+1\right\}_{j=1:p+1}, 0\right\} \end{array} \middle| \, z \right).$$

For this second test modules `PDFCompSymEqC0`, `CDFCompSymEqC0`, `PvalCompSymEqC0`, `QuantCompSymEqC0`, and `PvalDataCompSymEqC0` are available, with exactly the same arguments as the corresponding modules in Sects. 5.1.15 and 5.1.16.

5.1.21 *The Likelihood Ratio Statistic to Test Scalar Block Sphericity for Blocks of Two Variables [BSSph]*

Let us suppose that $\underline{X} \sim N_p(\mu, \Sigma)$ and that we want to test the hypothesis that Σ has a scalar block-spherical structure with groups of two variables. That is, that we want to test the null hypothesis

$$H_0 : \Sigma = \begin{bmatrix} \sigma_1^2 & 0 & & & & & & \\ 0 & \sigma_1^2 & & & & 0 & & \\ & & \ddots & & & & & \\ & & & \sigma_k^2 & 0 & & & \\ & & & 0 & \sigma_k^2 & & & \\ & & & & & \ddots & & \\ & & 0 & & & & \sigma_m^2 & 0 \\ & & & & & & 0 & \sigma_m^2 \end{bmatrix} \tag{5.339}$$

or, equivalently

$$H_0 : \Sigma = bdiag\left(\sigma_k^2 I_2,\, k = 1, \dots, m\right), \tag{5.340}$$

for $m = p/2$, which may be a hypothesis that one wants to test for example for the error structure of a 2^m factorial mixed model.

Then, as stated in Coelho and Marques (2009), one may split the null hypothesis in (5.339) or (5.340) as

$$H_0 \equiv H_{0b|a} \circ H_{0a}$$

where

$$H_{0a} : \Sigma = bdiag(\Sigma_k,\, k = 1, \dots, m)$$

is the null hypothesis of independence of m groups of two variables, where Σ_k ($k = 1, \dots, m$) are 2×2 symmetric positive-definite matrices, and

$$\begin{array}{c} H_{0b|a} : \Sigma_k = \sigma_k^2 I_2\ , k = 1, \dots, m \\ \text{(assuming } H_{0a}\text{)} \end{array} \tag{5.341}$$

is the composition of the m hypothesis of sphericity of the m matrices Σ_k, once having assumed H_{0a}.

Then, from Sect. 5.1.11, for a sample of size n, the $(2/n)$-th power of the l.r.t. statistic to test H_{0a} is

$$\Lambda_a = \frac{|A|}{\prod_{k=1}^m |A_{kk}|}$$

where A is the m.l.e. of Σ and A_{kk} is its k-th diagonal block of dimensions 2×2 ($k = 1, \dots, m$), and the $(2/n)$-th power of the l.r.t. statistic to test $H_{0b|a}$ is

$$\Lambda_{b|a} = \prod_{k=1}^m \Lambda_{bk|a} = \prod_{k=1}^m \frac{|A_{kk}|}{\left(tr\left(\frac{A_{kk}}{2}\right)\right)^2} = 2^{2m} \prod_{k=1}^m \frac{|A_{kk}|}{(tr\,(A_{kk}))^2},$$

where

$$\Lambda_{bk|a} = \frac{|A_{kk}|}{\left(tr\left(\frac{A_{kk}}{2}\right)\right)^2} \tag{5.342}$$

is the $(2/n)$-th power of the l.r. statistic to test the sphericity, that is, the diagonality and equality of diagonal elements, of Σ_k. We should note here that under H_{0a} the m

statistics $\Lambda_{bk|a}$ are independent, given the independence, under this null hypothesis, of the matrices A_{kk}.

Then, from Lemma 10.3.1 in Anderson (2003), the $(2/n)$-th power of the l.r.t. statistic to test H_0 in (5.339) or (5.340) is

$$\Lambda = \Lambda_a\, \Lambda_{b|a} = 2^{2m} \frac{|A|}{\left(\prod_{k=1}^{m} tr(A_{kk})\right)^2} \tag{5.343}$$

with

$$\Lambda_a \stackrel{\text{d}}{\equiv} \prod_{k=1}^{m-1} \prod_{j=1}^{2} Y_{jk}$$

where, for $n > 2m$,

$$Y_{jk} \sim Beta\left(\frac{n-2(m-k)-j}{2}, \frac{2(m-k)}{2}\right) \equiv Beta\left(\frac{n-j}{2} - (m-k), m-k\right)$$

form a set of $2(m-1)$ independent r.v.'s, and with

$$\Lambda_{b|a} \stackrel{\text{d}}{\equiv} \prod_{k=1}^{m} Y_k^*$$

where

$$Y_k^* \sim Beta\left(\frac{n-2}{2}, 1\right)$$

form a set of m independent r.v.'s, since for each statistic $\Lambda_{bk|a}$ in (5.342) we have

$$\Lambda_{bk|a} \stackrel{\text{d}}{\equiv} Beta\left(\frac{n-2}{2}, 1\right),$$

as it is shown in Appendix 16.

But then the exact distribution of Λ_a is given by Theorem 3.2, and its p.d.f. and c.d.f. by Corollary 4.2 with

$$m^* = m-1\,, \quad k_\nu = 2\,, \quad a_\nu = \frac{n}{2} - (m-\nu)\,, \quad n_\nu = 1\,, \quad m_\nu = 2(m-\nu)\,,$$

for $\nu = 1, \ldots, m-1$, and the exact distribution of $\Lambda_{b|a}$ is also given by Theorem 3.2, and its p.d.f. and c.d.f. by Corollary 4.2 with

$$m^* = m\,, \quad k_\nu = 1\,, \quad a_\nu = \frac{n}{2}, \quad n_\nu = 1\,, \quad m_\nu = 1\,, \quad \nu = 1, \ldots, m\,,$$

and, given the independence of Λ_a and $\Lambda_{b|a}$, which derives from the independence of Λ_a and the matrices A_{kk} $(k = 1, \ldots, m)$, we may say that

$$\Lambda \stackrel{d}{\equiv} \left\{ \prod_{k=1}^{m-1} \prod_{j=1}^{2} Y_{jk} \right\} \left\{ \prod_{k=1}^{m} Y_k^* \right\},$$

where all Beta distributed r.v.'s are independent, and that, as such, its distribution is given by Theorem 3.2 and its exact p.d.f. and c.d.f. are given by Corollary 4.2 with

$$m^* = 2m-1\,,$$

$$k_\nu = \begin{cases} 2\,,\ \nu = 1,\ldots,m-1 \\ 1\,,\ \nu = m,\ldots,2m-1\,, \end{cases} \qquad a_\nu = \begin{cases} \frac{n}{2} - (m-\nu)\,,\ \nu = 1,\ldots,m-1 \\ \frac{n}{2}\,, \qquad\qquad \nu = m,\ldots,2m-1\,, \end{cases}$$

$$n_\nu = 1, \nu = 1,\ldots,2m-1\,, \quad m_\nu = \begin{cases} 2(m-\nu)\,,\ \nu = 1,\ldots,m-1 \\ 1\,, \qquad\qquad \nu = m,\ldots,2m-1\,. \end{cases}$$

The exact p.d.f. and c.d.f. of Λ in (5.343) are thus given, in terms of the EGIG distribution representation, respectively by

$$f_\Lambda(z) = f^{EGIG}\left(z \,\middle|\, \{r_j\}_{j=1:2m-1};\ \left\{ \frac{n-1-j}{2} \right\}_{j=1:2m-1};\ 2m-1 \right)$$

and

$$F_\Lambda(z) = F^{EGIG}\left(z \,\middle|\, \{r_j\}_{j=1:2m-1};\ \left\{ \frac{n-1-j}{2} \right\}_{j=1:2m-1};\ 2m-1 \right),$$

where

$$r_j = \left\lfloor \frac{2m-j+1}{2} \right\rfloor, \qquad j = 1,\ldots,2m-1\,.$$

In terms of the Meijer G function, the p.d.f. and c.d.f. of Λ in (5.343) are given by

$$f_\Lambda(z) = \left\{ \prod_{k=1}^{m-1} \prod_{j=1}^{2} \frac{\Gamma\left(\frac{n-j}{2}\right)}{\Gamma\left(\frac{n-j}{2} - (m-k)\right)} \right\} \left\{ \prod_{k=1}^{m} \frac{\Gamma\left(\frac{n}{2}\right)}{\Gamma\left(\frac{n-2}{2}\right)} \right\}$$
$$\times\, G^{3m-2,0}_{3m-2,3m-2}\left(\begin{array}{c} \left\{ \left\{ \frac{n-j}{2} - 1 \right\}_{\substack{k=1:m-1\\ j=1:2}}, \left\{ \frac{n}{2} - 1 \right\}_{k=1:m} \right\} \\ \left\{ \left\{ \frac{n-j}{2} - (m-k) - 1 \right\}_{\substack{k=1:m-1\\ j=1:2}}, \left\{ \frac{n-2}{2} - 1 \right\}_{k=1:m} \right\} \end{array} \middle|\, z \right)$$

and

$$F_\Lambda(z) = \left\{ \prod_{k=1}^{m-1} \prod_{j=1}^{2} \frac{\Gamma\left(\frac{n-j}{2}\right)}{\Gamma\left(\frac{n-j}{2} - (m-k)\right)} \right\} \left\{ \prod_{k=1}^{m} \frac{\Gamma\left(\frac{n}{2}\right)}{\Gamma\left(\frac{n-2}{2}\right)} \right\}$$
$$\times G^{3m-2,1}_{3m-1,3m-1} \left(\begin{array}{c} \left\{ 1, \left\{ \frac{n-j}{2} \right\}_{\substack{k=1:m-1\\ j=1:2}}, \left\{ \frac{n}{2} \right\}_{k=1:m} \right\} \\ \left\{ \left\{ \frac{n-j}{2} - (m-k) \right\}_{\substack{k=1:m-1\\ j=1:2}}, \left\{ \frac{n-2}{2} \right\}_{k=1:m}, 0 \right\} \end{array} \middle| \, z \right).$$

Modules PDFBSSph, CDFBSSph, PvalBSSph, QuantBSSph, and PvalDataBSSph are available to implement this test. The plots in Figs 5.72, 5.73, and 5.74 were made using the first two of these modules and they illustrate a number of p.d.f.'s and c.d.f.'s of Λ for different values of m and n. While in Figs. 5.72 and 5.73 the idea is to see how the changes in n affect the shape of the p.d.f.'s and c.d.f.'s, in Fig. 5.74 we may see how the evolution in m affects the shape of the p.d.f.'s and c.d.f.'s.

The first three of those modules have three mandatory arguments which are the sample size n, the number of groups m and the running or computed value of the statistic, in this order, while the fourth module also has three mandatory arguments which are the sample size n, the number of groups m and the value of α for which the quantile is sought. These four modules have exactly the same optional arguments as the corresponding modules in Sect. 5.1.1. Module PvalDataBSSph has a single mandatory argument which is the name of the data file, with its optional arguments being exactly the same as those of the corresponding module in Sect. 5.1.1.

To exemplify the application of the present test we will use the data set on overtime hours from the police department of Madison, Wisconsin, used in Sect. 5.1.17, depicted in Table 5.8 of Johnson and Wichern (2014), to test if the population covariance matrix of "legal appearance," "meeting," "extraordinary event" and

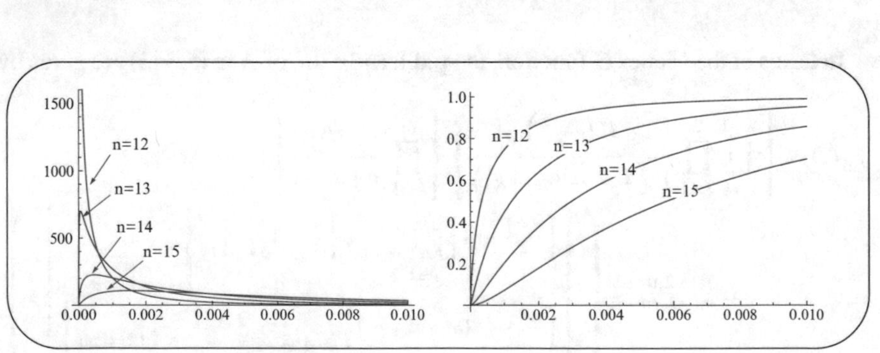

Fig. 5.72 Plots of p.d.f.'s and c.d.f.'s of Λ in (5.343) for $m = 5$ and $n = 12, 13, 14, 15$

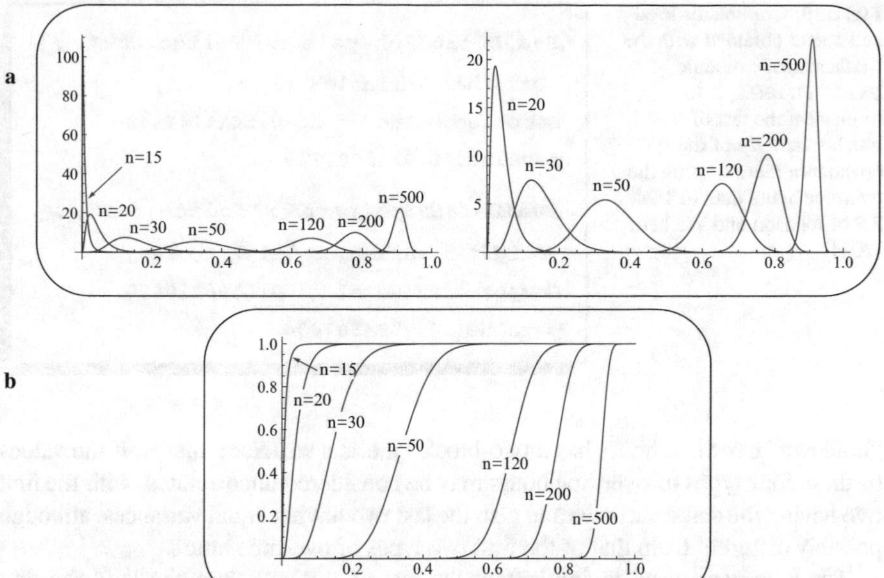

Fig. 5.73 Plots of p.d.f.'s and c.d.f.'s of Λ in (5.343) for $m = 5$ and $n = 15, 20, 30, 50, 120, 200$ and 500: (**a**) two plots of p.d.f.'s, the first one with and the second one without the p.d.f. for $n = 15$ (the plots have different vertical scales); (**b**) the corresponding c.d.f.'s

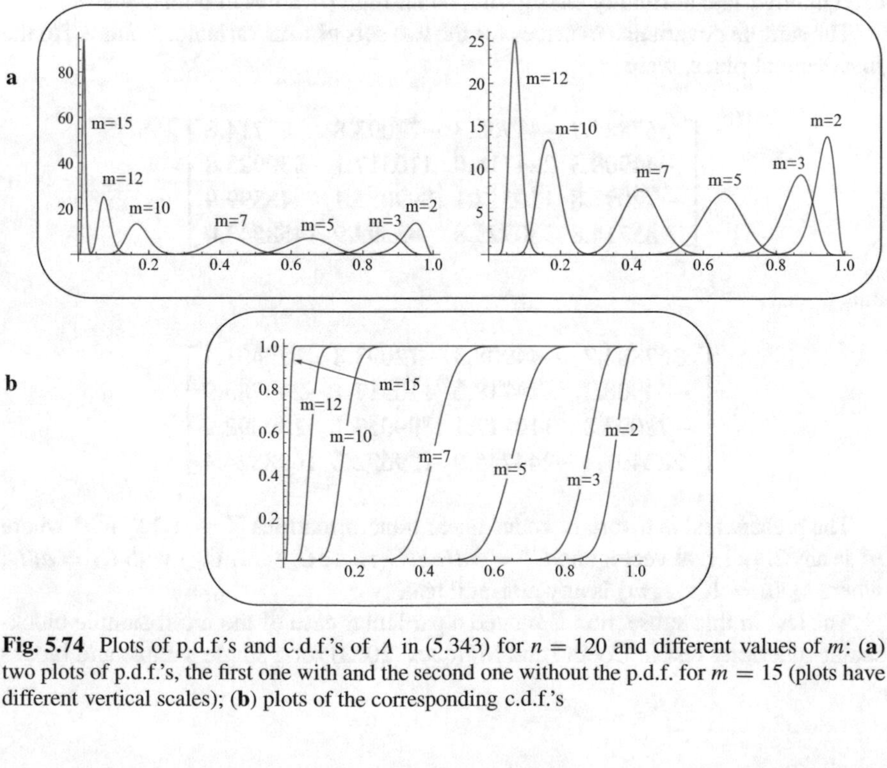

Fig. 5.74 Plots of p.d.f.'s and c.d.f.'s of Λ in (5.343) for $n = 120$ and different values of m: (**a**) two plots of p.d.f.'s, the first one with and the second one without the p.d.f. for $m = 15$ (plots have different vertical scales); (**b**) plots of the corresponding c.d.f.'s

```
PvalDataBSSph[path <> "Police.dat", 1]
 Original variables {1, 5, 2, 3}
Computed value of Λ: 0.5416442942
p-value: 0.4242887919
PvalDataBSSph[path <> "Police.dat", 1]
 Original variables {1, 5, 2, 4}
Computed value of Λ: 0.6596110490
p-value: 0.7034587434
```

Fig. 5.75 Commands used and output obtained with the Mathematica® module `PvalDataBSSph` to implement the test of block-sphericity of the covariance matrix, using the overtime hours data in Tables 5.8 of Johnson and Wichern (2014)

"holdover" overtime hours has a two-block spherical structure, that is, if the values of these four types of overtime hours may be considered uncorrelated, with the first two having the same variance and also the last two having equal variances, although possibly different from that of the first two types of overtime hours.

The command used to implement the test of this null hypothesis is the first command in Fig. 5.75. The large p-value obtained leads us to not reject the above null hypothesis. A similar hypothesis was also tested for a first group formed by the same two variables that formed the above first group, and a second group being now formed by the "extraordinary event" and "compensatory overtime allowed" hours. The result was similar, with an even larger p-value, with most of the more commonly used multivariate normality tests giving rather high p-values in both cases.

The sample covariance matrices for the two sets of four variables, rounded to the first decimal place, were

$$\begin{bmatrix} 367884.7 & -44908.3 & -72093.8 & 85714.8 \\ -44908.3 & 224718.0 & 110517.1 & 330923.8 \\ -72093.8 & 110517.1 & 1399053.1 & 43399.9 \\ 85714.8 & 330923.8 & 43399.9 & 1458543.0 \end{bmatrix}$$

and

$$\begin{bmatrix} 367884.7 & -44908.3 & -72093.8 & 222491.4 \\ -44908.3 & 224718.0 & 110517.1 & -244785.9 \\ -72093.8 & 110517.1 & 1399053.1 & 139692.2 \\ 222491.4 & -244785.9 & 139692.2 & 1698324.4 \end{bmatrix}.$$

The present test is invariant under linear transformations $\underline{X} \to C^*\underline{X} + \underline{b}^*$ where $\underline{b}^*$ is any $2m\times 1$ real vector and $C^* = bdiag(C_1, \ldots, C_k, \ldots, C_m)$ with $C_k = a_k I_2$, where a_k $(k = 1, \ldots, m)$ is any non-null real.

The test in this subsection is indeed a particular case of the multi-sample block-scalar sphericity test in Coelho and Marques (2013) for a single sample and blocks all of dimension 2×2. As such, for blocks of different dimensions or for blocks of dimensions larger than 2×2, the near-exact distributions in the above reference may be used.

5.1.22 *The Likelihood Ratio Test for Equality of Mean Vectors, Under the Assumption of Circularity of the Covariance Matrices [EqMeanVecCirc]*

Let us assume that for $k = 1, \ldots, q$, we have $\underline{X}_k \sim N_p(\underline{\mu}_k, \Sigma_k)$, where all matrices Σ_k are assumed to be equal and circular or circulant.

A $p \times p$ matrix Σ_{cp} is said to be circular or circulant if for $i, j = 1, \ldots, p$ and $k = 0, \ldots, p-1$,

$$\Sigma_{cp} = [\sigma_{ij}], \quad \text{with} \quad \sigma_{i,i+k} = \sigma_{i+k,i} = Cov(X_i, X_{i+k}) = \sigma_0^2 \rho_k\,, \tag{5.344}$$

where $\rho_0 = Corr(X_i, X_i) = 1$ and for $i = 1, \ldots, p-1$ and $k = 1, \ldots, p-i$, $\rho_k = Corr(X_i, X_{i+k}) = Corr(X_i, X_{i+p-k}) = \rho_{p-k}$. For example, for $p = 6$ and $p = 7$ we have respectively,

$$\Sigma_{c6} = \sigma_0^2 \begin{bmatrix} 1 & \rho_1 & \rho_2 & \rho_3 & \rho_2 & \rho_1 \\ \rho_1 & 1 & \rho_1 & \rho_2 & \rho_3 & \rho_2 \\ \rho_2 & \rho_1 & 1 & \rho_1 & \rho_2 & \rho_3 \\ \rho_3 & \rho_2 & \rho_1 & 1 & \rho_1 & \rho_2 \\ \rho_2 & \rho_3 & \rho_2 & \rho_1 & 1 & \rho_1 \\ \rho_1 & \rho_2 & \rho_3 & \rho_2 & \rho_1 & 1 \end{bmatrix}, \qquad \Sigma_{c7} = \sigma_0^2 \begin{bmatrix} 1 & \rho_1 & \rho_2 & \rho_3 & \rho_3 & \rho_2 & \rho_1 \\ \rho_1 & 1 & \rho_1 & \rho_2 & \rho_3 & \rho_3 & \rho_2 \\ \rho_2 & \rho_1 & 1 & \rho_1 & \rho_2 & \rho_3 & \rho_3 \\ \rho_3 & \rho_2 & \rho_1 & 1 & \rho_1 & \rho_2 & \rho_3 \\ \rho_3 & \rho_3 & \rho_2 & \rho_1 & 1 & \rho_1 & \rho_2 \\ \rho_2 & \rho_3 & \rho_3 & \rho_2 & \rho_1 & 1 & \rho_1 \\ \rho_1 & \rho_2 & \rho_3 & \rho_3 & \rho_2 & \rho_1 & 1 \end{bmatrix}.$$

See Sect. 5.2.1 for the l.r. test for circularity of a covariance matrix.

Let us then suppose that for $k = 1, \ldots, q$, we have a sample of size n_k from $\underline{X}_k$ ($k = 1, \ldots, q$), with $n = \sum_{k=1}^{q} n_k$, and that these q samples are independent.

Then the $(2/n)$-th power of the l.r.t. statistic to test the null hypothesis

$$\begin{aligned} &H_0 : \underline{\mu}_1 = \ldots \underline{\mu}_q \\ &\qquad \text{assuming } \Sigma_1 = \cdots = \Sigma_q (= \Sigma_{cp} \text{ non-specified})\,, \end{aligned} \tag{5.345}$$

where Σ_{cp} represents a circular matrix of order p, may be obtained in a number of different ways. One of them, which may indeed be the easiest one, is exactly from the l.r.t. statistic in Sect. 5.1.1. The l.r.t. statistic to test H_0 in (5.345) will be initially similar to the l.r.t. statistic in (5.2), to which we have to add the fact that now the matrices Σ_k are assumed to be circular. This means that we will need to be able to obtain the m.l.e.'s of a circular or circulant Σ_{cp} matrix both under H_0 in (5.345) as well as under the alternative hypothesis

$$\begin{aligned} H_1 : \exists j, j' \in \{1, \dots, q\} : \underline{\mu}_j \neq \underline{\mu}_{j'}, \\ \text{assuming } \Sigma_1 = \cdots = \Sigma_q (= \Sigma_{cp} \text{ non-specified}) . \end{aligned} \tag{5.346}$$

Well, it happens that from the definition of a circular or circulant matrix in (5.344) we see that any $p \times p$ circular or circulant matrix Σ_{cp} may be written as

$$\Sigma_{cp} = \sigma_0^2 \sum_{k=0}^{p-1} \rho_k W_k \tag{5.347}$$

where $\rho_0 = 1$, $\rho_k = \rho_{p-k}$, and

$$W_k = \begin{bmatrix} \underset{(p-k)\times k}{0} & I_{p-k} \\ I_k & \underset{k\times(p-k)}{0} \end{bmatrix},$$

with I_p an identity matrix of order p, I_0 an empty matrix, and $W_0 = I_p$. But it happens that if Σ_{cp} is a circular or circulant matrix, then also Σ_{cp}^{-1} is a circular or circulant matrix (Lv and Huang, 2007). As such Σ_{cp}^{-1} may also be written in the form (5.347), with ρ_k replaced by an appropriate linear combination of $\rho_0, \dots, \rho_p$. But then, on one hand, from (5.347) we may write

$$\widehat{\Sigma_{cp}} = \sum_{k=0}^{p-1} \widehat{\sigma_0^2 \rho_k}\, W_k ,$$

while, on the other hand, from the results in Szatrowski (1978, 1980), since the matrices W_k in (5.347) are not a function of the elements in Σ_{cp} but only a function of its dimensions, we can be assured that the m.l.e.'s of $\sigma_0^2 \rho_k$, both under H_0 in (5.345), as well as under H_1 in (5.346), may be obtained by averaging the "right" elements in the respective unstructured or usual m.l.e. of the covariance matrix. All we have to do is to figure out which and how many are the elements in this unstructured m.l.e. that correspond to a given ρ_k.

If we take into account which are the elements in Σ_{cp} that are equal to $\sigma_0^2 \rho_k$, for $k = 0, 1, \dots, \lfloor p/2 \rfloor$, and that there are $2p$ of each of these elements, except for $k = 0$ and, when p is even, also for $k = \lfloor p/2 \rfloor$, in which cases there are p of these elements, then, from the results in Szatrowski (1978, 1980), the m.l.e. of Σ_{cp}, under the alternative hypothesis in (5.346) is $A^* = [a^*_{ij}]$, with

$$a^*_{j+k,j} = a^*_{j,j+k} = \widehat{\sigma_0^2 \rho_k}\Big|_{H_1} = \frac{1}{2p} \sum_{\ell=1}^{p} \left(a_{\ell, mod^*(\ell+k,p)} + a_{mod^*(\ell+k,p),\ell} \right) , \tag{5.348}$$

for $k = 0, \ldots, p-1$ and $j = 1, \ldots, p-k$, where a_{ij} represents the (i,j)-th element of the matrix A in (5.3), and

$$mod^*(a,b) = \begin{cases} mod(a,b)\,, & mod(a,b) \neq 0 \\ b\,, & mod(a,b) = 0\,, \end{cases} \tag{5.349}$$

while the m.l.e. of Σ_{cp} under the null hypothesis in (5.345) is $C^* = [c^*_{ij}]$, with

$$c^*_{j+k,j} = c^*_{j,j+k} = \widehat{\sigma_0^2 \rho_k}\big|_{H_0} = \frac{1}{2p}\sum_{\ell=1}^{p}\Big(a_{\ell,mod^*(\ell+k,p)} + b_{\ell,mod^*(\ell+k,p)} + a_{mod^*(\ell+k,p),\ell} + b_{mod^*(\ell+k,p),\ell}\Big)\,, \tag{5.350}$$

also for $k = 0, \ldots, p-1$ and $j = 1, \ldots, p-k$, where a_{ij} and b_{ij} represent the elements in row i and column j of the matrices A and B in (5.3).

Then, the $(2/n)$-th power of the l.r.t. statistic to test H_0 in (5.345) may be written as

$$\Lambda = \frac{|A^*|}{|C^*|}\,. \tag{5.351}$$

However, this form of the statistic is not only not easy to compute as well as it is a form from which it is not easy to obtain its distribution.

But, it is possible to show that we may write, for $m = \lfloor p/2 \rfloor$,

$$\begin{aligned} |A^*| &= a^{**}_{11}\left(a^{**}_{1+m,1+m}\right)^{mod(p+1,2)} \prod_{j=2}^{p-m}\left(\frac{a^{**}_{jj} + a^{**}_{p-j+2,p-j+2}}{2}\right)^2 \\ &= v^*_1\left(v^*_{1+m,1+m}\right)^{mod(p+1,2)} \prod_{j=2}^{p-m} v^*_j \end{aligned} \tag{5.352}$$

and

$$\begin{aligned} |C^*| &= c^{**}_{11}\left(c^{**}_{1+m,1+m}\right)^{mod(p+1,2)} \prod_{j=2}^{p-m}\left(\frac{c^{**}_{jj} + c^{**}_{p-j+2,p-j+2}}{2}\right)^2 \\ &= v^{**}_1\left(v^{**}_{1+m,1+m}\right)^{mod(p+1,2)} \prod_{j=2}^{p-m} v^{**}_j \end{aligned} \tag{5.353}$$

where,

$$v_j^* = \begin{cases} a_{jj}^{**}\,, & j=1 \text{ and } j=m+1 \text{ if } p \text{ is even} \\ \left(a_{jj}^{**}+a_{p-j+2,p-j+2}^{**}\right)/2\,, & j=2,\dots,p-m,m+2,\dots,p\,, \end{cases} \tag{5.354}$$

and

$$v_j^{**} = \begin{cases} c_{jj}^{**}\,, & j=1 \text{ and } j=m+1 \text{ if } p \text{ is even} \\ \left(c_{jj}^{**}+c_{p-j+2,p-j+2}^{**}\right)/2\,, & j=2,\dots,p-m,m+2,\dots,p\,, \end{cases} \tag{5.355}$$

with $v_j^* = v_{p-j+2}^*$ and $v_j^{**} = v_{p-j+2}^{**}$ for $j=2,\dots,p-m$, and where a_{jj}^{**} and c_{jj}^{**} represent the j-th diagonal elements respectively of

$$A^{**} = UAU' \quad \text{and} \quad C^{**} = U(A+B)U'\,, \tag{5.356}$$

where A and B are defined as in (5.3) and the matrix U is an orthogonal symmetric matrix with (i,j)-th element

$$u_{ij} = \frac{1}{\sqrt{p}}\left\{\cos\left(2\pi(i-1)(j-1)/p\right)+\sin\left(2\pi(i-1)(j-1)/p\right)\right\}, \tag{5.357}$$

for $i, j \in \{1,\dots,p\}$.

As such, Λ in (5.351) may be written as

$$\Lambda = \frac{v_1^*}{v_1^{**}}\left(\frac{v_{1+m,1+m}^*}{v_{1+m,1+m}^{**}}\right)^{mod(p+1,2)}\underbrace{\prod_{j=2}^{p-m}\left(\frac{v_j^*}{v_j^{**}}\right)^2}_{\Lambda^*}, \tag{5.358}$$

or, in a more concise form, as

$$\Lambda = \prod_{j=1}^{p}\frac{v_j^*}{v_j^{**}}\,. \tag{5.359}$$

Then, in order to obtain the distribution of the l.r.t. statistic Λ in (5.358) or (5.359), one only has to take into account that

$$U\Sigma_{cp}U' = \Delta = diag(\delta_1,\delta_2,\dots,\delta_p)\,, \tag{5.360}$$

where $\delta_j = \delta_{p-j+2}$, for $j=2,\dots,p-m$, so that, since we know that the matrices A and B are independent, with

$$A \sim W_p(n-q,\Sigma_{cp}) \quad \text{and} \quad B \sim W_p(q-1,\Sigma_{cp})\,, \tag{5.361}$$

we have A^{**} and $B^{**} = UBU'$ independent, with

$$A^{**} \sim W_p(n-q, \Delta) \quad \text{and} \quad B^{**} \sim W_p(q-1, \Delta), \tag{5.362}$$

so that

$$C^{**} = A^{**} + B^{**} \sim W_p(n-1, \Delta). \tag{5.363}$$

As such, the diagonal elements of A^{**} are independent, as well as the diagonal elements of B^{**} and C^{**}, and for $n > q$ we have

$$\frac{a_{jj}^{**}}{\delta_j} \sim \chi^2_{n-q}, \qquad \frac{b_{jj}^{**}}{\delta_j} \sim \chi^2_{q-1}, \qquad \frac{c_{jj}^{**}}{\delta_j} = \frac{a_{jj}^{**}}{\delta_j} + \frac{b_{jj}^{**}}{\delta_j} \sim \chi^2_{n-1}, \tag{5.364}$$

for $j = 1, \ldots, p$.

But then, from (5.358) and (5.364) we see that, for $n > q$,

$$\Lambda \stackrel{d}{\equiv} Y_1 \left(Y^*\right)^{mod(p+1,2)} \underbrace{\prod_{j=2}^{p-m} Y_j^2}_{\Lambda^*} \tag{5.365}$$

where all r.v.'s are independent, with

$$Y_1 \stackrel{d}{\equiv} Y^* \sim Beta\left(\frac{n-q}{2}, \frac{q-1}{2}\right) \qquad \text{and} \qquad Y_j \sim Beta\left(n-q, q-1\right). \tag{5.366}$$

We may note that v_j^* and v_j^{**}, defined in (5.354) and (5.355), are the m.l.e.'s of the eigenvalues δ_j in (5.360), respectively under H_1 in (5.346) and H_0 in (5.345).

The exact distribution of Λ^* in (5.358) and (5.365) is thus given by Theorem 3.2 with

$$m^* = \left\lfloor \frac{p-1}{2} \right\rfloor, \quad n_\nu = 1, \quad m_\nu = q-1, \quad k_\nu = 2, \quad a_\nu = \frac{n-q}{2} + 1,$$

which is better seen if we identify the r.v.'s Y_j in (5.366) with the r.v.'s $Y_{\nu\ell}^{***}$ in (3.10), and as such, it is an EGIG distribution of depth $q-1$, with rate parameters $(n-q+\ell-1)/2$ $(\ell = 1, \ldots, q-1)$ and shape parameters all equal to $\lfloor p/2 \rfloor$. If one wants to accommodate also the distributions of Y_1 and Y^* in this setting, then one has to see the distribution of each one these two r.v.'s as given, for odd q, by Theorem 3.2 with

$$m^* = 1, \; n_1 = 1, \; m_1 = \frac{q-1}{2}, \; k_1 = 1, \; a_1 = \frac{n-q}{2} + 1,$$

their exact distribution being an EGIG distribution of depth $(q-1)/2$, with rate parameters $(n-q)/2+\ell$ $(\ell=1,\dots,(q-1)/2)$ and shape parameters all equal to 1, independent of the distribution of Λ^*.

The exact distribution of Λ in (5.358) and (5.365) is thus given, for odd q, by Theorem 3.2 and its exact p.d.f. and c.d.f. by Corollary 4.2, with $m^*=\left\lfloor\frac{p-1}{2}\right\rfloor+1+mod(p+1,2)$,

$$n_\nu=1\,,\quad a_\nu=\frac{n-q}{2}+1\,,\qquad \nu=1,\dots,m^*,$$

$$m_\nu=\begin{cases} q-1\,, & \nu=1,\dots,m^{**}\\ \frac{q-1}{2}\,, & \nu=m^{**}+1,\dots,m^*\end{cases}\quad \text{and}\quad k_\nu=\begin{cases} 2\,, & \nu=1,\dots,m^{**}\\ 1\,, & \nu=m^{**}+1,\dots,m^*\end{cases},$$

for $m^{**}=\lfloor(p-1)/2\rfloor$, so that the exact distribution of Λ is in this case an EGIG distribution of depth $q-1$ with rate parameters $(n-q+\ell-1)/2$ and shape parameters $\lfloor p/2\rfloor+1-(1+\text{mod}(p+1,2))\text{mod}(\ell-1,2)$ $(\ell=1,\dots,q-1)$, with p.d.f. and c.d.f. given respectively by

$$f_\Lambda(z)=f^{EGIG}\left(z\;\middle|\;\left\{\left\lfloor\frac{p}{2}\right\rfloor+1-(1+mod(p+1,2))mod(\ell-1,2)\right\}_{\ell=1:q-1};\right.$$
$$\left.\left\{\frac{n-q+\ell-1}{2}\right\}_{\ell=1:q-1};q-1\right)$$

and

$$F_\Lambda(z)=F^{EGIG}\left(z\;\middle|\;\left\{\left\lfloor\frac{p}{2}\right\rfloor+1-(1+mod(p+1,2))mod(\ell-1,2)\right\}_{\ell=1:q-1};\right.$$
$$\left.\left\{\frac{n-q+\ell-1}{2}\right\}_{\ell=1:q-1};q-1\right).$$

For any q, the p.d.f. and c.d.f. of Λ in (5.359) may be expressed in terms of Fox's H function, for $m=\lfloor p/2\rfloor$ and $p^*=p-m+mod(p+1,2)$, as

$$f_\Lambda(z)=\left\{\left(\frac{\Gamma\left(\frac{n-1}{2}\right)}{\Gamma\left(\frac{n-q}{2}\right)}\right)^{1+mod(p+1,2)}\prod_{j=2}^{p-m}\frac{\Gamma(n-1)}{\Gamma(n-q)}\right\}$$
$$\times H^{p^*,0}_{p^*,p^*}\left(\begin{array}{c}\left\{\left(\frac{n-3}{2},1\right)\right\}_{j=1:1+mod(p+1,2)},\{(n-3,2)\}_{j=2:p-m}\\ \left\{\left(\frac{n-q-2}{2},1\right)\right\}_{j=1:1+mod(p+1,2)},\{(n-q-2,2)\}_{j=2:p-m}\end{array}\;\middle|\;z\right)$$

and

$$F_{\Lambda}(z) = \left\{ \left(\frac{\Gamma\left(\frac{n-1}{2}\right)}{\Gamma\left(\frac{n-q}{2}\right)} \right)^{1+mod(p+1,2)} \prod_{j=2}^{p-m} \frac{\Gamma(n-1)}{\Gamma(n-q)} \right\}$$

$$\times H^{p^*,1}_{p^*+1,p^*+1} \left(\left. \begin{array}{l} \left\{ (1,1), \left\{ \left(\frac{n-1}{2}, 1 \right) \right\}_{j=1:1+mod(p+1,2)}, \{(n-1,2)\}_{j=2:p-m} \right\} \\ \left\{ \left\{ \left(\frac{n-q}{2}, 1 \right) \right\}_{j=1:1+mod(p+1,2)}, \{(n-q,2)\}_{j=2:p-m}, (0,1) \right\} \end{array} \right| z \right).$$

The test in this subsection is invariant under linear transformations $\underline{X}_k \to C\underline{X}_k + \underline{b}$, where C is any $p\times p$ nonrandom full-rank circular or circulant matrix and $\underline{b}$ is any $p\times 1$ real vector.

For this test modules `PDFEqMeanVecCirc`, `CDFEqMeanVecCirc`, `PvalEqMeanVecCirc`, `QuantEqMeanVecCirc`, and `PvalDataEqMeanVecCirc` are made available. These modules have exactly the same optional and non-optional arguments as the corresponding modules in Sect. 5.1.1, and module `PvalDataEqMeanVecCirc` accepts files with the same structure as the corresponding `PvalData` module in Sect. 5.1.1.

To exemplify the use and usefulness of the present test we use the data on response times for the words following tokens placed at five different locations in a sentence, reported in Tables 3.9.7 and 3.9.8 of Timm (2002) and used in Sects. 5.1.17, 5.1.18 and 5.2.1–5.2.3, as forming three independent samples that we will use to test the equality of the corresponding population mean vectors, accounting for the circularity of the covariance matrices, which, according to the results in Sects. 5.2.1 and 5.2.3, should not be rejected for all three populations (the one corresponding to the sample in Table 3.9.7 and the other two which samples are formed respectively by the low and high short-term memory subjects in Table 3.9.8 of Timm (2002)). The data were stored in the file `Response_times_all.dat`, which has the same structure as the data files used in Sect. 5.1.1.

The command used and the output obtained are shown in Fig. 5.76, where we may see that the test for the equality of covariance matrices gives a rather high near-exact p-value, while the p-value for the test of equality of the mean vectors has a rather low value. The second command in this same figure implements the test of equality of the mean vectors using the test in Sect. 5.1.1, that is, without accounting for the circularity of the covariance matrices. We may see how by taking into account the circular structure of the covariance matrices we are able to profit from a non-negligible gain in power (see Sect. 5.2.3 for the simultaneous test of circularity of the three population covariance matrices involved, with the non-rejection of this null hypothesis).

The material presented up to the end of the paragraph after expression (5.366) may also be found in Sect. 2.1 of Coelho (2018), in a somewhat less detailed way. In this reference, an approach dedicated to this particular test produced, for the case of odd q, equivalent results to those obtained in the present subsection, for

```
PvalDataEqMeanVecCirc[path <> "Response_times_all.dat", , , , , cov → 1]
 There are 3 samples, with sizes {11, 10, 10} and 5 variables
(p-value for the test of equality of covariance matrices: 0.5246128246)
Computed value of Λ: 0.4726507428
p-value: 0.01926051171
PvalDataEqMeanVecR[path <> "Response_times_all.dat"]
 There are 3 samples, with sizes {11, 10, 10} and 5 variables
Computed value of Λ: 0.4910667258
p-value: 0.04826267255
```

Fig. 5.76 Commands used and output obtained with the Mathematica® module `PvalDataEqMeanVecCirc` to implement the test of equality of mean vectors when a circular structure is assumed for the covariance matrices, using the data on response times in Tables 3.9.7 and 3.9.8 of Timm (2002); and comparison with the result obtained when the test that is used assumes no structure for the covariance matrices

the distribution of Λ in (5.359), (5.351) or (5.358), where an all-encompassing approach based on Theorem 3.2 and Corollary 4.2 was used. In this same reference the reader may also find near-exact distributions for Λ for the case of even q.

5.1.23 *The Likelihood Ratio Test for Simultaneous Nullity of Mean Vectors, Under the Assumption of Circularity of the Covariance Matrices [NullMeanVecCirc]*

Let us consider a similar setting to the one in the previous subsection. In this subsection we are concerned with testing the hypothesis

$$\begin{array}{c} H_0 : \underline{\mu}_1 = \ldots \underline{\mu}_q = \underline{0} \\ \text{assuming } \Sigma_1 = \cdots = \Sigma_q (= \Sigma_{cp} \text{ non-specified}) \end{array} \qquad (5.367)$$

where Σ_{cp} represents a circular matrix of order p.

Then, from the results in Szatrowski (1978, 1980), which may be applied given the reasons pointed out in the previous subsection, right after expression (5.346), while the m.l.e. of Σ_{cp} under the alternative hypothesis is the same as it is in the previous subsection, the m.l.e. of Σ_{cp} under the null hypothesis in (5.367) is now $\widetilde{C}^* = [\tilde{c}^*_{ij}]$, with

$$\tilde{c}^*_{j+k,j} = \tilde{c}^*_{j,j+k} = \widehat{\sigma_0^2 \rho_k}\big|_{H_0} = \frac{1}{2p} \sum_{\ell=1}^{p} \Big(a_{\ell,mod^*(\ell+k,p)} + b^*_{\ell,mod^*(\ell+k,p)} + a_{mod^*(\ell+k,p),\ell} + b^*_{mod^*(\ell+k,p),\ell} \Big),$$

for $j = 1, \dots, p-k$ and $k = 0, \dots, p-1$, and for $mod^*(a, b)$ given by (5.349), and where a_{ij} and b^*_{ij} represent respectively the elements in row i and column j of the matrix A in (5.3) and (5.356) and of the matrix

$$B^* = \sum_{k=1}^{q} n_k \underline{\overline{X}}_k \underline{\overline{X}}_k' \,. \tag{5.368}$$

As such, for $n = \sum_{k=1}^{q} n_k$, where n_k is the size of the sample from $\underline{X}_k$, the $(2/n)$-th power of the l.r.t. statistic to test H_0 in (5.367) may be written as

$$\Lambda = \frac{|A^*|}{|\widetilde{C}^*|} \tag{5.369}$$

where A^* is the matrix with elements given by (5.348), with $|A^*|$ given by (5.352), and

$$\begin{aligned} |\widetilde{C}^*| &= c^{***}_{11} \left(c^{***}_{1+m,1+m}\right)^{mod(p+1,2)} \prod_{j=2}^{p-m} \left(\frac{c^{***}_{jj} + c^{***}_{p-j+2,p-j+2}}{2} \right)^2 \\ &= v^{***}_1 \left(v^{***}_{1+m,1+m}\right)^{mod(p+1,2)} \prod_{j=2}^{p-m} v^{***}_j \end{aligned} \tag{5.370}$$

where c^{***}_{jj} is the j-th diagonal element of the matrix

$$C^{***} = U(A + B^*)U', \tag{5.371}$$

with A given by (5.3), B^* given by (5.368), and where U is the symmetric matrix with elements given by (5.357), and where, for $m = \lfloor p/2 \rfloor$, v^*_j are given by (5.354), and

$$v^{***}_j = \begin{cases} c^{***}_{jj} \,, & j = 1 \text{ and } j = m+1 \text{ if } p \text{ is even} \\ \left(c^{***}_{jj} + c^{***}_{p-j+2,p-j+2}\right)/2 \,, & j = 2, \dots, p-m, m+2, \dots, p \,, \end{cases} \tag{5.372}$$

with c^{***}_{jj} representing the j-th diagonal element of the matrix C^{***} in (5.371).

But, hence, Λ in (5.369) may be written as

$$\Lambda = \underbrace{\frac{v^*_1}{v^{***}_1} \left(\frac{v^*_{1+m,1+m}}{v^{***}_{1+m,1+m}} \right)^{mod(p+1,2)} \prod_{j=2}^{p-m} \left(\frac{v^*_j}{v^{***}_j} \right)^2}_{\Lambda^*} , \tag{5.373}$$

or, in a more concise form, as

$$\Lambda = \prod_{j=1}^{p} \frac{v_j^*}{v_j^{***}}\,. \tag{5.374}$$

Then, in order to obtain the distribution of the l.r.t. statistic Λ in (5.369), (5.373), or (5.374), one only has to observe that the matrices A and B^* are independent, with

$$A \sim W_p(n-q, \Sigma_{cp}) \quad \text{and} \quad B^* \sim W_p(q, \Sigma_{cp})\,,$$

so that $A^{**} = UAU'$ and $B^{***} = UB^*U'$ are independent, with

$$A^{**} \sim W_p(n-q, \Delta) \quad \text{and} \quad B^{***} \sim W_p(q, \Delta)\,,$$

for Δ given by (5.360), and thus,

$$C^{***} = A^{**} + B^{***} \sim W_p(n, \Delta)\,.$$

Therefore, given that the matrix Δ is a diagonal matrix, the diagonal elements of A^{**} are independent, as well as the diagonal elements of B^{***} and C^{***}, with, for $n > q$,

$$\frac{a_{jj}^{**}}{\delta_j} \sim \chi^2_{n-q}\,, \qquad \frac{b_{jj}^{***}}{\delta_j} \sim \chi^2_q\,, \qquad \frac{c_{jj}^{***}}{\delta_j} = \frac{a_{jj}^{**}}{\delta_j} + \frac{b_{jj}^{***}}{\delta_j} \sim \chi^2_n\,. \tag{5.375}$$

so that, from (5.373) and (5.375) we see that, for $n > q$,

$$\Lambda \stackrel{d}{\equiv} Y_1 \left(Y^*\right)^{mod(p+1,2)} \underbrace{\prod_{j=2}^{p-m} Y_j^2}_{\Lambda^*} \tag{5.376}$$

where all r.v.'s are independent, with

$$Y_1 \stackrel{d}{\equiv} Y^* \sim Beta\left(\frac{n-q}{2}, \frac{q}{2}\right) \quad \text{and} \quad Y_j \sim Beta\,(n-q, q)\,.$$

We may note that v_j^{***} and v_j^*, defined in (5.354) and (5.372), are the m.l.e.'s of the eigenvalues δ_j in (5.360), respectively under H_0 in (5.367) and under the alternative hypothesis, where the matrices Σ_k are still assumed to be circular.

The exact distribution of Λ^* in (5.373) and (5.376) is thus given by Theorem 3.2, with

$$m^* = \left\lfloor \frac{p-1}{2} \right\rfloor, \quad n_\nu = 1\,, \; m_\nu = q\,, \; k_\nu = 2\,, \; a_\nu = \frac{n-q}{2} + 1\,,$$

so that it is an EGIG distribution of depth q, with rate parameters $(n-q+\ell-1)/2$ ($\ell = 1, \ldots, q$) and shape parameters all equal to $\lfloor p/2 \rfloor$ (see the brief note in Appendix 16), while if one wants to accommodate also the distributions of Y_1 and Y^* in this setting, then one has to see the distribution of these two r.v.'s as given, for even q, by Theorem 3.2 with

$$m^* = 1, \;\; n_1 = 1, \;\; k_1 = 1, \;\; a_1 = \frac{n-q}{2} + 1, \;\; m_1 = \frac{q}{2},$$

their exact distribution being, for even q, an EGIG distribution of depth $(q-1)/2$, with rate parameters $(n-q)/2 + \ell$ ($\ell = 1, \ldots, (q-1)/2$) and shape parameters all equal to 1.

This yields for Λ in (5.373), (5.369), or (5.374), its exact distribution, for even q, given by Theorem 3.2 and Corollary 4.2 with $m^* = \left\lfloor \frac{p-1}{2} \right\rfloor + 1 + mod(p+1, 2)$,

$$n_\nu = 1\,, \quad a_\nu = \frac{n-q}{2} + 1\,, \qquad \nu = 1, \ldots, m^*\,,$$

$$m_\nu = \begin{cases} q\,, \; \nu = 1, \ldots, m^{**} \\ \frac{q}{2}\,, \; \nu = m^{**} + 1, \ldots, m^* \end{cases} \quad \text{and} \quad k_\nu = \begin{cases} 2\,, \; \nu = 1, \ldots, m^{**} \\ 1\,, \; \nu = m^{**} + 1, \ldots, m^* \end{cases},$$

for $m^{**} = \lfloor (p-1)/2 \rfloor$, which is an EGIG distribution of depth q with rate parameters $(n-q+\ell-1)/2$ and shape parameters $\lfloor p/2 \rfloor + 1 - (1 + \text{mod}(p+1, 2))\text{mod}(\ell-1, 2)$ ($\ell = 1, \ldots, q$), with p.d.f. and c.d.f. given respectively by

$$f_\Lambda(z) = f^{EGIG}\left(z \,\middle|\, \Big\{ \lfloor p/2 \rfloor + 1 - (1 + mod(p+1, 2))mod(\ell-1, 2) \Big\}_{\ell=1:q} ; \left\{ \frac{n-q+\ell-1}{2} \right\}_{\ell=1:q} ; q \right)$$

and

$$F_\Lambda(z) = F^{EGIG}\left(z \,\middle|\, \Big\{ \lfloor p/2 \rfloor + 1 - (1 + mod(p+1, 2))mod(\ell-1, 2) \Big\}_{\ell=1:q} ; \left\{ \frac{n-q+\ell-1}{2} \right\}_{\ell=1:q} ; q \right).$$

For any q, the p.d.f. and c.d.f. of Λ in (5.374) may be expressed in terms of Fox's H function, for $m = \lfloor p/2 \rfloor$ and $p^* = p - m + mod(p+1, 2)$, as

$$f_\Lambda(z) = \left\{ \left(\frac{\Gamma\left(\frac{n}{2}\right)}{\Gamma\left(\frac{n-q}{2}\right)} \right)^{1+mod(p+1,2)} \prod_{j=2}^{p-m} \frac{\Gamma(n)}{\Gamma(n-q)} \right\} \times H^{p^*,0}_{p^*,p^*}\left(\begin{array}{c} \left\{ \left\{ \left(\frac{n-2}{2}, 1\right) \right\}_{j=1:1+mod(p+1,2)}, \{(n-2,2)\}_{j=2:p-m} \right\} \\ \left\{ \left\{ \left(\frac{n-q-2}{2}, 1\right) \right\}_{j=1:1+mod(p+1,2)}, \{(n-q-2,2)\}_{j=2:p-m} \right\} \end{array} \middle| \, z \right)$$

and

$$F_\Lambda(z) = \left\{ \left(\frac{\Gamma\left(\frac{n}{2}\right)}{\Gamma\left(\frac{n-q}{2}\right)} \right)^{1+mod(p+1,2)} \prod_{j=2}^{p-m} \frac{\Gamma(n)}{\Gamma(n-q)} \right\} \times H^{p^*,1}_{p^*+1,p^*+1}\left(\begin{array}{c} \left\{ (1,1), \left\{ \left(\frac{n}{2}, 1\right) \right\}_{j=1:1+mod(p+1,2)}, \{(n,2)\}_{j=2:p-m} \right\} \\ \left\{ \left\{ \left(\frac{n-q}{2}, 1\right) \right\}_{j=1:1+mod(p+1,2)}, \{(n-q,2)\}_{j=2:p-m}, (0,1) \right\} \end{array} \middle| \, z \right).$$

This test is invariant under linear transformations $\underline{X}_k \to C\underline{X}_k$ where C is any $p \times p$ nonrandom full-rank circular or circulant matrix.

For this test modules `PDFNullMeanVecCirc`, `CDFNullMeanVecCirc`, `PvalNullMeanVecCirc`, `QuantNullMeanVecCirc`, and `PvalDataNullMeanVecCirc` are made available. All these modules have exactly the same optional and non-optional arguments as the corresponding modules in Sect. 5.1.1, with module `PvalDataNullMeanVecCirc` accepting data files with exactly the same structure as the corresponding `PvalData` module in the previous subsection and in Sect. 5.1.1.

The material presented up to the end of the paragraph after expression (5.376) may also be found in Sect. 2.2 of Coelho (2018), where an approach devoted to the present test produced for the case of even q equivalent results in terms of the distribution of Λ in (5.374), (5.369) or (5.373), to those obtained in the present subsection, where the authors use now a general approach. In this same reference the reader may also find near-exact distributions for Λ in (5.374) for the case of odd q.

5.1.24 The Likelihood Ratio Test for Equality of Mean Vectors, Under the Assumption of Compound Symmetry of the Covariance Matrices [EqMeanVecCS]

Let us assume that $\underline{X}_k \sim N_p(\underline{\mu}_k, \Sigma_k)$, $k = 1, \dots, q$, where the matrices Σ_k are assumed to be equal and compound symmetric.

A $p{\times}p$ positive-definite covariance matrix Σ_{CS} is said to be compound symmetric if, for $-\frac{a}{p-1} < b < a$, it can be written as

$$\Sigma_{CS} = (a-b)I_p + bE_{pp} = aI_p + b(E_{pp} - I_p)\,, \tag{5.377}$$

where I_p represents the identity matrix of order p and E_{pp} a $p{\times}p$ matrix of 1's.

Let us further suppose that we have a sample of size n_k from $\underline{X}_k$ $(k = 1, \dots, q)$ and that these q samples are independent, with $n = \sum_{k=1}^{q} n_k$, and that we want to test the null hypothesis of equality of the q mean vectors $\underline{\mu}_k$ $(k = 1, \dots, q)$, that is, the null hypothesis

$$\begin{aligned} H_0 : \underline{\mu}_1 &= \dots \underline{\mu}_q \\ \text{assuming } \Sigma_1 = \cdots &= \Sigma_q (= \Sigma_{CS} \text{ non-specified})\,, \end{aligned} \tag{5.378}$$

where Σ_{CS} represents a compound symmetric matrix of order p.

Then the $(2/n)$-th power of the l.r.t. statistic to test the null hypothesis in (5.378) is

$$\Lambda = \frac{a_{11}^{**}(a^{**})^{p-1}}{c_{11}^{**}(c^{**})^{p-1}}\,, \tag{5.379}$$

where

$$a^{**} = \frac{1}{p-1}\sum_{j=2}^{p} a_{jj}^{**} \quad \text{and} \quad c^{**} = \frac{1}{p-1}\sum_{j=2}^{p} c_{jj}^{**}\,, \tag{5.380}$$

with a_{jj}^{**} and c_{jj}^{**}, the diagonal elements of the matrices

$$A^{**} = UAU' \quad \text{and} \quad C^{**} = U(A+B)U' \tag{5.381}$$

where U is either the $p{\times}p$ orthogonal matrix with elements given by (5.357), or a Helmert matrix of order p, which is a $p{\times}p$ orthogonal matrix with first row equal to $\frac{1}{\sqrt{p}}E_{1p}$ and i-th row $(i = 2, \dots, p)$ equal to

$$\frac{1}{\sqrt{(i-1)i}}\Big[\underbrace{1, \dots, 1}_{i-1}, -(i-1)\,, \underbrace{0, \dots, 0}_{p-i}\Big] \quad (i = 2, \dots, p)\,.$$

We should note that although the use of a matrix U which is the $p\times p$ orthogonal matrix with elements given by (5.357), or which is a Helmert matrix of order p will yield different matrices A^{**} and different matrices C^{**}, these matrices will anyway have the same first element, that is, the same element in the first row and first column, and will, in either case, yield exactly the same computed value for Λ in (5.379).

As happened in Sect. 5.1.22, there are a number of ways in which the l.r.t. statistic in (5.379) may be obtained, one of them being exactly by establishing a parallel with the l.r.t. statistic in Sect. 5.1.1, and by seeing that the l.r.t. statistic to test H_0 in (5.378) will be initially similar to the l.r.t. statistic in (5.2), to which we have to add the fact that now the matrices Σ_k arc assumed to be compound symmetric.

Indeed since if Σ_{CS} is a compound symmetric matrix, then it may be written as in (5.377), which is of the form of (5.347) if we take $p=2$, $\sigma_0^2\rho_0=a$, $\sigma_0^2\rho_1=b$, $W_0=I_p$ and $W_1=E_{pp}-I_p$. Then also Σ_{CS}^{-1} is of the same form, with a replaced by $(a+(p-2)b)/(a^2+(p-2)ab-(p-1)b^2)$ and b replaced by $b/((p-1)b^2-a^2-(p-2)ab)$. Since the matrices W_0 and W_1 are only functions of the dimensions of the matrix Σ_{CS} and not of its contents, from Szatrowski (1978, 1980) it is then possible to define the m.l.e.'s of a and b, respectively under H_1 and H_0, by averaging the elements in the matrices A and $A+B$ that correspond to the parameters a and b in Σ_{CS}.

If we take into account that both under the alternative and under the null hypothesis all the diagonal elements of Σ_{CS} are equal among themselves and that also all the off-diagonal elements of Σ_{CS} are also equal among themselves, then the m.l.e. of Σ_{CS}, under the alternative hypothesis, is $A^*=[a^*_{ij}]$, with

$$a^*_{ii}=\widehat{a}\big|_{H_1}=\frac{1}{p}\sum_{\ell=1}^{p}a_{\ell\ell}\quad\text{and}\quad a^*_{ij}=\widehat{b}\big|_{H_1}=\frac{1}{p(p-1)}\sum_{\substack{k=1\\k\neq\ell}}^{p}\sum_{\ell=1}^{p}a_{k\ell}\ \ (i\neq j)\,, \tag{5.382}$$

while the m.l.e. of Σ_{CS} under the null hypothesis in (5.378) is $C^*=[c^*_{ij}]$, with

$$c^*_{ii}=\widehat{a}\big|_{H_0}=\frac{1}{p}\sum_{\ell=1}^{p}\left(a_{\ell\ell}+b_{\ell\ell}\right)\ \ \text{and}\ \ c^*_{ij}=\widehat{b}\big|_{H_0}=\frac{1}{p(p-1)}\sum_{\substack{k=1\\k\neq\ell}}^{p}\sum_{\ell=1}^{p}\left(a_{k\ell}+b_{k\ell}\right)\ \ (i\neq j), \tag{5.383}$$

where $a_{k\ell}$ and $b_{k\ell}$ represent the elements in row k and column ℓ of the matrices A and B in (5.3).

Then, the l.r.t. statistic to test H_0 in (5.378) may be written as

$$\Lambda=\frac{|A^*|}{|C^*|} \tag{5.384}$$

where it may be shown that

$$|A^*| = a_{11}^{**} \left(\frac{1}{p-1} \sum_{j=2}^{p} a_{jj}^{**} \right)^{p-1} \tag{5.385}$$

and

$$|C^*| = c_{11}^{**} \left(\frac{1}{p-1} \sum_{j=2}^{p} c_{jj}^{**} \right)^{p-1} \tag{5.386}$$

where, as in (5.379) and (5.380), a_{jj}^{**} and c_{jj}^{**} represent respectively the j-th diagonal element of the matrices A^{**} and C^{**} in (5.381). Thus, Λ in (5.384) may be written as in (5.379).

In order to obtain the distribution of the l.r.t. statistic Λ in (5.384) or (5.379) now one only has to observe that

$$U \Sigma_{CS} U' = \Delta = diag\big(a + (p-1)b, \underbrace{a-b, \dots, a-b}_{p-1}\big), \tag{5.387}$$

so that, since A and B are independent, now with distributions similar to the ones in (5.361), with Σ_{cp} replaced by Σ_{CS}, A^{**} and $B^{**} = UBU'$ are independent, with similar distributions to the ones in (5.362), and, as such, also C^{**} has a similar distribution to that in (5.363), now for Δ in (5.387).

Given the fact that the matrix Δ in (5.387) is a diagonal matrix, the diagonal elements of A^{**} are thus independent, as well as the diagonal elements of B^{**} and C^{**}, and, if $n > q$, with similar distributions to the ones in (5.364), for $\delta_1 = a + (p-1)b$ and $\delta_2 = \cdots = \delta_p = a - b$.

Therefore, from (5.384), (5.385), and (5.386), we see that, for $n > q$,

$$\Lambda \stackrel{d}{\equiv} Y_1 \, (Y_2)^{p-1} \tag{5.388}$$

where Y_1 and Y_2 are independent, with

$$Y_1 \sim Beta\left(\frac{n-q}{2}, \frac{q-1}{2}\right) \quad \text{and} \quad Y_2 \sim Beta\left(\frac{(n-q)(p-1)}{2}, \frac{(q-1)(p-1)}{2}\right). \tag{5.389}$$

The exact distribution of $(Y_2)^{p-1}$ in (5.388) is thus given, for odd q or odd p, by Theorem 3.2 with

$$m^* = 1, \;\; n_1 = 1 \;, \; m_1 = \frac{(q-1)(p-1)}{2}, \;\; k_1 = p-1 \,, \;\; a_1 = \frac{n-q}{2} + 1 \,,$$

which is better seen if we identify the r.v. Y_2 in (5.389) with the r.v.'s $Y_{\nu\ell}^{***}$ in (3.10).

As such, for odd q or odd p, the distribution of $(Y_2)^{p-1}$ is an EGIG distribution of depth $(q-1)(p-1)/2$, with rate parameters $\frac{(n-q)}{2}+\frac{\ell-1}{p-1}$ ($\ell=1,\dots,(q-1)(p-1)/2$) and shape parameters all equal to 1, while if one wants to accommodate also the distribution of Y_1 in this setting, then one has to see the distribution of this r.v. as given, for odd q, by Theorem 3.2 with

$$m^*=1,\ n_1=1,\ k_1=1,\ a_1=\frac{n-q}{2}+1,\ m_1=\frac{q-1}{2},$$

its exact distribution being an EGIG distribution of depth $(q-1)/2$, with rate parameters $(n-q)/2+\ell$ ($\ell=1,\dots,(q-1)/2$) and shape parameters all equal to 1.

This yields, for odd q, the exact distribution of Λ in (5.379) given by Theorem 3.2 and Corollary 4.2 with

$$m^*=2\,,\quad k_\nu=\{1,p-1\}\,,\quad a_\nu=\left\{\frac{n-q}{2}+1,\frac{n-q}{2}+1\right\},\quad n_\nu=\{1,1\}\,,$$

$$m_\nu=\left\{\frac{q-1}{2},\frac{(q-1)(p-1)}{2}\right\},$$

which is an EGIG distribution of depth $(q-1)(p-1)/2$, with rate parameters $\frac{(n-q)}{2}+\frac{\ell-1}{p-1}$ ($\ell=1,\dots,(q-1)(p-1)/2$) and shape parameters

$$r_\ell=\begin{cases}1\,, & \ell=1,\dots,(q-1)(p-1)/2\\ & \ell\neq(j-1)(p-1)+1 \text{ for } j=1,\dots,(q-1)/2\\ 2\,, & \ell=(j-1)(p-1)+1 \text{ for } j=1,\dots,(q-1)/2\,,\end{cases}\tag{5.390}$$

with p.d.f. and c.d.f. given respectively by

$$f_\Lambda(z)=f^{EGIG}\left(z\,\middle|\,\{r_\ell\}_{\ell=1:(q-1)(p-1)/2}\,;\ \left\{\frac{(n-q)}{2}+\frac{\ell-1}{p-1}\right\}_{\ell=1:(q-1)(p-1)/2};\ \frac{(q-1)(p-1)}{2}\right)\tag{5.391}$$

and

$$F_\Lambda(z)=F^{EGIG}\left(z\,\middle|\,\{r_\ell\}_{\ell=1:(q-1)(p-1)/2}\,;\ \left\{\frac{(n-q)}{2}+\frac{\ell-1}{p-1}\right\}_{\ell=1:(q-1)(p-1)/2};\ \frac{(q-1)(p-1)}{2}\right)\tag{5.392}$$

for r_ℓ given by (5.390).

We may note that, indeed as it was supposed to be, for $p = 2$ and $p = 3$, the test in this subsection is equivalent to the test in Sect. 5.1.22, and that while a_{11}^{**} and c_{11}^{**} are the m.l.e.'s of $a + (p-1)b$, respectively under

$$H_1 : \exists j, j' \in \{1, \dots, q\} : \underline{\mu}_j \neq \underline{\mu}_{j'}\,, \\ \text{assuming } \Sigma_1 = \cdots = \Sigma_q (= \Sigma_{CS} \text{ non-specified})\,, \tag{5.393}$$

and under H_0 in (5.378), a^{**} and c^{**} in (5.379) and (5.380) are the m.l.e.'s of $a - b$, respectively under H_1 in (5.393) and under H_0 in (5.378).

For any q and p, the p.d.f. and c.d.f. of Λ in (5.379) may be expressed in terms of Fox's H function as

$$f_\Lambda(z) = \frac{\Gamma\left(\frac{n-1}{2}\right)}{\Gamma\left(\frac{n-q}{2}\right)} \frac{\Gamma\left(\frac{(n-1)(p-1)}{2}\right)}{\Gamma\left(\frac{(n-q)(p-1)}{2}\right)} H^{2,0}_{2,2}\left(\begin{array}{c} \left\{\left(\frac{n-3}{2}, 1\right), \left(\frac{(n-2)(p-1)}{2}, p-1\right)\right\} \\ \left\{\left(\frac{n-q-2}{2}, 1\right), \left(\frac{(n-q-1)(p-1)}{2}, p-1\right)\right\} \end{array} \middle| \, z \right)$$

and

$$F_\Lambda(z) = \frac{\Gamma\left(\frac{n-1}{2}\right)}{\Gamma\left(\frac{n-q}{2}\right)} \frac{\Gamma\left(\frac{(n-1)(p-1)}{2}\right)}{\Gamma\left(\frac{(n-q)(p-1)}{2}\right)} H^{2,1}_{3,3}\left(\begin{array}{c} \left\{(1,1), \left(\frac{n-1}{2}, 1\right), \left(\frac{(n-1)(p-1)}{2}, p-1\right)\right\} \\ \left\{\left(\frac{n-q}{2}, 1\right), \left(\frac{(n-q)(p-1)}{2}, p-1\right), (0,1)\right\} \end{array} \middle| \, z \right).$$

The present test is invariant under linear transformations $\underline{X}_k \to C\underline{X}_k + \underline{b}$, where C is any nonrandom full-rank $p \times p$ compound symmetric matrix and $\underline{b}$ any $p \times 1$ real vector.

Modules `PDFEqMeanVecCS`, `CDFEqMeanVecCS`, `PvalEqMeanVecCS`, `QuantEqMeanVecCS`, and `PvalDataEqMeanVecCS` are available for the implementation of this test, and their optional and non-optional arguments are exactly the same as those of the corresponding modules in Sect. 5.1.1.

In order to illustrate the use of the present test we use once again the data on the response times in Tables 3.9.7 and 3.9.8 in Timm (2002), which were used in Sect. 5.1.22, and exactly the same data file. The command used and the output obtained are in Fig. 5.77.

The p-value obtained, being very similar to the p-value obtained in Fig. 5.76 when module `PvalDataEqMeanVecCirc` was used, that is, when the test carried out was the test for equality of mean vectors when a circular structure is assumed for the covariance matrices, shows that the assumption of compound symmetry of the covariance matrices may also be an acceptable one, the compound symmetry structure being a particular case of the circular structure.

The material presented up to the end of the paragraph that ends with expression (5.389) may also be found in Coelho (2017), where an approach devoted to the present test produced for the case of odd q equivalent results to those obtained in the present subsection, now using a general approach. In this same reference the reader may also find near-exact distributions for Λ in (5.379) for the case of even q.

```
PvalDataEqMeanVecCS[path <> "Response_times_all.dat"]
 There are 3 samples, with sizes {11, 10, 10} and 5 variables
Computed value of Λ: 0.4735703109
p-value: 0.01868752829
```

Fig. 5.77 Command used and output obtained with the Mathematica® module `PvalDataEqMeanVecCS` to implement the test of equality of mean vectors when a compound symmetric structure is assumed for the covariance matrices, using the data on response times in Tables 3.9.7 and 3.9.8 of Timm (2002)

5.1.25 The Likelihood Ratio Test for Simultaneous Nullity of Mean Vectors, Under the Assumption of Compound Symmetry of the Covariance Matrices [NullMeanVecCS]

Let us consider a similar setting to the one in the previous subsection and let us suppose we are interested in testing the hypothesis

$$\begin{array}{l} H_0 : \underline{\mu}_1 = \dots \underline{\mu}_q = \underline{0} \\ \qquad \text{assuming } \Sigma_1 = \cdots = \Sigma_q (= \Sigma_{CS} \text{ non-specified}) \end{array} \tag{5.394}$$

where Σ_{CS} represents a compound symmetric matrix of order p.

Then, for a sample of size n, the $(2/n)$-th power of the l.r.t. statistic to test this null hypothesis is

$$\Lambda = \frac{a_{11}^{**}(a^{**})^{p-1}}{c_{11}^{***}(c^{***})^{p-1}}, \tag{5.395}$$

where a_{11}^{**} and a^{**} are the same as in (5.379) and (5.380) and

$$c^{***} = \frac{1}{p-1}\sum_{j=2}^{p} c_{jj}^{***}, \tag{5.396}$$

with c_{jj}^{***} representing the j-th diagonal element of a matrix C^{***} with a similar definition to that of the matrix C^{***} in (5.371), with B^* being a matrix with a similar definition to that of the matrix B^* in (5.368) but now with the matrix U being either an orthogonal symmetric matrix with elements given by (5.357), or a Helmert matrix of order p (see the previous subsection for the definition of a Helmert matrix and some brief remarks on its use, compared to the use of the matrix with elements given by (5.357)).

The m.l.e. of Σ_{CS} under the alternative hypothesis remains the same as in the previous subsection, while the m.l.e. of Σ_{CS} under H_0 in (5.394) is $\widetilde{C}^* = [\tilde{c}^*_{ij}]$, with

$$\tilde{c}^*_{ii} = \widehat{a}\Big|_{H_0} = \frac{1}{p}\sum_{\ell=1}^{p}\left(a_{\ell\ell}+b^*_{\ell\ell}\right) \quad \text{and} \quad \tilde{c}^*_{ij} = \widehat{b}\Big|_{H_0} = \frac{1}{p(p-1)}\sum_{\substack{k=1\\k\neq\ell}}^{p}\sum_{\ell=1}^{p}\left(a_{k\ell}+b^*_{k\ell}\right)\ (i\neq j),$$

where $a_{k\ell}$ and $b^*_{k\ell}$ represent respectively the elements in row k and column ℓ of the matrices A in (5.3) and B^* in (5.368).

Therefore, the l.r.t. statistic to test H_0 in (5.394) may be written as

$$\Lambda = \frac{|A^*|}{|\widetilde{C}^*|} \tag{5.397}$$

where $|A^*|$ is given by (5.385) and

$$|\widetilde{C}^*| = c^{***}_{11}\left(\frac{1}{p-1}\sum_{j=2}^{p} c^{***}_{jj}\right)^{p-1} \tag{5.398}$$

where c^{***}_{jj} is the j-th diagonal element of the matrix C^{***} in (5.371), so that Λ in (5.397) may be written as in (5.395).

In order to obtain the distribution of the l.r.t. statistic Λ in (5.397) and (5.395), one only has to notice that, for $n > q$, now the matrices A and B^* are independent, with

$$A \sim W_p(n-q, \Sigma_{CS}) \quad \text{and} \quad B^* \sim W_p(q, \Sigma_{CS}),$$

so that A^{**} and $B^{***} = UB^*U'$ are independent, with

$$A^{**} \sim W_p(n-q, \Delta) \quad \text{and} \quad B^{***} \sim W_p(q, \Delta),$$

for Δ given by (5.387), and thus,

$$C^{***} = A^{**} + B^{***} \sim W_p(n, \Delta).$$

Therefore, the diagonal elements of A^{**} are independent, as well as the diagonal elements of B^{***} and C^{***}, with the same distributions as those in (5.375) and thus, from (5.395),

$$\Lambda \stackrel{d}{\equiv} Y_1\,(Y_2)^{p-1} \tag{5.399}$$

where Y_1 and Y_2 are independent, with

$$Y_1 \sim Beta\left(\frac{n-q}{2}, \frac{q}{2}\right) \quad \text{and} \quad Y_2 \sim Beta\left(\frac{(n-q)(p-1)}{2}, \frac{q(p-1)}{2}\right). \tag{5.400}$$

The exact distribution of $(Y_2)^{p-1}$ in (5.399) is thus given, for even q or odd p, by Theorem 3.2 with

$$m^* = 1,\ n_1 = 1\ ,\ m_1 = \frac{q(p-1)}{2}\ ,\ k_1 = p-1\ ,\ a_1 = \frac{n-q}{2} + 1\ ,$$

which is better seen if we identify the r.v. Y_2 in (5.400) with the r.v.'s $Y^{***}_{\nu\ell}$ in (3.10).

As such, for even q or odd p, the distribution of $(Y_2)^{p-1}$ is an EGIG distribution of depth $q(p-1)/2$, with rate parameters $\frac{(n-q)}{2} + \frac{\ell-1}{p-1}$ ($\ell = 1, \ldots, q(p-1)/2$) and shape parameters all equal to 1, while if one wants to accommodate also the distribution of Y_1 in this setting, then one has to see the distribution of this r.v. as given, for even q, by Theorem 3.2 with

$$m^* = 1,\ n_1 = 1,\ k_1 = 1,\ a_1 = \frac{n-q}{2} + 1,\ m_1 = \frac{q}{2},$$

its exact distribution being an EGIG distribution of depth $q/2$, with rate parameters $(n-q)/2 + \ell$ ($\ell = 1, \ldots, q/2$) and shape parameters all equal to 1.

This yields, under H_0 in (5.394), the exact distribution of Λ in (5.395) given by Theorem 3.2 and Corollary 4.2, for even q, with

$$m^* = 2\,, \quad k_\nu = \{1, p-1\}\,, \quad a_\nu = \left\{\frac{n-q}{2} + 1, \frac{n-q}{2} + 1\right\}, \quad n_\nu = \{1, 1\}\,,$$

$$m_\nu = \left\{\frac{q}{2}, \frac{q(p-1)}{2}\right\},$$

which is an EGIG distribution of depth $q(p-1)/2$, with rate parameters $\frac{(n-q)}{2} + \frac{\ell-1}{p-1}$ ($\ell = 1, \ldots, q(p-1)/2$) and shape parameters

$$r_\ell = \begin{cases} 1\,, & \ell = 1, \ldots, q(p-1)/2 \\ & \ell \neq (j-1)(p-1)+1 \ \text{ for } j = 1, \ldots, q/2 \\ 2\,, & \ell = (j-1)(p-1)+1 \ \text{ for } j = 1, \ldots, q/2\,, \end{cases} \tag{5.401}$$

with p.d.f. and c.d.f. given respectively by

$$f_\Lambda(z) = f^{EGIG}\left(z \,\middle|\, \left\{\frac{(n-q)}{2} + \frac{\ell-1}{p-1}\right\}_{\ell=1:\frac{q(p-1)}{2}} ; \{r_\ell\}_{\ell=1:\frac{q(p-1)}{2}} ; \frac{q(p-1)}{2}\right)$$

and

$$F_\Lambda(z) = f^{EGIG}\left(z \,\middle|\, \left\{\frac{(n-q)}{2} + \frac{\ell-1}{p-1}\right\}_{\ell=1:\frac{q(p-1)}{2}} ; \{r_\ell\}_{\ell=1:\frac{q(p-1)}{2}} ; \frac{q(p-1)}{2}\right)$$

for r_ℓ given by (5.401).

For any p and q, the p.d.f. and c.d.f. of Λ in (5.395) may be expressed in terms of Fox's H function as

$$f_\Lambda(z) = \frac{\Gamma\left(\frac{n}{2}\right)}{\Gamma\left(\frac{n-q}{2}\right)} \frac{\Gamma\left(\frac{n(p-1)}{2}\right)}{\Gamma\left(\frac{(n-q)(p-1)}{2}\right)} H^{2,0}_{2,2}\left(\begin{array}{c} \left\{\left(\frac{n-2}{2},1\right),\left(\frac{(n-1)(p-1)}{2},p-1\right)\right\} \\ \left\{\left(\frac{n-q-2}{2},1\right),\left(\frac{(n-q-1)(p-1)}{2},p-1\right)\right\} \end{array}\middle|\, z\right)$$

and

$$F_\Lambda(z) = \frac{\Gamma\left(\frac{n}{2}\right)}{\Gamma\left(\frac{n-q}{2}\right)} \frac{\Gamma\left(\frac{n(p-1)}{2}\right)}{\Gamma\left(\frac{(n-q)(p-1)}{2}\right)} H^{2,1}_{3,3}\left(\begin{array}{c} \left\{(1,1),\left(\frac{n}{2},1\right),\left(\frac{n(p-1)}{2},p-1\right)\right\} \\ \left\{\left(\frac{n-q}{2},1\right),\left(\frac{(n-q)(p-1)}{2},p-1\right),(0,1)\right\} \end{array}\middle|\, z\right).$$

This test is invariant under linear transformations $\underline{X}_k \to C\underline{X}_k$ where C is any nonrandom full-rank $p\times p$ compound symmetric matrix.

Modules `PDFNullMeanVecCS`, `CDFNullMeanVecCS`, `PvalNullMeanVecCS`, `QuantNullMeanVecCS`, and `PvalDataNullMeanVecCS` are available to implement the test and these have exactly the same arguments as the corresponding modules in Sect. 5.1.1, with module `PvalDataNullMeanVecCS` accepting the same type of data files as the corresponding module in Sect. 5.1.1.

5.1.26 *Brief Note on the Likelihood Ratio Test for Equality of Mean Vectors, Under the Assumption of Sphericity of the Covariance Matrices [EqMeanVecSph]*

If we assume that $\underline{X}_k \sim N_p(\underline{\mu}_k, \Sigma_k)$, $k = 1, \ldots, q$, where the covariance matrices are assumed to be all equal and spherical, that is, with $\Sigma_1 = \cdots = \Sigma_q = \sigma^2 I_p$, then, for a similar setup of independent samples as the one in the previous four subsections, with $n = n_1 + \cdots + n_q$, where n_k is the size of the sample from $\underline{X}_k$ $(k = 1, \ldots, q)$, and following similar steps to the ones taken in these same subsections, the $(2/n)$-th power of the l.r.t. statistic used to test the null hypothesis

$$\begin{array}{l} H_0 : \underline{\mu}_1 = \cdots = \underline{\mu}_q \\ \quad \text{assuming } \Sigma_1 = \cdots = \Sigma_q (= \sigma^2 I_p, \text{ with } \sigma^2 \text{ not specified}), \end{array} \tag{5.402}$$

is

$$\Lambda = \frac{|A^*|}{|C^*|} = \left(\frac{tr(A)}{tr(A+B)}\right)^p \tag{5.403}$$

where A and B are the matrices in (5.3) and A^* and C^* are diagonal matrices with all diagonal elements respectively equal to a^*_{ii} and c^*_{ii} in (5.382) and (5.383).

Again, following similar steps to the ones pursued in Sects. 5.1.22–5.1.25, with the only difference that now we do not need any U matrix, given that the population covariance matrices are already diagonal, it is easy to show that, for $n > q$,

$$\Lambda \stackrel{d}{\equiv} Y^p$$

where

$$Y \sim Beta\left(\frac{(n-q)p}{2}, \frac{p(q-1)}{2}\right). \tag{5.404}$$

As such, either for odd q or even p, the exact distribution of Λ in (5.403) is given by Theorem 3.2 with

$$m^* = 1\,,\;\; n_1 = 1\,,\;\; m_1 = \frac{p(q-1)}{2}\,,\;\; k_1 = p\,,\;\; a_1 = \frac{n-q}{2} + 1\,,$$

and therefore, in these cases, it is an EGIG distribution of depth $\frac{p(q-1)}{2}$, with rate parameters $\frac{n-q}{2} + \frac{\ell-1}{p}$ ($\ell = 1, \ldots, \frac{p(q-1)}{2}$) and shape parameters all equal to 1.

It happens though that in the case of this test one may take what may be a simpler approach which would be to use the statistic

$$F_1 = \frac{p(q-1)}{p(n-q)}\,\frac{\Lambda^{1/p}}{1-\Lambda^{1/p}} \tag{5.405}$$

which has an $F_{p(n-q),p(q-1)}$ distribution, rejecting H_0 in (5.402) if the computed value of the statistic in (5.405) exceeds the $1-\alpha$ quantile of an F distribution with $p(n-q)$ and $p(q-1)$ degrees of freedom, for a test of level α, or, equivalently, use the statistic

$$F_2 = \frac{p(n-q)}{p(q-1)}\,\frac{1-\Lambda^{1/p}}{\Lambda^{1/p}} \tag{5.406}$$

which has an $F_{p(q-1),p(n-q)}$ distribution, rejecting H_0 in (5.402) if the computed value of the statistic in (5.406) exceeds the $1-\alpha$ quantile of an F distribution with $p(q-1)$ and $p(n-q)$ degrees of freedom, for a test of level α. Given the spherical structure of the covariance matrices Σ_k ($k = 1, \ldots, q$) we are in a setup somewhat similar to the usual one-way ANOVA setting with one factor with q

levels, but where now the number of degrees of freedom of the F distributions appears multiplied by p, the number of variables in each vector $\underline{X}_k$.

For any p and q the p.d.f. of Λ in (5.403) may be expressed in terms of Fox's H function, or, alternatively, through the transformation of the p.d.f. of the Beta distribution as

$$f_\Lambda(z) = \frac{\Gamma\left(\frac{(n-1)p}{2}\right)}{\Gamma\left(\frac{(n-q)p}{2}\right)} H_{1,1}^{1,0}\left(\begin{array}{c}\left\{\left(\frac{(n-3)p}{2}, p\right)\right\} \\ \left\{\left(\frac{(n-q-2)p}{2}, p\right)\right\}\end{array}\middle| z\right)$$

or,

$$f_\Lambda(z) = \frac{1}{p\,B\left(\frac{(n-q)p}{2}, \frac{p(q-1)}{2}\right)} z^{\frac{n-q}{2}-1}\left(1 - z^{1/p}\right)^{\frac{p(q-1)}{2}-1}$$

and its c.d.f. as

$$F_\Lambda(z) = \frac{\Gamma\left(\frac{(n-1)p}{2}\right)}{\Gamma\left(\frac{(n-q)p}{2}\right)} H_{2,2}^{1,1}\left(\begin{array}{c}\left\{(1,1), \left(\frac{(n-1)p}{2}, p\right)\right\} \\ \left\{\left(\frac{(n-q)p}{2}, p\right), (0,1)\right\}\end{array}\middle| z\right)$$

or,

$$F_\Lambda(z) = \frac{B\left(\frac{(n-q)p}{2}, \frac{p(q-1)}{2}; z^{1/p}\right)}{B\left(\frac{(n-q)p}{2}, \frac{p(q-1)}{2}\right)},$$

where $B(a,b)$ and $B(a,b;y)$ represent respectively the complete and the incomplete Beta functions, with

$$B(a,b;y) = \int_0^y x^{a-1}(1-x)^{b-1}\,dx \quad \text{and} \quad B(a,b) = B(a,b;1)\,. \tag{5.407}$$

We may note here that the spherical structure may be seen as a particular case of the compound symmetric structure, which in turn may be seen as a particular case of the circular structure.

The present test is invariant under linear transformations $\underline{X}_k \to C\underline{X}_k + \underline{b}$ where $C = aI_p$ for any non-null real a and where $\underline{b}$ is any $p\times 1$ real vector.

Modules `PDFEqMeanVecSph`, `CDFEqMeanVecSph`, `PvalEqMeanVecSph`, `QuantEqMeanVecSph`, and `PvalDataEqMeanVecSph` are available to implement the test and these have exactly the same arguments as the corresponding modules in Sect. 5.1.1, with module `PvalDataNullMeanVecCS` accepting the same type of data files as the corresponding module in Sect. 5.1.1.

5.1.27 *Brief Note on the Likelihood Ratio Test for Simultaneous Nullity of Mean Vectors, Under the Assumption of Sphericity of the Covariance Matrices [NullMeanVecSph]*

Let us suppose a similar setup to the one in the previous subsection, where now we want to test the null hypothesis

$$\begin{array}{l} H_0 : \underline{\mu}_1 = \dots \underline{\mu}_q = \underline{0} \\ \quad \text{assuming } \Sigma_1 = \cdots = \Sigma_q (= \sigma^2 I_p, \text{ with } \sigma^2 \text{ not specified}) \,. \end{array} \tag{5.408}$$

Then the $(2/n)$-th power of the l.r.t. statistic to test this null hypothesis is

$$\Lambda = \left(\frac{tr(A)}{tr(A + B^*)} \right)^p , \tag{5.409}$$

where A is the matrix in (5.3) and B^* is the matrix in (5.368).

Following similar steps to the ones in the previous five subsections, we may see that, for $n > q$,

$$\Lambda \stackrel{\text{d}}{\equiv} Y^p$$

where

$$Y \sim Beta\left(\frac{(n-q)p}{2}, \frac{pq}{2} \right).$$

Therefore, for either even q or even p, the exact distribution of Λ in (5.409) is given by Theorem 3.2 with

$$m^* = 1 , \;\; n_1 = 1 , \;\; m_1 = \frac{pq}{2} , \;\; k_1 = p , \;\; a_1 = \frac{n-q}{2} + 1 ,$$

thus yielding, in these cases, an EGIG distribution of depth $\frac{pq}{2}$, with rate parameters $\frac{n-q}{2} + \frac{\ell-1}{p}$ $(\ell = 1, \dots, \frac{pq}{2})$ and shape parameters all equal to 1 as the exact distribution of Λ in (5.409).

As happened in the previous subsection, also for the test in the present subsection one may alternatively use the statistic

$$F_1 = \frac{pq}{p(n-q)} \frac{\Lambda^{1/p}}{1 - \Lambda^{1/p}}$$

which now has an $F_{p(n-q),pq}$ distribution, or, equivalently, use the statistic

$$F_2 = \frac{p(n-q)}{pq}\frac{1-\Lambda^{1/p}}{\Lambda^{1/p}}$$

which has an $F_{pq,p(n-q)}$ distribution.

Similar to what happens in the previous subsection, for any p and any q the p.d.f. of Λ in (5.409) may be expressed in terms of Fox's H function, or, alternatively, through the transformation of the p.d.f. of the Beta distribution as

$$f_\Lambda(z) = \frac{\Gamma\left(\frac{np}{2}\right)}{\Gamma\left(\frac{(n-q)p}{2}\right)} H^{1,0}_{1,1}\left(\left.\begin{array}{c}\left\{\left(\frac{(n-2)p}{2}, p\right)\right\} \\ \left\{\left(\frac{(n-q-2)p}{2}, p\right)\right\}\end{array}\right| z\right)$$

or,

$$f_\Lambda(z) = \frac{1}{p\,B\left(\frac{(n-q)p}{2}, \frac{pq}{2}\right)} z^{\frac{n-q}{2}-1}\left(1-z^{1/p}\right)^{\frac{pq}{2}-1}$$

and its c.d.f. as

$$F_\Lambda(z) = \frac{\Gamma\left(\frac{np}{2}\right)}{\Gamma\left(\frac{(n-q)p}{2}\right)} H^{1,1}_{2,2}\left(\left.\begin{array}{c}\left\{(1,1), \left(\frac{np}{2}, p\right)\right\} \\ \left\{\left(\frac{(n-q)p}{2}, p\right), (0,1)\right\}\end{array}\right| z\right)$$

or,

$$F_\Lambda(z) = \frac{B\left(\frac{(n-q)p}{2}, \frac{pq}{2}; z^{1/p}\right)}{B\left(\frac{(n-q)p}{2}, \frac{pq}{2}\right)},$$

where, once again $B(a,b)$ and $B(a,b;y)$ represent respectively the complete and the incomplete Beta functions defined in (5.407).

This test is invariant under linear transformations $\underline{X}_k \to C\underline{X}_k$ where $C = aI_p$ for any non-null real a.

Modules `PDFNullMeanVecSph`, `CDFNullMeanVecSph`, `PvalNullMeanVecSph`, `QuantNullMeanVecSph`, and `PvalDataNullMeanVecSph` are available to implement the test and these have exactly the same arguments as the corresponding modules in Sect. 5.1.1, with module `PvalDataNullMeanVecSph` accepting the same type of data files as the corresponding module in Sect. 5.1.1.

5.1.28 The Likelihood Ratio Test for Profile Parallelism Under the Assumption of Circularity of the Covariance Matrices [ProfParCirc]

Let us assume a setup similar to the one in Sect. 5.1.22, where now we want to test the parallelism of the q profiles, that is, the null hypothesis in (5.21), assuming now that the matrices Σ_k $(k = 1, \ldots, q)$ are circular, that is,

$$\Sigma_1 = \cdots = \Sigma_q = \Sigma_{cp},$$

where Σ_{cp} is defined in (5.344) in Sect. 5.1.22. Then, the l.r.t. statistic to test the null hypothesis of profile parallelism is

$$\Lambda = \frac{|CA^*C'|}{|CC^*C'|} \tag{5.410}$$

where C is the matrix in (5.20), A^* is the matrix with elements given by (5.348) and C^* is the matrix with elements given by (5.350).

It is not too hard to show that

$$|CA^*C'| = \prod_{j=2}^{p} v_j^* \quad \text{and} \quad |CC^*C'| = \prod_{j=2}^{p} v_j^{**}$$

where v_j^* and v_j^{**} are respectively given by (5.354) and (5.355), so that we may write

$$\Lambda = \prod_{j=2}^{p} \frac{v_j^*}{v_j^{**}}, \tag{5.411}$$

which, directly from the derivations in Sect. 5.1.22, show that, for $m = \lfloor p/2 \rfloor$ and $n > q$,

$$\Lambda \stackrel{d}{\equiv} (Y^*)^{\text{ mod } (p+1,2)} \prod_{j=2}^{p-m} Y_j^2$$

where the r.v.'s Y^* and Y_j $(j = 2, \ldots, p-m)$ are the r.v.'s in (5.366). This shows that for odd p, in which case Y^* does not exist, the exact distribution of Λ is given by Theorem 3.2 with

$$m^* = \frac{p-1}{2} \quad n_\nu = 1\,, \quad m_\nu = q-1\,, \quad k_\nu = 2\,, \quad a_\nu = \frac{n-q}{2} + 1\,.$$

This is an EGIG distribution of depth $q-1$, with rate parameters $(n-q+\ell-1)/2$ ($\ell=1,\dots,q-1$) and shape parameters all equal to $\lfloor p/2 \rfloor$, with p.d.f. and c.d.f. respectively given by

$$f_\Lambda(z) = f^{EGIG}\left(z \,\Big|\, \left\{\frac{p-1}{2}\right\}_{\ell=1:q-1};\, \left\{\frac{n-q+\ell-1}{2}\right\}_{\ell=1:q-1};\, q-1\right)$$

and

$$F_\Lambda(z) = F^{EGIG}\left(z \,\Big|\, \left\{\frac{p-1}{2}\right\}_{\ell=1:q-1};\, \left\{\frac{n-q+\ell-1}{2}\right\}_{\ell=1:q-1};\, q-1\right),$$

while if p is even, then if q is odd, the distribution of Y^* is given by Theorem 3.2 with

$$m^* = 1, \;\; n_1 = 1, \;\; m_1 = \frac{q-1}{2}, \;\; k_1 = 1, \;\; a_1 = \frac{n-q}{2} + 1\,,$$

its exact distribution being an EGIG distribution of depth $(q-1)/2$, with rate parameters $(n-q)/2 - 1 + \ell$ ($\ell = 1, \dots, (q-1)/2$) and shape parameters all equal to 1, independent of the distribution of $\prod_{j=2}^{p-m} Y_j^2$, this way yielding for Λ its exact distribution given by Theorem 3.2 and Corollary 4.2 with

$$m^* = \frac{p}{2}, \quad n_\nu = 1\,, \quad a_\nu = \frac{n-q}{2} + 1\,, \quad \nu = 1, \dots, m^*\,,$$

$$k_\nu = \begin{cases} 2, \; \nu = 1, \dots, m^*-1 \\ 1, \; \nu = m^* \end{cases}, \quad m_\nu = \begin{cases} q-1, \; \nu = 1, \dots, m^*-1 \\ \frac{q-1}{2}, \quad \nu = m^* \end{cases},$$

which is an EGIG distribution of depth $q-1$ with rate parameters $(n-q+\ell-1)/2$ and shape parameters $\frac{p}{2} - \text{mod}(\ell-1, 2)$ ($\ell = 1, \dots, q-1$), with p.d.f. and c.d.f. given respectively by

$$f_\Lambda(z) = f^{EGIG}\left(z \,\Big|\, \left\{\frac{p}{2} - \text{mod}(\ell-1,2)\right\}_{\ell=1:q-1};\, \left\{\frac{n-q+\ell-1}{2}\right\}_{\ell=1:q-1};\, q-1\right)$$

and

$$F_\Lambda(z) = F^{EGIG}\left(z \,\Big|\, \left\{\frac{p}{2} - \text{mod}(\ell-1,2)\right\}_{\ell=1:q-1};\, \left\{\frac{n-q+\ell-1}{2}\right\}_{\ell=1:q-1};\, q-1\right).$$

Thus, Theorem 3.2 and Corollary 4.2 yield the exact p.d.f. and c.d.f. of Λ in (5.411), under the null hypothesis of parallelism of the profiles, for odd p, or for even p and odd q, expressed in terms of the EGIG p.d.f. and c.d.f. as

$$f_\Lambda(z) = f^{EGIG}\left(z \,\middle|\, \left\{ \left\lfloor \frac{p}{2} \right\rfloor - \text{mod}(p+1,2)\text{mod}(\ell-1,2) \right\}_{\ell=1:q-1}; \left\{ \frac{n-q+\ell-1}{2} \right\}_{\ell=1:q-1}; q-1 \right)$$

and

$$F_\Lambda(z) = F^{EGIG}\left(z \,\middle|\, \left\{ \left\lfloor \frac{p}{2} \right\rfloor - \text{mod}(p+1,2)\ \text{mod}(\ell-1,2) \right\}_{\ell=1:q-1}; \left\{ \frac{n-q+\ell-1}{2} \right\}_{\ell=1:q-1}; q-1 \right).$$

Alternatively, for any p and q, the p.d.f. and c.d.f. of Λ in (5.359) may be expressed in terms of Fox's H function, for $m = \lfloor p/2 \rfloor$ and $p^* = p - m - 1 + mod(p+1, 2)$, as

$$f_\Lambda(z) = \left\{ \left(\frac{\Gamma\left(\frac{n-1}{2}\right)}{\Gamma\left(\frac{n-q}{2}\right)} \right)^{mod(p+1,2)} \prod_{j=2}^{p-m} \frac{\Gamma(n-1)}{\Gamma(n-q)} \right\} \times H^{p^*,0}_{p^*,p^*}\left(\begin{array}{c} \left\{ \left\{ \left(\frac{n-3}{2}, 1\right) \right\}_{j=1:mod(p+1,2)}, \{(n-3,2)\}_{j=2:p-m} \right\} \\ \left\{ \left\{ \left(\frac{n-q-2}{2}, 1\right) \right\}_{j=1:mod(p+1,2)}, \{(n-q-2,2)\}_{j=2:p-m} \right\} \end{array} \middle|\, z \right)$$

and

$$F_\Lambda(z) = \left\{ \left(\frac{\Gamma\left(\frac{n-1}{2}\right)}{\Gamma\left(\frac{n-q}{2}\right)} \right)^{mod(p+1,2)} \prod_{j=2}^{p-m} \frac{\Gamma(n-1)}{\Gamma(n-q)} \right\} \times H^{p^*,1}_{p^*+1,p^*+1}\left(\begin{array}{c} \left\{ (1,1), \left\{ \left(\frac{n-1}{2}, 1\right) \right\}_{j=1:mod(p+1,2)}, \{(n-1,2)\}_{j=2:p-m} \right\} \\ \left\{ \left\{ \left(\frac{n-q}{2}, 1\right) \right\}_{j=1:mod(p+1,2)}, \{(n-q,2)\}_{j=2:p-m}, (0,1) \right\} \end{array} \middle|\, z \right).$$

We may note here that, as happens with the test in Sect. 5.1.3, also in (5.410) the matrix C may be replaced by any matrix $C^+ = I^*_{p-1}C$, where I^*_{p-1} is any permutation matrix of order $p - 1$.

Modules `PDFProfParCirc`, `CDFProfParCirc`, `PvalProfParCirc`, `QuantProfParCirc`, and `PvalDataProfParCirc` are made available for the implementation of the test. These modules have exactly the same optional and non-optional arguments as the corresponding modules in Sect. 5.1.1, and module `PvalDataProfParCirc` accepts files with the same structure as the corresponding `PvalData` module in Sect. 5.1.1.

As an example of application of the present test, we will use the data on response times in Tables 3.9.7 and 3.9.8 of Timm (2002), already used in Sect. 5.1.22, to test the parallelism of the three profiles of the five response times for the population that corresponds to the sample in Table 3.9.7 of Timm (2002) and for the populations of low and high long-term memory capacity. A plot of the sample profiles may be seen in Fig. 5.78 (although there may seem to be some inversion of the values, those are indeed the values obtained from the data), while the command used and the output obtained with the Mathematica® version of module `PvalDataProfParCirc` may be examined in Fig. 5.79.

From the high p-value obtained in Fig. 5.79 we see that we should not reject the hypothesis of parallelism of the three profiles, assuming a circular structure for the covariance matrices, which as already remarked in Sect. 5.1.22 seems to be a much plausible assumption, given the results obtained in Sects. 5.2.1 and 5.2.3 for the tests of circularity of the covariance matrices. The decision of not rejecting the parallelism of the profiles seems plausible in the light of the plot in Fig. 5.78.

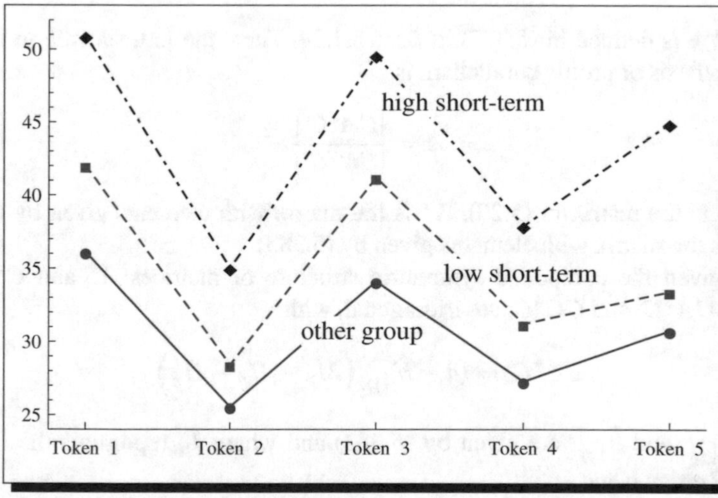

Fig. 5.78 Plot of the sample profiles for the five response times for the three subpopulations corresponding to the data in Tables 3.9.7 and 3.9.8 of Timm (2002)

```
PvalDataProfParCirc[path <> "Response_times_all.dat"]
 There are 3 samples, with sizes {11, 10, 10} and 5 variables
Computed value of Λ: 0.8509248421
p-value: 0.7993711273
```

Fig. 5.79 Command used and output obtained with the Mathematica® module PvalDataProfParCirc to implement the test of profile parallelism when a circular structure is assumed for the covariance matrices, using the data on response times in Tables 3.9.7 and 3.9.8 of Timm (2002)

As expected, the present test is invariant under linear transformations of the type $\underline{X}_k \to C\underline{X}_k + \underline{b}$ where $\underline{b}$ is any $p\times 1$ real vector and C is any $p\times p$ full-rank nonrandom circular or circulant matrix.

5.1.29 *Brief Note on the Likelihood Ratio Test for Profile Parallelism Under the Assumption of Compound Symmetry of the Covariance Matrices [ProfParCS]*

Let us assume a setup similar to the one in Sect. 5.1.24, where now we want to test the parallelism of the q profiles, that is, the null hypothesis in (5.21), assuming now that the matrices Σ_k $(k = 1, \dots, q)$ are compound symmetric, that is,

$$\Sigma_1 = \cdots = \Sigma_q = \Sigma_{CS}\,,$$

where Σ_{CS} is defined in (5.377) in Sect. 5.1.24. Then, the l.r.t. statistic to test the null hypothesis of profile parallelism is

$$\Lambda = \frac{|CA^*C'|}{|CC^*C'|} \tag{5.412}$$

where C is the matrix in (5.20), A^* is the matrix with elements given by (5.382) and C^* is the matrix with elements given by (5.383).

But, given the compound symmetric structure of matrices A^* and C^*, both matrices CA^*C' and CC^*C' are tridiagonal, with

$$CA^*C' = (\hat{a} - \hat{b})_{|H_1}\left(2I_p - H_p - H_p'\right)$$

where $\hat{a}_{|H_1}$ and $\hat{b}_{|H_1}$ are given by (5.382) and where I_p represents the identity matrix of order p and

$$\underset{(p\times p)}{H_p} = \left[\begin{array}{c|c} \multicolumn{2}{c}{\underline{0}_{1\times p}} \\ \hline I_{p-1} & \underline{0}_{(p-1)\times 1} \end{array}\right] \tag{5.413}$$

and with

$$CC^*C' = (\hat{a} - \hat{b})_{|H_0}\left(2I_p - H_p - H_p'\right)$$

where $\hat{a}_{\,|\,H_0}$ and $\hat{b}_{\,|\,H_0}$ are given by (5.383).

As such, the eigenvalues of CA^*C' are (see Seber (2008, Sec. 8.11))

$$2(\hat{a} - \hat{b})_{|H_1}\left(1 - \cos\frac{j\pi}{p}\right), \quad j = 1, \ldots, p-1$$

and the eigenvalues of CC^*C' are

$$2(\hat{a} - \hat{b})_{|H_0}\left(1 - \cos\frac{j\pi}{p}\right), \quad j = 1, \ldots, p-1\,,$$

so that

$$|CA^*C'| = \left(2(\hat{a} - \hat{b})_{|H_1}\right)^{p-1}\prod_{j=1}^{p-1}\left(1 - \cos\frac{j\pi}{p}\right) = p\left((\hat{a} - \hat{b})_{|H_1}\right)^{p-1}$$

and

$$|CC^*C'| = \left(2(\hat{a} - \hat{b})_{|H_0}\right)^{p-1}\prod_{j=1}^{p-1}\left(1 - \cos\frac{j\pi}{p}\right) = p\left((\hat{a} - \hat{b})_{|H_0}\right)^{p-1}.$$

But, then we may write

$$\Lambda = \left(\frac{(\hat{a} - \hat{b})_{|H_1}}{(\hat{a} - \hat{b})_{|H_0}}\right)^{p-1} = \left(\frac{\frac{1}{p-1}\left(tr(A^{**}) - a_{11}^{**}\right)}{\frac{1}{p-1}\left(tr(C^{**}) - c_{11}^{**}\right)}\right)^{p-1} = \left(\frac{tr(A^{**}) - a_{11}^{**}}{tr(C^{**}) - c_{11}^{**}}\right)^{p-1}, \tag{5.414}$$

where A^{**} and C^{**} are the matrices in (5.381). But then, from (5.364), where now we have $\delta_2 = \cdots = \delta_p$, we know that

$$tr(A^{**}) - a_{11}^{**} = \sum_{j=2}^{p} a_{jj}^{**} \sim \chi^2_{(p-1)(n-q)}\,,$$

while

$$tr(B^{**}) - b_{11}^{**} = \sum_{j=2}^{p} b_{jj}^{**} \sim \chi^2_{(p-1)(q-1)}\,,$$

independent of $tr(A^{**}) - a_{11}^{**}$, so that

$$tr(C^{**}) - c_{11}^{**} = tr(A^{**}) + tr(B^{**}) - a_{11}^{**} - b_{11}^{**} = \sum_{j=2}^{p} c_{jj}^{**} \sim \chi^2_{(p-1)n} \, .$$

As such, for $n > q$,

$$\Lambda \stackrel{d}{\equiv} (Y_2)^{p-1}$$

where Y_2 is the r.v. in (5.389), and thus the exact distribution of Λ is given, for odd q or odd p, by Theorem 3.2 with

$$m^* = 1, \;\; n_1 = 1 \;, \; m_1 = \frac{(q-1)(p-1)}{2} \;, \;\; k_1 = p-1 \;, \;\; a_1 = \frac{n-q}{2} + 1 \;,$$

which, as it happens in Sect. 5.1.24, is better seen if we identify the r.v. Y_2 with the r.v.'s $Y_{\nu\ell}^{***}$ in (3.10). This way, for odd q or odd p, the distribution of Λ is an EGIG distribution of depth $(q-1)(p-1)/2$, with rate parameters $\frac{(n-q)}{2} + \frac{\ell-1}{p-1}$ ($\ell = 1, \dots, (q-1)(p-1)/2$) and shape parameters all equal to 1, with p.d.f. and c.d.f. respectively given by (5.391) and (5.392), with $r_\ell = 1$, for $\ell = 1, \dots, (q-1)(p-1)/2$.

As happens in Sects. 5.1.26 and 5.1.27, also for the present test, one may take what may be a simpler approach which would be to use the statistic

$$F_1 = \frac{(p-1)(q-1)}{(p-1)(n-q)} \, \frac{\Lambda^{1/(p-1)}}{1 - \Lambda^{1/(p-1)}}$$

which has an $F_{(p-1)(n-q),(p-1)(q-1)}$ distribution, or, equivalently, the statistic

$$F_2 = \frac{(p-1)(n-q)}{(p-1)(q-1)} \, \frac{1 - \Lambda^{1/(p-1)}}{\Lambda^{1/(p-1)}}$$

which has an $F_{(p-1)(q-1),(p-1)(n-q)}$ distribution.

For any p and q, the p.d.f. and c.d.f. of Λ in (5.412) or (5.414), may be expressed in terms of Fox's H function as

$$f_\Lambda(z) = \frac{\Gamma\left(\frac{(n-1)(p-1)}{2}\right)}{\Gamma\left(\frac{(n-q)(p-1)}{2}\right)} \, H_{1,1}^{1,0}\left(\begin{array}{c} \left\{\left(\frac{(n-3)(p-1)}{2}, p-1\right)\right\} \\ \left\{\left(\frac{(n-q-2)(p-1)}{2}, p-1\right)\right\} \end{array} \middle| \, z \right)$$

and

$$F_\Lambda(z) = \frac{\Gamma\left(\frac{(n-1)(p-1)}{2}\right)}{\Gamma\left(\frac{(n-q)(p-1)}{2}\right)} H_{2,2}^{1,1}\left(\begin{array}{c} \left\{(1,1), \left(\frac{(n-1)(p-1)}{2}, p-1\right)\right\} \\ \left\{\left(\frac{(n-q)(p-1)}{2}, p-1\right), (0,1)\right\} \end{array} \middle| \, z \right).$$

Alternatively, this p.d.f. and this c.d.f. may be respectively given by

$$f_\Lambda(z) = \frac{1}{(p-1)\, B\left(\frac{(n-q)(p-1)}{2}, \frac{(q-1)(p-1)}{2}\right)}\, z^{\frac{n-q}{2}-1}\left(1 - z^{1/(p-1)}\right)^{\frac{(q-1)(p-1)}{2}-1}$$

and

$$F_\Lambda(z) = \frac{B\left(\frac{(n-q)(p-1)}{2}, \frac{(q-1)(p-1)}{2}; z^{1/(p-1)}\right)}{B\left(\frac{(n-q)(p-1)}{2}, \frac{(q-1)(p-1)}{2}\right)},$$

where, $B(a,b)$ and $B(a,b;y)$ represent respectively the complete and the incomplete Beta functions defined in (5.407).

Also as it happens in Sects. 5.1.3 and 5.1.28, once again in (5.412) one may replace matrix C by any matrix $C^+ = I_{p-1}^* C$ where I_{p-1}^* is any permutation matrix of order $p-1$.

Modules `PDFProfParCS`, `CDFProfParCS`, `PvalProfParCS`, `QuantProfParCS`, and `PvalDataProfParCS` are available for the implementation of this test. These modules have exactly the same optional and non-optional arguments as the corresponding modules in Sect. 5.1.1, and module `PvalDataProfParCS` accepts files with the same structure as the corresponding `PvalData` module in Sect. 5.1.1.

This test is invariant under linear transformations of the type $\underline{X}_k \to C\underline{X}_k + \underline{b}$ where $\underline{b}$ is any $p\times 1$ real vector and C is any $p\times p$ full-rank nonrandom compound symmetric matrix.

An application of the test in this subsection is done using the same data set on response times used in the previous subsection. The command used and the output obtained are in Fig. 5.80 and we may see that the p-value obtained is quite close to the one obtained in Fig. 5.79 for the test in the previous subsection. This fact was actually expected, given the closeness of the p-values obtained in Sects. 5.1.22 and 5.1.24 for the same data set, showing that, as already stated in Sect. 5.1.24, the compound symmetric structure may also be a plausible one for the covariance matrices of the populations from which the samples were taken.

```
PvalDataProfParCS[path <> "Response_times_all.dat"]
 There are 3 samples, with sizes {11, 10, 10} and 5 variables
Computed value of Λ: 0.8525803634
p-value: 0.8009705115
```

Fig. 5.80 Command used and output obtained with the Mathematica® module `PvalDataProfParCS` to implement the test of profile parallelism when a compound symmetric structure is assumed for the covariance matrices, using the data on response times in Tables 3.9.7 and 3.9.8 of Timm (2002)

5.1.30 *Brief Note on the Likelihood Ratio Test for Profile Parallelism Under the Assumption of Sphericity of the Covariance Matrices [ProfParSph]*

Let us assume a setup similar to the one in Sect. 5.1.26, where now we want to test the parallelism of the q profiles, that is, the null hypothesis in (5.21), assuming that the matrices Σ_k $(k = 1, \dots, q)$ are spherical, that is,

$$\Sigma_1 = \cdots = \Sigma_q = \sigma^2 I_p \text{ (with } \sigma^2 \text{ unspecified)}\,.$$

Then, the l.r.t. statistic to test the null hypothesis of profile parallelism is

$$\Lambda = \frac{|CA^*C'|}{|CC^*C'|} \tag{5.415}$$

where C is the matrix in (5.20), and A^* and C^* are the same as in Sect. 5.1.26, that is, A^* is a diagonal matrix with all diagonal elements equal to $\frac{1}{p}tr(A)$ and C^* is a diagonal matrix with all diagonal elements equal to $\frac{1}{p}tr(A+B)$, with A and B being the matrices in (5.3), and where once again, as happens in Sects. 5.1.3, 5.1.28 and 5.1.29, the matrix C may be replaced by any $C^+ = I^*_{p-1}C$ matrix, where I^*_{p-1} is any permutation matrix of order $p-1$.

Then, given the diagonal structure of matrices A^* and C^*, both matrices CA^*C' and CC^*C' are tridiagonal, with

$$CA^*C' = \hat{\sigma^2}\,|_{H_1}\left(2I_p - H_p - H'_p\right)$$

where $\hat{\sigma^2}\,|_{H_1} = \frac{1}{p}tr(A)$, I_p represents the identity matrix of order p and H_p is the matrix in (5.413), and with

$$CC^*C' = \hat{\sigma^2}\,|_{H_0}\left(2I_p - H_p - H'_p\right)$$

where $\hat{\sigma^2}\,|_{H_0} = \frac{1}{p}tr(A+B)$.

As such, the eigenvalues of CA^*C' are (see Seber (2008, Sec. 8.11))

$$2\hat{\sigma^2}_{|H_1}\left(1-\cos\frac{j\pi}{p}\right),\quad j=1,\dots,p-1$$

and the eigenvalues of CC^*C' are

$$2\hat{\sigma^2}_{|H_0}\left(1-\cos\frac{j\pi}{p}\right),\quad j=1,\dots,p-1\,,$$

so that

$$|CA^*C'|=\left(2\hat{\sigma^2}_{|H_1}\right)^{p-1}\prod_{j=1}^{p-1}\left(1-\cos\frac{j\pi}{p}\right)=p\left(\hat{\sigma^2}_{|H_1}\right)^{p-1}$$

and

$$|CC^*C'|=\left(2\hat{\sigma^2}_{|H_0}\right)^{p-1}\prod_{j=1}^{p-1}\left(1-\cos\frac{j\pi}{p}\right)=p\left(\hat{\sigma^2}_{|H_0}\right)^{p-1}.$$

But, then we may write

$$\Lambda=\left(\frac{\hat{\sigma^2}_{|H_1}}{\hat{\sigma^2}_{|H_0}}\right)^{p-1}=\left(\frac{\frac{1}{p}tr(A)}{\frac{1}{p}tr(A+B)}\right)^{p-1}=\left(\frac{tr(A)}{tr(A+B)}\right)^{p-1},\tag{5.416}$$

where A and B are the matrices in (5.3). But then, given that the population covariance matrices are assumed equal and spherical, we know that, for $n>q$,

$$tr(A)\sim\chi^2_{p(n-q)}\,,$$

while

$$tr(B)\sim\chi^2_{p(q-1)}\,,$$

independent of $tr(A)$. As such,

$$\Lambda\stackrel{\mathrm{d}}{\equiv}Y^{p-1}$$

where Y is the r.v. in (5.404).

Thus, either for odd q or even p, the exact distribution of Λ in (5.403) is given by Theorem 3.2 with

$$m^* = 1\,,\;\; n_1 = 1\,,\;\; m_1 = \frac{p(q-1)}{2}\,,\;\; k_1 = p-1\,,\;\; a_1 = \frac{(n-q)p}{2(p-1)} + 1\,,$$

and therefore, in these cases, this distribution is, as given by Corollary 4.2, an EGIG distribution of depth $\frac{p(q-1)}{2}$, with rate parameters $\frac{(n-q)p}{2(p-1)} + \frac{\ell-1}{p-1}$ ($\ell = 1, \ldots, \frac{p(q-1)}{2}$) and shape parameters all equal to 1, with p.d.f. and c.d.f. given respectively by

$$f_\Lambda(z) = f^{EGIG}\left(z \,\middle|\, \{1\}_{\ell=1:\frac{p(q-1)}{2}}\,;\, \left\{ \frac{(n-q)p}{2(p-1)} + \frac{\ell-1}{p-1} \right\}_{\ell=1:\frac{p(q-1)}{2}}\,;\, \frac{p(q-1)}{2} \right)$$

and

$$F_\Lambda(z) = f^{EGIG}\left(z \,\middle|\, \{1\}_{\ell=1:\frac{p(q-1)}{2}}\,;\, \left\{ \frac{(n-q)p}{2(p-1)} + \frac{\ell-1}{p-1} \right\}_{\ell=1:\frac{p(q-1)}{2}}\,;\, \frac{p(q-1)}{2} \right).$$

However, as happens in Sects. 5.1.26, 5.1.27, and 5.1.29, also for the present test, one may take what may be a simpler approach which would be to use the statistic

$$F_1 = \frac{p(q-1)}{p(n-q)}\, \frac{\Lambda^{1/(p-1)}}{1-\Lambda^{1/(p-1)}}$$

which has an $F_{p(n-q),p(q-1)}$ distribution, or, equivalently, use the statistic

$$F_2 = \frac{p(n-q)}{p(q-1)}\, \frac{1-\Lambda^{1/(p-1)}}{\Lambda^{1/(p-1)}}$$

which has an $F_{p(q-1),p(n-q)}$ distribution.

For any p and q, the p.d.f. and c.d.f. of Λ in (5.415) or (5.416), may be expressed in terms of Fox's H function as

$$f_\Lambda(z) = \frac{\Gamma\left(\frac{(n-1)p}{2}\right)}{\Gamma\left(\frac{(n-q)p}{2}\right)} H^{1,0}_{1,1}\left(\begin{array}{c} \left\{\left(\frac{(n-3)p}{2}+1,\, p-1\right)\right\} \\ \left\{\left(\frac{(n-q-2)p}{2}+1,\, p-1\right)\right\} \end{array} \middle|\, z \right)$$

and

$$F_\Lambda(z) = \frac{\Gamma\left(\frac{(n-1)p}{2}\right)}{\Gamma\left(\frac{(n-q)p}{2}\right)} H^{1,1}_{2,2}\left(\begin{array}{c} \left\{(1,1),\, \left(\frac{(n-1)p}{2},\, p-1\right)\right\} \\ \left\{\left(\frac{(n-q)p}{2},\, p-1\right),\, (0,1)\right\} \end{array} \middle|\, z \right).$$

or, alternatively, through the transformation of the Beta p.d.f. and c.d.f. as

$$f_\Lambda(z) = \frac{1}{(p-1)\,B\left(\frac{(n-q)p}{2}, \frac{(q-1)p}{2}\right)}\, z^{\frac{(n-q)*p}{2(p-1)}-1}\left(1 - z^{1/(p-1)}\right)^{\frac{(q-1)p}{2}-1}$$

and

$$F_\Lambda(z) = \frac{B\left(\frac{(n-q)p}{2}, \frac{(q-1)p}{2}; z^{1/(p-1)}\right)}{B\left(\frac{(n-q)p}{2}, \frac{(q-1)p}{2}\right)},$$

where, $B(a,b)$ and $B(a,b;y)$ represent respectively the complete and the incomplete Beta functions defined in (5.407).

Modules `PDFProfParSph`, `CDFProfParSph`, `PvalProfParSph`, `QuantProfParSph`, and `PvalDataProfParSph` are available for the implementation of the test. These modules have exactly the same arguments as the corresponding modules in Sect. 5.1.1, and module `PvalDataProfParSph` accepts files with the same structure as the corresponding `PvalData` module in Sect. 5.1.1.

The present test is invariant under linear transformations of the type $\underline{X}_k \to C\underline{X}_k + \underline{b}$ where $\underline{b}$ is any $p{\times}1$ real vector and $C = aI_p$, where a is any non-null real.

5.2 Likelihood Ratio Test Statistics Whose Distributions Correspond to the Product in Theorem 3.3 and That Have p.d.f. and c.d.f. Given by Corollary 4.3

Parallel to the discussion in Sect. 5.1, products of independent Beta r.v.'s in Theorem 3.3 yield the exact distributions of a number of l.r. statistics. Some of these are described next.

5.2.1 The Likelihood Ratio Statistic to Test Circularity of the Covariance Matrix [CircOddp]

Let us suppose that $\underline{X} \sim N_p(\underline{\mu}, \Sigma)$ and that we are interested in testing the null hypothesis of circularity of the covariance matrix. See Sect. 5.1.22 for the definition of a circular covariance matrix.

For a sample of size n, the $(2/n)$-th power of the l.r.t. statistic to test the null hypothesis

$$H_0 : \Sigma = \Sigma_{cp}, \qquad (5.417)$$

where the structure for Σ_{cp} is given in (5.344) in Sect. 5.1.22, is (see Olkin and Press (1969, Sec. 3.3))

$$\Lambda = n^p \frac{|A|}{\prod_{j=1}^{p} v_j^*} = \frac{|V|}{\prod_{j=1}^{p} v_j^*} \tag{5.418}$$

where A is the m.l.e. of Σ, and, for $m = \lfloor \frac{p}{2} \rfloor$,

$$v_j^* = \begin{cases} v_{jj} & j = 1, \text{ and also } j = m+1 \text{ if } p \text{ is even} \\ (v_{jj} + v_{p-j+2,p-j+2})/2 & j = 2, \ldots, p-m, m+2, \ldots, p\,, \end{cases} \tag{5.419}$$

with $v_j^* = v_{p-j+2}^*$ $(j = 2, \ldots, m)$, and where v_{jj} is the j-th diagonal element of $V = nUAU'$, where $U = [u_{ij}]$ $(i, j = 1, \ldots, p)$ is a $p\times p$ orthogonal matrix, with

$$u_{ij} = \frac{1}{\sqrt{p}} \left\{\cos\left(2\pi(j-1)(i-1)/p\right) + \sin\left(2\pi(j-1)(i-1)/p\right)\right\}.$$

We should note that the second expression for Λ in (5.418) agrees with the first expression in (3.5) of Olkin and Press (1969), while the second expression in (3.5) of this reference appears to have a typographical error. In it the diagonal elements of the matrix V should be replaced by the diagonal elements of the matrix S. The first expression for Λ in (5.418) may be indeed easily derived from the second one, taking into account that, given the fact that U is an orthogonal matrix, $|V| = |nUAU'| = n^p|UAU'| = n^p|A|$.

For an alternative way to obtain the second expression for the l.r. statistic in (5.418), please refer to Appendix 17.

Under H_0 in (5.417) (see Olkin and Press (1969, Sec. 3.3)),

$$\Lambda \stackrel{d}{\equiv} \prod_{j=1}^{p-1} Y_j \tag{5.420}$$

where, for $n > p$,

$$Y_j \sim \begin{cases} Beta\left(\frac{n-1-j}{2}, \frac{j}{2}\right) & j = 1, \ldots, m \\ Beta\left(\frac{n-1-j}{2}, \frac{j+1}{2}\right) & j = m+1, \ldots, p-1\,, \end{cases} \tag{5.421}$$

form a set of $p-1$ independent r.v.'s, so that the exact p.d.f. and c.d.f. of Λ in (5.418) are, for $m=\lfloor p/2\rfloor$, given in terms of the Meijer G function by

$$f_\Lambda(z)=\left\{\prod_{j=1}^{m}\frac{\Gamma\left(\frac{n-1}{2}\right)}{\Gamma\left(\frac{n-1-j}{2}\right)}\right\}\left\{\prod_{j=m+1}^{p-1}\frac{\Gamma\left(\frac{n}{2}\right)}{\Gamma\left(\frac{n-1-j}{2}\right)}\right\}\times G^{\,p-1,0}_{\,p-1,p-1}\left(\begin{array}{c}\left\{\left\{\frac{n-1}{2}-1\right\}_{j=1:m},\left\{\frac{n}{2}-1\right\}_{j=m+1:p-1}\right\}\\ \left\{\frac{n-1-j}{2}-1\right\}_{j=1:p-1}\end{array}\middle|\, z\right)$$

and

$$F_\Lambda(z)=\left\{\prod_{j=1}^{m}\frac{\Gamma\left(\frac{n-1}{2}\right)}{\Gamma\left(\frac{n-1-j}{2}\right)}\right\}\left\{\prod_{j=m+1}^{p-1}\frac{\Gamma\left(\frac{n}{2}\right)}{\Gamma\left(\frac{n-1-j}{2}\right)}\right\}\times G^{\,p-1,1}_{\,p,p}\left(\begin{array}{c}\left\{1,\left\{\frac{n-1}{2}\right\}_{j=1:m},\left\{\frac{n}{2}\right\}_{j=m+1:p-1}\right\}\\ \left\{\left\{\frac{n-1-j}{2}\right\}_{j=1:p-1},0\right\}\end{array}\middle|\, z\right).$$

For odd p, the exact distribution of Λ is given by Theorem 3.3 and its p.d.f. and c.d.f. are given by Corollary 4.3, with

$$m^*=1\,,\quad n_1=2\,,\quad k_1=\frac{p-1}{2}\,,\quad a_1=\frac{n-1}{2}\quad\text{and}\quad s_1=0\,.$$

The exact p.d.f. and c.d.f. of Λ in (5.418) are thus given, in terms of the EGIG distribution representation, respectively by

$$f_\Lambda(z)=f^{EGIG}\left(z\,\middle|\,\{r_j\}_{j=2:p};\left\{\frac{n-j}{2}\right\}_{j=2:p};p-1\right)$$

and

$$F_\Lambda(z)=F^{EGIG}\left(z\,\middle|\,\{r_j\}_{j=2:p};\left\{\frac{n-j}{2}\right\}_{j=2:p};p-1\right),$$

where

$$r_j=1+\left\lfloor\frac{p-j}{2}\right\rfloor,\qquad j=2,\ldots,p\,. \tag{5.422}$$

The same distribution was obtained by Marques and Coelho (2013), using a process specially dedicated to this statistic, while for even p one may use the near-exact distributions developed for Λ in this reference.

This test is invariant under linear transformations of the same type as those that assure the invariance of the test in Sect. 5.1.28, that is, linear transformations $\underline{X} \rightarrow C\underline{X} + \underline{b}$ where $\underline{b}$ is any $p\times 1$ real vector and C is any $p\times p$ nonrandom full-rank circular or circulant matrix.

For the implementation of this test modules `PDFCircOddp`, `CDFCircOddp`, `PvalCircOddp`, `QuantCircOddp`, and `PvalDataCircOddp` are made available by the authors. The first three of these modules have three mandatory arguments which are the sample size, the number of variables and the running or computed value of the statistic, in this order, while the fourth module also has three mandatory arguments which are the sample size, the number of variables and the value of α for which the quantile is sought. These four modules have exactly the same optional arguments as the corresponding modules in Sect. 5.1.1.

The plots in Fig. 5.81 were obtained with module `PDFCircOddp`. Part **a** of this figure illustrates how increasing values of n reshape the p.d.f., while part **b** shows how different values of p affect the p.d.f., for fixed values of n.

a

b

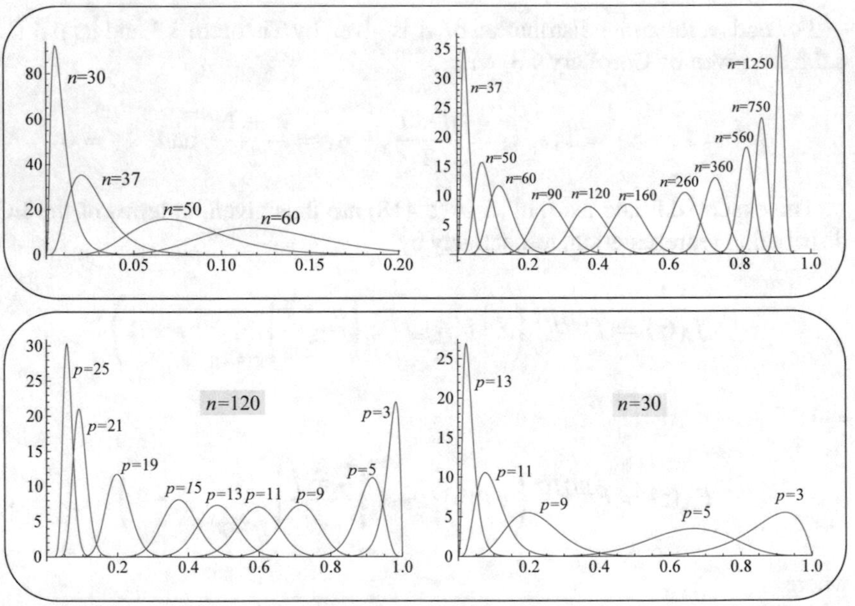

Fig. 5.81 Plots of p.d.f.'s of Λ in (5.418) for the test of circularity of the covariance matrix, for different values of p and n: (**a**) plots of p.d.f.'s for $p = 15$ and different values of n (the two plots have different horizontal and vertical scales); (**b**) plots of p.d.f.'s for $n = 120$ and $n = 30$ and different values of p (the two plots have different vertical scales)

Module `PvalDataCircOddp` has a single mandatory argument which is the name of the data file, with its optional arguments being exactly the same as those of the corresponding module in Sect. 5.1.1.

An application of this test is done by testing the circularity of the covariance matrix of the five response times for the study on response times described in Sect. 5.1.17 and in Exercise 3.9.2 of Timm (2002). This hypothesis seems to be a plausible one, given the type of variables involved in this study. The data is presented in Table 3.9.7 of the above reference, the last five columns of which are stored in file `Response_times.dat`. We also test separately the circularity of the covariance matrices of the response times for the low and high short term memory groups using the data reported in Table 3.9.8 of Timm (2002), which were separately stored in files `Response_times_low.dat` and `Response_times_high.dat` for each one of the two groups of observations. The commands used and the output obtained are in Fig. 5.82 and we may see that for all three covariance matrices, given the rather large p-values obtained, we are lead to not reject the null hypothesis of circularity, with all three samples giving quite high p-values for all multivariate normality tests, except the Shapiro-Wilk test (these are not shown).

One should not be surprised by the extremely high p-value for the first test in Fig. 5.82, since the sample covariance matrix for the five response times in file `Response_times.dat`, rounded to four decimal places, is

$$\begin{bmatrix} 65.0909 & 33.6455 & 47.5909 & 36.7727 & 25.4273 \\ 33.6455 & 46.0727 & 28.9455 & 40.3364 & 28.3636 \\ 47.5909 & 28.9455 & 60.6909 & 37.3727 & 41.1273 \\ 36.7727 & 40.3364 & 37.3727 & 62.8182 & 31.6818 \\ 25.4273 & 28.3636 & 41.1273 & 31.6818 & 58.2182 \end{bmatrix}.$$

```
PvalDataCircOddp[path <> "Response_times.dat"]
Computed value of Λ: 0.5910806436
p-value: 0.9822093724
PvalDataCircOddp[path <> "Response_times_low.dat"]
Computed value of Λ: 0.1644851847
p-value: 0.4434696464
PvalDataCircOddp[path <> "Response_times_high.dat"]
Computed value of Λ: 0.1480759539
p-value: 0.3898165880
```

Fig. 5.82 Commands used and output obtained with the Mathematica® module `PvalDataCircOddp` to implement the test of circularity of the covariance matrix, using the response times data in Tables 3.9.7 and 3.9.8 of Timm (2002)

5.2.2 *The Likelihood Ratio Statistic to Test Simultaneously the Circularity of the Covariance Matrix and the Equality of the Means (for an Odd Number of Variables) [CircMeanOddp]*

Let us suppose that $\underline{X} \sim N_p(\underline{\mu}, \Sigma)$ and that we are interested in testing simultaneously if Σ is circular and all the mean values in $\underline{\mu}$ are equal, that is, that, using the notation in the previous subsection, we are interested in testing the null hypothesis

$$H_0 : \Sigma = \Sigma_{cp},\ \underline{\mu} = aE_{p1}, \quad \text{for some unspecified } a \in \mathbb{R} \tag{5.423}$$

where E_{p1} represents a vector of dimension p with all elements equal to 1.

Then, for a sample of size n, the $(2/n)$-th power of the l.r.t. statistic to test H_0 in (5.423) is, using the first expression in (3.5) of Olkin and Press (1969), given by

$$\Lambda = n^p \frac{|A|}{v_1^*} \prod_{j=2}^{p} \frac{1}{v_j^* + w_j} = \frac{|V|}{v_1^*} \prod_{j=2}^{p} \frac{1}{v_j^* + w_j}, \tag{5.424}$$

where m, A, and V are defined as in the previous subsection, with v_j^* $(j = 1, \ldots, p)$ given by (5.419), and where

$$w_j = \begin{cases} y_j^2 & j = 1, \text{ and also } j = m+1 \text{ if } p \text{ is even,} \\ y_j^2 + y_{p-j+2}^2 & j = 2, \ldots, p-m, m+2, \ldots, p \end{cases}$$

with $\underline{Y} = [y_j] = \sqrt{n}\, U \overline{\underline{X}}$, where U is the matrix defined in the previous subsection, with running element given by (5.357) and $\overline{\underline{X}}$ is the vector of sample means.

According to Olkin and Press (1969, Sec. 5.2), for $n > p$,

$$\Lambda \equiv \prod_{j=2}^{p} Y_j \tag{5.425}$$

where

$$Y_j \sim \begin{cases} Beta\left(\frac{n-j}{2}, \frac{j}{2}\right), & j = 2, \ldots, m+1 \\ Beta\left(\frac{n-j}{2}, \frac{j+1}{2}\right), & j = m+2, \ldots, p, \end{cases} \tag{5.426}$$

are a set of $p - 1$ independent r.v.'s.

Thus, for odd p the distribution of Λ in (5.424) is given by Theorem 3.3 and its p.d.f. and c.d.f. given by Corollary 4.3, with

$$m^* = 1\,, \quad n_1 = 2\,, \quad k_1 = \frac{p-1}{2}\,, \quad a_1 = \frac{n-1}{2}\,, \quad s_1 = 1\,.$$

The exact p.d.f. and c.d.f. of Λ in (5.424) are thus given, for odd p, in terms of the EGIG distribution representation, respectively by

$$f_\Lambda(z) = f^{EGIG}\left(z \,\middle|\, \{r_j\}_{j=1:p}; \left\{\frac{n-j}{2}\right\}_{j=1:p}; p\right) \tag{5.427}$$

and

$$F_\Lambda(z) = F^{EGIG}\left(z \,\middle|\, \{r_j\}_{j=1:p}; \left\{\frac{n-j}{2}\right\}_{j=1:p}; p\right), \tag{5.428}$$

where

$$r_j = \frac{p-1}{2} - \left\lfloor \frac{|j-2|}{2} \right\rfloor, \quad j = 1, \ldots, p\,, \tag{5.429}$$

a result that confirms the ones in Coelho et al. (2013) and (2016, Sect. 2) for this statistic.

For a closed finite form representation of the p.d.f. and c.d.f of Λ in (5.424) for even p we refer the reader to Sect. 5.3.1.

For general p the p.d.f. and c.d.f. of Λ are given in terms of the Meijer G function as

$$f_\Lambda(z) = \left\{\prod_{j=2}^{m+1} \frac{\Gamma\left(\frac{n}{2}\right)}{\Gamma\left(\frac{n-j}{2}\right)}\right\} \left\{\prod_{j=m+2}^{p} \frac{\Gamma\left(\frac{n+1}{2}\right)}{\Gamma\left(\frac{n-j}{2}\right)}\right\} \times G^{p-1,0}_{p-1,p-1}\left(\begin{array}{c} \left\{\frac{n-2}{2}\right\}_{j=2:m+1}, \left\{\frac{n-1}{2}\right\}_{j=m+2:p} \\ \left\{\frac{n-j}{2}-1\right\}_{j=2:p} \end{array} \,\middle|\, z\right)$$

and

$$F_\Lambda(z) = \left\{\prod_{j=2}^{m+1} \frac{\Gamma\left(\frac{n}{2}\right)}{\Gamma\left(\frac{n-j}{2}\right)}\right\} \left\{\prod_{j=m+2}^{p} \frac{\Gamma\left(\frac{n+1}{2}\right)}{\Gamma\left(\frac{n-j}{2}\right)}\right\} \times G^{p-1,1}_{p,p}\left(\begin{array}{c} 1, \left\{\frac{n}{2}\right\}_{j=2:m+1}, \left\{\frac{n+1}{2}\right\}_{j=m+2:p} \\ \left\{\frac{n-j}{2}\right\}_{j=2:p}, 0 \end{array} \,\middle|\, z\right).$$

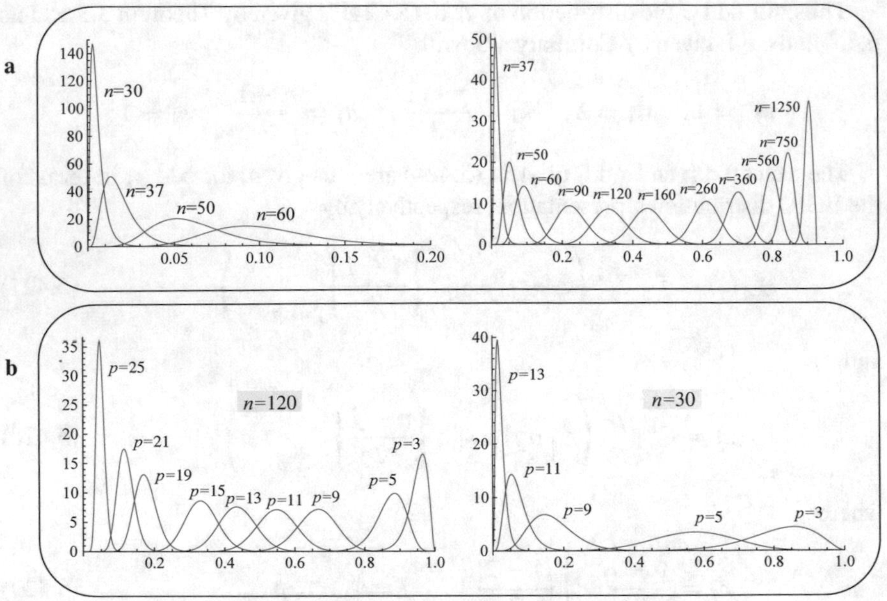

Fig. 5.83 Plots of p.d.f.'s of Λ in (5.424) for the simultaneous test of circularity of the covariance matrix and equality of means, for different values of p and n: (**a**) plots of p.d.f.'s for $p = 15$ and different values of n (the two plots have different horizontal and vertical scales); (**b**) plots of p.d.f.'s for $n = 120$ and $n = 30$ and different values of p

This test is invariant under linear transformations $\underline{X} \to C\underline{X} + bE_{p1}$ where b is any real and C is any $p \times p$ nonrandom full-rank circular or circulant matrix.

As for all other tests, modules `PDFCircMeanOddp`, `CDFCircMeanOddp`, `PvalCircMeanOddp`, `QuantCircMeanOddp`, and `PvalDataCircMeanOddp` are available, with exactly the same mandatory and optional arguments as the corresponding modules in the previous subsection and with the `PvalData` module taking data files of the same type as the `PvalData` module in that subsection. Plots in Fig. 5.83 were obtained with the first of these modules and illustrate how the shape of p.d.f.'s of Λ in (5.424) evolves for $p = 15$ and increasing values of n and for fixed n and decreasing values of p.

As an example of application of this test we use once again the data on response times in Tables 3.9.7 and 3.9.8 of Timm (2002), already used in Sects. 5.1.22, 5.1.24, 5.1.28 and in the previous subsection. The commands used and the output obtained are reported in Fig. 5.84.

The results obtained show that we should not reject the null hypothesis of equality of the means and circularity of the covariance matrix for the five response times for all three populations, that is, for the population of undiscriminated respondents which sample is in Table 3.9.7 of Timm (2002) as well as for the low and high short-term memory capacity respondents populations. If by looking at the plot in Fig. 5.78 the result concerning the equality of the means for the five response times may sur-

```
PvalDataCircMeanOddp[path <> "Response_times.dat"]
Computed value of Λ: 0.3611844437
p-value: 0.9373015928
PvalDataCircMeanOddp[path <> "Response_times_low.dat"]
Computed value of Λ: 0.1019273456
p-value: 0.4337944754
PvalDataCircMeanOddp[path <> "Response_times_high.dat"]
Computed value of Λ: 0.06333264805
p-value: 0.2389066061
```

Fig. 5.84 Commands used and output obtained with the Mathematica® module `PvalDataCircMeanOddp` to implement the simultaneous test of circularity of the covariance matrix and equality of means, using the response times data in Tables 3.9.7 and 3.9.8 of Timm (2002)

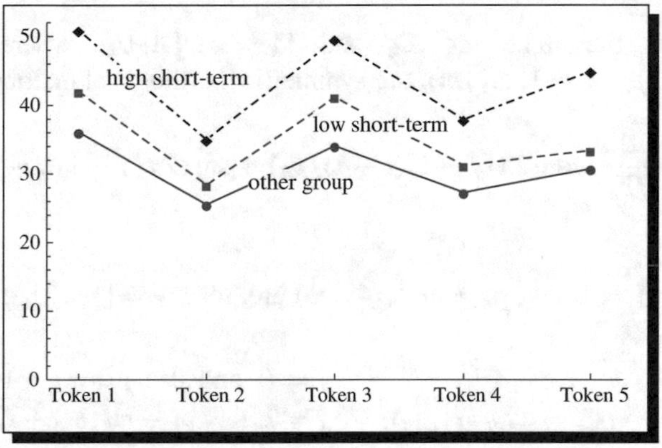

Fig. 5.85 Plot of the sample profiles for the five response times for the three subpopulations corresponding to the data in Tables 3.9.7 and 3.9.8 of Timm (2002)

prise us a bit, we should take into account that this plot has its vertical axis cut, which may give us the wrong message. If the whole vertical scale is taken into account, it should look like the plot in Fig. 5.85. Then it seems that all three p-values obtained in Fig. 5.84, and even their relative values, make absolute sense, moreover since the samples being rather small give us less power in rejecting the null hypotheses.

5.2.3 The Likelihood Ratio Statistic to Test the Simultaneous Circularity of m Covariance Matrices [CircS]

Let us suppose we have m populations, $\underline{X}_k$ $(k = 1, \dots, m)$, with $\underline{X}_k \sim N_{p_k}(\underline{\mu}_k, \Sigma_k)$ and that, based on m independent samples, one from each population, we are interested in testing the hypothesis

$$H_0 : \Sigma_k = \Sigma_{cp_k}\,, \quad k = 1, \dots, m \tag{5.430}$$

where Σ_{cp_k} represents a circular matrix of order $p_k \times p_k$, that is, to test simultaneously the circularity of all the m covariance matrices Σ_k $(k = 1, \dots, m)$. Let us further suppose that all samples have dimension $n > \max_{k\in\{1,\dots,m\}} p_k$. Then, the $(2/n)$-th power of the l.r.t. statistic to test H_0 in (5.430) is

$$\Lambda = n^{\sum_{k=1}^m p_k} \prod_{k=1}^m \frac{|A_k|}{\prod_{j=1}^{p_k} v_{jk}^*} = \prod_{k=1}^m \frac{|V_k|}{\prod_{j=1}^{p_k} v_{jk}^*}\,, \tag{5.431}$$

where A_k is the m.l.e. of Σ_k and $V_k = n\, U_k' A_k U_k$, where $U_k = [u_{ijk}]$ $(i, j = 1, \dots, p_k;\ k = 1, \dots, m)$, are symmetrical orthogonal matrices, with

$$u_{ijk} = \frac{1}{\sqrt{p_k}} \{\cos(2\pi(j-1)(i-1)/p_k) + \sin(2\pi(j-1)(i-1)/p_k)\} \tag{5.432}$$

and the v_{jk}^*'s $(j = 1, \dots, p_k;\ k = 1, \dots, m)$ are, for $m_k = \lfloor p_k/2 \rfloor$, given by

$$v_{jk}^* = \begin{cases} v_{jjk} & j = 1, \text{ and also } j = m_k + 1 \text{ if } p_k \text{ is even}, \\ (v_{jjk} + v_{(p_k-j+2),(p_k-j+2),k})/2 & j = 2, \dots, p_k - m_k, m_k + 2, \dots, p_k, \end{cases} \tag{5.433}$$

where v_{jjk} is the j-th diagonal element of V_k.

Under H_0 in (5.430),

$$\Lambda \stackrel{d}{\equiv} \prod_{k=1}^m \prod_{j=1}^{p_k-1} Y_{jk} \tag{5.434}$$

where, for $k = 1, \dots, m$,

$$Y_{jk} \sim \begin{cases} Beta\left(\frac{n-1-j}{2}, \frac{j}{2}\right) & j = 1, \dots, m_k \\ Beta\left(\frac{n-1-j}{2}, \frac{j+1}{2}\right) & j = m_k + 1, \dots, p_k - 1\,, \end{cases} \tag{5.435}$$

form m sets of $p_k - 1$ independent r.v.'s ($k = 1, \ldots, m$), so that if all p_k are odd, the exact distribution of Λ in (5.431) is given by Theorem 3.3 and its p.d.f. and c.d.f. are given by Corollary 4.3, with

$$m^* = m\,, \quad n_\nu = 2\,, \quad k_\nu = \frac{p_\nu - 1}{2}\,, \quad a_\nu = \frac{n-1}{2} \text{ and } s_\nu = 0 \quad (\nu = 1, \ldots, m^*)\,. \tag{5.436}$$

Then, if for some odd p, $p_k = p$ ($k = 1, \ldots, m$), the exact p.d.f. and c.d.f. of Λ in (5.431) are given, in terms of the EGIG distribution representation, respectively by

$$f_\Lambda(z) = f^{EGIG}\left(z \,\middle|\, \{r_j^*\}_{j=2:p};\, \left\{\frac{n-j}{2}\right\}_{j=2:p};\, p-1\right)$$

and

$$F_\Lambda(z) = F^{EGIG}\left(z \,\middle|\, \{r_j^*\}_{j=2:p};\, \left\{\frac{n-j}{2}\right\}_{j=2:p};\, p-1\right),$$

where $r_j^* = m r_j$, with r_j given by (5.422), while if at least one of the p_k is different, but they remain all odd, the exact p.d.f. and c.d.f. of Λ in (5.431) are given respectively by

$$f_\Lambda(z) = f^{EGIG}\left(z \,\middle|\, \{r_j^*\}_{j=2:\max\{p_k\}};\, \left\{\frac{n-j}{2}\right\}_{j=2:\max\{p_k\}};\, \max\{p_k\}-1\right)$$

and

$$F_\Lambda(z) = F^{EGIG}\left(z \,\middle|\, \{r_j^*\}_{j=2:\max\{p_k\}};\, \left\{\frac{n-j}{2}\right\}_{j=2:\max\{p_k\}};\, \max\{p_k\}-1\right),$$

where

$$r_j^* = \sum_{k=1}^{m} r_{jk}^*$$

with

$$r_{jk}^* = \begin{cases} r_{jk}\,, & j = 2, \ldots, p_k \\ 0\,, & j = 1 + p_k, \ldots, \max\{p_k\} \end{cases}$$

where, for $j = 2, \ldots, p_k$,

$$r_{jk} = 1 + \left\lfloor \frac{p_k - j}{2} \right\rfloor . \tag{5.437}$$

In case the sample from $\underline{X}_k$ has size $n_k > p_k$ ($k = 1, \ldots, m$), with possible different values for the n_k's, then we will have to replace $n^{\sum_{k=1}^{m} p_k}$ by $\prod_{k=1}^{m} n_k^{p_k}$ in (5.431). However, the statistic in (5.431) will no longer be the $(2/n)$-th power of the l.r.t. statistic to test the hypothesis in (5.430), but rather, it will be the product of the $(2/n_k)$-th powers of the l.r.t. statistics to test each sub-hypothesis

$$H_{0k} : \Sigma_k = \Sigma_{cp_k}$$

for a given $k \in \{1, \ldots, m\}$. Anyway, we may still use the statistic in (5.431) to test the null hypothesis in (5.430), with its exact distribution given by (5.434) and (5.435) with n replaced by n_k. As such, in case all p_k are odd, the exact distribution of Λ in (5.431) will be given by Theorem 3.3 and its exact p.d.f. and c.d.f. given by Corollary 4.3, with the parameters in (5.436) with n replaced by n_k, yielding a representation of the exact p.d.f. and c.d.f. of Λ in (5.431), in terms of the EGIG distribution representation, respectively given by

$$f_\Lambda(z) = f^{EGIG}\left(z \;\middle|\; \overset{\approx}{\left\{ \{r_{jk}\}_{\substack{k=1:m\\ j=2:p_k}} \right\}} ; \overset{\sim}{\left\{ \left\{ \frac{n_k - j}{2} \right\}_{\substack{k=1:m\\ j=2:p_k}} \right\}} ; \leq \sum_{k=1}^{m} (p_k - 1) \right),$$

and

$$F_\Lambda(z) = F^{EGIG}\left(z \;\middle|\; \overset{\approx}{\left\{ \{r_{jk}\}_{\substack{k=1:m\\ j=2:p_k}} \right\}} ; \overset{\sim}{\left\{ \left\{ \frac{n_k - j}{2} \right\}_{\substack{k=1:m\\ j=2:p_k}} \right\}} ; \leq \sum_{k=1}^{m} (p_k - 1) \right),$$

where the r_{jk} are given by (5.437) and where we used the same notation used in Corollary 4.3 and Sect. 2.2 for the vectors of shape and rate parameters, where

$$\overset{\sim}{\left\{ \left\{ \frac{n_k - j}{2} \right\}_{\substack{k=1:m\\ j=2:p_k}} \right\}}$$

denotes the set of unique values of $\frac{n_k - j}{2}$ for all $j = 2, \ldots, p_k$ and $k = 1, \ldots, m$, and

$$\overset{\approx}{\left\{ \{r_{jk}\}_{\substack{k=1:m\\ j=2:p_k}} \right\}}$$

denotes the corresponding "contraction" of the set of shape parameters, obtained by adding the shape parameters r_{jk} that correspond to equal values of $\frac{n_k-j}{2}$.

For any set of p_k $(k = 1, \ldots, m)$, the p.d.f. and c.d.f. of Λ in (5.431), under H_0 in (5.430), are given in terms of the Meijer G function, for different n_k and for $m_k = \lfloor p_k/2 \rfloor$ and $p = \sum_{k=1}^{m}(p_k - 1)$, by

$$f_\Lambda(z) = \left\{\prod_{k=1}^{m}\left\{\prod_{j=1}^{m_k}\frac{\Gamma\left(\frac{n_k-1}{2}\right)}{\Gamma\left(\frac{n_k-1-j}{2}\right)}\right\}\left\{\prod_{j=m_k+1}^{p_k-1}\frac{\Gamma\left(\frac{n_k}{2}\right)}{\Gamma\left(\frac{n_k-1-j}{2}\right)}\right\}\right\} \times G_{p,p}^{p,0}\left(\begin{array}{c}\left\{\left\{\frac{n_k-3}{2}\right\}_{\substack{j=1:m_k\\ k=1:m}}, \left\{\frac{n_k-2}{2}\right\}_{\substack{j=m_k+1:p_k-1\\ k=1:m}}\right\}\\ \left\{\frac{n_k-3-j}{2}\right\}_{\substack{j=1:p_k-1\\ k=1:m}}\end{array}\middle|\, z\right)$$

and

$$F_\Lambda(z) = \left\{\prod_{k=1}^{m}\left\{\prod_{j=1}^{m_k}\frac{\Gamma\left(\frac{n_k-1}{2}\right)}{\Gamma\left(\frac{n_k-1-j}{2}\right)}\right\}\left\{\prod_{j=m_k+1}^{p_k-1}\frac{\Gamma\left(\frac{n_k}{2}\right)}{\Gamma\left(\frac{n_k-1-j}{2}\right)}\right\}\right\} \times G_{p+1,p+1}^{p,1}\left(\begin{array}{c}\left\{1, \left\{\frac{n_k-1}{2}\right\}_{\substack{j=1:m_k\\ k=1:m}}, \left\{\frac{n_k}{2}\right\}_{\substack{j=m_k+1:p_k-1\\ k=1:m}}\right\}\\ \left\{\left\{\frac{n_k-1-j}{2}\right\}_{\substack{j=1:p_k-1\\ k=1:m}}, 0\right\}\end{array}\middle|\, z\right).$$

For the test in this subsection modules `PDFCircS`, `CDFCircS`, `PvalCircS`, `QuantCircS`, and `PvalDataCircS` are made available. The first four of these modules have as their first two mandatory arguments a list with the m sample sizes n_k and a list with the p_k values $(k = 1, \ldots, m)$, with the first three of these modules having then as third mandatory argument the computed value of the statistic for which the p.d.f. or the c.d.f. is to be computed or for which the p-value is intended to be computed, while the fourth module has as its third mandatory argument the α-value for which the quantile is sought. Module `PvalDataCircS` has as its only mandatory argument the name of the data file with the m samples. This file, given the possibility of existence of samples with different numbers of variables, and as such with different variables, has to have a structure similar to that of the files in Fig. 5.7. The optional arguments for all these modules are exactly the same as those for the corresponding modules in Sect. 5.1.1.

To exemplify the use of this test we again use the data in Tables 3.9.7 and 3.9.8 of Timm (2002), to test simultaneously the circularity of the three population covariance matrices for the undiscriminated respondents and the low and high short-term memory capacity populations whose samples are reported in the above

```
PvalDataCircS[path <> "Response_times_all.dat"]
 There are 3 samples, with sizes {11, 10, 10} and {5, 5, 5} variables
Computed value of Λ: 0.01439653785
p-value: 0.7648539951

PvalDataCircS[path <> "Response_times_all.dat", 1]
 There are 3 samples, with sizes {11, 10, 10} and {5, 5, 5} variables
 Original variables kept {{3, 4, 5}, {1, 2, 3}, {3, 4, 5}}
 There are now 3 samples, with sizes {11, 10, 10} and {3, 3, 3} variables
Computed value of Λ: 0.5037735696
p-value: 0.9455527355
```

Fig. 5.86 Commands used and output obtained with the Mathematica® module `PvalDataCircS` to implement the simultaneous test of circularity of the three covariance matrices for the three populations whose samples of response times are reported in Tables 3.9.7 and 3.9.8 of Timm (2002)

mentioned tables. The first command in Fig. 5.86 tests the hypothesis of circularity of the covariance matrix for all five response times for the three populations, while the second command tests the hypothesis of circularity of the covariance matrix for the last three response times for the populations of undiscriminated respondents and high short-term memory capacity respondents, and for the first three response times for the population of low short-term memory capacity respondents, to illustrate the use of the first optional argument of module `PvalDataCircS`, which allows for the selection of different sets of variables for each population. The quite high p-values obtained show that in both cases we should not reject these null hypotheses of circularity of all three population covariance matrices. The result for the first test clearly matches the results obtained in Sect. 5.2.1.

The present test is invariant under linear transformations $\underline{X}_k \to C_k\underline{X}_k + \underline{b}_k$ where $\underline{b}_k$ is any $p{\times}1$ real vector and C_k is any $p_k{\times}p_k$ nonrandom full-rank circular or circulant matrix.

5.2.4 *The Likelihood Ratio Statistic to Test Simultaneously the Circularity of the Covariance Matrices and the Equality of Means in m Subsets with an Odd Number of Variables [CircMeansOddp]*

As in the previous subsection, let us suppose we have m populations $\underline{X}_k$ $(k = 1, \dots, m)$, with $\underline{X}_k \sim N_{p_k}(\underline{\mu}_k, \Sigma_k)$ and that we want to test simultaneously for all m populations the circularity of their covariance matrices and the equality of all means in each vector $\underline{\mu}_k$, that is, that we want to test the null hypothesis

$$H_0 : \bigwedge_{k=1}^{m} \Sigma_k = \Sigma_{cp_k}, \ \underline{\mu}_k = a_k\, E_{p_k 1} \quad \text{(for some unspecified } \underline{a}_k \in \mathbb{R}), \tag{5.438}$$

where $E_{p_k 1}$ is a vector of 1's, of dimension p_k.

Let us suppose we have m independent samples, one from each $\underline{X}_k$ and that all samples have common dimension $n > \max_{k\in\{1,\dots,m\}} p_k$.

Then, the $(2/n)$-th power of the l.r.t. statistic to test H_0 is

$$\Lambda = n^{\sum_{k=1}^{m} p_k} \prod_{k=1}^{m} \frac{|A_k|}{v_{1k}^* \prod_{j=2}^{p_k} (w_{jk} + v_{jk}^*)} = \prod_{k=1}^{m} \frac{|V_k|}{v_{1k}^* \prod_{j=2}^{p_k} (w_{jk} + v_{jk}^*)}, \tag{5.439}$$

where A_k is, as in the previous subsection, the m.l.e. of Σ_k, and m_k and v_{jk}^* are defined as in the previous subsection, $V_k = nU_k' A_k U_k$, and, for $m_k = \lfloor p_k/2 \rfloor$,

$$w_{jk} = \begin{cases} y_{jk}^2 & j = 1, \text{ and also } j = m_k + 1 \text{ if } p_k \text{ is even,} \\ y_{jk}^2 + y_{p-j+2,k}^2 & j = 2, \dots, p_k - m_k, m_k + 2, \dots, p_k \end{cases} \tag{5.440}$$

with $\underline{Y}_k = [y_{jk}] = \sqrt{n}\, U_k\, \overline{\underline{X}}_k$, where U_k is the matrix defined in the previous subsection, with elements given by (5.432) and $\overline{\underline{X}}_k$ is the vector of sample means for the sample from $\underline{X}_k$.

Then

$$\Lambda \stackrel{d}{\equiv} \prod_{k=1}^{m} \prod_{j=2}^{p_k} Y_{jk} \tag{5.441}$$

where, for $k = 1, \dots, m$,

$$Y_{jk} \sim \begin{cases} Beta\left(\frac{n-j}{2}, \frac{j}{2}\right), & j = 2, \dots, m_k + 1 \\ Beta\left(\frac{n-j}{2}, \frac{j+1}{2}\right), & j = m_k + 2, \dots, p_k, \end{cases} \tag{5.442}$$

form m sets of $p_k - 1$ independent r.v.'s ($k = 1, \dots, m$), so that if all p_k are odd, the exact distribution of Λ in (5.431) is given by Theorem 3.3 and its p.d.f. and c.d.f. are given by Corollary 4.3, with

$$m^* = m\,, \ \ n_\nu = 2\,, \ \ k_\nu = \frac{p_\nu - 1}{2}\,, \ \ a_\nu = \frac{n-1}{2} \ \text{ and } s_\nu = 1 \quad (\nu = 1, \dots, m^*)\,.$$

Then, if $p_k = p$ $(k = 1, \dots, m)$, the exact p.d.f. and c.d.f. of Λ in (5.439) are given, in terms of the EGIG distribution representation, respectively by

$$f_\Lambda(z) = f^{EGIG}\left(z \,\Big|\, \{r_j^*\}_{j=1:p};\; \left\{\frac{n-j}{2}\right\}_{j=1:p};\; p\right) \tag{5.443}$$

and

$$F_\Lambda(z) = F^{EGIG}\left(z \,\Big|\, \{r_j^*\}_{j=1:p};\; \left\{\frac{n-j}{2}\right\}_{j=1:p};\; p\right), \tag{5.444}$$

where $r_j^* = m r_j$, with r_j given by (5.429), while if at least some of the p_k are different, the exact p.d.f. and c.d.f. of Λ in (5.431) are given respectively by

$$f_\Lambda(z) = f^{EGIG}\left(z \,\Big|\, \{r_j^*\}_{j=1:\max\{p_k\}};\; \left\{\frac{n-j}{2}\right\}_{j=1:\max\{p_k\}};\; \max\{p_k\}\right) \tag{5.445}$$

and

$$F_\Lambda(z) = F^{EGIG}\left(z \,\Big|\, \{r_j^*\}_{j=1:\max\{p_k\}};\; \left\{\frac{n-j}{2}\right\}_{j=1:\max\{p_k\}};\; \max\{p_k\}\right), \tag{5.446}$$

where

$$r_j^* = \sum_{k=1}^{m} r_{jk}^* \tag{5.447}$$

with

$$r_{jk}^* = \begin{cases} r_{jk}\,, & j = 1, \dots, p_k \\ 0\,, & j = 1 + p_k, \dots, \max\{p_k\} \end{cases} \tag{5.448}$$

where, for $j = 1, \dots, p_k$, and $k = 1, \dots, m$,

$$r_{jk} = \frac{p_k - 1}{2} - \left\lfloor \frac{|j-2|}{2} \right\rfloor. \tag{5.449}$$

In case the sample from $\underline{X}_k$ has size $n_k > p_k$ $(k = 1, \dots, m)$, with possible different values for the n_k's, as happened in the previous subsection, we will have to replace in the expression for Λ in (5.439), $n^{\sum_{k=1}^m p_k}$ by $\prod_{k=1}^m n_k^{p_k}$. Although the statistic with this expression is no longer the $(2/n)$-th power of the l.r.t. statistic to test the hypothesis in (5.438), it will be the product of the $(2/n_k)$-th powers of the l.r.t. statistics to test each sub-hypothesis

$$H_{0k}: \Sigma_k = \Sigma_{cp_k},\ \underline{\mu}_k = a_k E_{p_k 1} \quad \text{(for some unspecified } a_k \in \mathbb{R}) \tag{5.450}$$

for a given $k \in \{1, \ldots, m\}$. As such, we may still use this statistic to test the null hypothesis in (5.438), with its exact distribution given by (5.441) and (5.442) with n replaced by n_k. In the case in which all p_k are odd, the exact distribution of Λ in (5.439) will be given by Theorem 3.3 and its exact p.d.f. and c.d.f. given by Corollary 4.3, with the parameters in (5.436) with n replaced by n_ν, for $\nu = 1, \ldots, m$, yielding a representation of the exact p.d.f. and c.d.f. of Λ in (5.439), in terms of the EGIG distribution representation, respectively given by

$$f_\Lambda(z) = f^{EGIG}\left(z \,\middle|\, \widetilde{\widetilde{\left\{\{r_{jk}\}_{\substack{k=1:m\\ j=1:p_k}}\right\}}};\ \widetilde{\left\{\left\{\frac{n_k-j}{2}\right\}_{\substack{k=1:m\\ j=1:p_k}}\right\}};\ \le \sum_{k=1}^{m} p_k\right), \tag{5.451}$$

and

$$F_\Lambda(z) = F^{EGIG}\left(z \,\middle|\, \widetilde{\widetilde{\left\{\{r_{jk}\}_{\substack{k=1:m\\ j=1:p_k}}\right\}}};\ \widetilde{\left\{\left\{\frac{n_k-j}{2}\right\}_{\substack{k=1:m\\ j=1:p_k}}\right\}};\ \le \sum_{k=1}^{m} p_k\right), \tag{5.452}$$

where the r_{jk} are given by (5.449) and where we used the same notation used in Corollary 4.3 and Sect. 2.2 for the vectors of shape and rate parameters, where

$$\widetilde{\left\{\left\{\frac{n_k-j}{2}\right\}_{\substack{k=1:m\\ j=1:p_k}}\right\}}$$

denotes the set of unique values of $\frac{n_k+1-j}{2}$ for all $j = 1, \ldots, p_k$ and $k = 1, \ldots, m$, and

$$\widetilde{\widetilde{\left\{\{r_{jk}\}_{\substack{k=1:m\\ j=1:p_k}}\right\}}}$$

denotes the corresponding "contraction" of the set of shape parameters r_{jk}, obtained by adding the shape parameters that correspond to equal values of $\frac{n_k-j}{2}$.

For the case of even p_k's, see Sect. 5.3.2 and for the case of both even and odd p_k's see Sect. 5.3.3 in the next section.

For any set of p_k ($k = 1, \ldots, m$), the p.d.f. and c.d.f. of Λ in (5.431), under H_0 in (5.430), are given in terms of the Meijer G function, for different n_k and for $m_k = \lfloor p_k/2 \rfloor$ and $p = \sum_{k=1}^{m}(p_k - 1)$, by

$$f_\Lambda(z) = \left\{ \prod_{k=1}^{m} \left\{ \prod_{j=2}^{m_k+1} \frac{\Gamma\left(\frac{n_k}{2}\right)}{\Gamma\left(\frac{n_k-j}{2}\right)} \right\} \left\{ \prod_{j=m_k+2}^{p_k} \frac{\Gamma\left(\frac{n_k+1}{2}\right)}{\Gamma\left(\frac{n_k-j}{2}\right)} \right\} \right\} \times G_{p,p}^{p,0}\left(\begin{array}{c} \left\{ \left\{ \frac{n_k-2}{2} \right\}_{\substack{j=2:m_k+1 \\ k=1:m}}, \left\{ \frac{n_k-1}{2} \right\}_{\substack{j=m_k+2:p_k \\ k=1:m}} \right\} \\ \left\{ \frac{n_k-2-j}{2} \right\}_{\substack{j=2:p_k \\ k=1:m}} \end{array} \middle| \, z \right)$$

and

$$F_\Lambda(z) = \left\{ \prod_{k=1}^{m} \left\{ \prod_{j=2}^{m_k+1} \frac{\Gamma\left(\frac{n_k}{2}\right)}{\Gamma\left(\frac{n_k-j}{2}\right)} \right\} \left\{ \prod_{j=m_k+2}^{p_k} \frac{\Gamma\left(\frac{n_k+1}{2}\right)}{\Gamma\left(\frac{n_k-j}{2}\right)} \right\} \right\} \times G_{p+1,p+1}^{p,1}\left(\begin{array}{c} \left\{ 1, \left\{ \frac{n_k}{2} \right\}_{\substack{j=2:m_k+1 \\ k=1:m}}, \left\{ \frac{n_k+1}{2} \right\}_{\substack{j=m_k+2:p_k \\ k=1:m}} \right\} \\ \left\{ \left\{ \frac{n_k-j}{2} \right\}_{\substack{j=2:p_k \\ k=1:m}}, 0 \right\} \end{array} \middle| \, z \right).$$

To implement this test modules `PDFCircMeansOddp`, `CDFCircMeansOddp`, `PvalCircMeansOddp`, `QuantCircMeansOddp`, and `PvalDataCircMeansOddp` are available, with exactly the same arguments as the corresponding modules in the previous subsection.

As an example of application of the present test we will first test simultaneously the circularity of the covariance matrices and the equality of the mean values for the five response times for the three populations of undiscriminated respondents and low and high short-term memory capacity respondents whose samples are found in Tables 3.9.7 and 3.9.8 of Timm (2002). The command used and the output obtained are in Fig. 5.87 and the quite high p-value obtained shows that we should not reject this null hypothesis, a result that perfectly matches the results obtained in Sect. 5.2.2. With the second command in this same figure we test a similar hypothesis but now for the five response times for the population of undiscriminated respondents and for the first, third and fifth response times for the populations of low and high short-term memory capacity.

The p-value obtained is now even higher, which is due to the fact that, as it may be observed in Figs. 5.78 and 5.85, the mean values for the first and third response times are indeed very close for these two latter populations. This second command in Fig. 5.87 illustrates the use of the first optional argument of module

```
PvalDataCircMeansOddp[path <> "Response_times_all.dat"]
 There are 3 samples, with sizes {11, 10, 10} and {5, 5, 5} variables
Computed value of Λ: 0.002331564307
p-value: 0.5917051958
PvalDataCircMeansOddp[path <> "Response_times_all.dat", 1]
 There are 3 samples, with sizes {11, 10, 10} and {5, 5, 5} variables
 Original variables kept {{1, 2, 3, 4, 5}, {1, 3, 5}, {1, 3, 5}}
 There are now 3 samples, with sizes {11, 10, 10} and {5, 3, 3} variables
Computed value of Λ: 0.05845406646
p-value: 0.7375709426
```

Fig. 5.87 Commands used and output obtained with the Mathematica® module `PvalDataCircMeansOddp` to implement the simultaneous test of circularity of the covariance matrix and equality of means for the three populations whose samples of response times are reported in Tables 3.9.7 and 3.9.8 of Timm (2002)

`PvalDataCircMeansOddp`, which allows for the selection of samples and/or the selection of variables, with the possibility of selecting different sets of variables for the different populations.

The test in this subsection is invariant under linear transformations $\underline{X}_k \to C_k\underline{X}_k + b_k E_{p1}$ where b_k is any real and C_k is any $p_k \times p_k$ nonrandom full-rank circular or circulant matrix.

5.3 Likelihood Ratio Test Statistics Whose Distributions Correspond to a Multiplication of the Products in Theorem 3.1 or 3.2 and in Theorem 3.3 and That Have p.d.f. and c.d.f. Given by Corollary 4.4 or 4.5

Testing some of the hypotheses already addressed in the previous subsections for situations with a different parity of the number of variables involved or certain more complex hypotheses may call for a hybrid approach based on a combination of the results in at least two of the Theorems in Chap. 3, with the distributions of the statistics given by Corollary 4.4 or 4.5 in Chap. 4.

5.3.1 The Likelihood Ratio Statistic to Test Simultaneously the Circularity of the Covariance Matrix and the Equality of the Means (for an Even Number of Variables) [CircMeanEvenp]

Let us assume the setup of Sect. 5.2.2, and let us consider the l.r.t. statistic in (5.424) and its distribution in (5.425) and (5.426).

For even p, the distribution of Λ in (5.424) is given by a combination of the products in Theorem 3.1 or 3.2 and Theorem 3.3, and its p.d.f. and c.d.f. are thus given by Corollary 4.4 or 4.5, with

$$m^* = 1\,,\ n_1 = 1\,,\ k_1 = 2\,,\ a_1 = \tfrac{n}{2}\,,\ m_1 = 1\,,\ \text{and}$$
$$m^{**} = 1\,,\ n_1^* = 2\,,\ k_1^* = \tfrac{p}{2} - 1\,,\ a_1^* = \tfrac{n-2}{2}\,,\ s_1^* = 2\,.$$

Interestingly enough is the fact that in terms of the EGIG representation, the exact p.d.f. and c.d.f. of Λ are still respectively given by (5.427) and (5.428), now with

$$r_j = \begin{cases} \frac{p}{2} - 1 & j = 1 \\ \frac{p}{2} - \left\lfloor \frac{j-1}{2} \right\rfloor & j = 2, \dots, p\,. \end{cases} \tag{5.453}$$

For references where the same distribution was obtained using an alternative approach see Coelho et al. (2013) and (2016, Sect. 2).

For the case of odd p, and for the expression of the p.d.f. and c.d.f. in terms of the Meijer G function, see Sect. 5.2.2 in the previous section.

For the implementation of the present test modules `PDFCircMeanEvenp`, `CDFCircMeanEvenp`, `PvalCircMeanEvenp`, `QuantCircMeanEvenp`, and `PvalDataCircMeanEvenp` are made available, which have exactly the same arguments as their counterparts in Sect. 5.2.2.

To illustrate the use of the present test we will once again use the response times data in Tables 3.9.7 and 3.9.8 of Timm (2002), already used in Sect. 5.2.2 and described in Sect. 5.1.17 for the undiscriminated respondents population. We will be testing, for each of the three populations of undiscriminated, low and high short-term memory capacity respondents, the null hypothesis of circularity of the covariance matrices and equality of the means for the response times for the last four token words.

The commands used are in Fig. 5.88 and the results obtained show that we should not reject this null hypothesis for all three populations.

We may note that although in Sect. 5.2.2 we had already tested similar hypotheses for all five response times and we did not reject these null hypotheses, that does not directly imply that we would not reject similar hypotheses for the first four of these response times since the circularity of the covariance matrices for the five response

```
PvalDataCircMeanEvenp[path <> "Response_times.dat", 1]
 Original variables {2, 3, 4, 5}
Computed value of Λ: 0.7512369813
p-value: 0.9917433951

PvalDataCircMeanEvenp[path <> "Response_times_low.dat", 1]
 Original variables {2, 3, 4, 5}
Computed value of Λ: 0.2371223066
p-value: 0.3779868376

PvalDataCircMeanEvenp[path <> "Response_times_high.dat", 1]
 Original variables {2, 3, 4, 5}
Computed value of Λ: 0.3383930791
p-value: 0.6185795547
```

Fig. 5.88 Commands used and output obtained with the Mathematica® module `PvalDataCircMeanEvenp` to implement the simultaneous test of circularity of the covariance matrix and equality of means for the response times for the 2nd, 3rd, 4th and 5th token words, whose samples are in Tables 3.9.7 and 3.9.8 of Timm (2002)

times does not necessarily imply the circularity of the covariance matrices for the last four of these response times.

The very high p-value obtained for the undiscriminated respondents may come from the fact that the sample covariance matrix for the response times for the last four token words for these respondents, which rounded to four decimal places is equal to

$$\begin{bmatrix} 46.0727 & 28.9455 & 40.3364 & 28.3636 \\ 28.9455 & 60.6909 & 37.3727 & 41.1273 \\ 40.3364 & 37.3727 & 62.8182 & 31.6818 \\ 28.3636 & 41.1273 & 31.6818 & 58.2182 \end{bmatrix},$$

has a structure which is quite compliant with the circular structure for $p = 4$ variables, which is

$$\sigma^2 \begin{bmatrix} 1 & \rho_1 & \rho_2 & \rho_1 \\ \rho_1 & 1 & \rho_1 & \rho_2 \\ \rho_2 & \rho_1 & 1 & \rho_1 \\ \rho_1 & \rho_2 & \rho_1 & 1 \end{bmatrix}.$$

The test in this subsection is invariant under linear transformations similar to those under which the test in Sect. 5.2.2 is invariant.

5.3.2 The l.r.t. Statistic to Test Simultaneously the Circularity of the Covariance Matrices and the Equality of Means in m Subsets with an Even Number of Variables [CircMeansEvenp]

Let us assume the setup of Sect. 5.2.4, and let us consider the l.r.t. statistic in (5.439) and its distribution in (5.441) and (5.442), thus assuming that all samples have a common dimension $n > \max_{k\in\{1,\dots,m\}}\{p_k\}$.

If all p_k are even, the distribution of Λ in (5.439) is given by a combination of the products in Theorems 3.2 and 3.3, and its p.d.f. and c.d.f. given by Corollary 4.5, with

$$m^* = m\,,\;\; n_\nu = 1\;,\;\; k_\nu = 2\,,\;\; a_\nu = \frac{n}{2}\,,\;\; m_\nu = 1\,,\quad (\nu = 1,\dots,m^*)\,, \qquad (5.454)$$

and

$$m^{**} = m\,,\;\; n^*_\nu = 2\,,\;\; k^*_\nu = \frac{p_\nu}{2} - 1\,,\;\; a^*_\nu = \frac{n-2}{2}\,,\;\; s^*_\nu = 2\,,\quad (\nu = 1,\dots,m^{**}) \qquad (5.455)$$

so that if $p_k = p$ $(k = 1,\dots,m)$, the exact p.d.f. and c.d.f. of Λ in (5.439) are still respectively given by (5.443) and (5.444), now with r_j given by (5.453), while if at least one of the p_k is different, the exact p.d.f. and c.d.f. of Λ in (5.439) are given respectively by (5.445) and (5.446), still with r^*_j and r^*_{jk} respectively given by (5.447) and (5.448) but now with

$$r_{jk} = \begin{cases} \frac{p_k}{2} - 1\,, & j = 1 \\ \frac{p_k}{2} - \left\lfloor \frac{j-1}{2} \right\rfloor\,, & j = 2,\dots,p_k\,. \end{cases} \qquad (5.456)$$

In case the sample from $\underline{X}_k$ has size $n_k > p_k$ $(k = 1,\dots,m)$, with possibly different values for the n_k's, the statistic in (5.439), with $n^{\sum_{k=1}^m p_k}$ replaced by $\prod_{k=1}^m n_k^{p_k}$, as happened in Sect. 5.2.4, is no longer the $(2/n)$-th power of the l.r.t. statistic to test the hypothesis in (5.438), but it will be the product of the $(2/n_k)$-th powers of the l.r.t. statistics to test each sub-hypothesis in (5.450) for a given $k \in \{1,\dots,m\}$. Anyway, as in Sect. 5.2.4, we may still use the statistic in (5.439), with $n^{\sum_{k=1}^m p_k}$ replaced by $\prod_{k=1}^m n_k^{p_k}$, to test the null hypothesis in (5.438), with its exact distribution given by (5.441) and (5.442) with n replaced by n_k. As such, in case all p_k are even, the exact distribution of Λ in (5.439), with $n^{\sum_{k=1}^m p_k}$ replaced by $\prod_{k=1}^m n_k^{p_k}$, will be given by a combination of the products in Theorem 3.2 and Theorem 3.3, and its exact p.d.f. and c.d.f. given by Corollary 4.5, with the parameters in (5.454) and (5.455) with n replaced by n_ν, for $\nu = 1,\dots,m$, yielding a representation of the exact p.d.f. and c.d.f. for Λ in (5.439), in terms of the EGIG distribution representation, respectively given by (5.451) and (5.452), where the sets of shape and rate parameters are defined as in Sect. 5.2.4, with the only difference that the r_{jk} are now given by (5.456) above.

For the case of odd p_k and for the representation of the p.d.f. and c.d.f. of Λ in terms of Meijer G functions, see Sect. 5.2.4 in the previous section.

Modules available to implement this test are modules `PDFCircMeansEvenp`, `CDFCircMeansEvenp`, `PvalCircMeansEvenp`, `QuantCircMeansEvenp`, and `PvalDataCircMeansEvenp`. These modules have exactly the same arguments as the corresponding modules in Sect. 5.2.4.

As an example of application of the present test we use once again the response time data in Tables 3.9.7 and 3.9.8 of Timm (2002) to test the circularity of the covariance matrix and the equality of means for the last four response times simultaneously for the three populations of undiscriminated, low and high-term memory capability respondents. The command used is the first command in Fig. 5.89, and the high p-value obtained shows that we should not reject this null hypothesis, with all multivariate normality tests giving rather high p-values, except the Shapiro-Wilk test for the set of four response times.

A second command in the same figure tests a similar hypothesis but now for the last four response times for the undiscriminated respondents and for the last two response times for the other two populations, in order to illustrate the possibility of having different numbers of variables for different populations. The high p-value obtained shows once again that this hypothesis should not be rejected, once again with all multivariate normality tests giving rather high p-values, except the Shapiro-Wilk test for the set of four response times.

The test in this subsection is invariant under linear transformations similar to those under which the test in Sect. 5.2.4 is invariant.

```
PvalDataCircMeansEvenp[path <> "Response_times_all.dat", 1]
 There are 3 samples, with sizes {11, 10, 10} and {5, 5, 5} variables
 Original variables kept {{2, 3, 4, 5}, {2, 3, 4, 5}, {2, 3, 4, 5}}
 There are now 3 samples, with sizes {11, 10, 10} and {4, 4, 4} variables
Computed value of Λ: 0.06027966666
p-value: 0.8579677013
PvalDataCircMeansEvenp[path <> "Response_times_all.dat", 1]
 There are 3 samples, with sizes {11, 10, 10} and {5, 5, 5} variables
 Original variables kept {{2, 3, 4, 5}, {4, 5}, {4, 5}}
 There are now 3 samples, with sizes {11, 10, 10} and {4, 2, 2} variables
Computed value of Λ: 0.6011097946
p-value: 0.9936962480
```

Fig. 5.89 Commands used and output obtained with the Mathematica® module `PvalDataCircMeansEvenp` to implement the simultaneous test of circularity of the covariance matrix and equality of means for the response times for the last four token words, for the three populations whose samples of response times are reported in Tables 3.9.7 and 3.9.8 of Timm (2002)

5.3.3 The l.r.t. Statistic to Test Simultaneously the Circularity of the Covariance Matrices and the Equality of Means in m Subsets of Variables, Some with an Odd and the Other with an Even Number of Variables [CircMeans]

Let as assume a similar setup to the one in the previous subsection, where now we have q_1 of the m sets of variables with an odd number of variables and the remaining $q_2 = m - q_1$ sets with an even number of variables. Without any loss of generality and for the sake of simplicity in the notation, let the first q_1 sets, indexed with $k = 1, \ldots, q_1$, be the sets with an odd number of variables.

Then, if we are interested in testing a null hypothesis as the one in (5.438) and if the m samples have a common dimension $n > \max\{p_k\}$, the $(2/n)$-th power of the l.r.t. statistic to test H_0 in (5.438) is the statistic in (5.439) and its exact distribution will be given by a combination of the products in Theorem 3.1 or 3.2 and Theorem 3.3, and its p.d.f. and c.d.f. given by Corollary 4.5, with

$$m^* = q_2\,,\;\; n_\nu = 1\,,\;\; k_\nu = 2\,,\;\; a_\nu = \frac{n}{2},\;\; m_\nu = 1\,,\;\;\; \nu = 1, \ldots, q_2\,, \qquad (5.457)$$

and

$$\begin{gathered} m^{**} = m\,, \quad n^*_\nu = 2\,, \qquad (\nu = 1, \ldots, m^{**}), \quad \text{and} \\ a^*_\nu = \begin{cases} \frac{n-1}{2}, & \nu = 1, \ldots, q_1 \\ \frac{n-2}{2}, & \nu = q_1 + 1, \ldots, m \end{cases}, \\ k^*_\nu = \begin{cases} \frac{p_\nu - 1}{2}, & \nu = 1, \ldots, q_1 \\ \frac{p_\nu}{2} - 1\,, & \nu = q_1 + 1, \ldots, m \end{cases}, \quad s^*_\nu = \begin{cases} 1\,, & \nu = 1, \ldots, q_1 \\ 2\,, & \nu = q_1 + 1, \ldots, m\,, \end{cases} \end{gathered} \qquad (5.458)$$

so that its exact p.d.f. and c.d.f. are respectively given by (5.445) and (5.446), still with r^*_j and r^*_{jk} respectively given by (5.447) and (5.448) but now with r_{jk} given by (5.449) for $k = 1, \ldots, q_1$ and by (5.456) for $k = q_1 + 1, \ldots, m$.

In case the sample from $\underline{X}_k$ has size $n_k > p_k$ $(k = 1, \ldots, m)$, with possibly different values for the n_k's, the statistic in (5.439), as in Sect. 5.2.4, is no longer the $(2/n)$-th power of the l.r.t. statistic to test the hypothesis in (5.438), but it is the product of the $(2/n_k)$-th powers of the l.r.t. statistics to test each sub-hypothesis in (5.450). Anyway, as in Sect. 5.2.4, we may still use the statistic in (5.439) to test the null hypothesis in (5.438), with its exact distribution given by Theorems 3.3 and 3.2 and Corollary 4.5, with the parameters in (5.457) and (5.458), with n replaced by n_k, respectively for $\nu = 1, \ldots, m^*$ and $\nu = 1, \ldots, m^{**}$.

In this case, the exact p.d.f. and c.d.f. of Λ are given, in terms of the EGIG distribution, respectively by (5.451) and (5.452), where the sets of shape and rate parameters are defined as in Sect. 5.2.4, with the only difference that the r_{jk} are now given by (5.449) for $k = 1, \ldots, q_1$ and (5.456) for $k = q_1 + 1, \ldots, m$.

For a representation of the p.d.f. and c.d.f. of the statistic Λ in terms of Meijer G functions see Sect. 5.2.4.

```
PvalDataCircMeans[path <> "Response_times_all.dat", 1]
 There are 3 samples, with sizes {11, 10, 10} and {5, 5, 5} variables
 Original variables kept {{1, 2, 3, 4, 5}, {2, 3, 4, 5}, {2, 3, 4, 5}}
 There are now 3 samples, with sizes {11, 10, 10} and {5, 4, 4} variables
Computed value of Λ: 0.02898163750
p-value: 0.8432949097
```

Fig. 5.90 Command used and output obtained with the Mathematica® module `PvalDataCircMeans` to implement the simultaneous test of circularity of the covariance matrix and equality of means for all five response times for the population of undiscriminated respondents and for the response times for the last four token words, for the other two populations of low and high short-term memory capacity respondents whose samples of response times are reported in Tables 3.9.7 and 3.9.8 of Timm (2002)

For the implementation of the present test modules `PDFCircMeans`, `CDFCircMeans`, `PvalCircMeans`, `QuantCircMeans`, and `PvalDataCircMeans` are made available. These have exactly the same arguments as the corresponding modules in the previous subsection and in Sect. 5.2.4.

As an example of application of this test we use once again the response time data in Tables 3.9.7 and 3.9.8 of Timm (2002) to test the circularity of the covariance matrix and the equality of means simultaneously for all five response times for the population of undiscriminated respondents and for the last four response times for the other two populations of low and high short-term memory capability respondents. The command used and the output obtained are in Fig. 5.90, and the high p-value obtained shows that we should not reject this null hypothesis.

The test in this subsection is invariant under linear transformations similar to those under which the tests in Sects. 5.2.2 and 5.2.4 are invariant.

5.3.4 *The l.r.t. Statistic to Test Simultaneously the Independence of m Sets of Variables, the Circularity of Their Covariance Matrices and the Equality of Means, When All Sets Have an Even Number of Variables [IndCircMeans]*

Let us assume a similar setup to the one in Sect. 5.1.11, where the random vector $\underline{X} \sim N_p(\underline{\mu}, \Sigma)$ is split into m subvectors $\underline{X}_k$, the k-th of which has p_k variables $(k = 1, \dots, m)$, inducing in Σ a partition such as the one in (5.232).

Let us suppose then that we have a sample of size $n(> p)$ from $\underline{X}$ and that we are interested in testing the hypothesis

$$\begin{aligned} H_0 : \Sigma &= bdiag(\Sigma_{kk},\ k = 1, \dots, m) \\ \text{with } \Sigma_{kk} &= \Sigma_{cp_k} \text{ and } \underline{\mu}_k = a_k E_{p_k 1} \end{aligned} \qquad (5.459)$$

where Σ_{cp_k}, a circular matrix of order $p_k \times p_k$, and $a_k \in \mathbb{R}$ are unspecified, and $E_{p_k 1}$ represents a vector of 1's of dimension p_k.

Since the null hypothesis in (5.459) may be decomposed as

$$H_{0b|a} \circ H_{0a}$$

where

$$H_{0a} : \Sigma = bdiag(\Sigma_{kk},\ k = 1, \ldots, m)$$

is the null hypothesis of independence of the m subvectors $\underline{X}_k$ $(k = 1, \ldots, m)$, and

$$H_{0b|a} : \bigwedge_{k=1}^{m} \left\{ \Sigma_{kk} = \Sigma_{cp_k} \right\} \text{ and } \bigwedge_{k=1}^{m} \left\{ \underline{\mu}_k = a_k E_{p_k 1} \right\}$$

is the null hypothesis of circularity of the m variance-covariance matrices $\Sigma_{kk} = Var(\underline{X}_k)$ and equality of the means in each expected value vector $\underline{\mu}_k = E(\underline{X}_k)$ $(k = 1, \ldots, m)$ where we have $\underline{X}_k \sim N_{p_k}(\underline{\mu}_k, \Sigma_{kk})$.

Then, since the $(2/n)$-th power of the l.r.t. statistic to test H_{0a} is the statistic in (5.234), where A is the m.l.e. of Σ and A_{kk} its k-th diagonal block of dimensions $p_k \times p_k$, and the $(2/n)$-th power of the l.r.t. statistic to test $H_{0b|a}$ is the statistic in (5.439), where A_k is to be replaced by A_{kk}, using Lemma 10.3.1 in Anderson (2003), the $(2/n)$-th power of the l.r.t. statistic to test H_0 in (5.459) is

$$\Lambda = n^p |A| \prod_{k=1}^{m} \frac{1}{v_{1k}^* \prod_{j=2}^{p_k} \left(w_{jk} + v_{jk}^*\right)} = |V| \prod_{k=1}^{m} \frac{1}{v_{1k}^* \prod_{j=2}^{p_k} \left(w_{jk} + v_{jk}^*\right)} \tag{5.460}$$

where A is the m.l.e. of Σ, $V = n\,UAU'$, with $U = [u_{ij}]$, where u_{ij} is given by (5.357), and the v_{jk}^* and w_{jk} are defined as in (5.433) and (5.440), respectively in Sects. 5.2.3 and 5.2.4.

Then,

$$\Lambda \stackrel{d}{\equiv} \left\{ \prod_{k=1}^{m-1} \prod_{j=1}^{p_k} Y_{jk} \right\} \left\{ \prod_{k=1}^{m} \prod_{j=2}^{p_k} Y_{jk}^* \right\} \tag{5.461}$$

where, given the independence of the l.r.t. statistic to test H_{0a} and the m matrices A_{kk}, we have, for $m_k = p_k/2$ and $q_k = p_k + \cdots + p_m$,

$$Y_{jk} \sim Beta\left(\frac{n - q_k - j}{2}, \frac{q_k}{2}\right) \quad \text{and} \quad Y_{jk}^* \sim \begin{cases} Beta\left(\frac{n-j}{2}, \frac{j}{2}\right), & j = 2, \ldots, m_k + 1 \\ Beta\left(\frac{n-j}{2}, \frac{j+1}{2}\right), & j = m_k + 2, \ldots, p_k \end{cases} \tag{5.462}$$

forming a set of $\sum_{k=1}^{m-1} p_k + \sum_{k=1}^{m}(p_k - 1) = p - p_m + p - m = 2p - p_m - m$ independent r.v.'s. As such, the exact distribution of Λ in (5.460) is given by a combination of the products in Theorems 3.2 and 3.3, and its p.d.f. and c.d.f. given by Corollary 4.5, with

$$m^* = 2m-1\,,\quad k_\nu = 2\;(\nu = 1,\ldots,2m-1)\,,\quad n_\nu = \begin{cases} \frac{p_\nu}{2}\,, & \nu = 1,\ldots,m-1 \\ 1\,, & \nu = m,\ldots,2m-1\,, \end{cases}$$

$$m_\nu = \begin{cases} q_\nu\,, & \nu = 1,\ldots,m-1 \\ 1\,, & \nu = m,\ldots,2m-1\,, \end{cases} \qquad a_\nu = \begin{cases} \frac{m-q_\nu}{2}\,, & \nu = 1,\ldots,m-1 \\ \frac{n}{2}\,, & \nu = m,\ldots,2m-1\,, \end{cases}$$

and

$$m^{**} = m\,,\;\; n_\nu^* = 2\,,\;\; a_\nu^* = \frac{n-2}{2}\,,\;\; k_\nu^* = \frac{p_\nu}{2} - 1\,,\;\; s_\nu^* = 2\,,\quad (\nu = 1,\ldots,m^{**})$$

for

$$q_\nu = p_{\nu+1} + \cdots + p_m\,, \tag{5.463}$$

so that the exact p.d.f. and c.d.f. of Λ in (5.460) are given, in terms of the EGIG distribution representation, respectively by

$$f_\Lambda(z) = f^{EGIG}\left(z\,\middle|\,\{r_j^*\}_{j=1:p};\left\{\frac{n-j}{2}\right\}_{j=1:p};p\right) \tag{5.464}$$

and

$$F_\Lambda(z) = F^{EGIG}\left(z\,\middle|\,\{r_j^*\}_{j=1:p};\left\{\frac{n-j}{2}\right\}_{j=1:p};p\right), \tag{5.465}$$

where

$$r_j^* = \begin{cases} \sum_{k=1}^{m} r_{jk}^*\,, & j = 1,2 \\ r_{j-2} + \sum_{k=1}^{m} r_{jk}^*\,, & j = 3,\ldots,p \end{cases} \tag{5.466}$$

with r_j given by (5.241)–(5.242) and

$$r_{jk}^* = \begin{cases} r_{jk}\,, & j = 1,\ldots,p_k \\ 0\,, & j = p_k+1,\ldots,p \end{cases} \tag{5.467}$$

where r_{jk} are given by (5.456).

For any p_k the p.d.f. and c.d.f. of Λ in (5.460) may be written in terms of the Meijer G function, for $p^* = 2p - p_m - m$ as

$$f_\Lambda(z) = \left\{ \prod_{k=1}^{m-1} \prod_{j=1}^{p_k} \frac{\Gamma\left(\frac{n-j}{2}\right)}{\Gamma\left(\frac{n-q_k-j}{2}\right)} \right\} \left\{ \prod_{k=1}^{m} \left\{ \prod_{j=2}^{m_k+1} \frac{\Gamma\left(\frac{n}{2}\right)}{\Gamma\left(\frac{n-j}{2}\right)} \right\} \left\{ \prod_{j=m_k+2}^{p_k} \frac{\Gamma\left(\frac{n+1}{2}\right)}{\Gamma\left(\frac{n-j}{2}\right)} \right\} \right\}$$
$$\times G^{p^*,0}_{p^*,p^*} \left(\begin{array}{c} \left\{ \left\{\frac{n-j-2}{2}\right\}_{\substack{j=1:p_k\\k=1:m-1}}, \left\{\frac{n-2}{2}\right\}_{\substack{j=2:m_k+1\\k=1:m}}, \left\{\frac{n-1}{2}\right\}_{\substack{j=m_k+2:p_k\\k=1:m}} \right\} \\ \left\{ \left\{\frac{n-q_k-j-2}{2}\right\}_{\substack{j=1:p_k\\k=1:m-1}}, \left\{\frac{n-j-2}{2}\right\}_{\substack{j=2:p_k\\k=1:m}} \right\} \end{array} \middle| \, z \right)$$

and

$$F_\Lambda(z) = \left\{ \prod_{k=1}^{m-1} \prod_{j=1}^{p_k} \frac{\Gamma\left(\frac{n-j}{2}\right)}{\Gamma\left(\frac{n-q_k-j}{2}\right)} \right\} \left\{ \prod_{k=1}^{m} \left\{ \prod_{j=2}^{m_k+1} \frac{\Gamma\left(\frac{n}{2}\right)}{\Gamma\left(\frac{n-j}{2}\right)} \right\} \left\{ \prod_{j=m_k+2}^{p_k} \frac{\Gamma\left(\frac{n+1}{2}\right)}{\Gamma\left(\frac{n-j}{2}\right)} \right\} \right\}$$
$$\times G^{p^*,1}_{p^*+1,p^*+1} \left(\begin{array}{c} \left\{ 1, \left\{\frac{n-j}{2}\right\}_{\substack{j=1:p_k\\k=1:m-1}}, \left\{\frac{n}{2}\right\}_{\substack{j=2:m_k+1\\k=1:m}}, \left\{\frac{n+1}{2}\right\}_{\substack{j=m_k+2:p_k\\k=1:m}} \right\} \\ \left\{ \left\{\frac{n-q_k-j}{2}\right\}_{\substack{j=1:p_k\\k=1:m-1}}, \left\{\frac{n-j}{2}\right\}_{\substack{j=2:p_k\\k=1:m}} 0 \right\} \end{array} \middle| \, z \right).$$

Modules PDFIndCircMeans, CDFIndCircMeans, PvalIndCircMeans, QuantIndCircMeans, and PvalDataIndCircMeans are made available for the implementation of this test. These modules have exactly the same arguments as the corresponding modules in Sect. 5.1.11.

As an example of application of this test we will use the data on response times for the five token words in a given sentence in Tables 3.9.7 and 3.9.8 of Timm (2002) (see this reference and Sect. 5.1.17 for a more detailed description of the data in these tables). Since we need a sample of a given size on several sets of variables, we decided to use the data on the first ten individuals for the five response times in Table 3.9.7 of Timm (2002) and all ten observations for the five response times for the two groups of individuals in Table 3.9.8 of the same reference, as if they would refer to three sets of five response times for three different sentences, collected on the same ten individuals, making up this way a sample of size ten for an overall set of 15 variables (the 15 response times, five for each sentence). Then we may be interested in testing the independence for example of the response times for the last four token words of the first sentence and the response times for the last two of the other two sentences, at the same time that we also test for the circularity of the covariance matrices and the equality of the mean response times. The command used and the output obtained are in Fig. 5.91 and the p-value obtained would lead us to not reject the null hypothesis being tested. We should note that we are aware of the fact that we are using a very small sample, with size ten, for a total of eight

```
PvalDataIndCircMeans[path <> "Response_times_1set.dat", {4, 2, 2}, 1]
 Original variables {2, 3, 4, 5, 9, 10, 14, 15}
Computed value of Λ: 0.001373151688
p-value: 0.4457293088
```

Fig. 5.91 Command used and output obtained with the Mathematica® module PvalDataIndCircMeans to implement the simultaneous test of independence, circularity of the covariance matrix and equality of means, using a data set obtained from a transformation of the data sets on response times in Tables 3.9.7 and 3.9.8 of Timm (2002) (see text)

variables involved in the analysis, which may give us a very limited power for rejection of the null hypothesis. We may also note that the test of circularity for the 2×2 covariance matrices of the last two token words for the second and third sentences is indeed a test of equality of two variances, based on two paired samples.

The present test is invariant under linear transformations of the type $\underline{X} \to C\underline{X} + \underline{b}$, where $C = bdiag(C_1, \ldots, C_k, \ldots, C_m)$ and $\underline{b} = [\underline{b}_1', \ldots, \underline{b}_k', \ldots, \underline{b}_m']'$, with $\underline{b}_k = a_k E_{p_k 1}$, where a_k is any real and with C_k any nonrandom full-rank $p_k \times p_k$ circular or circulant matrix.

5.3.5 The l.r.t. Statistic to Test Simultaneously the Independence of m Sets of Variables, the Circularity of Their Covariance Matrices and the Equality of Means, When All but One of the Sets Have an Even Number of Variables [IndCircMeans1Odd]

Let us consider a similar setup to the one in the previous subsection, where now all but one of the subvectors $\underline{X}_k$ have an even number of variables. Without any loss of generality, let the set with an odd number of variables be the m-th set. Let us then suppose that we are interested in testing the null hypothesis in (5.459). The $(2/n)$-th power of the l.r.t. statistic to test this hypothesis is the statistic Λ in (5.460), still with a similar distribution to that given by (5.461) and (5.462), now with $m_k = p_k/2$ for $k = 1, \ldots, m-1$ and $m_m = (p_m - 1)/2$, so that the exact distribution of Λ in (5.460) is now given by a combination of the products in Theorem 3.3 and Theorem 3.1 or 3.2, and its p.d.f. and c.d.f. given by Corollary 4.4 or 4.5, with

$$m^{**} = m\,,\ n_\nu^* = 2\,,\ a_\nu^* = \frac{n-1}{2}\,,\qquad (\nu = 1, \ldots, m^{**})$$

$$k_\nu^* = \begin{cases} \frac{p_\nu}{2} - 1\,, & \nu = 1, \ldots, m-1 \\ \frac{p_m - 1}{2}\,, & \nu = m\,, \end{cases} \qquad s_\nu^* = \begin{cases} 2\,, & \nu = 1, \ldots, m \\ 1\,, & \nu = m\,, \end{cases}$$

and

$$m^* = 2m - 2\,, \quad k_\nu = 2\,, \quad n_\nu = \begin{cases} 1\,, & \nu = 1, \ldots, m-1 \\ \frac{p_\nu}{2}\,, & \nu = m, \ldots, 2m-2\,, \end{cases}$$

$$m_\nu = \begin{cases} 1\,, & \nu = 1, \ldots, m-1 \\ q_\nu\,, & \nu = m, \ldots, 2m-2\,, \end{cases} \qquad a_\nu = \begin{cases} \frac{n+1}{2}\,, & \nu = 1, \ldots, m-1 \\ \frac{n-m_\nu}{2}\,, & \nu = m, \ldots, 2m-2\,, \end{cases}$$

for q_ν given by (5.463).

The exact p.d.f. and c.d.f. of Λ in (5.460) are thus given, in terms of the EGIG distribution representation, respectively by (5.464) and (5.465) with r_j^* and r_{jk}^* given by (5.466) and (5.467), now with r_{jk} given by (5.456) for $k = 1, \ldots, m-1$, and by (5.449) for $k = m$.

For a representation of the p.d.f. and c.d.f. of the l.r.t. statistic in terms of Meijer G functions see the previous subsection.

Modules `PDF IndCircMeans1Odd`, `CDF IndCircMeans1Odd`, `Pval IndCircMeans1Odd`, `Quant IndCircMeans1Odd`, and `PvalData IndCircMeans1Odd` are made available for the implementation of this test. These modules have exactly the same arguments as the corresponding modules in the previous subsection and in Sect. 5.1.11.

As an example of application of this test we will use the same data we used in the previous subsection, that is, the data for the first ten observations in Table 3.9.7 of Timm (2002) and the data in Table 3.9.8 of this same reference as if they would refer to the response times for three different sentences, collected on the same 10 individuals, making a sample of size 10 for an overall set of 15 variables, to test the hypothesis of independence of the set of response times for the last three token words of the first sentence and the response times for the last two token words of the other two sentences, at the same time that we also test for the circularity of the covariance matrices and the equality of the mean response times. The command used and the output obtained are in Fig. 5.92, with the p-value obtained leading us to not reject this null hypothesis.

```
PvalDataIndCircMeans1Odd["<Path>/Response_times_1set.dat", {3, 2, 2}, 1]
 Original variables {3, 4, 5, 9, 10, 14, 15}
Computed value of Λ: 0.006981259028
p-value: 0.3159993467
```

Fig. 5.92 Command used and output obtained with the Mathematica® module `PvalDataIndCircMeans1Odd` to implement the simultaneous test of independence, circularity of the covariance matrix and equality of means, using a data set obtained from a transformation of the data sets on response times in Tables 3.9.7 and 3.9.8 of Timm (2002) (see text)

We may note that although in the presentation of the test the set with an odd number of variables was placed, without any loss of generality, as the last set, in the practical implementation of the test and namely when using module `PvalData IndCircMeans1Odd` the set with an odd number of variables may appear in any position among the m sets of variables.

The test in this subsection is invariant under linear transformations similar to those under which the test in the previous subsection is invariant.

5.3.6 *Testing for a Two-Block Independent Circular-Spherical Covariance Structure [IndCircSph]*

Let us suppose a spatial setup where a point of interest is at the center of a regular polygon with p vertices and let us suppose that a variable of interest is measured at these p points as well as at two other points situated far apart from this set of p points, and also far apart from each other. Such setup may occur for example in seismological studies where a point of interest is at the center of a regular polygon with p vertices, where a variable of interest is measured and then two other points placed at similar but quite large distances, say one on each side, of this set of p points, where the "same" variable is measured. Or, the point of interest being considered may be a tree, taken as the center of a regular polygon, at whose p vertices the soil moisture content at a given depth is measured, while this soil moisture content is also measured at the same depth at two other points in opposite sides of the tree, laying much farther from the tree. A similar setup may occur in a time frame, if we consider for example the price of some good or the value of some variable whose covariance structure is thought to be cyclic in time, as for example along the days of some reference week, while values of this "same" variable are also considered 1 year before and 1 year after this reference week. Then one may be interested in testing the hypothesis that the covariance structure of the measurements on these $p+2$ points has a structure that is circular or cyclical for the p points that are at the vertices of the regular polygon or that refer to the days in that reference week and that the variances at the other two points are similar, but that these two points lie so far apart from each other and from the other p points that the covariances between the measurements at these two points as well as the covariances of the measurements at these two points and those taken at the set of the p points are all null.

For example, for $p=7$, this would be equivalent to testing that the covariance matrix Σ for the set of $p+2$ variables measured (since we will have to see the measurements taken at each of the $p+2$ points as a different random variable) would have a structure like

$$\Sigma = \left[\begin{array}{ccccccc|cc} a & b & c & d & d & c & b & 0 & 0 \\ b & a & b & c & d & d & c & 0 & 0 \\ c & b & a & b & c & d & d & 0 & 0 \\ d & c & b & a & b & c & d & 0 & 0 \\ d & d & c & b & a & b & c & 0 & 0 \\ c & d & d & c & b & a & b & 0 & 0 \\ b & c & d & d & c & b & a & 0 & 0 \\ \hline 0 & 0 & 0 & 0 & 0 & 0 & 0 & e & 0 \\ 0 & 0 & 0 & 0 & 0 & 0 & 0 & 0 & e \end{array}\right]. \tag{5.468}$$

If we take $\underline{X} = \left[\underline{X}_1, \underline{X}_2\right]' \sim N_{p+2}(\underline{\mu}, \Sigma)$, where $\underline{X}_1$ is $p{\times}1$ and $\underline{X}_2$ is $2{\times}1$, with

$$\underline{\mu} = \left[\underline{\mu}_1, \underline{\mu}_2\right]' \quad \text{with} \quad \underline{\mu}_1 = E(\underline{X}_1) \quad \text{and} \quad \underline{\mu}_2 = E(\underline{X}_2),$$

then to test the null hypothesis that the covariance matrix Σ has a structure as that in (5.468) is the same as to test

$$H_0 \equiv H_{0b|a} \circ H_{0a}$$

where

$$H_{0a} : \Sigma = bdiag(\Sigma_1, \Sigma_2), \text{ (where } \Sigma_1 \text{ is } p{\times}p \text{ and } \Sigma_2 \text{ is } 2{\times}2) \tag{5.469}$$

and

$$H_{0b|a} : \Sigma_1 \text{ is circular and } \Sigma_2 = \sigma^2 I_2 (\sigma^2 \text{ non-specified}) \\ \text{assuming } H_{0a}.$$

The $(2/n)$-th power of the l.r.t. statistic to test H_{0a} is

$$\Lambda_a = \frac{|A|}{|A_1|\,|A_2|} \tag{5.470}$$

where A is the m.l.e. of Σ and A_1 and A_2, its first and second diagonal blocks, respectively of dimensions $p{\times}p$ and $2{\times}2$, are the m.l.e.'s of Σ_1 and Σ_2, with (see Sect. 5.1.9)

$$\Lambda_a \equiv Y_1^* \, Y_2^* \tag{5.471}$$

where, for $n > p+2$,

$$Y_1^* \sim Beta\left(\frac{n-p-1}{2}, \frac{p}{2}\right) \quad \text{and} \quad Y_2^* \sim Beta\left(\frac{n-p-2}{2}, \frac{p}{2}\right) \tag{5.472}$$

are two independent r.v.'s. The $(2/n)$-th power of the l.r.t. statistic to test $H_{0b|a}$, once H_{0a} is assumed, is the product of two independent l.r.t. statistics, one to test the circularity of Σ_1 and the other to test the sphericity of Σ_2, with (see Sects. 5.2.1 and 2 in Marques et al. (2011))

$$\Lambda_{b|a} = n^p \frac{|A_1|}{\prod_{j=1}^{p} v_j^*} \frac{|A_2|}{\left(tr\left(\frac{1}{2}A_2\right)\right)^2},$$

where v_j^* is given by (5.419), with $V = nUA_1U'$, being U a matrix with dimensions $p\times p$ with elements given by (5.357).

From Sects. 5.2.1 and 2 in Marques et al. (2011), we know that, for $m = \lfloor p/2 \rfloor$,

$$\Lambda_{b|a} \stackrel{d}{\equiv} \underbrace{\left\{\prod_{j=1}^{p-1} Y_j\right\}}_{Y^*} Y \tag{5.473}$$

where

$$Y_j \sim \begin{cases} Beta\left(\frac{n-1-j}{2}, \frac{j}{2}\right), & j = 1, \dots, m \\ Beta\left(\frac{n-1-j}{2}, \frac{j+1}{2}\right), & j = m+1, \dots, p-1 \end{cases} \quad \text{and} \quad Y \sim Beta\left(\frac{n-2}{2}, 1\right) \tag{5.474}$$

form a set of p independent r.v.'s.

As such, the $(2/n)$-th power of the overall l.r.t. statistic will be

$$\Lambda = \Lambda_a \, \Lambda_{b|a} = 2^2 n^p \frac{|A|}{\left\{\prod_{j=1}^{p} v_j^*\right\} (tr(A_2))^2}, \tag{5.475}$$

with its exact distribution, for odd p, given by a combination of two independent products of independent Beta r.v.'s, one given by Theorem 3.2 and the other one given by Theorem 3.3, with the exact p.d.f. and c.d.f. of Λ given by Corollary 4.5, with

$$m^* = 2, \; n_\nu = \{1, 1\}, \; k_\nu = \{2, 1\}, \; a_\nu = \left\{\frac{n-p}{2}, \frac{n}{2}\right\}, \; m_\nu = \{p, 1\}$$

which refers to the distribution of Λ_a together with the distribution of Y in (5.473), and

$$m^{**}=1\,,\quad n_1^*=2\,,\quad k_1^*=\frac{p-1}{2}\,,\quad a_1^*=\frac{n-1}{2}\,,\quad s_1^*=0$$

which takes care of the distribution of Y^* in (5.473). Thus, the exact p.d.f. and c.d.f. of Λ are given in terms of the EGIG p.d.f. and c.d.f. as

$$f_\Lambda(z)=f^{EGIG}\left(z\,\middle|\,\{r_j^*\}_{j=2:p+2};\left\{\frac{n-j}{2}\right\}_{j=2:p+2};p+1\right) \tag{5.476}$$

and

$$F_\Lambda(z)=F^{EGIG}\left(z\,\middle|\,\{r_j^*\}_{j=2:p+2};\left\{\frac{n-j}{2}\right\}_{j=2:p+2};p+1\right), \tag{5.477}$$

where

$$r_j^*=\begin{cases}2+\left\lfloor\frac{p-j}{2}\right\rfloor, & j=2,\dots,p\\ 1, & j=p+1,p+2\,.\end{cases} \tag{5.478}$$

In terms of the Meijer G function, either for even or odd p, the p.d.f. and c.d.f. of the statistic Λ in (5.475), for $m=\lfloor p/2\rfloor$, is written as

$$\begin{aligned}f_\Lambda(z)=&\frac{\Gamma\left(\frac{n}{2}\right)}{\Gamma\left(\frac{n-2}{2}\right)}\left\{\prod_{j=1}^{2}\frac{\Gamma\left(\frac{n-j}{2}\right)}{\Gamma\left(\frac{n-p-j}{2}\right)}\right\}\left\{\prod_{j=1}^{m}\frac{\Gamma\left(\frac{n-1}{2}\right)}{\Gamma\left(\frac{n-1-j}{2}\right)}\right\}\left\{\prod_{j=m+1}^{p-1}\frac{\Gamma\left(\frac{n}{2}\right)}{\Gamma\left(\frac{n-1-j}{2}\right)}\right\}\\ &\times G^{p+2,0}_{p+2,p+2}\left(\begin{array}{c}\frac{n-2}{2},\left\{\frac{n-j-2}{2}\right\}_{j=1:2},\left\{\frac{n-3}{2}\right\}_{j=1:m},\left\{\frac{n-2}{2}\right\}_{j=m+1:p-1}\\ \frac{n-4}{2},\left\{\frac{n-p-j-2}{2}\right\}_{j=1:2},\left\{\frac{n-3-j}{2}\right\}_{j=1:p-1}\end{array}\middle|\,z\right)\end{aligned}$$

and

$$\begin{aligned}F_\Lambda(z)=&\frac{\Gamma\left(\frac{n}{2}\right)}{\Gamma\left(\frac{n-2}{2}\right)}\left\{\prod_{j=1}^{2}\frac{\Gamma\left(\frac{n-j}{2}\right)}{\Gamma\left(\frac{n-p-j}{2}\right)}\right\}\left\{\prod_{j=1}^{m}\frac{\Gamma\left(\frac{n-1}{2}\right)}{\Gamma\left(\frac{n-1-j}{2}\right)}\right\}\left\{\prod_{j=m+1}^{p-1}\frac{\Gamma\left(\frac{n}{2}\right)}{\Gamma\left(\frac{n-1-j}{2}\right)}\right\}\\ &\times G^{p+2,1}_{p+3,p+3}\left(\begin{array}{c}1,\frac{n}{2},\left\{\frac{n-j}{2}\right\}_{j=1:2},\left\{\frac{n-1}{2}\right\}_{j=1:m},\left\{\frac{n}{2}\right\}_{j=m+1:p-1}\\ \frac{n-2}{2},\left\{\frac{n-p-j}{2}\right\}_{j=1:2},\left\{\frac{n-1-j}{2}\right\}_{j=1:p-1},0\end{array}\middle|\,z\right).\end{aligned}$$

Modules PDFIndCircSph, CDFIndCircSph, PvalIndCircSph, QuantIndCircSph, and PvalDataIndCircSph are available for the implementation of this test. These modules use similar arguments to the corresponding modules in Sect. 5.1.11, with the only difference that instead of the vector with the p_k values one has now a simple argument with the value of p. Also, the PvalData module uses a data file with a similar structure of that used in that subsection, that is, a data file of a sample of size n for $p+2$ variables, but it has as its only mandatory argument the name of the data file, keeping the same optional arguments as the corresponding module in Sect. 5.1.11.

For an application of the present test we use again the response time data in Tables 3.9.7 and 3.9.8 of Timm (2002). We analyze separately the undiscriminated respondents, low and high-term memory capability respondents to test if we can consider that the variances of the response times for the second, third and fourth token words are equal and also their covariances are equal, while we consider these response times independent of those for the first and fifth token words, whose variances are assumed equal and whose covariance is considered null. We should note here that for $p = 3$ the test of circularity is exactly the same as the test for compound symmetry, since for $p = 3$ the two structures are the same.

The rather high p-values reported in Fig. 5.93 indicate that we should not reject this null hypothesis for all three groups of respondents. If one may think that this fact may be due to the fact that we would not reject the hypothesis of sphericity for the whole covariance matrix for the five response times, this is not the case,

```
PvalDataIndCircSph[path <> "Response_times.dat", 1]
 Original variables {2, 3, 4, 1, 5}
Computed value of Λ: 0.1275949109
p-value: 0.2014553179
PvalDataIndCircSph[path <> "Response_times_low.dat", 1]
 Original variables {2, 3, 4, 1, 5}
Computed value of Λ: 0.1460165321
p-value: 0.3829495965
PvalDataIndCircSph[path <> "Response_times_high.dat", 1]
 Original variables {2, 3, 4, 1, 5}
Computed value of Λ: 0.06470570606
p-value: 0.1119136152
```

Fig. 5.93 Commands used and output obtained with the Mathematica® version of module PvalDataIndCircSph to implement the simultaneous test of independence, circularity of the covariance matrix for the first set with three variables and sphericity of the covariance matrix for the set with two variables, using the data sets obtained from Tables 3.9.7 and 3.9.8 of Timm (2002) (see text)

at least for the undiscriminated respondents group and for the high-term memory capability group, for which the p-value for the sphericity test gives a value around 0.03, being the case that for the low-term memory capability group the p-value for the sphericity test for the whole covariance matrix of the five response times is indeed rather high, around 0.48.

Another example with quite interesting results comes out of the analysis of the covariance matrix of the raw scores for six psychological tests administered to 20 successful engineers and 20 successful pilots reported in Table 1 of Travers (1939) and also in Table 5.6 of Rencher and Christensen (2012). If we consider testing the hypothesis of equality of the variances for the scores of the Intelligence (Test 33 of the Nat. Inst. Ind. Psychol.), Dotting and Perseveration tests, equality of their covariances and the independence of the scores of these three tests from the scores of the tests for Form Relations (test from the Nat. Inst. Ind. Psychol.) and Dynamometer, which are assumed to be independent and have equal variances, for each of the two professional groups we see from the results in Fig. 5.94 that we should not reject this null hypothesis for either one of the two professional groups (data for engineers is in file `Travers_1.dat` and data for pilots in file `Travers_2.dat`), which is in itself an interesting conclusion, given the type of tests involved.

Once again, in case one may wonder if this result is not a simple consequence of a spherical structure for the whole covariance matrix of the scores for the five tests, we may assure that this is not the case since the p-values for the tests of sphericity for the covariance matrix of the scores for the five tests are quite low, both for the engineers (p-value = 0.000388) and for the pilots (p-value = 1.50588×10^{-7}), leading us to reject the sphericity hypothesis for both population covariance matrices.

```
PvalDataIndCircSph[path <> "Travers_1.dat", 1]
 Original variables {1, 4, 6, 2, 3}
Computed value of Λ: 0.5505237715
p-value: 0.6149160226
PvalDataIndCircSph[path <> "Travers_2.dat", 1]
 Original variables {1, 4, 6, 2, 3}
Computed value of Λ: 0.4516992314
p-value: 0.3459442510
```

Fig. 5.94 Commands used and output obtained with the Mathematica® version of module `PvalDataIndCircSph` to implement the simultaneous test of independence, circularity of the covariance matrix for the first set with three variables and sphericity of the covariance matrix for the set with two variables, using the data sets obtained from Table 1 in Travers (1939) or from Table 5.6 in Rencher and Christensen (2012) (see text)

As a side note, for all five examples addressed all the multivariate normality tests, except the Shapiro-Wilk test, gave very high p-values.

In case there are more than two "far away" points, or the number of vertices p is even, although a similar testing procedure can be used, it will not be possible to encompass the exact distribution of Λ in the setup defined by Theorems 3.1–3.3, and the exact distribution of Λ is no longer representable as an EGIG distribution. In this case we will have to resort to the use of near-exact distributions (see Marques and Coelho (2008, 2013), Coelho et al. (2010, Sects. 3,5) and Chap. 7).

The present test is invariant under linear transformations $\underline{X} \rightarrow C\underline{X} + \underline{b}$ where $\underline{b}$ is any $(p+2)\times 1$ real vector and $C = bdiag(C_1, C_2)$, with C_1 any nonrandom full-rank $p\times p$ circular matrix and $C_2 = aI_2$, where a is any non-null real.

5.3.7 *Testing for a Two-Block Independent Circular-Spherical Covariance Structure and Equality of Means [IndCircSphEqMean]*

Let us suppose a similar setup to the one in the previous subsection, where now, besides testing the covariance structure one wants to test if the means for the p points that define the vertices of the regular polygon are all equal.

In this case our null hypothesis may be expressed as

$$H_0 \equiv H_{0b|a} \circ H_{0a} \tag{5.479}$$

where H_{0a} is the null hypothesis in (5.469), with associated l.r.t. statistic Λ_a in (5.470), and

$$H_{0b|a} : \Sigma_1 \text{ is circular and } \underline{\mu}_1 = bE_{p1};\ \Sigma_2 = \sigma^2 I_2 (b \text{ and } \sigma^2 \text{ non-specified}) \text{ assuming } H_{0a}\,, \tag{5.480}$$

where E_{p1} represents a vector of ones of dimension p.

As in the previous subsection, once H_{0a} is assumed, the test for $H_{0b|a}$ is the composition of two independent tests, the test for the circularity of Σ_1 and the equality of means in $\underline{\mu}_1$, and the test for sphericity of Σ_2. From Sects. 5.2.2 and 5.3.1 and from Sect. 2 in Marques et al. (2011), for a sample of size n, the $(2/n)$-th power of the l.r.t. statistic to test $H_{0b|a}$ is

$$\Lambda_{b|a} = n^p \frac{|A_1|}{v_1^*} \frac{|A_2|}{\left(tr\left(\frac{1}{2}A_2\right)\right)^2} \prod_{j=2}^{p} \frac{1}{v_j^* + w_j}$$

with

$$\Lambda_{b|a} \stackrel{d}{\equiv} \left\{ \prod_{j=2}^{p} Y_j \right\} Y$$

where, for $m = \lfloor p/2 \rfloor$,

$$Y_j \sim \begin{cases} Beta\left(\frac{n-j}{2}, \frac{j}{2}\right), & j = 2, \ldots, m+1 \\ Beta\left(\frac{n-j}{2}, \frac{j+1}{2}\right), & j = m+2, \ldots, p \end{cases} \quad \text{and} \quad Y \sim Beta\left(\frac{n-2}{2}, 1\right). \tag{5.481}$$

The $(2/n)$-th power of the l.r.t. statistic to test H_0 is thus

$$\Lambda = \Lambda_a \, \Lambda_{b|a} = 2^2 \, n^p \, \frac{|A|}{v_1^* \, (tr\,(A_2))^2} \prod_{j=2}^{p} \frac{1}{v_j^* + w_j}, \tag{5.482}$$

with

$$\Lambda \stackrel{d}{\equiv} Y_1^* Y_2^* \left\{ \prod_{j=2}^{p} Y_j \right\} Y$$

where all r.v.'s are independent and where Y_1^* and Y_2^* are the r.v.'s in (5.471) and (5.472).

For odd p, the exact distribution of Λ is thus given by a combination of Theorems 3.2 and 3.3 and its exact p.d.f. and c.d.f. are given by Corollary 4.5, in terms of the EGIG p.d.f. and c.d.f., with

$$m^* = 2\,, \;\; n_\nu = \{1, 1\}\,, \;\; k_\nu = \{2, 1\}\,, \;\; a_\nu = \left\{ \frac{n-p}{2}, \frac{n}{2} \right\}\,, \;\; m_\nu = \{p, 1\} \;\; (\nu = 1, 2)$$

and

$$m^{**} = 1\,, \quad n_1^* = 2\,, \quad k_1^* = \frac{p-1}{2}\,, \quad a_1^* = \frac{n-1}{2}\,, \quad s_1^* = 1$$

respectively as

$$f_\Lambda(z) = f^{EGIG}\left(z \,\middle|\, \{r_j^*\}_{j=1:p+2}; \left\{ \frac{n-j}{2} \right\}_{j=1:p+2}; p+2 \right) \tag{5.483}$$

and

$$F_\Lambda(z) = F^{EGIG}\left(z \,\middle|\, \{r_j^*\}_{j=1:p+2};\ \left\{\frac{n-j}{2}\right\}_{j=1:p+2};\ p+2 \right), \tag{5.484}$$

where

$$r_j^* = \begin{cases} \frac{p-1}{2}, & j=1 \\ \frac{p+1}{2} - \left\lfloor \frac{j-2}{2} \right\rfloor, & j=2,\ldots,p \\ 1, & j=p+1, p+2, \end{cases} \tag{5.485}$$

while for even p the exact p.d.f. and c.d.f. of Λ are given by the same corollary with

$$m^* = 3\,,\ n_\nu = \{1,1,1\}\,,\ k_\nu = \{2,1,2\}\,,\ a_\nu = \left\{\frac{n-p}{2}, \frac{n}{2}, \frac{n}{2}\right\}\,,\ m_\nu = \{p,1,1\}$$

and

$$m^{**} = 1\,,\quad n_1^* = 2\,,\quad k_1^* = \frac{p}{2} - 1\,,\quad a_1^* = \frac{n-2}{2}\,,\quad s_1^* = 2$$

with expressions still given by (5.483) and (5.484) now with

$$r_j^* = \begin{cases} \frac{p}{2} - 1, & j=1 \\ \frac{p}{2} + 1 - \left\lfloor \frac{j-1}{2} \right\rfloor, & j=2,\ldots,p \\ 1, & j=p+1, p+2\,. \end{cases} \tag{5.486}$$

Both for odd or even p, the p.d.f. and c.d.f. of Λ in (5.482) may be written in terms of the Meijer G function as

$$f_\Lambda(z) = \frac{\Gamma\left(\frac{n}{2}\right)}{\Gamma\left(\frac{n-2}{2}\right)} \left\{ \prod_{j=1}^{2} \frac{\Gamma\left(\frac{n-j}{2}\right)}{\Gamma\left(\frac{n-p-j}{2}\right)} \right\} \left\{ \prod_{j=2}^{m+1} \frac{\Gamma\left(\frac{n}{2}\right)}{\Gamma\left(\frac{n-j}{2}\right)} \right\} \left\{ \prod_{j=m+2}^{p} \frac{\Gamma\left(\frac{n+1}{2}\right)}{\Gamma\left(\frac{n-j}{2}\right)} \right\}$$
$$\times G^{p+2,0}_{p+2,p+2}\left(\begin{matrix} \left\{\frac{n-2}{2}, \left\{\frac{n-j-2}{2}\right\}_{j=1:2}, \left\{\frac{n-2}{2}\right\}_{j=2:m+1}, \left\{\frac{n-1}{2}\right\}_{j=m+2:p}\right\} \\ \left\{\frac{n-4}{2}, \left\{\frac{n-p-j-2}{2}\right\}_{j=1:2}, \left\{\frac{n-2-j}{2}\right\}_{j=2:p}\right\} \end{matrix} \,\middle|\, z \right)$$

and

$$F_\Lambda(z) = \frac{\Gamma\left(\frac{n}{2}\right)}{\Gamma\left(\frac{n-2}{2}\right)} \left\{ \prod_{j=1}^{2} \frac{\Gamma\left(\frac{n-j}{2}\right)}{\Gamma\left(\frac{n-p-j}{2}\right)} \right\} \left\{ \prod_{j=2}^{m+1} \frac{\Gamma\left(\frac{n}{2}\right)}{\Gamma\left(\frac{n-j}{2}\right)} \right\} \left\{ \prod_{j=m+2}^{p} \frac{\Gamma\left(\frac{n+1}{2}\right)}{\Gamma\left(\frac{n-j}{2}\right)} \right\}$$
$$\times G^{p+2,1}_{p+3,p+3}\left(\begin{matrix} \left\{1, \frac{n}{2}, \left\{\frac{n-j}{2}\right\}_{j=1:2}, \left\{\frac{n}{2}\right\}_{j=2:m+1}, \left\{\frac{n+1}{2}\right\}_{j=m+2:p}\right\} \\ \left\{\frac{n-2}{2}, \left\{\frac{n-p-j}{2}\right\}_{j=1:2}, \left\{\frac{n-j}{2}\right\}_{j=2:p}, 0\right\} \end{matrix} \,\middle|\, z \right).$$

Modules PDFIndCircSphEqMean, CDFIndCircSphEqMean, PvalIndCircSphEqMean, QuantIndCircSphEqMean, and PvalDataIndCircSphEqMean are available for the implementation of this test, and these modules have exactly the same arguments as the modules in the two previous subsections.

To exemplify the present test we will once again use the data set used in Sects. 5.3.4 and 5.3.5 to test whether we may assume that the five response times for the first sentence and the last two response times for the second sentence are independent, with the covariance matrix for the set of five response times being circular and their means all equal, while the covariance matrix for the set of two response times is spherical, that is, while these two response times may be considered to have equal variances and to be independent. We will also carry out a similar test where we will use for the first sentence only the first four response times. The commands used and the output obtained with the Mathematica® version of module PvalDataIndCircSphEqMean may be viewed in Fig. 5.95, with the p-values obtained leading us to not reject any of the two null hypotheses.

In case one also wants to test whether the means of the two "far away" points are the same, then the distribution of Λ will no longer be an EGIG distribution, but in case the variable that is being considered is a variable whose value may vanish as we get far away from the central point of interest, then it may make sense to test if the means for the two "far away" points are null, that is to test if $\underline{\mu}_2 = \underline{0}_{(2\times 1)}$.

We would then be considering a null hypothesis of the type

$$H_0 \equiv H_{0b|a} \circ H_{0a}$$

```
PvalDataIndCircSphEqMean[path <> "Response_times_1set.dat", 1]
 Original variables {1, 2, 3, 4, 5, 9, 10}
Computed value of Λ: 0.007750275406
p-value: 0.4064726808
PvalDataIndCircSphEqMean[path <> "Response_times_1set.dat", 1]
 Original variables {1, 2, 3, 4, 9, 10}
Computed value of Λ: 0.1878514708
p-value: 0.9481901958
```

Fig. 5.95 Command used and output obtained with the Mathematica® version of module PvalDataIndCircSphEqMean to implement the simultaneous test of independence, circularity of the covariance matrix and equality of means for the first set of five or four variables and sphericity of the covariance matrix for the set with two variables, using a data set obtained from a transformation of the data sets on response times in Tables 3.9.7 and 3.9.8 of Timm (2002) (see text)

where, for non-specified b and σ^2,

$$H_{0b|a}: \Sigma_1 \text{ is circular and } \underline{\mu}_1 = bE_{p1};\ \Sigma_2 = \sigma^2 I_2 \text{ and } \underline{\mu}_2 = \underline{0}_{(2\times 1)}$$
assuming H_{0a} .

From Sects. 5.2.2 and 5.3.1, the $(2/n)$-th power of the l.r.t. statistic to test $H_{0b|a}$ is

$$\Lambda_{b|a} = n^p \frac{|A_1|}{v_1^*} \frac{|A_2|}{\left(tr\left(\frac{1}{2}X_2X_2'\right)\right)^2} \prod_{j=2}^{p} \frac{1}{v_j^* + w_j},$$

where X_2 is the $2\times n$ matrix of the sample values for the measurements taken on the two "far away" points, with

$$\Lambda_{b|a} \stackrel{d}{\equiv} \left\{\prod_{j=2}^{p} Y_j\right\} Y\,(Y^*)^2$$

where Y_j $(j = 1, \ldots, p)$ and Y are the r.v.'s in (5.481) and

$$Y^* \sim Beta\,(n-1, 1)$$

are all independent r.v.'s.

The $(2/n)$-th power of the l.r.t. statistic to test H_0 is thus

$$\Lambda = 2^2\, n^p \frac{|A|}{v_1^*\left(tr\left(X_2X_2'\right)\right)^2} \prod_{j=2}^{p} \frac{1}{v_j^* + w_j},$$

with

$$\Lambda \stackrel{d}{\equiv} Y_1^* Y_2^* \left\{\prod_{j=2}^{p} Y_j\right\} Y\,(Y^*)^2$$

and its exact distribution is given by a combination of Theorems 3.3 and 3.2 and for odd p its exact p.d.f. and c.d.f. are given by Corollary 4.4, with

$$m^{**} = 1\,, \quad n_1^* = 2\,, \quad k_1^* = \frac{p-1}{2}\,, \quad a_1^* = \frac{n-1}{2}\,, \quad s_1^* = 1$$

and

$$m^* = 3\,,\;\; n_\nu = \{1,1,1\}\,,\;\; k_\nu = \{2,1,2\}\,,\;\; a_\nu = \left\{\frac{n-p}{2}, \frac{n}{2}, \frac{n+1}{2}\right\}\,,$$
$$m_\nu = \{p,1,1\}$$

and for even p with

$$m^{**} = 1\,, \quad n_1^* = 2\,, \quad k_1^* = \frac{p}{2} - 1\,, \quad a_1^* = \frac{n-2}{2}\,, \quad s_1^* = 2$$

and

$$m^* = 4\,,\;\; n_\nu = \{1,1,1,1\}\,,\;\; k_\nu = \{2,1,2,2\}\,,\;\; a_\nu = \left\{\frac{n-p}{2}, \frac{n}{2}, \frac{n+1}{2}, \frac{n}{2}\right\}\,,$$
$$m_\nu = \{p,1,1,1\}\,.$$

Its exact p.d.f. and c.d.f. are thus given respectively by (5.483) and (5.484), with

$$r_1 = \begin{cases} \frac{p+1}{2} & \text{odd } p \\ \frac{p}{2} & \text{even } p \end{cases}$$

and the remaining r_j, $j = 2, \ldots, p+2$, given by (5.485) for odd p or by (5.486) for even p.

The p.d.f and c.d.f. of Λ may also be expressed in terms of Fox H function, for $p^* = p+3$ and $p^{**} = p+4$, as

$$f_\Lambda(z) = (n-1)\,\frac{\Gamma\left(\frac{n}{2}\right)}{\Gamma\left(\frac{n-2}{2}\right)} \left\{\prod_{j=1}^{2} \frac{\Gamma\left(\frac{n-j}{2}\right)}{\Gamma\left(\frac{n-p-j}{2}\right)}\right\} \left\{\prod_{j=2}^{m+1} \frac{\Gamma\left(\frac{n}{2}\right)}{\Gamma\left(\frac{n-j}{2}\right)}\right\} \left\{\prod_{j=m+2}^{p} \frac{\Gamma\left(\frac{n+1}{2}\right)}{\Gamma\left(\frac{n-j}{2}\right)}\right\}$$
$$\times H^{p^*,0}_{p^*,p^*}\left(\begin{array}{c} \left\{\left(\frac{n-2}{2},1\right), (n-2,2), \left\{\left(\frac{n-j-2}{2},1\right)\right\}_{j=1:2}, \left\{\left(\frac{n-2}{2},1\right)\right\}_{j=2:m+1}, \left\{\left(\frac{n-1}{2},1\right)\right\}_{j=m+2:p}\right\} \\ \left\{\left(\frac{n-4}{2},1\right), (n-3,2), \left\{\left(\frac{n-p-j-2}{2},1\right)\right\}_{j=1:2}, \left\{\left(\frac{n-2-j}{2},1\right)\right\}_{j=2:p}\right\} \end{array} \middle|\, z \right)$$

and

$$F_\Lambda(z) = (n-1)\,\frac{\Gamma\left(\frac{n}{2}\right)}{\Gamma\left(\frac{n-2}{2}\right)} \left\{\prod_{j=1}^{2} \frac{\Gamma\left(\frac{n-j}{2}\right)}{\Gamma\left(\frac{n-p-j}{2}\right)}\right\} \left\{\prod_{j=2}^{m+1} \frac{\Gamma\left(\frac{n}{2}\right)}{\Gamma\left(\frac{n-j}{2}\right)}\right\} \left\{\prod_{j=m+2}^{p} \frac{\Gamma\left(\frac{n+1}{2}\right)}{\Gamma\left(\frac{n-j}{2}\right)}\right\}$$
$$\times H^{p^*,1}_{p^{**},p^{**}}\left(\begin{array}{c} \left\{(1,1), \left(\frac{n}{2},1\right), (n,2), \left\{\left(\frac{n-j}{2},1\right)\right\}_{j=1:2}, \left\{\left(\frac{n}{2},1\right)\right\}_{j=2:m+1}, \left\{\left(\frac{n+1}{2},1\right)\right\}_{j=m+2:p}\right\} \\ \left\{\left(\frac{n-2}{2},1\right), (n-1,2), \left\{\left(\frac{n-p-j}{2},1\right)\right\}_{j=1:2}, \left\{\left(\frac{n-j}{2},1\right), (0,1)\right\}_{j=2:p}\right\} \end{array} \middle|\, z \right).$$

Modules PDFIndCircSphEqMean0, CDFIndCircSphEqMean0, PvalIndCircSphEqMean0, QuantIndCircSphEqMean0, and PvalDataIndCircSphEqMean0 are available for the implementation of this test, and these modules have exactly the same arguments as the modules for the previous test and the modules in the two previous subsections.

To exemplify the implementation of the present test we will once more use the data set used for the previous test and in Sects. 5.3.4 and 5.3.5 to test whether we may assume that the five response times for the first sentence and the last two response times for the second sentence are independent, with the covariance matrix for the set of five response times being circular and their means all equal, while the covariance matrix for the set of two response times is spherical and their means null. We will also carry out a similar test where we will use for the first sentence only the first four response times. The commands used and the output obtained with the Mathematica® version of module PvalDataIndCircSphEqMean0 may be viewed in Fig. 5.96. The extremely low p-values obtained indicate that we should reject our null hypotheses, which, taking into account the results from the previous test, may be seen as due to the fact that the mean values for the two last response times for the second sentence cannot be considered as null.

The first test in the present subsection is invariant under linear transformations $\underline{X} \rightarrow C\underline{X} + \underline{b}$ where C is defined as in the previous subsection and $\underline{b} = [\underline{b}_1', \underline{b}_2']'$ with $\underline{b}_1 = aE_{p1}$, where a is any real, and $\underline{b}_2$ any $2{\times}1$ real vector, and the second test is invariant under similar linear transformations with $\underline{b}_2 = \underline{0}$, a null $2{\times}1$ vector.

For the case where there are more than two "far away" points, near-exact distributions based on the results in Coelho et al. (2016, Sects. 2,6) and (2010, Sects. 3,5) may be used.

```
PvalDataIndCircSphEqMean0[path <> "Response_times_1set.dat", 1]
 Original variables {1, 2, 3, 4, 5, 9, 10}
Computed value of Λ: 9.809770821 × 10^-7
p-value: 6.846257388 × 10^-6

PvalDataIndCircSphEqMean0[path <> "Response_times_1set.dat", 1]
 Original variables {1, 2, 3, 4, 9, 10}
Computed value of Λ: 0.00003683406833
p-value: 7.452853003 × 10^-6
```

Fig. 5.96 Command used and output obtained with the Mathematica® version of module PvalDataIndCircSphEqMean0 to implement the simultaneous test of independence, circularity of the covariance matrix and equality of means for the first set of five or four variables and sphericity of the covariance matrix and nullity of means for the set with two variables, using a data set obtained from a transformation of the data sets on response times in Tables 3.9.7 and 3.9.8 of Timm (2002) (see a detailed description of the data set in Sect. 5.3.4)

Appendix 1: On the Distribution of $\Lambda = \frac{|A|}{|A+B|}$ When A and B Are Independent Real Wishart Matrices (and on the Independence of Λ and $A + B$)

If the matrix A is a $p\times p$ matrix with a (real) Wishart distribution with $n-q$ degrees of freedom and parameter matrix Σ and B is an independent $p\times p$ matrix with a (real) Wishart (or pseudo-Wishart) distribution with $q-1$ degrees of freedom and parameter matrix Σ, that is, if

$$A \sim W_p(n-q, \Sigma) \quad \text{and} \quad B \sim W_p(q-1, \Sigma) \tag{5.487}$$

are two independent matrices, then (see Theorem 7.3.2 in Anderson (2003))

$$A + B \sim W_p(n-1, \Sigma) \tag{5.488}$$

with (see Theorem 3.2.15 in Muirhead (2005))

$$|A| \stackrel{d}{\equiv} |\Sigma| \prod_{j=1}^{p} Z_j \quad \text{and} \quad |A+B| \stackrel{d}{\equiv} |\Sigma| \prod_{j=1}^{p} W_j$$

where, for $j = 1, \ldots, p$,

$$Z_j \sim \chi^2_{n-q+1-j} \quad \text{and} \quad W_j \sim \chi^2_{n-j}$$

are two sets of p independent r.v.'s, with

$$W_j = Z_j + W_j^*$$

where Z_j is independent of W_j^*, with

$$W_j^* \sim \chi^2_{q-1}, \quad j = 1, \ldots, p,$$

forming a set of p independent r.v.'s.

As such

$$\Lambda = \frac{|A|}{|A+B|} \stackrel{d}{\equiv} \prod_{j=1}^{p} Y_j \tag{5.489}$$

where

$$Y_j = \frac{Z_j}{Z_j + W_j^*} \sim Beta\left(\frac{n-q+1-j}{2}, \frac{q-1}{2}\right), \quad j = 1, \ldots, p \tag{5.490}$$

are a set of p independent r.v.'s (see Theorem 8.4.1 in Anderson (2003)).

Alternatively, one may obtain the h-th moment of Λ directly from the joint distribution of A and B. We know that if $A^* \sim W_p(n, \Sigma)$, then its p.d.f. is given by (see for example Theorem 3.2.1 in Sect. 3.2 of Muirhead (2005) or expression (1) in Sect. 7.2 of Anderson (2003))

$$f_{A^*}(\mathcal{A}) = \frac{e^{-\frac{1}{2}tr(\Sigma^{-1}\mathcal{A})}|\mathcal{A}|^{(n-p-1)/2}}{2^{np/2}\,\Gamma_p\left(\frac{n}{2}\right)|\Sigma|^{n/2}}, \tag{5.491}$$

where $\Gamma_p(\cdot)$ represents the (real) multivariate Gamma function, which is a convenient shorthand notation, with (see Def. 2.1.10 and Theorem 2.1.12 in Muirhead (2005))

$$\Gamma_p(a) = \pi^{p(p-1)/4}\prod_{j=1}^{p}\Gamma\left(a-\frac{j-1}{2}\right). \tag{5.492}$$

From (5.491) it is easy to obtain the h-th moment of $|A^*|$, since, for $h > (p+1-n)/2$,

$$\begin{aligned}
E\left(|A^*|^h\right) &= \int_{\mathcal{A}>0}|\mathcal{A}|^h\, f_{A^*}(\mathcal{A})\, d\mathcal{A}\\
&= \int_{\mathcal{A}>0}|\mathcal{A}|^h\,\frac{e^{-\frac{1}{2}tr(\Sigma^{-1}\mathcal{A})}|\mathcal{A}|^{(n-p-1)/2}}{2^{np/2}\,\Gamma_p\left(\frac{n}{2}\right)|\Sigma|^{n/2}}\, d\mathcal{A}\\
&= \frac{2^{(n+2h)p/2}\,\Gamma_p\left(\frac{n+2h}{2}\right)|\Sigma|^{n/2+h}}{2^{np/2}\,\Gamma_p\left(\frac{n}{2}\right)|\Sigma|^{n/2}}\underbrace{\int_{\mathcal{A}>0}\underbrace{\frac{e^{-\frac{1}{2}tr(\Sigma^{-1}\mathcal{A})}|\mathcal{A}|^{(n+2h-p-1)/2}}{2^{(n+2h)p/2}\,\Gamma_p\left(\frac{n+2h}{2}\right)|\Sigma|^{(n+2h)/2}}}_{\text{p.d.f. of } W_p(n+2h,\Sigma)}\, d\mathcal{A}}_{=1}\\
&= 2^{hp}\,|\Sigma|^h\,\frac{\Gamma_p\left(\frac{n}{2}+h\right)}{\Gamma_p\left(\frac{n}{2}\right)}
\end{aligned} \tag{5.493}$$

where the notation $\int_{\mathcal{A}>0}$ indicates that the integral is taken over the space of all positive-definite symmetric matrices.

Then, assuming that A and B are independent, with the distributions in (5.487), and using (5.493) and (5.492), we have

$$
\begin{aligned}
E\left[\left(\frac{|A|}{|A+B|}\right)^h\right] &= E\left(\frac{|A|^h}{|A+B|^h}\right)\\
&= \int_{\mathcal{A}>0}\int_{\mathcal{B}>0} \frac{|\mathcal{A}|^h}{|\mathcal{A}+\mathcal{B}|^h}\, \frac{e^{-\frac{1}{2}tr(\Sigma^{-1}\mathcal{A})}|\mathcal{A}|^{(n-q-p-1)/2}}{2^{(n-q)p/2}\,\Gamma_p\left(\frac{n-q}{2}\right)|\Sigma|^{(n-q)/2}}\\
&\qquad\qquad \times \frac{e^{-\frac{1}{2}tr(\Sigma^{-1}\mathcal{B})}|\mathcal{B}|^{(q-p-2)/2}}{2^{(q-1)p/2}\,\Gamma_p\left(\frac{q-1}{2}\right)|\Sigma|^{(q-1)/2}}\, d\mathcal{B}\, d\mathcal{A}\\
&= 2^{hp}\,|\Sigma|^h \frac{\Gamma_p\left(\frac{n-q}{2}+h\right)}{\Gamma_p\left(\frac{n-q}{2}\right)}\\
&\qquad \int_{\mathcal{A}>0}\int_{\mathcal{B}>0} \frac{1}{|\mathcal{A}+\mathcal{B}|^h}\, \underbrace{\frac{e^{-\frac{1}{2}tr(\Sigma^{-1}\mathcal{A})}|\mathcal{A}|^{(n-q+2h-p-1)/2}}{2^{(n-q+2h)p/2}\,\Gamma_p\left(\frac{n-q+2h}{2}\right)|\Sigma|^{(n-q+2h)/2}}}_{\text{p.d.f. of } W_p(n-q+2h,\Sigma)}\\
&\qquad\qquad \times \underbrace{\frac{e^{-\frac{1}{2}tr(\Sigma^{-1}\mathcal{B})}|\mathcal{B}|^{(q-p-2)/2}}{2^{(q-1)p/2}\,\Gamma_p\left(\frac{q-1}{2}\right)|\Sigma|^{(q-1)/2}}}_{\text{p.d.f. of } W_p(q-1,\Sigma)}\, d\mathcal{B}\, d\mathcal{A}\\
&= 2^{hp}\,|\Sigma|^h \frac{\Gamma_p\left(\frac{n-q}{2}+h\right)}{\Gamma_p\left(\frac{n-q}{2}\right)} \underbrace{E\left(|A+B|^{-h}\right)}_{\text{with } A+B\sim W_p(n-1+2h,\Sigma)}\\
&= 2^{hp}\,|\Sigma|^h \frac{\Gamma_p\left(\frac{n-q}{2}+h\right)}{\Gamma_p\left(\frac{n-q}{2}\right)}\, 2^{-hp}\,|\Sigma|^{-h} \frac{\Gamma_p\left(\frac{n-1}{2}+h-h\right)}{\Gamma_p\left(\frac{n-1}{2}+h\right)}\\
&= \prod_{j=1}^{p} \frac{\Gamma\left(\frac{n-q}{2}+h-\frac{j-1}{2}\right)}{\Gamma\left(\frac{n-q}{2}-\frac{j-1}{2}\right)}\, \frac{\Gamma\left(\frac{n-1}{2}-\frac{j-1}{2}\right)}{\Gamma\left(\frac{n-1}{2}+h-\frac{j-1}{2}\right)} = \prod_{j=1}^{p} E\left(Y_j^h\right) \quad \left(h>\frac{p+q-n-1}{2}\right)
\end{aligned}
\tag{5.494}
$$

where the r.v.'s Y_j are the r.v.'s in (5.490). Since $0 < \Lambda < 1$, the set of moments in (5.494) completely identifies the distribution of Λ, and as such we can infer (5.489)–(5.490).

Yet another way to obtain $E(\Lambda^h)$ would be to use Lemma 10.4.1 of Anderson (2003) or Theorem 5 of Jensen (1988), which, in simple terms, state that $\Lambda = \frac{|A|}{|A+B|}$ is independent of $A+B$, so that from

$$|A| = \Lambda\,|A+B|$$

we may write

$$E\left(|A|^h\right) = E\left(\Lambda^h\right) E\left(|A+B|^h\right),$$

and thus from (5.493) and the distributions of A and $A+B$ in (5.487) and (5.488) we have

$$E\left(\Lambda^h\right) = \frac{E\left(|A|^h\right)}{E\left(|A+B|^h\right)} = 2^{hp}\,|\Sigma|^h \frac{\Gamma_p\left(\frac{n-q}{2}+h\right)}{\Gamma_p\left(\frac{n-q}{2}\right)}\, 2^{-hp}\,|\Sigma|^{-h} \frac{\Gamma_p\left(\frac{n-1}{2}\right)}{\Gamma_p\left(\frac{n-1}{2}+h\right)}$$

which is exactly the expression before the last in (5.494).

It is possible to show that, without any restrictions on either p or q,

$$\prod_{j=1}^{p} Y_j \stackrel{d}{\equiv} \prod_{k=1}^{q-1} Y_k^*,$$

where

$$Y_k^* \sim Beta\left(\frac{n-p-k}{2}, \frac{p}{2}\right), \quad k = 1, \ldots, q-1 \qquad (5.495)$$

are a set of q independent r.v.'s (see for example Coelho (1999)).

This is so, since from (5.494) we may write

$$\begin{aligned}
E\left(\Lambda^h\right) &= \prod_{j=1}^{p} \frac{\Gamma\left(\frac{n-q+1-j}{2}+h\right)\Gamma\left(\frac{n-j}{2}\right)}{\Gamma\left(\frac{n-q+1-j}{2}\right)\Gamma\left(\frac{n-j}{2}+h\right)}\\
&= \left\{\prod_{j=q}^{p+q-1} \frac{\Gamma\left(\frac{n-j}{2}+h\right)}{\Gamma\left(\frac{n-j}{2}\right)}\right\}\left\{\prod_{j=1}^{p} \frac{\Gamma\left(\frac{n-j}{2}\right)}{\Gamma\left(\frac{n-j}{2}+h\right)}\right\}\\
&= \left\{\prod_{j=1}^{p+q-1} \frac{\Gamma\left(\frac{n-j}{2}+h\right)}{\Gamma\left(\frac{n-j}{2}\right)}\right\}\left\{\prod_{j=1}^{q-1} \frac{\Gamma\left(\frac{n-j}{2}\right)}{\Gamma\left(\frac{n-j}{2}+h\right)}\right\}\left\{\prod_{j=1}^{p} \frac{\Gamma\left(\frac{n-j}{2}\right)}{\Gamma\left(\frac{n-j}{2}+h\right)}\right\}\\
&= \left\{\prod_{j=p+1}^{p+q-1} \frac{\Gamma\left(\frac{n-j}{2}+h\right)}{\Gamma\left(\frac{n-j}{2}\right)}\right\}\left\{\prod_{j=1}^{q-1} \frac{\Gamma\left(\frac{n-j}{2}\right)}{\Gamma\left(\frac{n-j}{2}+h\right)}\right\}\\
&= \prod_{j=1}^{q-1} \frac{\Gamma\left(\frac{n-p-j}{2}+h\right)\Gamma\left(\frac{n-j}{2}\right)}{\Gamma\left(\frac{n-p-j}{2}\right)\Gamma\left(\frac{n-j}{2}+h\right)} = \prod_{k=1}^{q-1} E\left((Y_k^*)^h\right),
\end{aligned}$$

where Y_k^* $(k = 1, \ldots, q-1)$ are the r.v.'s in (5.495).

Appendix 2: On the Equivalence of the Statistic Λ and the Hotelling T^2 Statistic for $q = 2$

First of all let us note that for $q = 2$ we have, from (5.4),

$$\underline{\overline{X}} = \tfrac{1}{n}\left(n_1\underline{\overline{X}}_1 + n_2\underline{\overline{X}}_2\right) = \tfrac{n_1}{n}\underline{\overline{X}}_1 + \tfrac{n_2}{n}\underline{\overline{X}}_2 ,$$

so that, for $n = n_1 + n_2$,

$$\begin{aligned}
n_1 & \left(\underline{\overline{X}}_1 - \underline{\overline{X}}\right)\left(\underline{\overline{X}}_1 - \underline{\overline{X}}\right)' + n_2\left(\underline{\overline{X}}_2 - \underline{\overline{X}}\right)\left(\underline{\overline{X}}_2 - \underline{\overline{X}}\right)' \\
&= n_1\left(\underline{\overline{X}}_1 - \tfrac{n_1}{n}\underline{\overline{X}}_1 - \tfrac{n_2}{n}\underline{\overline{X}}_2\right)\left(\underline{\overline{X}}_1 - \tfrac{n_1}{n}\underline{\overline{X}}_1 - \tfrac{n_2}{n}\underline{\overline{X}}_2\right)' \\
&\qquad + n_2\left(\underline{\overline{X}}_2 - \tfrac{n_1}{n}\underline{\overline{X}}_1 - \tfrac{n_2}{n}\underline{\overline{X}}_2\right)\left(\underline{\overline{X}}_2 - \tfrac{n_1}{n}\underline{\overline{X}}_1 - \tfrac{n_2}{n}\underline{\overline{X}}_2\right)' \\
&= n_1\left(\tfrac{n-n_1}{n}\underline{\overline{X}}_1 - \tfrac{n_2}{n}\underline{\overline{X}}_2\right)\left(\tfrac{n-n_1}{n}\underline{\overline{X}}_1 - \tfrac{n_2}{n}\underline{\overline{X}}_2\right)' \\
&\qquad + n_2\left(\tfrac{n-n_2}{n}\underline{\overline{X}}_2 - \tfrac{n_2}{n}\underline{\overline{X}}_2\right)\left(\tfrac{n-n_2}{n}\underline{\overline{X}}_2 - \tfrac{n_2}{n}\underline{\overline{X}}_2\right)' \\
&= n_1\left(\tfrac{n_2}{n}\right)^2\left(\underline{\overline{X}}_1 - \underline{\overline{X}}_2\right)\left(\underline{\overline{X}}_1 - \underline{\overline{X}}_2\right)' + n_2\left(\tfrac{n_1}{n}\right)^2\left(\underline{\overline{X}}_1 - \underline{\overline{X}}_2\right)\left(\underline{\overline{X}}_1 - \underline{\overline{X}}_2\right)' \\
&= \tfrac{n_1n_2^2 + n_2n_1^2}{n^2}\left(\underline{\overline{X}}_1 - \underline{\overline{X}}_2\right)\left(\underline{\overline{X}}_1 - \underline{\overline{X}}_2\right)' = \tfrac{n_1n_2}{n_1+n_2}\left(\underline{\overline{X}}_1 - \underline{\overline{X}}_2\right)\left(\underline{\overline{X}}_1 - \underline{\overline{X}}_2\right)' .
\end{aligned} \tag{5.496}$$

Then, since if the matrix H is partitioned as

$$H = \begin{bmatrix} H_{11} & H_{12} \\ H_{21} & H_{22} \end{bmatrix}$$

we have

$$|H| = |H_{11}|\,|H_{22.1}| = |H_{22}|\,|H_{11.2}| ,$$

where

$$H_{11.2} = H_{11} - H_{12}H_{22}^{-1}H_{21} , \qquad \text{and} \qquad H_{22.1} = H_{22} - H_{21}H_{11}^{-1}H_{12} .$$

Let us, following a similar approach to the one used in Sect. 4.2 of Morrison (2005), take the matrix

$$H = \begin{bmatrix} -1 & \sqrt{\tfrac{n_1n_2}{n_1+n_2}}\left(\underline{\overline{X}}_1 - \underline{\overline{X}}_2\right)' \\ \sqrt{\tfrac{n_1n_2}{n_1+n_2}}\left(\underline{\overline{X}}_1 - \underline{\overline{X}}_2\right) & A \end{bmatrix}$$

where A is the matrix in (5.3) for $q = 2$.

Then we have

$$|H| = -\left|A + \frac{n_1 n_2}{n_1+n_2}\left(\underline{X}_1 - \underline{X}_2\right)\left(\underline{X}_1 - \underline{X}_2\right)'\right| \tag{5.497}$$

$$= |A|\left(-1 - \frac{n_1 n_2}{n_1+n_2}\left(\underline{X}_1 - \underline{X}_2\right)' A^{-1}\left(\underline{X}_1 - \underline{X}_2\right)\right), \tag{5.498}$$

so that by equating (5.497) and (5.498) we have

$$\left|A + \frac{n_1 n_2}{n_1+n_2}\left(\underline{X}_1 - \underline{X}_2\right)\left(\underline{X}_1 - \underline{X}_2\right)'\right| = |A|\Big(1 + \underbrace{\frac{n_1 n_2}{n_1+n_2}\left(\underline{X}_1 - \underline{X}_2\right)' A^{-1}\left(\underline{X}_1 - \underline{X}_2\right)}_{T^2}\Big)$$

or, taking into account the definition of the statistic T^2 in (5.14), the equality in (5.496) and the definition of the matrix B in (5.3) for $q = 2$,

$$1 + T^2 = \frac{\left|A + \frac{n_1 n_2}{n_1+n_2}\left(\underline{X}_1 - \underline{X}_2\right)\left(\underline{X}_1 - \underline{X}_2\right)'\right|}{|A|}$$

$$= \frac{\left|A + n_1\left(\underline{X}_1 - \underline{X}\right)\left(\underline{X}_1 - \underline{X}\right)' + n_2\left(\underline{X}_2 - \underline{X}\right)\left(\underline{X}_2 - \underline{X}\right)'\right|}{|A|} = \frac{|A+B|}{|A|} = \frac{1}{\Lambda}$$

or

$$T^2 = \frac{1}{\Lambda} - 1 = \frac{1-\Lambda}{\Lambda}.$$

Appendix 3: On the Covariance Matrix of CXD, Where X Is Random and C and D Are Nonrandom

Let X be a $p{\times}n$ random matrix with $Var(X) = Var(vec(X)) = F \otimes \Sigma$, where F is $n{\times}n$ and Σ is $p{\times}p$. Then, using the following relations, involving Kronecker products,

$$vec(CXD) = (D' \otimes C)vec(X),$$

$$(D \otimes C)' = D' \otimes C' \qquad \text{and} \qquad (C \otimes D)(G \otimes H) = CG \otimes DH,$$

we have, for nonrandom matrices C and D,

$$\begin{aligned} Var(CXD) &= Var(vec(CXD)) = Var((D' \otimes C)vec(X)) \\ &= (D' \otimes C)\, Var(vec(X))\, (D' \otimes C) \\ &= (D' \otimes C)(F \otimes \Sigma)(D \otimes C') \\ &= (D'F \otimes C\Sigma)(D \otimes C') = D'FD \otimes C\Sigma C'. \end{aligned}$$

Appendix 4: Some Properties of the Matrix D in (5.69)

It may be easily shown that for D in (5.69) we have $DMM'D' = DMM' = MM'D'$, since from the definition of the matrix D we may write

$$DMM' = \left(I_q - \tfrac{1}{n}MM'E_{qq}\right)MM' = MM' - \tfrac{1}{n}MM'E_{qq}MM' \\ = MM'\left(I_q - \tfrac{1}{n}E_{qq}MM'\right) = MM'D' \tag{5.499}$$

while, on the other hand, the matrix D is idempotent, since

$$\begin{aligned} DD &= \left(I_q - \tfrac{1}{n}MM'E_{qq}\right)\left(I_q - \tfrac{1}{n}MM'E_{qq}\right) \\ &= I_q - \tfrac{1}{n}MM'E_{qq} - \tfrac{1}{n}MM'E_{qq} + \tfrac{1}{n^2}MM'\underbrace{E_{qq}MM'E_{qq}}_{=nE_{qq}} \\ &= I_q - \tfrac{1}{n}MM'E_{qq} = D\,, \end{aligned}$$

where we have used the fact that $MM' = diag(n_k, k = 1, \ldots, q)$ and that as such $E_{qq}MM'E_{qq} = nE_{qq}$.

Now using (5.499) and the idempotent property of D we may write

$$DMM'D' = DDMM' = DMM' = MM'D'\,.$$

Appendix 5: On the Equality of A_0 and $A_1 + B^*$ in (5.85) and (5.84) for $B^* = \frac{1}{n}\beta DHH'D'\beta'$ with D and H in (5.76) and (5.77)

Let us consider the split of M in (5.79). Then, we will have

$$MM' = \begin{bmatrix} M_1M_1' & M_1M_2' \\ M_2M_1' & M_2M_2' \end{bmatrix} = \begin{bmatrix} M_{11} & M_{12} \\ M_{21} & M_{22} \end{bmatrix} \tag{5.500}$$

which, using Result 2 in Appendix 6, namely (5.506), for the inverse of a partitioned matrix, implies

$$(MM')^{-1} = \begin{bmatrix} M_{11}^{-1} + M_{11}^{-1}M_{12}M_{22.1}^{-1}M_{21}M_{11}^{-1} & -M_{11}^{-1}M_{12}M_{22.1}^{-1} \\ -M_{22.1}^{-1}M_{21}M_{11}^{-1} & M_{22.1}^{-1} \end{bmatrix}$$

for $M_{22.1}$ given by (5.78).

Then we have

$$\begin{aligned}\beta &= XM'(MM')^{-1}\\ &= X\left[\, M_1' \mid M_2' \,\right]\begin{bmatrix} M_{11}^{-1} + M_{11}^{-1}M_{12}M_{22.1}^{-1}M_{21}M_{11}^{-1} & -M_{11}^{-1}M_{12}M_{22.1}^{-1} \\ -M_{22.1}^{-1}M_{21}M_{11}^{-1} & M_{22.1}^{-1}\end{bmatrix}\\ &= X\Big[M_1'M_{11}^{-1} + M_1'M_{11}^{-1}M_{12}M_{22.1}^{-1}M_{21}M_{11}^{-1} - M_2'M_{22}^{-1}M_{21}M_{11}^{-1} \mid \\ &\qquad M_2'M_{22.1}^{-1} - M_1'M_{11}^{-1}M_{12}M_{22.1}^{-1}\Big] = \left[\, \beta_1 \mid \beta_2 \,\right]\end{aligned} \tag{5.501}$$

so that

$$\begin{aligned}\beta DHH'D'\beta' &= \left[\, \beta_1 \mid \beta_2 \,\right]\begin{bmatrix} 0 & 0 \\ 0 & I_{q2}\end{bmatrix}\begin{bmatrix} 0 & 0 \\ 0 & M_{22.1}\end{bmatrix}\begin{bmatrix} 0 & 0 \\ 0 & I_{q2}\end{bmatrix}\begin{bmatrix} \beta_1' \\ \beta_2'\end{bmatrix}\\ &= \left[\, 0 \mid \beta_2 \,\right]\begin{bmatrix} 0 & 0 \\ 0 & M_{22.1}\end{bmatrix}\begin{bmatrix} 0 \\ \beta_2'\end{bmatrix} = \beta_2 M_{22.1}\beta_2'\\ &= \left(XM_2' - XM_1'M_{11}^{-1}M_{12}\right)\left(M_{22.1}^{-1}M_2X' - M_{22.1}^{-1}M_{21}M_{11}^{-1}M_1X'\right)\\ &= X_2M_2'M_{22.1}^{-1}M_2X' - XM_2'M_{22.1}^{-1}M_{21}M_{11}^{-1}M_1X'\\ &\quad -XM_1'M_{11}^{-1}M_{12}M_{22.1}^{-1}M_2X' + XM_1'M_{11}^{-1}M_{12}M_{22.1}^{-1}M_{21}M_{11}^{-1}M_1X'.\end{aligned}$$

Then, since

$$A_1 = \tfrac{1}{n}\left(XX' - XM'(MM')^{-1}MX\right)$$

where

$$\begin{aligned}XM'(MM')^{-1}MX = \beta MX &= \left[\, \beta_1 \mid \beta_2 \,\right]\begin{bmatrix} M_1 \\ M_2\end{bmatrix}X\\ &= \beta_1 M_1 X + \beta_2 M_2 X,\end{aligned}$$

it follows from (5.501) that we have

$$\begin{aligned}XM'(MM')^{-1}MX = \beta MX &= \beta_1 M_1 X + \beta_2 M_2 X\\ &= XM_1'M_{11}^{-1}M_1X + XM_1'M_{11}^{-1}M_{12}M_{22.1}^{-1}M_{21}M_{11}^{-1}M_1X'\\ &\quad -XM_2'M_{22.1}^{-1}M_{21}M_{11}^{-1}M_1X' + XM_2'M_{22.1}^{-1}M_2X'\\ &\quad -XM_1'M_{11}^{-1}M_{12}M_{22.1}^{-1}M_2X'.\end{aligned}$$

Therefore,

$$XM'(MM')^{-1}MX = XM_1'M_{11}^{-1}M_1X + \beta DHH'D'\beta',$$

and as such, we have for A_0 and A_1 in (5.85) and (5.84),

$$A_0 = A_1 + \frac{1}{n}\beta DHH'D'\beta',$$

confirming that the l.r.t. statistics in (5.83) and (5.58) are the same.

Appendix 6: Results on the Inverse and the Determinant of Compound and Partitioned Matrices

In this appendix the authors state two results on matrix inverses which are available in essentially equivalent forms in several books, but which are shown here in simpler and/or somewhat more general forms. A result on the determinant of a partitioned matrix is also presented.

Result 1: Let A and B be respectively a $p\times p$ and $q\times q$ symmetric invertible matrices and let C be a $p\times q$ and D a $q\times p$ matrix. Then

$$P = A \pm CBD \Longrightarrow P^{-1} = A^{-1} \mp A^{-1}C(B \pm DA^{-1}B)^{-1}DA^{-1}. \tag{5.502}$$

Proof Just check that for the above definitions of P and P^{-1} we have $PP^{-1} = P^{-1}P = I_p$. □

This result is a slightly more general and simpler version of the result in Theorem A.5.1 in Muirhead (2005).

Two other results we will use are the following ones, on the inverse and the determinant of a partitioned matrix.

Result 2: Let A be a symmetric positive-definite matrix partitioned as

$$A = \begin{bmatrix} A_{11} & A_{12} \\ A_{21} & A_{22} \end{bmatrix}, \tag{5.503}$$

where A_{11} and A_{22} are square matrices. Then, if A^{-1} exists, we have, for

$$A_{11.2} = A_{11} - A_{12}A_{22}^{-1}A_{21}, \quad \text{and} \quad A_{22.1} = A_{22} - A_{21}A_{11}^{-1}A_{12}, \tag{5.504}$$

$$A^{-1} = \begin{bmatrix} A_{11.2}^{-1} & -A_{11}^{-1}A_{12}A_{22.1}^{-1} \\ -A_{22.1}^{-1}A_{21}A_{11}^{-1} & A_{22.1}^{-1} \end{bmatrix} \tag{5.505}$$

$$= \begin{bmatrix} A_{11}^{-1} + A_{11}^{-1}A_{12}A_{22.1}^{-1}A_{21}A_{11}^{-1} & -A_{11}^{-1}A_{12}A_{22.1}^{-1} \\ -A_{22.1}^{-1}A_{21}A_{11}^{-1} & A_{22.1}^{-1} \end{bmatrix} \tag{5.506}$$

$$= \begin{bmatrix} A_{11}^{-1} & 0 \\ 0 & A_{22.1}^{-1} \end{bmatrix} \begin{bmatrix} A_{11} + A_{12}A_{22.1}^{-1}A_{21} & -A_{12} \\ -A_{21} & A_{22.1} \end{bmatrix} \begin{bmatrix} A_{11}^{-1} & 0 \\ 0 & A_{22.1}^{-1} \end{bmatrix}, \quad (5.507)$$

or,

$$A^{-1} = \begin{bmatrix} A_{11.2}^{-1} & -A_{11.2}^{-1}A_{12}A_{22}^{-1} \\ -A_{22}^{-1}A_{21}A_{11.2}^{-1} & A_{22.1}^{-1} \end{bmatrix} \quad (5.508)$$

$$= \begin{bmatrix} A_{11.2}^{-1} & -A_{11.2}^{-1}A_{12}A_{22}^{-1} \\ -A_{22}^{-1}A_{21}A_{11.2}^{-1} & A_{22}^{-1} + A_{22}^{-1}A_{21}A_{11.2}^{-1}A_{12}A_{22}^{-1} \end{bmatrix} \quad (5.509)$$

$$= \begin{bmatrix} A_{11.2}^{-1} & 0 \\ 0 & A_{22}^{-1} \end{bmatrix} \begin{bmatrix} A_{11} + A_{12}A_{22.1}^{-1}A_{21} & -A_{12} \\ -A_{21} & A_{22.1} \end{bmatrix} \begin{bmatrix} A_{11.2}^{-1} & 0 \\ 0 & A_{22}^{-1} \end{bmatrix}, \quad (5.510)$$

Proof First of all let us establish a couple of results. From the definition of $A_{11.2}$ in (5.504) we have

$$A_{11} = A_{11.2} + A_{12}A_{22}^{-1}A_{21},$$

which, right multiplying by A_{11}^{-1}, gives

$$I = A_{11.2}A_{11}^{-1} + A_{12}A_{22}^{-1}A_{21}A_{11}^{-1}$$

which in turn, left multiplying by $A_{11.2}^{-1}$, gives

$$A_{11.2}^{-1} = A_{11}^{-1} + A_{11}^{-1}A_{12}A_{22.1}^{-1}A_{21}A_{11}^{-1}. \quad (5.511)$$

Then, we may write

$$\begin{aligned} A_{11.2}^{-1}A_{12}A_{22}^{-1} &= \left(A_{11}^{-1} + A_{11}^{-1}A_{12}A_{22.1}^{-1}A_{21}A_{11}^{-1}\right)A_{12}A_{22}^{-1} \\ &= A_{11}^{-1}A_{12}A_{22}^{-1} + A_{11}^{-1}A_{12}A_{22.1}^{-1}A_{21}A_{11}^{-1}A_{12}A_{22}^{-1} \\ &= A_{11}^{-1}A_{12}\left(A_{22}^{-1} + A_{22.1}^{-1}A_{21}A_{11}^{-1}A_{12}A_{22}^{-1}\right) \\ &= A_{11}^{-1}A_{12}A_{22.1}^{-1}. \end{aligned}$$

This equality establishes the equivalence between (5.505) and (5.508) and from it, by transposing and then left multiplying by A_{22}, and noticing that since both $A_{11.2}$ and $A_{22.1}$ are symmetric matrices, also $A_{11.2}^{-1}$ and $A_{22.1}^{-1}$ are symmetric matrices, we obtain

$$A_{22}^{-1}A_{21}A_{11.2}^{-1} = A_{22.1}^{-1}A_{21}A_{11}^{-1} \iff A_{21}A_{11.2}^{-1} = A_{22}A_{22.1}^{-1}A_{21}A_{11}^{-1}. \quad (5.512)$$

In (5.512) the indexes 1 and 2 may indeed be reversed and the equality will still hold. Such an equality may be obtained from the second equality in (5.512) also by left multiplying it by A_{22}^{-1} and right multiplying by A_{11}, and then transposing, obtaining

$$A_{22.1}^{-1}A_{21} = A_{22}^{-1}A_{21}A_{11.2}^{-1}A_{11} \iff A_{12}A_{22.1}^{-1} = A_{11}A_{11.2}^{-1}A_{12}A_{22}^{-1}. \qquad (5.513)$$

Then, the proof that the matrix in (5.505) is the inverse of A may be obtained by just checking that for this definition of A^{-1} we have $AA^{-1} = A^{-1}A = I$. In doing this we will have to use in the upper left corners of

$$AA^{-1} = \left[\begin{array}{c|c} A_{11}A_{11.2}^{-1} - A_{12}A_{22.1}^{-1}A_{21}A_{11}^{-1} & -A_{12}A_{22.1}^{-1} + A_{12}A_{22.1}^{-1} \\ \hline A_{21}A_{11.2}^{-1} - A_{22}A_{22.1}^{-1}A_{21}A_{11}^{-1} & -A_{21}A_{11}^{-1}A_{12}A_{22.1}^{-1} + A_{22}A_{22.1}^{-1} \end{array}\right]$$

and

$$A^{-1}A = \left[\begin{array}{c|c} A_{11.2}^{-1}A_{11} - A_{11}^{-1}A_{12}A_{22.1}^{-1}A_{21} & A_{11.2}^{-1}A_{12} - A_{11}^{-1}A_{12}A_{22.1}^{-1}A_{22} \\ \hline -A_{22.1}^{-1}A_{21} + A_{22.1}^{-1}A_{21} & -A_{22.1}^{-1}A_{21}A_{11}^{-1}A_{12} + A_{22.1}^{-1}A_{22} \end{array}\right]$$

the expression for $A_{11.2}^{-1}$ in (5.511), and in the lower right corners the definition of $A_{22.1}$ in (5.504), while in the lower left corner of AA^{-1} and in the upper right corner of $A^{-1}A$ we will have to use (5.512). Then (5.506) is obtained from (5.505) by applying Result 1 to obtain $A_{11.2}^{-1}$, and (5.507) is easily obtained from (5.506) by using only matrix multiplication rules.

As already mentioned, (5.508) is obtained from (5.505) through the use of (5.512), and then (5.509) is obtained from (5.508) by applying Result 1 to the lower right submatrix in (5.508) and finally (5.510) is easily obtained from (5.509) by using matrix multiplication rules. □

Result 3: Let A be the matrix in (5.503). If A_{11} is non-singular, then

$$|A| = |A_{11}|\,|A_{22} - A_{21}A_{11}^{-1}A_{12}|, \qquad (5.514)$$

and if A_{22} is non-singular,

$$|A| = |A_{22}|\,|A_{11} - A_{12}A_{22}^{-1}A_{21}|. \qquad (5.515)$$

Proof If A_{11} is non-singular, let

$$B = \begin{bmatrix} A_{11} & 0 \\ A_{21} & I \end{bmatrix}, \qquad C = \begin{bmatrix} I & A_{11}^{-1}A_{12} \\ 0 & A_{22} - A_{21}A_{11}^{-1}A_{12} \end{bmatrix}.$$

Then, $A = BC$, so that $|A| = |B||C|$, where B is upper block-triangular and C is lower block-triangular. As such, $|B| = |A_{11}|$ and $|C| = |A_{11} - A_{21}A_{11}^{-1}A_{12}|$, so that we have (5.514).

If A_{22} is non-singular, let

$$B = \begin{bmatrix} I & A_{12} \\ 0 & A_{22} \end{bmatrix}, \qquad C = \begin{bmatrix} A_{11} - A_{12}A_{22}^{-1}A_{21} & 0 \\ A_{22}^{-1}A_{21} & I \end{bmatrix},$$

so that we have $A = BC$, with $|B| = |A_{22}|$ and $|C| = |A_{11} - A_{12}A_{22}^{-1}A_{21}|$, and, as such, $|A| = |B||C|$, given by (5.515). □

Appendix 7: The p.d.f. and the Covariance Matrix of a Complex Normal Random Vector

As mentioned in Sect. 5.1.5, the complex Normal distribution we consider in this book is the one used in Wooding (1956), Goodman (1957, 1963a,b), James (1964, Sec. 8), Khatri (1965), Gupta (1971), Krishnaiah et al. (1976), Fang et al. (1982), Brillinger (2001, Sec. 4.2), and Anderson (2003, probl. 2.64), with

$$\underline{X}_k \sim CN_p(\underline{\mu}_k, \Sigma)$$

denoting the fact that the random vector $\underline{X}_k$ has a p-variate complex Normal distribution with expected value vector $\underline{\mu}_k$ and covariance matrix Σ. The p.d.f. of $\underline{X}_k$ is given by

$$f_{\underline{X}_k}(\underline{x}) = \pi^{-p}\,|\Sigma|^{-1}\,e^{-(\underline{x}-\underline{\mu}_k)^{\#}\Sigma^{-1}(\underline{x}-\underline{\mu}_k)}$$

where "#" denotes the transpose of the complex conjugate. If we take $\underline{Y}_k \in \mathbb{R}^p$ and $\underline{Z}_k \in \mathbb{R}^p$ to represent respectively the real and imaginary parts of $\underline{X}_k$, i.e. with

$$\underline{X}_k = \underline{Y}_k + \mathrm{i}\,\underline{Z}_k\,,$$

the complex Normal distribution we will use requires

$$\begin{bmatrix} \underline{Y}_k \\ \underline{Z}_k \end{bmatrix} \sim N_{2p}\left(\begin{bmatrix} \underline{\mu}_{\underline{Y}_k} \\ \underline{\mu}_{\underline{Z}_k} \end{bmatrix}, \begin{bmatrix} \Gamma & \Phi \\ -\Phi & \Gamma \end{bmatrix}\right), \tag{5.516}$$

that is, it requires $\underline{Y}_k$ and $\underline{Z}_k$ to be jointly and thus also marginally normally distributed vectors, with

$$Var(\underline{Y}_k) = Var(\underline{Z}_k) = \Gamma\,,$$

and

$$Cov(\underline{Y}_k, \underline{Z}_k) = \Phi\,, \qquad Cov(\underline{Z}_k, \underline{Y}_k) = \Phi' = -\Phi\,,$$

where Γ is a real positive-definite matrix and Φ a real skew-symmetric matrix, thus with $\Phi' = -\Phi$.

Then, with these prerequisites, we have, for

$$\underline{Y}_k^* = \underline{Y}_k - \underline{\mu}_{\underline{Y}_k}\,, \quad \underline{Z}_k^* = \underline{Z}_k - \underline{\mu}_{\underline{Z}_k}\,,$$

and

$$\underline{X}_k^* = \underline{X}_k - \underline{\mu}_k = \underline{Y}_k^* + \mathrm{i}\,\underline{Z}_k^*\,,$$

where

$$\underline{\mu}_k = E(\underline{X}_k) = \underline{\mu}_{\underline{Y}_k} + \mathrm{i}\,\underline{\mu}_{\underline{Z}_k}\,,$$

the covariance matrix of $\underline{X}_k$ given by

$$\begin{aligned}\Sigma = Var(\underline{X}_k) &= E\left[\underline{X}_k^*(\overline{\underline{X}_k^*})'\right]\\ &= E\left[\left(\underline{Y}_k^* + \mathrm{i}\underline{Z}_k^*\right)\left(\underline{Y}_k^* - \mathrm{i}\underline{Z}_k^*\right)'\right]\\ &= E\left[\underline{Y}_k^*(\underline{Y}_k^*)' - \mathrm{i}^2\underline{Z}_k^*(\underline{Z}_k^*)' - \mathrm{i}\,\underline{Y}_k^*(\underline{Z}_k^*)' + \mathrm{i}\,\underline{Z}_k^*(\underline{Y}_k^*)'\right]\\ &= Var(\underline{Y}_k) + Var(\underline{Z}_k) - \mathrm{i}\,Cov(\underline{Y}_k, \underline{Z}_k) + \mathrm{i}\,Cov(\underline{Z}_k, \underline{Y}_k)\\ &= 2\Gamma - 2\mathrm{i}\Phi\,,\end{aligned}$$

which is an Hermitian, positive-definite matrix.

Appendix 8: On the Distribution of $\Lambda = \frac{|A|}{|A+B|}$ When A and B Are Independent Complex Wishart Matrices

In this appendix it is shown how we can obtain the distribution of

$$\Lambda = \frac{|A|}{|A+B|} \tag{5.517}$$

when A and B have independent complex Wishart distributions (Goodman, 1963a,b).

Although this proof bears a close resemblance to the one in Appendix 1 for the real case, we decided to consider it since there are indeed some subtle differences, namely in the definitions of the complex Wishart p.d.f. and the complex multivariate Gamma function, which may at first sight cause some confusion.

Let the matrix A be a $p\times p$ matrix with a complex Wishart distribution with $n-q$ degrees of freedom and parameter matrix Σ and B an independent $p\times p$ complex Wishart matrix with $q-1$ degrees of freedom and parameter matrix Σ, that is, let

$$A \sim CW_p(n-q, \Sigma) \quad \text{and} \quad B \sim CW_p(q-1, \Sigma) \tag{5.518}$$

be two independent matrices.

Then (see Goodman (1963a, Sec. 1))

$$A + B \sim CW_p(n-1, \Sigma) \tag{5.519}$$

with (see Goodman (1963b))

$$|A| \stackrel{\text{d}}{\equiv} |\Sigma| \prod_{j=1}^{p} Z_j \quad \text{and} \quad |A+B| \stackrel{\text{d}}{\equiv} |\Sigma| \prod_{j=1}^{p} (Z_j + W_j)$$

where, for $j = 1, \ldots, p$,

$$Z_j \sim \chi^2_{2(n-q+1-j)} \quad \text{and} \quad W_j \sim \chi^2_{2(q-1)}$$

are two independent sets of p independent r.v.'s.

As such

$$\frac{|A|}{|A+B|} \stackrel{\text{d}}{\equiv} \prod_{j=1}^{p} Y_j \tag{5.520}$$

where

$$Y_j = \frac{Z_j}{Z_j + W_j} \sim Beta\,(n-q+1-j, q-1)\,, \quad j = 1, \ldots, p \tag{5.521}$$

are a set of p independent r.v.'s.

Alternatively, as in the real case, it is not hard to obtain the h-th moment of Λ in (5.517) directly from the joint distribution of A and B. It is known that if

$A^* \sim CW_p(n, \Sigma)$, then its pdf is given by (see Goodman (1963a); James (1964, Sec. 8))

$$f_{A^*}(\mathcal{A}) = \frac{e^{-tr(\Sigma^{-1}\mathcal{A})}|\mathcal{A}|^{n-p}}{\Gamma_p(n)\,|\Sigma|^n}, \tag{5.522}$$

where

$$\Gamma_p(a) = \pi^{p(p-1)/2}\prod_{j=1}^{p}\Gamma(a-j+1) \tag{5.523}$$

is the complex multivariate Gamma function of order p (see James (1964, Sec. 8)).

The h-th moment of $|A^*|$ may be easily obtained from (5.522), for $h > p+1-n$, as

$$\begin{aligned} E\left(|A^*|^h\right) &= \int_{\mathcal{A}>0} |\mathcal{A}|^h f_{A^*}(\mathcal{A})\,d\mathcal{A} = \int_{\mathcal{A}>0} |\mathcal{A}|^h \frac{e^{-tr(\Sigma^{-1}\mathcal{A})}|\mathcal{A}|^{n-p}}{\Gamma_p(n)\,|\Sigma|^n}\,d\mathcal{A} \\ &= \frac{\Gamma_p(n+h)\,|\Sigma|^{n+h}}{\Gamma_p(n)\,|\Sigma|^n} \underbrace{\int_{\mathcal{A}>0} \underbrace{\frac{e^{-tr(\Sigma^{-1}\mathcal{A})}|\mathcal{A}|^{n+h-p-1}}{\Gamma_p(n+h)\,|\Sigma|^{n+h}}}_{\text{pdf of } CW_p(n+h,\,\Sigma)} d\mathcal{A}}_{=1} \\ &= |\Sigma|^h \frac{\Gamma_p(n+h)}{\Gamma_p(n)} \end{aligned} \tag{5.524}$$

where the notation $\mathcal{A} > 0$ indicates that the integral is taken over the space of all positive-definite Hermitian matrices.

Then, assuming that A and B are independent, with the distributions in (5.518), and using (5.524) and (5.523), it follows that, for $h > p+q-n$,

$$\begin{aligned} E\left(\Lambda^h\right) &= E\left(\frac{|A|^h}{|A+B|^h}\right) \\ &= \int_{\mathcal{A}>0}\int_{\mathcal{B}>0} \frac{|\mathcal{A}|^h}{|\mathcal{A}+\mathcal{B}|^h}\,\frac{e^{-tr(\Sigma^{-1}\mathcal{A})}|\mathcal{A}|^{n-q-p}}{\Gamma_p\,(n-q)\,|\Sigma|^{n-q}}\,\frac{e^{-tr(\Sigma^{-1}\mathcal{B})}|\mathcal{B}|^{q-p-1}}{\Gamma_p\,(q)\,|\Sigma|^q}\,d\mathcal{B}\,d\mathcal{A} \\ &= |\Sigma|^h \frac{\Gamma_p\,(n-q+h)}{\Gamma_p\,(n-q)} \int_{\mathcal{A}>0}\int_{\mathcal{B}>0} \frac{1}{|\mathcal{A}+\mathcal{B}|^h} \underbrace{\frac{e^{-tr(\Sigma^{-1}\mathcal{A})}|\mathcal{A}|^{n-q+h-p}}{\Gamma_p\,(n-q+h)\,|\Sigma|^{n-q+h}}}_{\text{pdf of } CW_p(n-q+h,\,\Sigma)} \\ &\qquad \times \underbrace{\frac{e^{-tr(\Sigma^{-1}\mathcal{B})}|\mathcal{B}|^{q-p-1}}{\Gamma_p\,(q)\,|\Sigma|^q}}_{\text{pdf of } CW_p(q-1,\,\Sigma)}\,d\mathcal{B}\,d\mathcal{A} \\ &= |\Sigma|^h \frac{\Gamma_p\,(n-q+h)}{\Gamma_p\,(n-q)}\,E\left(|A+B|^{-h}\right), \end{aligned}$$

where $A + B \sim CW_p(n + h - 1, \Sigma)$, so that, from (5.524) we may write, for $h > p + q - n$,

$$E\left(\Lambda^h\right) = |\Sigma|^h \frac{\Gamma_p(n-q+h)}{\Gamma_p(n-q)} |\Sigma|^{-h} \frac{\Gamma_p(n-1+h-h)}{\Gamma_p(n-1+h)} \tag{5.525}$$

$$= \prod_{j=1}^{p} \frac{\Gamma(n-q+h-j+1)}{\Gamma(n-q-j+1)} \frac{\Gamma(n-j)}{\Gamma(n+h-j)} = \prod_{j=1}^{p} E\left(Y_j^h\right), \tag{5.526}$$

where the r.v.'s Y_j are the r.v.'s in (5.521). Since $0 < \Lambda < 1$, the set of moments in (5.494) completely identifies the distribution of Λ, and as such it is possible to infer the result in (5.520).

Similar to what happens in the real case in Appendix 1, yet another way to obtain $E(\Lambda^h)$ would be to use Lemma 10.4.1 of Anderson (2003) or Theorem 5 of Jensen (1988), which, as mentioned in Appendix 1, state, in simple terms, that $\Lambda = \frac{|A|}{|A+B|}$ is independent of $A + B$, the m.l.e. of Σ under H_0, so that from

$$|A| = \Lambda\, |A + B|$$

we may write

$$E\left(|A|^h\right) = E\left(\Lambda^h\right) E\left(|A + B|^h\right).$$

Consequently, from (5.524) and the distributions of A and $A + B$ in (5.518) and (5.519) we may write

$$E\left(\Lambda^h\right) = \frac{E\left(|A|^h\right)}{E\left(|A+B|^h\right)} = |\Sigma|^h \frac{\Gamma_p(n-q+h)}{\Gamma_p(n-q)} |\Sigma|^{-h} \frac{\Gamma_p(n-1+h-h)}{\Gamma_p(n-1+h)}$$

which is exactly the expression in (5.525).

Once again, similar to what happens in the real case in Appendix 1, it is possible to show that, without any restrictions on either p or q,

$$\prod_{j=1}^{p} Y_j \stackrel{\mathrm{d}}{\equiv} \prod_{k=1}^{q-1} Y_k^*,$$

where

$$Y_k^* \sim Beta\,(n-p-k, p), \quad k = 1, \ldots, q-1 \tag{5.527}$$

are a set of q independent r.v.'s.

This is so, since from (5.526) we may write

$$\begin{aligned}
E\left(\Lambda^h\right) &= \prod_{j=1}^{p} \frac{\Gamma(n-q+1-j+h)\,\Gamma(n-j)}{\Gamma(n-q+1-j)\,\Gamma(n-j+h)} \\
&= \left\{\prod_{j=q}^{p+q-1} \frac{\Gamma(n-j+h)}{\Gamma(n-j)}\right\}\left\{\prod_{j=1}^{p} \frac{\Gamma(n-j)}{\Gamma(n-j+h)}\right\} \\
&= \left\{\prod_{j=1}^{p+q-1} \frac{\Gamma(n-j+h)}{\Gamma(n-j)}\right\}\left\{\prod_{j=1}^{q-1} \frac{\Gamma(n-j)}{\Gamma(n-j+h)}\right\}\left\{\prod_{j=1}^{p} \frac{\Gamma(n-j)}{\Gamma(n-j+h)}\right\} \\
&= \left\{\prod_{j=p+1}^{p+q-1} \frac{\Gamma(n-j+h)}{\Gamma(n-j)}\right\}\left\{\prod_{j=1}^{q-1} \frac{\Gamma(n-j)}{\Gamma(n-j+h)}\right\} \\
&= \prod_{j=1}^{q-1} \frac{\Gamma(n-p-j+h)\,\Gamma(n-j)}{\Gamma(n-p-j)\,\Gamma(n-j+h)} = \prod_{k=1}^{q-1} E\left((Y_k^*)^h\right),
\end{aligned}$$

where Y_k^* $(k=1,\dots,q-1)$ are the r.v.'s in (5.527).

Appendix 9: On the Equivalence of the Test of Independence of $\underline{X}_1$ and $\underline{X}_2$ and the Test of the Simultaneous Nullity of the Canonical Correlations

Since the matrices Σ_{11} and Σ_{22} are symmetric and positive-definite there exist symmetric positive-definite matrices $\Sigma_{11}^{1/2}$ and $\Sigma_{22}^{1/2}$ such that

$$\Sigma_{11}^{-1/2}\Sigma_{11}^{-1/2} = \Sigma_{11}^{-1} \quad \text{and} \quad \Sigma_{22}^{-1/2}\Sigma_{22}^{-1/2} = \Sigma_{22}^{-1}.$$

Let then $\underline{u}_{1\alpha}^*$ be the normalized eigenvector of the symmetric positive-definite or positive-semi-definite matrix

$$\Sigma_{11}^{-1/2}\Sigma_{12}\Sigma_{22}^{-1}\Sigma_{21}\Sigma_{11}^{-1/2}, \tag{5.528}$$

associated with the α-th largest non-null eigenvalue ρ_α^2 $(\alpha=1,\dots,k)$, with $k=\min(p_1,p_2)$ and $\rho_1^2 \geq \rho_2^2 \geq \cdots \geq \rho_{p_1}^2$, that is, let

$$\Sigma_{11}^{-1/2}\Sigma_{12}\Sigma_{22}^{-1}\Sigma_{21}\Sigma_{11}^{-1/2}\underline{u}_{1\alpha}^* = \rho_\alpha^2\underline{u}_{1\alpha}^*, \quad \text{with} \quad \underline{u}_{1\alpha}^{*\prime}\underline{u}_{1\alpha}^* = 1 \quad (\alpha=1,\dots,k). \tag{5.529}$$

Note that since Σ_{11} and Σ_{22} are full-rank matrices, the matrix in (5.528) will be positive-definite if $p_1 < p_2$ and positive-semi-definite if $p_1 > p_2$, in which case it will have p_2 positive eigenvalues and $p_1 - p_2$ null ones.

Then if we define the matrices

$$U_1^* = \left[\underline{u}_{11}^* \,|\, \underline{u}_{12}^* \,|\, \dots \,|\, \underline{u}_{1k}^*\right] \quad \text{and} \quad R = diag(\rho_\alpha, \alpha = 1, \dots, k)\,,$$

where each ρ_α is the positive square root of ρ_α^2 ($\alpha = 1, \dots, k$), we may then write

$$\Sigma_{11}^{-1/2}\Sigma_{12}\Sigma_{22}^{-1}\Sigma_{21}\Sigma_{11}^{-1/2}U_1^* = U_1^* R^2\,,$$

or, given the normalization of the vectors $\underline{u}_{1\alpha}^*$,

$$U_1^{*\prime}\Sigma_{11}^{-1/2}\Sigma_{12}\Sigma_{22}^{-1}\Sigma_{21}\Sigma_{11}^{-1/2}U_1^* = R^2\,. \tag{5.530}$$

Subsequently take $\underline{u}_{1\alpha} = \Sigma_{11}^{-1/2}\underline{u}_{1\alpha}^*$ and define

$$U_1 = \left[\underline{u}_{11} \,|\, \underline{u}_{12} \,|\, \dots \,|\, \underline{u}_{1k}\right] = \Sigma_{11}^{-1/2}U_1^*\,.$$

Then, from (5.530) we may write

$$U_1'\Sigma_{12}\Sigma_{22}^{-1}\Sigma_{21}U_1 = R^2\,.$$

Let then $\underline{u}_{2\alpha}$ be the counterparts of the vectors $\underline{u}_{1\alpha}$ by interchanging the indexes 1 and 2, that is, let $\underline{u}_{2\alpha} = \Sigma_{22}^{-1/2}\underline{u}_{2\alpha}^*$, where $\underline{u}_{2\alpha}^*$ are the eigenvectors of the matrix

$$\Sigma_{22}^{-1/2}\Sigma_{21}\Sigma_{11}^{-1/2}\Sigma_{12}\Sigma_{22}^{-1/2}\,, \tag{5.531}$$

associated with the non-null eigenvalues ρ_α^2 ($\alpha = 1, \dots, k$), with

$$\underline{u}_{2\alpha}^{*\prime}\underline{u}_{2\alpha}^* = 1 \iff \underline{u}_{2\alpha}'\Sigma_{22}\underline{u}_{2\alpha} = 1\,.$$

That the eigenvalues or latent roots of the matrices in (5.528) and (5.531) are the same is easy to establish for example by left multiplying the first equality in (5.529) by $\Sigma_{22}^{-1/2}\Sigma_{21}\Sigma_{11}^{-1/2}$, obtaining

$$\Sigma_{22}^{-1/2}\Sigma_{21}\Sigma_{11}^{-1/2}\Sigma_{12}\Sigma_{22}^{-1}\Sigma_{21}\Sigma_{11}^{-1/2}\underline{u}_{1\alpha}^* = \rho_\alpha^2\Sigma_{22}^{-1/2}\Sigma_{21}\Sigma_{11}^{-1/2}\underline{u}_{1\alpha}^*\,,$$

which shows that $\Sigma_{22}^{-1/2}\Sigma_{21}\Sigma_{11}^{-1/2}\underline{u}_{1\alpha}^*$ is an eigenvector of $\Sigma_{22}^{-1/2}\Sigma_{21}\Sigma_{11}^{-1/2}\Sigma_{12}\Sigma_{22}^{-1/2}$ associated with the eigenvalue ρ_α^2 ($\alpha = 1, \dots, p_2$). Indeed the eigenvectors $\Sigma_{22}^{-1/2}\Sigma_{21}\Sigma_{11}^{-1/2}\underline{u}_{1\alpha}^*$ are homothetic to the unitary eigenvectors of the matrix in (5.531) associated with the eigenvalues ρ_α^2, which are the eigenvectors

$$\underline{u}_{2\alpha}^* = \frac{1}{\rho_\alpha} \Sigma_{22}^{-1/2} \Sigma_{21} \Sigma_{11}^{-1/2} \underline{u}_{1\alpha}^* , \tag{5.532}$$

with

$$\begin{aligned} \Sigma_{22}^{-1/2} \Sigma_{21} \Sigma_{11}^{-1} \Sigma_{12} \Sigma_{22}^{-1/2} \underline{u}_{2\alpha}^* &= \frac{1}{\rho_\alpha} \Sigma_{22}^{-1/2} \Sigma_{21} \Sigma_{11}^{-1/2} \underbrace{\Sigma_{11}^{-1/2} \Sigma_{12} \Sigma_{22}^{-1} \Sigma_{21} \Sigma_{11}^{-1/2} \underline{u}_{1\alpha}^*}_{=\rho_\alpha^2 \underline{u}_{1\alpha}^*} \\ &= \rho_\alpha \underbrace{\Sigma_{22}^{-1/2} \Sigma_{21} \Sigma_{11}^{-1/2} \underline{u}_{1\alpha}^*}_{\rho_\alpha \underline{u}_{2\alpha}^*} = \rho_\alpha^2 \underline{u}_{2\alpha}^* , \end{aligned}$$

where we used (5.529) and (5.532), and thus also with

$$\underline{u}_{2\alpha}^{*\prime} \underline{u}_{2\alpha}^* = \frac{1}{\rho_\alpha^2} \underline{u}_{1\alpha}^{*\prime} \underbrace{\Sigma_{11}^{-1/2} \Sigma_{12} \Sigma_{22}^{-1} \Sigma_{21} \Sigma_{11}^{-1/2} \underline{u}_{1\alpha}^*}_{=\rho_\alpha^2 \underline{u}_{1\alpha}^*} = \frac{1}{\rho_\alpha^2} \rho_\alpha^2 \underline{u}_{1\alpha}^{*\prime} \underline{u}_{1\alpha}^* = 1 ,$$

where we used again (5.529).

Then, if we define

$$U_2^* = \left[\underline{u}_{21}^* \,\middle|\, \underline{u}_{22}^* \,\middle|\, \dots \,\middle|\, \underline{u}_{2k}^* \right]$$

and take $\underline{u}_{2\alpha} = \Sigma_{22}^{-1/2} \underline{u}_{2\alpha}^*$ and then, according to (5.532), define

$$U_2 = \left[\underline{u}_{21} \,\middle|\, \underline{u}_{22} \,\middle|\, \dots \,\middle|\, \underline{u}_{2k} \right] = \Sigma_{22}^{-1/2} U_2^* = \Sigma_{22}^{-1} \Sigma_{21} \Sigma_{11}^{-1/2} U_1^* R^{-1} = \Sigma_{22}^{-1} \Sigma_{21} U_1 R^{-1} ,$$

we have

$$U_1' \Sigma_{11} U_1 = I_k , \quad U_2' \Sigma_{22} U_2 = I_k$$

and, using (5.530),

$$U_1' \Sigma_{12} U_2 = U_1^{*\prime} \Sigma_{11}^{-1/2} \Sigma_{12} \Sigma_{22}^{-1} \Sigma_{21} \Sigma_{11}^{-1/2} U_1^* R^{-1} = R^2 R^{-1} = R .$$

Appendix 10: Some Relations Involving Matrices M, M^*, and M^+ in (5.68), (5.166), and (5.175)

Concerning the matrices M and M^+ we need to show that

$$M'(MM')^{-1}M = M^{+\prime}(M^+M^{+\prime})^{-1}M^+ . \tag{5.533}$$

From (5.175) we easily see that

$$M^+M^{+\prime} = \left[\begin{array}{c|c} n & n_2 \; \dots \; n_q \\ \hline n_2 & \\ \vdots & M^*M^{*\prime} \\ n_q & \end{array}\right],$$

where

$$M^*M^{*\prime} = diag(n_2, \dots, n_q)\,. \tag{5.534}$$

Then, from (5.503) to (5.505),

$$(M^+M^{+\prime})^{-1} = \left[\begin{array}{c|c} * & (**)' \\ \hline ** & *** \end{array}\right]$$

where, noticing that $n = \sum_{k=1}^{q} n_k$, we have

$$* = \left(n - \left[n_2, \dots, n_q\right] diag\left(\tfrac{1}{n_2}, \dots, \tfrac{1}{n_q}\right) \begin{bmatrix} n_2 \\ \vdots \\ n_q \end{bmatrix} \right)^{-1} = \Big(n-(n-n_1)\Big)^{-1} = \frac{1}{n_1}$$

and

$$** = -diag\left(\tfrac{1}{n_2}, \dots, \tfrac{1}{n_q}\right) \begin{bmatrix} n_2 \\ \vdots \\ n_q \end{bmatrix} \frac{1}{n_1} = -\frac{1}{n_1} E_{q-1,1}\,,$$

and, noticing that

$$\left[\, n_2, \dots, n_q \right] = E_{1n} M^{*\prime}\,, \tag{5.535}$$

while, at the same time the matrices E_{pq} have the property that for any $p, q, n \in \mathbb{N}$, $E_{pq}E_{qn} = qE_{pn}$, we have

$$\begin{aligned} *** &= \left(M^*M^{*\prime} - \begin{bmatrix} n_2 \\ \vdots \\ n_q \end{bmatrix} \tfrac{1}{n}\left[n_2, \dots, n_q\right] \right)^{-1} = \left(M^*M^{*\prime} - \tfrac{1}{n} M^* E_{n1} E_{1n} M^{*\prime} \right)^{-1} \\ &= \left(M^*M^{*\prime} - \tfrac{1}{n} M^* E_{nn} M^{*\prime} \right)^{-1} = \left(M^*(I_n - \tfrac{1}{n} E_{nn}) M^{*\prime} \right)^{-1}. \end{aligned} \tag{5.536}$$

Therefore,

$$(M^+M^{+\prime})^{-1} = \left[\begin{array}{c|c} \frac{1}{n_1} & -\frac{1}{n_1}E_{1,q-1} \\ \hline -\frac{1}{n_1}E_{q-1,1} & \left(M^*(I_n - \frac{1}{n}E_{nn})M^{*\prime}\right)^{-1} \end{array}\right]$$

and then

$$\begin{aligned} M^{+\prime}(M^+M^{+\prime})^{-1} &= \left[\begin{array}{c|c} E_{n1} & M^{*\prime} \end{array}\right] \left[\begin{array}{c|c} \frac{1}{n_1} & -\frac{1}{n_1}E_{1,q-1} \\ \hline -\frac{1}{n_1}E_{q-1,1} & \left(M^*(I_n - \frac{1}{n}E_{nn})M^{*\prime}\right)^{-1} \end{array}\right] \\ &= \left[\begin{array}{c|c} \frac{1}{n_1}E_{n1} - \frac{1}{n_1}M^{*\prime}E_{q-1,1} & -\frac{1}{n_1}E_{n,q-1} + M^{*\prime}\left(M^*(I_n - \frac{1}{n}E_{nn})M^{*\prime}\right)^{-1} \end{array}\right] \end{aligned}$$

so that

$$\begin{aligned} &M^{+\prime}(M^+M^{+\prime})^{-1}M^+ \\ &= \left[\begin{array}{c|c} \frac{1}{n_1}E_{n1} - \frac{1}{n_1}M^{*\prime}E_{q-1,1} & -\frac{1}{n_1}E_{n,q-1} + M^{*\prime}\left(M^*(I_n - \frac{1}{n}E_{nn})M^{*\prime}\right)^{-1} \end{array}\right] \left[\begin{array}{c} E_{1n} \\ \hline M^* \end{array}\right] \\ &= \frac{1}{n_1}E_{nn} - \frac{1}{n_1}M^{*\prime}E_{q-1,n} - \frac{1}{n_1}E_{n,q-1}M^* + M^{*\prime}\left(M^*(I_n - \frac{1}{n}E_{nn})M^{*\prime}\right)^{-1}M^* \end{aligned} \tag{5.537}$$

where we need to identify the relationship between $M^{*\prime}\left(M^*(I_n - \frac{1}{n}E_{nn})M^{*\prime}\right)^{-1}M^*$ and $M^{*\prime}(M^*M^{*\prime})^{-1}M^*$. In order to do this we will rewrite $\left(M^*(I_n - \frac{1}{n}E_{nn})M^{*\prime}\right)^{-1}$ in a similar form to that in the second expression of (5.536) and then use (5.502) with $A = M^*M^{*\prime}$, $C = M^*E_{n1}$, $B = I_1$ and $D = E_{1n}M^{*\prime}$. We may thus write

$$\begin{aligned} &\left(M^*\left(I_n - \frac{1}{n}E_{nn}\right)M^{*\prime}\right)^{-1} = \left(M^*M^{*\prime} - \frac{1}{n}M^*E_{nn}M^{*\prime}\right)^{-1} = \left(M^*M^{*\prime} - \frac{1}{n}M^*E_{n1}E_{1n}M^{*\prime}\right)^{-1} \\ &= (M^*M^{*\prime})^{-1} \\ &\quad +(M^*M^{*\prime})^{-1}\frac{1}{n}M^*E_{n1}\left(I_1 - \frac{1}{n}E_{1n}M^{*\prime}(M^*M^{*\prime})^{-1}M^*E_{n1}\right)^{-1}E_{1n}M^{*\prime}(M^*M^{*\prime})^{-1} \end{aligned}$$

where, taking into account (5.534) and (5.535),

$$\left(I_1 - \frac{1}{n}\underbrace{E_{1n}M^{*\prime}(M^*M^{*\prime})^{-1}M^*E_{n1}}_{=E_{1,q-1}M^*E_{n1}=n-n_1}\right)^{-1} = \left(1 - \frac{n-n_1}{n}\right)^{-1} = \frac{n}{n_1}$$

so that we may finally write

$$\left(M^*\left(I_n - \tfrac{1}{n}E_{nn}\right)M^{*\prime}\right)^{-1} = (M^*M^{*\prime})^{-1} + \tfrac{1}{n}\tfrac{n}{n_1}(M^*M^{*\prime})^{-1}E_{n1}E_{1n}M^{*\prime}(M^*M^{*\prime})^{-1}$$
$$= (M^*M^{*\prime})^{-1} + \tfrac{1}{n_1}E_{q-1,q-1}\,.$$

Therefore, from (5.537) we have

$$M^{+\prime}(M^+M^{+\prime})^{-1}M^+$$
$$= \tfrac{1}{n_1}E_{nn} - \tfrac{1}{n_1}M^{*\prime}E_{q-1,n} - \tfrac{1}{n_1}E_{n,q-1}M^* + M^{*\prime}(M^*M^{*\prime})^{-1}M^* + \tfrac{1}{n_1}M^{*\prime}E_{q-1,q-1}M^*,$$

where, for M_1 defined in (5.166),

$$\frac{1}{n_1}E_{nn} - \frac{1}{n_1}M^{*\prime}E_{q-1,n} - \frac{1}{n_1}E_{n,q-1}M^* + \frac{1}{n_1}M^{*\prime}E_{q-1,q-1}M^* = \frac{1}{n_1}M_1'M_1$$

since

$$M^{*\prime}E_{q-1,n} = \left[\begin{array}{c|c} 0_{n_1\times n_1} & 0_{n_1\times(n-n_1)} \\ \hline E_{n-n_1,n_1} & E_{n-n_1,n-n_1} \end{array}\right],$$

$$E_{n,q-1}M^* = (M^{*\prime}E_{q-1,n})' = \left[\begin{array}{c|c} 0_{n_1\times n_1} & E_{n_1,n-n_1} \\ \hline 0_{(n-n_1)\times n_1} & E_{n-n_1,n-n_1} \end{array}\right]$$

and

$$M^{*\prime}E_{q-1,1} = \left[\begin{array}{c} 0_{n_1\times 1} \\ \hline E_{n-n_1,1} \end{array}\right]$$

so that

$$M^{*\prime}E_{q-1,q-1}M^* = M^{*\prime}E_{q-1,1}E_{1,q-1}M^* = \left[\begin{array}{c} 0_{n_1\times 1} \\ \hline E_{n-n_1,1} \end{array}\right]\left[\begin{array}{c|c} 0_{1\times n_1} & E_{1,n-n_1} \end{array}\right]$$
$$= \left[\begin{array}{c|c} 0_{n_1\times n_1} & 0_{n_1\times n-n_1} \\ \hline 0_{(n-n_1)\times n_1} & E_{n-n_1,n-n_1} \end{array}\right]$$

and as such we finally have

$$E_{nn} - M^{*\prime}E_{q-1,n} - E_{n,q-1}M^* + M^{*\prime}E_{q-1,q-1}M^* = \left[\begin{array}{c|c} E_{n_1,n_1} & 0_{n_1\times n-n_1} \\ \hline 0_{(n-n_1)\times n_1} & 0_{(n-n_1)\times(n-n_1)} \end{array}\right] = M_1'M_1\,,$$

and we may thus write

$$M^{+\prime}(M^+M^{+\prime})^{-1}M^+ = \frac{1}{n_1}M_1'M_1 + M^{*\prime}(M^*M^{*\prime})^{-1}M^*\,.$$

On the other hand, since

$$MM' = diag(n_k, k=1,\dots,q) = \left[\begin{array}{c|c} n_1 & 0_{1\times q-1} \\ \hline 0_{(q-1)\times 1} & M^*M^{*\prime} \end{array}\right],$$

which implies that

$$(MM')^{-1} = \left[\begin{array}{c|c} 1/n_1 & 0_{1\times q-1} \\ \hline 0_{(q-1)\times 1} & (M^*M^{*\prime})^{-1} \end{array}\right],$$

from (5.166) we may easily write

$$\begin{aligned} M'(MM')^{-1}M &= \left[\, M_1' \,\middle|\, M^{*\prime} \,\right] \left[\begin{array}{c|c} 1/n_1 & 0_{1\times q-1} \\ \hline 0_{(q-1)\times 1} & (M^*M^{*\prime})^{-1} \end{array}\right] \left[\begin{array}{c} M_1 \\ \hline M^* \end{array}\right] \\ &= \left[\, \tfrac{1}{n_1}M_1' \,\middle|\, M^{*\prime}(M^*M^{*\prime})^{-1} \,\right] \left[\begin{array}{c} M_1 \\ \hline M^* \end{array}\right] \\ &= \tfrac{1}{n_1}M_1'M_1 + M^{*\prime}(M^*M^{*\prime})^{-1}M^*\,, \end{aligned}$$

so that we have indeed the equality in (5.533), with

$$M'(MM')^{-1}M = M^{+\prime}(M^+M^{+\prime})^{-1}M^+ = bdiag\left(\frac{1}{n_k}E_{n_kn_k}, k=1,\dots,q\right).$$

Appendix 11: On the Equivalence of the Hypotheses $H_0 : \Sigma_{12} = 0$ and $H_0 : \beta = 0$ in Sect. 5.1.9, the Distribution of $\widehat{\beta}$, and Some Further Results

In this appendix we will first determine $\hat{\beta}$ and then its distribution. Then we will show that $\beta = \Sigma_{12}\Sigma_{22}^{-1}$, so that the null hypothesis $H_0 : \Sigma_{12} = 0$ is indeed equivalent to the null hypothesis $H_0 : \beta = 0$. From these results we will then obtain the necessary results to identify the distributions of Λ_{H_1,H_0^*}. Λ_{H_0,H_0^*} and Λ_{H_1,H_0}.

Let the matrix Σ be partitioned as

$$\Sigma = \begin{bmatrix} \Sigma_{11} & \Sigma_{12} \\ \Sigma_{21} & \Sigma_{22} \end{bmatrix} \tag{5.538}$$

and let A be the m.l.e. of Σ, which can be expressed as

$$A = \tfrac{1}{n} X^* X^{*\prime} = \tfrac{1}{n} X(I_n - \tfrac{1}{n} E_{nn})X' \tag{5.539}$$

where X is the $p{\times}n$ matrix of a sample of size n from $\underline{X}$, for $p = p_1 + p_2$, and where

$$X^* = X(I_n - \tfrac{1}{n} E_{nn})$$

is the centered sample matrix for the sample means, where $I_n - \frac{1}{n} E_{nn}$ is a symmetric and idempotent matrix.

From Corollary 3.2.2 in Muirhead (2005) we know that

$$A \sim W_p(n-1, \tfrac{1}{n}\Sigma)\,.$$

Let us consider for A a similar partition to the one of Σ in (5.538),

$$A = \begin{bmatrix} A_{11} & A_{12} \\ A_{21} & A_{22} \end{bmatrix}$$

and let us consider the matrices

$$A_{11.2} = A_{11} - A_{12}A_{22}^{-1}A_{21} \quad \text{and} \quad \Sigma_{11.2} = \Sigma_{11} - \Sigma_{12}\Sigma_{22}^{-1}\Sigma_{21}\,.$$

Then, from Theorem 3.2.10 in Muirhead (2005) we know that

$$A_{11.2} \sim W_{p_1}\left(n - 1 - p_2, \tfrac{1}{n}\Sigma_{11.2}\right)$$

and

$$A_{12}|A_{22} \sim N_{p_1\times p_2}\left(\Sigma_{12}\Sigma_{22}^{-1}A_{22}, \tfrac{1}{n}A_{22}\otimes\Sigma_{11.2}\right) \tag{5.540}$$

and that $A_{11.2}$ and $A_{12}|A_{22}$ are independent.

But then, from (5.540), using simple rules for the expected value and the rule for the covariance of a product matrix in Appendix 3,

$$A_{12}A_{22}^{-1}|A_{22} \sim N_{p_1\times p_2}\left(\Sigma_{12}\Sigma_{22}^{-1}, \tfrac{1}{n}A_{22}^{-1}\otimes\Sigma_{11.2}\right) \tag{5.541}$$

so that under the null hypothesis $H_0 : \Sigma_{12} = 0$ we have

$$A_{11.2} \sim W_{p_1}\left(n-1-p_2, \tfrac{1}{n}\Sigma_{11}\right)$$

and

$$A_{12}A_{22}^{-1}|A_{22} \sim N_{p_1\times p_2}\left(0, \tfrac{1}{n}A_{22}^{-1}\otimes\Sigma_{11}\right). \tag{5.542}$$

Now, taking $\underline{X}_1$ as the set of response variables and $\underline{X}_2$ as the set of explanatory variables, let us consider the model

$$\underline{X}_1 = \beta\underline{X}_2 + \mathcal{E}, \tag{5.543}$$

or, in terms of centered sample matrices,

$$X_1^* = \beta X_2^* + \mathcal{E}, \tag{5.544}$$

where

$$X_1^* = X_1(I_n - \tfrac{1}{n}E_{nn}), \qquad X_2^* = X_2(I_n - \tfrac{1}{n}E_{nn})$$

are the centered sample matrices, with X_1 $(p_1\times n)$ and X_2 $(p_2\times n)$ the sample matrices for the sample of size n from $\underline{X}_1$ and $\underline{X}_2$ respectively.

From (5.544) we have

$$E(X_1^*|X_2^*) = \beta$$

or yet,

$$\hat{X}_1^*|X_2^* = \hat{\beta}X_2^*$$

so that the $p_1 \times n$ matrix of residuals $\hat{\mathcal{E}}$ will be given by

$$\hat{\mathcal{E}} = X_1^* - \hat{X}_1^* | X_2^* = X_1^* - \hat{\beta} X_2^*$$

so that the SSPRE (sum of squares and products of residuals) matrix will be given by

$$\begin{aligned} SSPRE &= \hat{\mathcal{E}}\hat{\mathcal{E}}' = (X_1^* - \hat{\beta} X_2^*)(X_1^* - \hat{\beta} X_2^*)' \\ &= X_1^* X_1^{*\prime} - X_1^* X_2^{*\prime} \hat{\beta}' - \hat{\beta} X_2^* X_1^{*\prime} + \hat{\beta} X_2^* X_2^{*\prime} \hat{\beta}' \end{aligned}$$

with

$$\frac{\partial SSPRE}{\partial \hat{\beta}} = -2X_2^* X_1^{*\prime} + 2X_2^* X_2^{*\prime} \hat{\beta}' .$$

Thus,

$$\begin{aligned} \frac{\partial SSPRE}{\partial \hat{\beta}} = 0 &\Longrightarrow X_2^* X_2^{*\prime} \hat{\beta}' = X_2^* X_1^{*\prime} \\ &\Longleftrightarrow \hat{\beta}' = (X_2^* X_2^{*\prime})^{-1} X_2^* X_1^{*\prime} \\ &\Longleftrightarrow \hat{\beta} = X_1^* X_2^{*\prime} (X_2^* X_2^{*\prime})^{-1} = A_{12} A_{22}^{-1} , \end{aligned} \qquad (5.545)$$

so that we have the distribution of $\hat{\beta} | A_{22}$ given by (5.541), or, under $H_0 : \Sigma_{12} = 0$, by (5.542).

But, $\hat{\beta}$ is a centered estimator of β, since from (5.545) above and the definition of the matrix A in (5.539), we may easily obtain

$$\begin{aligned} E(\hat{\beta}|X_2^*) = E(\hat{\beta}|A_{22}) &= E(X_1^*|X_2^*) X_2^{*\prime} (X_2^* X_2^{*\prime})^{-1} \\ &= \beta X_2^* X_2^{*\prime} (X_2^* X_2^{*\prime})^{-1} = \beta \end{aligned} \qquad (5.546)$$

so that from (5.546) and (5.541) it is clear that $\beta = \Sigma_{12} \Sigma_{22}^{-1}$, which, since $\Sigma_{22} \neq 0$, shows that

$$H_0 : \Sigma_{12} = 0 \iff H_0 : \beta = 0 . \qquad (5.547)$$

We may note that from (5.545) we also have

$$A_{12} A_{22}^{-1} A_{21} = \hat{\beta} A_{22} \hat{\beta}'$$

and that under $H_0 : \Sigma_{12} = 0$, from (5.542),

$$\hat{\beta} A_{22}^{1/2} \sim N_{p_1 \times p_2}\big(0, \tfrac{1}{n} I_{p_2} \otimes \Sigma_{11}\big) ,$$

so that, under $H_0 : \Sigma_{12} = 0$,

$$\hat{\beta} A_{22} \hat{\beta}' \sim W_p\left(p_2, \tfrac{1}{n}\Sigma_{11}\right).$$

Consequently, we may write the statistic Λ_{H_1,H_0^*} in (5.193), to test the fit of the model in (5.543) or (5.544) or to test the null hypotheses in (5.547) as

$$\Lambda_{H_1,H_0^*} = \frac{|A_{11.2}|}{|A_{11}|} = \frac{|A_{11.2}|}{|A_{11.2} + \hat{\beta} A_{22} \hat{\beta}'|},$$

where $A_{11.2}$ and $\hat{\beta} A_{22} \hat{\beta}'$ are independent, given the independence between $A_{11.2}$ and $A_{12}|A_{22}$. From the results in Appendix 1, Λ_{H_1,H_0^*} has thus the distribution given by (5.178) and (5.179), with a rejection of the null hypothesis in (5.547) being equivalent to saying that the model in (5.543) or (5.544) "fits" or is appropriate for the data.

Next, if we consider splitting the subvector $\underline{X}_2$ into two subvectors, $\underline{X}_{21}$ with p_{21} variables and $\underline{X}_{22}$ with p_{22} variables, with $p_2 = p_{21} + p_{22}$, this will induce on X_2^* and $X^* = X(I_n - \frac{1}{n}E_{nn})$ the partition

$$\underset{p\times n}{X^*} = \begin{bmatrix} X_1^* \\ X_2^* \end{bmatrix} \begin{matrix} p_1 \\ p_2 \end{matrix} = \begin{bmatrix} X_1^* \\ X_{21}^* \\ X_{22}^* \end{bmatrix} \begin{matrix} p_1 \\ p_{21} \\ p_{22} \end{matrix} \tag{5.548}$$

and as such, it will induce on A the partition

$$A = \left[\begin{array}{c|c} A_{11} & A_{12} \\ \hline A_{21} & A_{22} \end{array}\right] = \left[\begin{array}{c|c:c} A_{11} & A_{12(1)} & A_{12(2)} \\ \hline A_{21(1)} & A_{22(11)} & A_{22(12)} \\ \hdashline A_{21(2)} & A_{22(21)} & A_{22(22)} \end{array}\right] \tag{5.549}$$

where

$$A_{12} = \left[\, A_{12(1)} \,\middle|\, A_{12(2)} \,\right] \quad \text{and} \quad A_{21} = A_{12}' = \begin{bmatrix} A_{21(1)} \\ A_{21(2)} \end{bmatrix}$$

with

$$A_{12(i)} = X_1(I_n - \tfrac{1}{n}E_{nn})X_{2i}' = X_1^* X_{2i}^{*\prime} \quad \text{and} \quad A_{21(i)} = X_{2i}(I_n - \tfrac{1}{n}E_{nn})X_1' = X_{2i}^* X_1^{*\prime},$$

for $i \in \{1, 2\}$, and

$$A_{22(ij)} = X_{2i}(I_n - \tfrac{1}{n}E_{nn})X_{2j}' = X_{2i}^* X_{2j}^{*\prime}, \qquad i, j \in \{1, 2\}.$$

Let us now consider the model

$$\underline{X}_1 = \beta_1 \underline{X}_{21} + \mathcal{E}^* \qquad (5.550)$$

or, in terms of centered sample matrices,

$$X_1^* = \beta_1 X_{21}^* + \mathcal{E}^* \,, \qquad (5.551)$$

with

$$E(X_1^* | X_{21}^*) = \beta_1 X_{21}^*$$

where

$$X_{21}^* = X_{21}(I_n - \tfrac{1}{n} E_{nn}) \,.$$

Then we may restrict ourselves to considering the submatrix in the upper left corner of A in (5.549) delimited by a dotted line, and we may draw a parallel with the model in (5.544), while considering a similar split for Σ to that of A in (5.549). With this in mind, we may easily see that we have

$$\hat{\beta}_1 = A_{12(1)} A_{22(11)}^{-1}$$

with

$$\hat{\beta}_1 | A_{22(11)} \sim N_{p_1 \times p_{21}} \big(\Sigma_{12(1)} \Sigma_{22(11)}^{-1}, \tfrac{1}{n} A_{22(11)} \otimes \Sigma_{11.2(1)} \big) \qquad (5.552)$$

where

$$\Sigma_{11.2(1)} = \Sigma_{11} - \Sigma_{12(1)} \Sigma_{22(11)}^{-1} \Sigma_{21(1)} \,,$$

and hence the null hypothesis $H_0 : \beta_1 = 0$ is equivalent to the null hypothesis $H_0 : \Sigma_{12(1)} = 0$, since we also have

$$E\big(\hat{\beta}_1 | A_{22(11)}\big) = E\big(A_{12(1)} A_{22(11)}^{-1} | A_{22(11)}\big) = E\big(X_1^* X_{21}^{*\prime} (X_{21}^* X_{21}^{*\prime})^{-1} | X_{21}^*\big)$$
$$= \beta_1 X_{21}^* X_{21}^{*\prime} (X_{21}^* X_{21}^{*\prime})^{-1} = \beta_1 \,.$$

The statistic Λ_{H_0, H_0^*} in (5.195), to test the fit of the model in (5.550) or (5.551), or to test the null hypothesis $H_0 : \beta_1 = 0 \Leftrightarrow H_0 : \Sigma_{12(1)} = 0$, is thus given by

$$\Lambda_{H_0, H_0^*} = \frac{|A_{11.2(1)}|}{|A_{11}|} = \frac{|A_{11.2(1)}|}{|A_{11.2(1)} + \hat{\beta}_1 A_{22(11)}^{-1} \hat{\beta}_1'|} \qquad (5.553)$$

where

$$A_{11.2(1)} = A_{11} - A_{12(1)}A_{22(11)}^{-1}A_{21(1)} \sim W_{p_1}\left(n-1-p_{21}, \tfrac{1}{n}\Sigma_{11.2(1)}\right),$$

which under $H_0 : \Sigma_{12(1)} = 0$ reduces to

$$A_{11.2(1)} \sim W_{p_1}\left(n-1-p_{21}, \tfrac{1}{n}\Sigma_{11}\right),$$

and where, from (5.552), under $H_0 : \Sigma_{12(1)} = 0 \Leftrightarrow H_0 : \beta_1 = 0$,

$$\hat{\beta}_1 A_{22(11)}^{1/2} \sim N_{p_1 \times p_{21}}\left(0, \tfrac{1}{n}I_{p_{21}} \otimes \Sigma_{11}\right)$$

so that

$$\hat{\beta}_1 A_{22(11)}^{-1}\hat{\beta}_1' \sim W_{p_1}\left(p_{21}, \tfrac{1}{n}\Sigma_{11}\right).$$

From the results in Appendix 1, the exact distribution of Λ_{H_0,H_0^*} in (5.553) above is given by (5.180) and (5.181), with a rejection of the null hypothesis being equivalent to saying that the model in (5.550) or (5.551) "fits," or is appropriate for the data.

Then, if we write the model in (5.544) as

$$X_1^* = \beta_1 X_{21}^* + \beta_2 X_{22}^* + \mathcal{E}, \tag{5.554}$$

considering the split of X_2^* in (5.548) and

$$\underset{p_1\times p_2}{\beta} = \left[\; \underset{p_1\times p_{21}}{\beta_1} \;\middle|\; \underset{p_1\times p_{22}}{\beta_2} \;\right],$$

the $(2/n)$-th power of the l.r.t. statistic to test the null hypothesis $H_0^{**} : \beta_2 = 0$, for a sample of size n, in the model (5.554), that is, to test the hypothesis $\beta_2 = 0$ in the presence of β_1 in the model, is

$$\Lambda_{H_1,H_0} = \frac{\Lambda_{H_1,H_0^*}}{\Lambda_{H_0,H_0^*}} = \frac{|A_{11.2}|}{|A_{11.2(1)}|}.$$

We will now show that we can write

$$A_{11.2(1)} = A_{11.2} + \hat{\beta}_2 A_{22(2.1)} \hat{\beta}_2,$$

where

$$A_{22(2.1)} = A_{22(22)} - A_{22(21)}A_{22(11)}^{-1}A_{22(12)}, \tag{5.555}$$

so that we may write

$$\Lambda_{H_1,H_0} = \frac{|A_{11.2}|}{|A_{11.2} + \hat{\beta}_2 A_{22(2.1)} \hat{\beta}_2|}. \tag{5.556}$$

Indeed, by considering the definition of $A_{22(2.1)}$ in (5.555), we may write, from (5.506) in Appendix 6,

$$A_{22}^{-1} = \begin{bmatrix} A_{22(11)}^{-1} + A_{22(11)}^{-1} A_{22(12)} A_{22(2.1)}^{-1} A_{22(21)} A_{22(11)}^{-1} & -A_{22(11)}^{-1} A_{22(12)} A_{22(2.1)}^{-1} \\ -A_{22(2.1)}^{-1} A_{22(21)} A_{22(11)}^{-1} & A_{22(2.1)}^{-1} \end{bmatrix} \tag{5.557}$$

so that

$$\begin{aligned} \hat{\beta} &= A_{12} A_{22}^{-1} = \left[\; A_{12(1)} \mid A_{12(2)} \; \right] A_{22}^{-1} \\ &= \left[A_{12(1)} A_{22(11)}^{-1} + A_{12(1)} A_{22(11)}^{-1} A_{22(12)} A_{22(2.1)}^{-1} A_{22(21)} A_{22(11)}^{-1} \right. \\ &\qquad \left. -A_{12(2)} A_{22(2.1)}^{-1} A_{22(21)} A_{22(11)}^{-1} \mid A_{12(2)} A_{22(2.1)}^{-1} - A_{12(1)} A_{22(11)}^{-1} A_{22(12)} A_{22(2.1)}^{-1} \right] \\ &= \Big[\underset{p_1 \times p_{21}}{\hat{\beta}_1} \mid \underset{p_1 \times p_{22}}{\hat{\beta}_2} \Big]. \end{aligned} \tag{5.558}$$

But then, from (5.558), we have

$$\begin{aligned} A_{12} A_{22}^{-1} A_{21} &= \hat{\beta} \begin{bmatrix} A_{21(1)} \\ A_{21(2)} \end{bmatrix} \\ &= A_{12(1)} A_{22(11)}^{-1} A_{21(1)} + A_{12(1)} A_{22(11)}^{-1} A_{22(12)} A_{22(2.1)}^{-1} A_{22(21)} A_{22(11)}^{-1} A_{21(1)} \\ &\quad -A_{12(2)} A_{22(2.1)}^{-1} A_{22(21)} A_{22(11)}^{-1} A_{21(1)} \\ &\quad -A_{12(1)} A_{22(11)}^{-1} A_{22(12)} A_{22(2.1)}^{-1} A_{21(2)} + A_{12(2)} A_{22(2.1)}^{-1} A_{21(2)} \\ &= A_{12(1)} A_{22(11)}^{-1} A_{21(1)} \\ &\quad +\big(A_{12(2)} - A_{12(1)} A_{22(11)}^{-1} A_{22(12)}\big) A_{22(2.1)}^{-1} \big(A_{12(2)} - A_{12(1)} A_{22(11)}^{-1} A_{22(12)}\big)' \\ &= A_{12(1)} A_{22(11)}^{-1} A_{21(1)} + \hat{\beta}_2 A_{22(2.1)} \hat{\beta}_2', \end{aligned} \tag{5.559}$$

which shows that we may indeed write Λ_{H_1,H_0} as in (5.556).

In order to obtain the distribution of Λ_{H_1,H_0}, one has to see that from the distribution of $\hat{\beta}|A_{22}$ given by (5.541), and the expression for A_{22}^{-1} in (5.557), taking for Σ_{22}^{-1} a similar definition of that of A_{22}^{-1} in (5.557), we obtain

$$\hat{\beta}_2 | A_{22(2.1)} \sim N_{p_1 \times p_{22}} \left(\Sigma_{12} \begin{bmatrix} -\Sigma_{22(11)}^{-1} \Sigma_{22(12)} \Sigma_{22(2.1)}^{-1} \\ \Sigma_{22(2.1)}^{-1} \end{bmatrix}, \frac{1}{n} A_{22(2.1)}^{-1} \otimes \Sigma_{11.2} \right),$$

where

$$\Sigma_{12}\begin{bmatrix} -\Sigma_{22(11)}^{-1}\Sigma_{22(12)}\Sigma_{22(2.1)}^{-1} \\ \Sigma_{22(2.1)}^{-1} \end{bmatrix} = \left[\, \Sigma_{12(1)} \mid \Sigma_{12(2)} \,\right]\begin{bmatrix} -\Sigma_{22(11)}^{-1}\Sigma_{22(12)}\Sigma_{22(2.1)}^{-1} \\ \Sigma_{22(2.1)}^{-1} \end{bmatrix}$$
$$= \Sigma_{12(2)}\Sigma_{22(2.1)}^{-1} - \Sigma_{12(1)}\Sigma_{22(11)}^{-1}\Sigma_{22(12)}\Sigma_{22(2.1)}^{-1}$$
$$= \left(\Sigma_{12(2)} - \Sigma_{12(1)}\Sigma_{22(11)}^{-1}\Sigma_{22(12)}\right)\Sigma_{22(2.1)}^{-1}\,. \qquad (5.560)$$

We should be aware of the fact that the null hypothesis we are testing now is the null hypothesis

$$H_0^{**}: \beta_2 = 0 \ \text{ in the model (5.554)} \qquad (5.561)$$

and not just $\beta_2 = 0$. Testing this latter hypothesis is equivalent to testing the fit of the model $X_1^* = \beta_2 X_{22}^* + \mathcal{E}^*$.

To test the null hypothesis in (5.561) will then be equivalent to testing the nullity of $E(\hat{\beta}_2)$ in the model (5.554), which, from (5.560), is seen to be equivalent to test the null hypothesis

$$H_0^{**}: \Sigma_{12(2)} - \Sigma_{12(1)}\Sigma_{22(11)}^{-1}\Sigma_{22(12)} = 0\,. \qquad (5.562)$$

Then, under this null hypothesis in (5.561) or (5.562), we have

$$\hat{\beta}_2|A_{22(2.1)} \sim N_{p_1\times p_{22}}\left(0, \tfrac{1}{n}A_{22(2.1)}^{-1} \otimes \Sigma_{11.2_{|H_0^{**}}}\right)$$

or

$$\hat{\beta}_2 A_{22(2.1)}^{1/2} \sim N_{p_1\times p_{22}}\left(0, \tfrac{1}{n}I_{p_{22}} \otimes \Sigma_{11.2_{|H_0^{**}}}\right),$$

which implies

$$\hat{\beta}_2 A_{22(2.1)}\hat{\beta}_2' \sim W_{p_1}\left(p_{22}, \tfrac{1}{n}\Sigma_{11.2_{|H_0^{**}}}\right). \qquad (5.563)$$

Also under H_0^{**} in (5.561) or (5.562),

$$A_{11.2} \sim W_{p_1}\left(n-1-p_2, \frac{1}{n}\Sigma_{11.2_{|H_0^{**}}}\right). \qquad (5.564)$$

From the expression of Λ_{H_1,H_0} in (5.556) and the distributions of $A_{11.2}$ and $\hat{\beta}_2 A_{22(2.1)}\hat{\beta}_2'$ in (5.564) and (5.563), and the results in Appendix 1, we may then obtain the distribution of Λ_{H_1,H_0} as given by (5.182) through (5.184).

Although, as may be seen from Appendix 1, the exact expression for $\Sigma_{11.2_{|H_0^{**}}}$ is not important, since the distribution of Λ_{H_1,H_0} will not depend on it, this expression is not hard to obtain. In the definition of $\Sigma_{11.2}$ as

$$\Sigma_{11.2} = \Sigma_{11} - \Sigma_{12}\Sigma_{22}^{-1}\Sigma_{21}$$

one has to notice that $\Sigma_{12}\Sigma_{22}^{-1}\Sigma_{21}$ has a similar structure to that of $A_{12}A_{22}^{-1}A_{21}$ in (5.559), once all A's are replaced by Σ's. Then, given the null hypothesis in (5.561) or (5.562), we have

$$\Sigma_{12}\Sigma_{22}^{-1}\Sigma_{21} = \Sigma_{12(1)}\Sigma_{22(11)}^{-1}\Sigma_{21(1)}\,,$$

given the nullity of the matrix in (5.562). In this way we obtain,

$$\Sigma_{11.2_{|H_0^{**}}} = \Sigma_{11} - \Sigma_{12(1)}\Sigma_{22(11)}^{-1}\Sigma_{21(1)}\,.$$

Appendix 12: On the Distribution of the l.r.t. Statistic to Test Independence of Several Sets of Real Variables

Let Λ be given by (5.234). For a sample of size n, A in (5.234) has a $W_p(n-1,\frac{1}{n}\Sigma)$ distribution. The diagonal block structure of Σ under H_0 in (5.233), entails the independence of the matrices A_{kk}, so that the h-th moment of Λ under this null hypothesis will be given by

$$\begin{aligned}
E\left(\Lambda^h\right) &= \int_{\mathcal{A}>0} \frac{|\mathcal{A}|^h}{\prod_{k=1}^m |\mathcal{A}_{kk}|^h} \frac{e^{-\frac{1}{2}tr(n\Sigma^{-1}\mathcal{A})}|\mathcal{A}|^{(n-2-p)/2}}{2^{(n-1)p/2}\Gamma_p(\frac{n-1}{2})\,|\frac{1}{n}\Sigma|^{(n-1)/2}}\,d\mathcal{A}\\
&= \frac{\Gamma_p(\frac{n-1}{2}+h)}{\Gamma_p(\frac{n-1}{2})}\,2^{hp}\,|\tfrac{1}{n}\Sigma|^h\\
&\qquad \int_{\mathcal{A}>0} \frac{1}{\prod_{k=1}^m |\mathcal{A}_{kk}|^h}\,\underbrace{\frac{e^{-\frac{1}{2}tr(n\Sigma^{-1}\mathcal{A})}|\mathcal{A}|^{(n-2+2h-p)/2}}{2^{(n-1+2h)p/2}\,\Gamma_p(\frac{n-1}{2}+h)\,|\frac{1}{n}\Sigma|^{(n-1+2h)/2}}}_{\text{pdf of } W_p(n-1+2h,\,\frac{1}{n}\Sigma)}\,d\mathcal{A}\\
&= \frac{\Gamma_p(\frac{n-1}{2}+h)}{\Gamma_p(\frac{n-1}{2})}\,2^{hp}\,|\tfrac{1}{n}\Sigma|^h \prod_{k=1}^m E\left(|A_{kk}|^{-h}\right)
\end{aligned} \tag{5.565}$$

where $A_{kk} \sim W(n-1+2h,\frac{1}{n}\Sigma_{kk})$. But then, from (5.493) and from the fact that under H_0 in (5.233), $|\frac{1}{n}\Sigma| = \prod_{k=1}^m |\frac{1}{n}\Sigma_{kk}|$, using the definition of the multivariate

Gamma function in (5.492), and taking into account that $p = \sum_{k=1}^{m} p_k$, we may write, for $h > (p-n)/2$,

$$E\left(\Lambda^h\right) = \frac{\Gamma_p(\frac{n-1}{2}+h)}{\Gamma_p(\frac{n-1}{2})} 2^{hp} |\tfrac{1}{n}\Sigma|^h \prod_{k=1}^{m} \frac{\Gamma_{p_k}(\frac{n-1}{2}+h-h)}{\Gamma_{p_k}(\frac{n-1}{2}+h)} 2^{-hp_k} |\tfrac{1}{n}\Sigma_{kk}|^{-h} \quad (5.566)$$

$$= \frac{\Gamma_p(\frac{n-1}{2}+h)}{\Gamma_p(\frac{n-1}{2})} \prod_{k=1}^{m} \frac{\Gamma_{p_k}(\frac{n-1}{2})}{\Gamma_{p_k}(\frac{n-1}{2}+h)} \quad (5.567)$$

$$= \left\{\prod_{j=1}^{p} \frac{\Gamma(\frac{n-j}{2}+h)}{\Gamma(\frac{n-j}{2})}\right\} \left\{\prod_{k=1}^{m}\prod_{j=1}^{p_k} \frac{\Gamma(\frac{n-j}{2})}{\Gamma(\frac{n-j}{2}+h)}\right\}$$

$$= \left\{\prod_{j=1+p_m}^{p} \frac{\Gamma(\frac{n-j}{2}+h)}{\Gamma(\frac{n-j}{2})}\right\} \left\{\prod_{k=1}^{m-1}\prod_{j=1}^{p_k} \frac{\Gamma(\frac{n-j}{2})}{\Gamma(\frac{n-j}{2}+h)}\right\}$$

$$= \left\{\prod_{k=1}^{m-1}\prod_{j=1}^{p_k} \frac{\Gamma\left(\frac{n-q_k-j}{2}+h\right)}{\Gamma\left(\frac{n-q_k-j}{2}\right)}\right\} \left\{\prod_{k=1}^{m-1}\prod_{j=1}^{p_k} \frac{\Gamma\left(\frac{n-j}{2}\right)}{\Gamma\left(\frac{n-j}{2}+h\right)}\right\}$$

for $q_k = p_{k+1} + \ldots + p_m$. For an equivalent way to handle the above products see Appendix 14 after expression (5.576).

We may thus write

$$E\left(\Lambda^h\right) = \prod_{k=1}^{m-1}\prod_{j=1}^{p_k} \frac{\Gamma\left(\frac{n-j}{2}\right)\Gamma\left(\frac{n-q_k-j}{2}+h\right)}{\Gamma\left(\frac{n-q_k-j}{2}\right)\Gamma\left(\frac{n-j}{2}+h\right)} = \prod_{k=1}^{m-1}\prod_{j=1}^{p_k} E\left(Y_{jk}^h\right), \quad (5.568)$$

where

$$Y_{jk} \sim Beta\left(\frac{n-q_k-j}{2}, \frac{q_k}{2}\right),$$

are a set of independent r.v.'s.

But then, since $0 < \Lambda < 1$, its distribution is determined by the set of its moments and consequently

$$\Lambda \stackrel{d}{\equiv} \prod_{k=1}^{m-1}\prod_{j=1}^{p_k} Y_{jk}.$$

An alternative proof may be obtained by writing

$$\Lambda = \prod_{k=1}^{m-1} \frac{|A^*_{kk}|}{|A_{kk}|\,|A^*_{k+1,k+1}|} \tag{5.569}$$

where

$$A^*_{kk} = \begin{bmatrix} A_{kk} & A_{k,k+1} & \cdots & A_{km} \\ A_{k+1,k} & A_{k+1,k+1} & \cdots & A_{k+1,m} \\ \vdots & \vdots & \ddots & \vdots \\ A_{mk} & A_{m,k+1} & \cdots & A_{mm} \end{bmatrix} \tag{5.570}$$

with $A^*_{11} = A$, and where A_{kk} is the k-th diagonal block of A, of dimensions $p_k \times p_k$.

In (5.569) each statistic

$$\Lambda_k = \frac{|A^*_{kk}|}{|A_{kk}|\,|A^*_{k+1,k+1}|} \qquad (k = 1, \ldots, m-1) \tag{5.571}$$

is indeed the l.r.t. statistic to test the subhypothesis

$$H_{0k} : \bigwedge_{\ell=k+1}^{m} \Sigma_{k\ell} = 0$$

which is the null hypothesis of independence of $\underline{X}_k$ and the superset formed by joining $\underline{X}_{k+1}, \ldots, \underline{X}_m$, and where $\bigwedge_{\ell=k+1}^{m}$ represents a conjunction.

Note that one may then write the null hypothesis in (5.233) as

$$H_0 \equiv H_{0,m-1|\{m-2,\ldots,1\}} \circ \cdots \circ H_{03|\{2,1\}} \circ H_{02|1} \circ H_{01} \tag{5.572}$$

where $H_{0k|\{k-1,\ldots,1\}}$ represents H_{0k}, assuming that $H_{0,k-1}, \ldots, H_{01}$ hold.

But then, from Lemma 10.4.1 in Anderson (2003) or Theorem 5 in Jensen (1988) it follows that not only, under H_0, Λ is independent of $A_{11}, \ldots, A_{mm}$ but also each l.r.t. statistic Λ_k in (5.571) is, under H_{0k}, independent of A_{kk} and $A^*_{k+1,k+1}$. Then, since Λ_{k+1} is only a function of $A^*_{k+1,k+1}$, this assures that Λ_k and Λ_{k+1} are independent, under H_{0k}, which makes all $m-1$ l.r.t. statistics Λ_k $(k = 1, \ldots, m-1)$ in (5.571) independent, under H_0 in (5.233).

As such,

$$E\left(\Lambda^h\right) = \prod_{k=1}^{m-1} E\left(\Lambda_k^h\right)$$

where $E\left(\Lambda_k^h\right)$ may then be obtained in two different ways.

One possible way to obtain $E(\Lambda_k^h)$ is through a similar procedure to the one used to obtain $E(\Lambda^h)$ in the beginning of this Appendix, from the fact that $A_{kk}^* \sim W_{p_k+q_k}(n-1, \frac{1}{n}\Sigma_{kk}^*)$, for a similar partitioning and notation on Σ to that used in (5.570) for the matrix A, and where, under H_{0k}, $|\Sigma_{kk}^*| = |\Sigma_{kk}||\Sigma_{k+1,k+1}^*|$, so that from (5.565)–(5.567) we have

$$
\begin{aligned}
E\left(\Lambda_k^h\right) &= \frac{\Gamma_{p_k+q_k}(\frac{n-1}{2}+h)}{\Gamma_{p_k+q_k}(\frac{n-1}{2})}\,\frac{\Gamma_{p_k}(\frac{n-1}{2})}{\Gamma_{p_k}(\frac{n-1}{2}+h)}\,\frac{\Gamma_{q_k}(\frac{n-1}{2})}{\Gamma_{q_k}(\frac{n-1}{2}+h)}\\
&= \left\{\prod_{j=1}^{p_k+q_k}\frac{\Gamma(\frac{n-j}{2}+h)}{\Gamma(\frac{n-j}{2})}\right\}\left\{\prod_{j=1}^{p_k}\frac{\Gamma(\frac{n-j}{2})}{\Gamma(\frac{n-j}{2}+h)}\right\}\left\{\prod_{j=1}^{q_k}\frac{\Gamma(\frac{n-j}{2})}{\Gamma(\frac{n-j}{2}+h)}\right\}\\
&= \left\{\prod_{j=q_k+1}^{p_k+q_k}\frac{\Gamma(\frac{n-j}{2}+h)}{\Gamma(\frac{n-j}{2})}\right\}\left\{\prod_{j=1}^{p_k}\frac{\Gamma(\frac{n-j}{2})}{\Gamma(\frac{n-j}{2}+h)}\right\}\\
&= \prod_{j=1}^{p_k}\frac{\Gamma\left(\frac{n-j}{2}\right)\Gamma\left(\frac{n-q_k-j}{2}+h\right)}{\Gamma\left(\frac{n-q_k-j}{2}\right)\Gamma\left(\frac{n-j}{2}+h\right)},
\end{aligned}
\tag{5.573}
$$

so that $E(\Lambda^h)$ comes immediately in the form given in (5.568).

The other way to obtain $E(\Lambda_k^h)$ is by writing

$$
|A_{kk}^*| = |A_{kk}|\,|A_{k+1,k+1}^* - A_{k+1}^* A_{kk}^{-1} A_{k+1}^{*\prime}|
$$

where

$$
A_{k+1}^{*\prime} = \left[A_{k,k+1} \;\cdots\; A_{km}\right]
$$

is the first row block of A_{kk}^*, without the block A_{kk}.

Then,

$$
\Lambda_k = \frac{|A_{k+1,k+1}^* - A_{k+1}^* A_{kk}^{-1} A_{k+1}^{*\prime}|}{|A_{k+1,k+1}^*|}
$$

where, under H_{0k}, $(A_{k+1,k+1}^* - A_{k+1}^* A_{kk}^{-1} A_{k+1}^{*\prime}) \sim W_{q_k}(n-1-p_k, \Sigma_{k+1,k+1}^*)$ and $A_{k+1}^* A_{kk}^{-1} A_{k+1}^{*\prime} \sim W_{q_k}(p_k, \Sigma_{k+1,k+1}^*)$ are two independent matrices, so that $A_{k+1,k+1}^* \sim W_{q_k}(n-1, \Sigma_{k+1,k+1}^*)$, and $E(\Lambda_k^h)$ may be obtained from (5.494) in Appendix 1, with the due adaptations, as

$$
E\left(\Lambda_k^h\right) = \prod_{j=1}^{q_k}\frac{\Gamma\left(\frac{n-p_k-j}{2}+h\right)\Gamma\left(\frac{n-j}{2}\right)}{\Gamma\left(\frac{n-p_k-j}{2}\right)\Gamma\left(\frac{n-j}{2}+h\right)},
$$

which is equivalent to (5.573), since

$$E\left(\Lambda_k^h\right) = \prod_{j=1}^{q_k} \frac{\Gamma(\frac{n-p_k-j}{2}+h)\,\Gamma(\frac{n-j}{2})}{\Gamma(\frac{n-p_k-j}{2})\,\Gamma(\frac{n-j}{2}+h)}$$
$$= \left\{\prod_{j=p_k+1}^{p_k+q_k} \frac{\Gamma(\frac{n-j}{2}+h)}{\Gamma(\frac{n-j}{2})}\right\}\left\{\prod_{j=1}^{q_k} \frac{\Gamma(\frac{n-j}{2})}{\Gamma(\frac{n-j}{2}+h)}\right\}$$
$$= \left\{\prod_{j=1}^{p_k+q_k} \frac{\Gamma(\frac{n-j}{2}+h)}{\Gamma(\frac{n-j}{2})}\right\}\left\{\prod_{j=1}^{p_k} \frac{\Gamma(\frac{n-j}{2})}{\Gamma(\frac{n-j}{2}+h)}\right\}\left\{\prod_{j=1}^{q_k} \frac{\Gamma(\frac{n-j}{2})}{\Gamma(\frac{n-j}{2}+h)}\right\},$$

which is the same as the second expression in (5.573).

Appendix 13: On the Independence of the $k-1$ Statistics in (5.268)

Let

$$W_{jk} = -\log \Lambda_{jk|(j+1,\dots,k-1)} \quad \text{and} \quad W_k = -\log \Lambda_{k(1,\dots,k-1)}.$$

Then, in order to show that the $k-1$ factors $\Lambda_{jk|(j+1\dots,k-1)}$ in (5.268) are independent, all we need to do is to show that $\Phi_{W_k}(t) = \prod_{j=1}^{k-1} \Phi_{W_{jk}}(t)$, where $\Phi_{W_k}(t)$ and $\Phi_{W_{jk}}(t)$ are respectively the c.f.'s of W_k and W_{jk}.

From the facts that

$$\Lambda_{k(1,\dots,k-1)} \stackrel{d}{\equiv} \prod_{\ell=1}^{p_k} Y_\ell \quad \text{and} \quad \Lambda_{jk|(j+1,\dots,k-1)} \stackrel{d}{\equiv} \prod_{\ell=1}^{p_k} Y_{j\ell}^*$$

where

$$Y_\ell \sim Beta\left(\frac{n-\sum_{\nu=1}^{k-1} p_\nu - \ell}{2}, \frac{\sum_{\nu=1}^{k-1} p_\nu}{2}\right)$$

are p_k independent r.v.'s, and, for $\ell = 1, \dots, p_k$,

$$Y_{j\ell}^* \sim Beta\left(\frac{n-p_j-\sum_{\nu=j+1}^{k-1} p_\nu - \ell}{2}, \frac{p_j}{2}\right)$$

are also p_k independent r.v.'s, we have

$$E\left(\Lambda^h_{jk|(j+1,\dots,k-1)}\right)=\prod_{\ell=1}^{p_k}\frac{\Gamma\left(\left(n-\sum_{\nu=j+1}^{k-1}p_\nu-\ell\right)/2\right)}{\Gamma\left(\left(n-\sum_{\nu=j}^{k-1}p_\nu-\ell\right)/2\right)}\frac{\Gamma\left(\left(n-\sum_{\nu=j}^{k-1}p_\nu-\ell\right)/2+h\right)}{\Gamma\left(\left(n-\sum_{\nu=j+1}^{k-1}p_\nu-\ell\right)/2+h\right)},$$

so that, by exchanging the order of the products and canceling out terms in the product in j, we may write

$$\begin{aligned}\prod_{j=1}^{k-1}\Phi_{W_{jk}}(t)&=\prod_{j=1}^{k-1}E\left(e^{\mathrm{i}tW_{jk}}\right)=\prod_{j=1}^{k-1}E\left(\Lambda^{-\mathrm{i}t}_{jk|(j+1,\dots,k-1)}\right)\\&=\prod_{\ell=1}^{p_k}\prod_{j=1}^{k-1}\frac{\Gamma\left(\left(n-\sum_{\nu=j+1}^{k-1}p_\nu-\ell\right)/2\right)}{\Gamma\left(\left(n-\sum_{\nu=j}^{k-1}p_\nu-\ell\right)/2\right)}\frac{\Gamma\left(\left(n-\sum_{\nu=j}^{k-1}p_\nu-\ell\right)/2-\mathrm{i}t\right)}{\Gamma\left(\left(n-\sum_{\nu=j+1}^{k-1}p_\nu-\ell\right)/2-\mathrm{i}t\right)}\\&=\prod_{\ell=1}^{p_k}\frac{\Gamma\left((n-\ell)/2\right)\,\Gamma\left(\left(n-\sum_{\nu=1}^{k-1}p_\nu-\ell\right)/2-\mathrm{i}t\right)}{\Gamma\left(\left(n-\sum_{\nu=1}^{k-1}p_\nu-\ell\right)/2\right)\,\Gamma\left((n-\ell)/2-\mathrm{i}t\right)}\\&=\prod_{\ell=1}^{p_k}E\left(Y_\ell^{-\mathrm{i}t}\right)=E\left(\Lambda^{-\mathrm{i}t}_{k(1,\dots,k-1)}\right)=\Phi_{W_k}(t)\,.\end{aligned}$$

Appendix 14: On the Distribution of the l.r.t. Statistic to Test Independence of Several Sets of Complex Variables

Let Λ be given by (5.234), now for complex Normal r.v.'s. For a sample of size n, A in (5.234) has a $CW_p(n-1,\frac{1}{n}\Sigma)$ distribution. Then, using a procedure similar to the one used in the real case, under H_0 in (5.233), given the independence of the matrices A_{kk}, the h-th moment of Λ will be given by

$$\begin{aligned}E\left(\Lambda^h\right)&=\int_{\mathcal{A}>0}\frac{|\mathcal{A}|^h}{\prod_{k=1}^m|\mathcal{A}_{kk}|^h}\frac{e^{-tr(n\Sigma^{-1}\mathcal{A})}|\mathcal{A}|^{n-1-p}}{\Gamma_p(n-1)\,|\frac{1}{n}\Sigma|^{n-1}}\,d\mathcal{A}\\&=\frac{\Gamma_p(n-1+h)}{\Gamma_p(n-1)}\,|\tfrac{1}{n}\Sigma|^h\int_{\mathcal{A}>0}\frac{1}{\prod_{k=1}^m|\mathcal{A}_{kk}|^h}\underbrace{\frac{e^{-tr(n\Sigma^{-1}\mathcal{A})}|\mathcal{A}|^{n-1+h-p}}{\Gamma_p(n-1+h)\,|\frac{1}{n}\Sigma|^{n-1+h}}}_{\text{pdf of }CW_p\left(n-1+h,\,\frac{1}{n}\Sigma\right)}\,d\mathcal{A}\\&=\frac{\Gamma_p(n-1+h)}{\Gamma_p(n-1)}\,|\tfrac{1}{n}\Sigma|^h\prod_{k=1}^m E\left(|A_{kk}|^{-h}\right)\end{aligned}\qquad(5.574)$$

for $A_{kk} \sim CW(n-1+h, \frac{1}{n}\Sigma_{kk})$. But then, from (5.524) and from the fact that under H_0 in (5.233), $|\Sigma| = \prod_{k=1}^{m} |\Sigma_{kk}|$, using (5.523), we may write, for $h > p - n$,

$$E\left(\Lambda^h\right) = \frac{\Gamma_p(n-1+h)}{\Gamma_p(n-1)} |\tfrac{1}{n}\Sigma|^h \prod_{k=1}^{m} \frac{\Gamma_{p_k}(n-1+h-h)}{\Gamma_{p_k}(n-1+h)} |\tfrac{1}{n}\Sigma_{kk}|^{-h} \quad (5.575)$$

$$= \frac{\Gamma_p(n-1+h)}{\Gamma_p(n-1)} \prod_{k=1}^{m} \frac{\Gamma_{p_k}(n-1)}{\Gamma_{p_k}(n-1+h)} \quad (5.576)$$

$$= \left\{\prod_{j=1}^{p} \frac{\Gamma(n+h-j)}{\Gamma(n-j)}\right\} \left\{\prod_{k=1}^{m}\prod_{j=1}^{p_k} \frac{\Gamma(n-j)}{\Gamma(n+h-j)}\right\}$$

$$= \left\{\prod_{j=p_1+1}^{p} \frac{\Gamma(n+h-j)}{\Gamma(n-j)}\right\} \left\{\prod_{k=2}^{m}\prod_{j=1}^{p_k} \frac{\Gamma(n-j)}{\Gamma(n+h-j)}\right\}$$

$$= \left\{\prod_{k=2}^{m}\prod_{j=1}^{p_k} \frac{\Gamma(n+h-p_k^*-j)}{\Gamma(n-p_k^*-j)}\right\} \left\{\prod_{k=2}^{m}\prod_{j=1}^{p_k} \frac{\Gamma(n-j)}{\Gamma(n+h-j)}\right\}$$

for $p_k^* = p_1 + \ldots + p_{k-1}$. Then, by reversing the indexing of the sets, $p_1 \to p_m$, $p_2 \to p_{m-1}, \ldots, p_{m-1} \to p_2$, $p_m \to p_1$, it is possible to write, for $q_k = p_{k+1} + \cdots + p_m$,

$$E\left(\Lambda^h\right) = \prod_{k=1}^{m-1}\prod_{j=1}^{p_k} \frac{\Gamma(n-j)\,\Gamma(n-q_k-j+h)}{\Gamma(n-q_k-j)\,\Gamma(n-j+h)} = \prod_{k=1}^{m-1}\prod_{j=1}^{p_k} E\left(Y_{jk}^h\right), \quad (5.577)$$

where

$$Y_{jk} \sim Beta\,(n-q_k-j, q_k)\ ,$$

are a set of independent r.v.'s.

But then, since $0 < \Lambda < 1$, its distribution is determined by the set of its moments and consequently

$$\Lambda \stackrel{d}{\equiv} \prod_{k=1}^{m-1}\prod_{j=1}^{p_k} (Y_{jk})^n\ .$$

We may note that expression (5.577) matches the expressions for the moments of Λ in Krishnaiah et al. (1976) and Fang et al. (1982).

As in the real case, an alternative proof may be obtained by writing

$$\Lambda = \prod_{k=1}^{m-1} \frac{|A^*_{kk}|}{|A_{kk}|\,|A^*_{k+1,k+1}|} \tag{5.578}$$

where

$$A^*_{kk} = \begin{bmatrix} A_{kk} & A_{k,k+1} & \cdots & A_{km} \\ A_{k+1,k} & A_{k+1,k+1} & \cdots & A_{k+1,m} \\ \vdots & \vdots & \ddots & \vdots \\ A_{mk} & A_{m,k+1} & \cdots & A_{mm} \end{bmatrix} \tag{5.579}$$

with $A^*_{11} = A$, and where A_{kk} is the k-th diagonal block of A, of dimensions $p_k \times p_k$.

In (5.578) each statistic

$$\Lambda_k = \frac{|A^*_{kk}|}{|A_{kk}|\,|A^*_{k+1,k+1}|} \qquad (k = 1, \ldots, m-1) \tag{5.580}$$

is indeed the l.r.t. statistic to test the subhypothesis

$$H_{0k} : \bigwedge_{\ell=k+1}^{m} \Sigma_{k\ell} = 0$$

which is the null hypothesis of independence of $\underline{X}_k$ and the superset formed by joining $\underline{X}_{k+1}, \ldots, \underline{X}_m$, and where $\bigwedge_{\ell=k+1}^{m}$ represents a conjunction.

Note that one may then write the null hypothesis in (5.233) as

$$H_0 \equiv H_{0,m-1|\{m-2,\ldots,1\}} \circ \cdots \circ H_{03|\{2,1\}} \circ H_{02|1} \circ H_{01} \tag{5.581}$$

where $H_{0k|\{k-1,\ldots,1\}}$ represents H_{0k}, assuming that $H_{0,k-1}, \ldots, H_{01}$ hold.

But then, as in the real case, from Lemma 10.4.1 in Anderson (2003) or Theorem 5 in Jensen (1988) it follows that not only, under H_0, Λ is independent of $A_{11}, \ldots, A_{mm}$ but also each l.r.t. statistic Λ_k in (5.580) is, under H_{0k}, independent of A_{kk} and $A^*_{k+1,k+1}$. Then, since Λ_{k+1} is only built on $A^*_{k+1,k+1}$, this assures that Λ_k and Λ_{k+1} are independent, under H_{0k}, which makes all $m-1$ l.r.t. statistics Λ_k $(k = 1, \ldots, m-1)$ in (5.580) independent, under H_0 in (5.233).

As such,

$$E\left(\Lambda^h\right) = \prod_{k=1}^{m-1} E\left(\Lambda_k^h\right)$$

where $E\left(\Lambda_k^h\right)$ may then be obtained in two different ways.

One possible way to obtain $E(\Lambda_k^h)$ is through a similar procedure to the one used to obtain $E(\Lambda^h)$ in the beginning of this Appendix, from the fact that $A_{kk}^* \sim CW_{p_k+q_k}(n-1, \frac{1}{n}\Sigma_{kk}^*)$, for a similar partitioning and notation on Σ to that used in (5.579) for the matrix A, and where, under H_{0k}, $|\Sigma_{kk}^*| = |\Sigma_{kk}||\Sigma_{k+1,k+1}^*|$, so that from (5.574)–(5.576) we have

$$\begin{aligned}
E\left(\Lambda_k^h\right) &= \frac{\Gamma_{p_k+q_k}(n-1+h)}{\Gamma_{p_k+q_k}(n-1)}\,\frac{\Gamma_{p_k}(n-1)}{\Gamma_{p_k}(n-1+h)}\,\frac{\Gamma_{q_k}(n-1)}{\Gamma_{q_k}(n-1+h)} \\
&= \left\{\prod_{j=1}^{p_k+q_k}\frac{\Gamma(n+h-j)}{\Gamma(n-j)}\right\}\left\{\prod_{j=1}^{p_k}\frac{\Gamma(n-j)}{\Gamma(n+h-j)}\right\}\left\{\prod_{j=1}^{q_k}\frac{\Gamma(n-j)}{\Gamma(n+h-j)}\right\} \\
&= \left\{\prod_{j=q_k+1}^{p_k+q_k}\frac{\Gamma(n+h-j)}{\Gamma(n-j)}\right\}\left\{\prod_{j=1}^{p_k}\frac{\Gamma(n-j)}{\Gamma(n+h-j)}\right\} \\
&= \prod_{j=1}^{p_k}\frac{\Gamma(n-j)\,\Gamma(n+h-q_k-j)}{\Gamma(n-q_k-j)\,\Gamma(n+h-j)}\,,
\end{aligned} \tag{5.582}$$

so that $E(\Lambda^h)$ comes immediately in the form given in (5.577).

The other way to obtain $E(\Lambda_k^h)$ is by writing

$$|A_{kk}^*| = |A_{kk}|\,|A_{k+1,k+1}^* - A_{k+1}^* A_{kk}^{-1} A_{k+1}^{*\prime}|$$

where

$$A_{k+1}^{*\prime} = \left[A_{k,k+1}\ \cdots\ A_{km}\right]$$

is the first row block of A_{kk}^*, without the block A_{kk}.

Then,

$$\Lambda_k = \frac{|A_{k+1,k+1}^* - A_{k+1}^* A_{kk}^{-1} A_{k+1}^{*\prime}|}{|A_{k+1,k+1}^*|}$$

where, under H_{0k}, $(A_{k+1,k+1}^* - A_{k+1}^* A_{kk}^{-1} A_{k+1}^{*\prime}) \sim CW_{q_k}(n-1-p_k, \Sigma_{k+1,k+1}^*)$ and $A_{k+1}^* A_{kk}^{-1} A_{k+1}^{*\prime} \sim CW_{q_k}(p_k, \Sigma_{k+1,k+1}^*)$ are two independent matrices, so that $A_{k+1,k+1}^* \sim CW_{q_k}(n-1, \Sigma_{k+1,k+1}^*)$, and $E(\Lambda_k^h)$ may be obtained from (5.526) in Appendix 8, with the due adaptations, as

$$E\left(\Lambda_k^h\right) = \prod_{j=1}^{q_k}\frac{\Gamma(n-p_k+h-j)\,\Gamma(n-j)}{\Gamma(n-p_k-j)\,\Gamma(n+h-j)}\,,$$

which is equivalent to (5.582), since

$$
\begin{aligned}
E\left(\Lambda_k^h\right) &= \prod_{j=1}^{q_k} \frac{\Gamma(n-p_k+h-j)\,\Gamma(n-j)}{\Gamma(n-p_k-j)\,\Gamma(n+h-j)} \\
&= \left\{ \prod_{j=p_k+1}^{p_k+q_k} \frac{\Gamma(n+h-j)}{\Gamma(n-j)} \right\} \left\{ \prod_{j=1}^{q_k} \frac{\Gamma(n-j)}{\Gamma(n+h-j)} \right\} \\
&= \left\{ \prod_{j=1}^{p_k+q_k} \frac{\Gamma(n+h-j)}{\Gamma(n-j)} \right\} \left\{ \prod_{j=1}^{p_k} \frac{\Gamma(n-j)}{\Gamma(n+h-j)} \right\} \left\{ \prod_{j=1}^{q_k} \frac{\Gamma(n-j)}{\Gamma(n+h-j)} \right\},
\end{aligned}
$$

which is the same as the second expression in (5.582).

Appendix 15: Showing That the Matrix $\left(\Gamma \otimes I_p\right) \Sigma \left(\Gamma \otimes I_p\right)'$ Is Block-Diagonal If and Only If Σ has the Structure in (5.280)

That the covariance matrix $\left(\Gamma \otimes I_p\right) \Sigma \left(\Gamma \otimes I_p\right)'$ is block-diagonal if and only if the matrix Σ has the structure in (5.280) may be shown from the fact that if

$$
\Sigma = \begin{bmatrix} \Sigma_{11} & \Sigma_{12} \\ \Sigma_{21} & \Sigma_{22} \end{bmatrix} \quad \text{with} \quad \Sigma_{21} = \Sigma_{12}',
$$

then

$$
\begin{aligned}
\left(\Gamma \otimes I_p\right) \Sigma \left(\Gamma \otimes I_p\right)' &= \left[\begin{array}{c|c} \frac{1}{\sqrt{2}} I_p & \frac{1}{\sqrt{2}} I_p \\ \hline \frac{1}{\sqrt{2}} I_p & -\frac{1}{\sqrt{2}} I_p \end{array}\right] \begin{bmatrix} \Sigma_{11} & \Sigma_{12} \\ \Sigma_{21} & \Sigma_{22} \end{bmatrix} \left[\begin{array}{c|c} \frac{1}{\sqrt{2}} I_p & \frac{1}{\sqrt{2}} I_p \\ \hline \frac{1}{\sqrt{2}} I_p & -\frac{1}{\sqrt{2}} I_p \end{array}\right] \\
&= \left[\begin{array}{c|c} \frac{1}{\sqrt{2}} (\Sigma_{11}+\Sigma_{21}) & \frac{1}{\sqrt{2}} (\Sigma_{12}+\Sigma_{22}) \\ \hline \frac{1}{\sqrt{2}} (\Sigma_{11}-\Sigma_{21}) & \frac{1}{\sqrt{2}} (\Sigma_{12}-\Sigma_{22}) \end{array}\right] \left[\begin{array}{c|c} \frac{1}{\sqrt{2}} I_p & \frac{1}{\sqrt{2}} I_p \\ \hline \frac{1}{\sqrt{2}} I_p & -\frac{1}{\sqrt{2}} I_p \end{array}\right] \\
&= \left[\begin{array}{c|c} \frac{1}{2} (\Sigma_{11}+\Sigma_{21}+\Sigma_{12}+\Sigma_{22}) & \frac{1}{2} (\Sigma_{11}+\Sigma_{21}-\Sigma_{12}-\Sigma_{22}) \\ \hline \frac{1}{2} (\Sigma_{11}-\Sigma_{21}+\Sigma_{12}-\Sigma_{22}) & \frac{1}{2} (\Sigma_{11}-\Sigma_{21}-\Sigma_{12}+\Sigma_{22}) \end{array}\right].
\end{aligned}
$$

In order this matrix to be block-diagonal we need the two off-diagonal blocks to be null, that is, we need their difference and their sum to be null, i.e., we need

$$\begin{cases} \cancel{\Sigma_{11}} + \Sigma_{21} - \Sigma_{12} - \cancel{\Sigma_{22}} - \cancel{\Sigma_{11}} + \Sigma_{21} - \Sigma_{12} + \cancel{\Sigma_{22}} = 0 \\ \Sigma_{11} + \cancel{\Sigma_{21}} - \cancel{\Sigma_{12}} - \Sigma_{22} + \Sigma_{11} - \cancel{\Sigma_{21}} + \cancel{\Sigma_{12}} - \Sigma_{22} = 0 \end{cases}$$

$$\iff \begin{cases} 2\Sigma_{21} - 2\Sigma_{12} = 0 \\ 2\Sigma_{11} - 2\Sigma_{22} = 0 \end{cases} \iff \begin{cases} \Sigma_{21} = \Sigma_{12} \\ \Sigma_{11} = \Sigma_{22} \end{cases}.$$

Appendix 16: On the Distribution of the l.r.t. Statistic $\Lambda_{bk|a}$ in (5.342)

Given that $A_{kk} \sim W_2\left(n-1, \frac{1}{n}\Sigma_{kk}\right)$, we have, from the p.d.f. of the Wishart distribution in (5.491),

$$\begin{aligned} E\left(\Lambda_{bk|a}^h\right) &= 2^{2h} \int_{\mathcal{A}_{kk}>0} \frac{|\mathcal{A}_{kk}|^h}{(tr(\mathcal{A}_{kk}))^{2h}} \frac{e^{-\frac{1}{2}tr\left(n\Sigma_{kk}^{-1}\mathcal{A}_{kk}\right)}|\mathcal{A}_{kk}|^{(n-4)/2}}{2^{n-1}\,\Gamma_2\left(\frac{n-1}{2}\right)|\frac{1}{n}\Sigma_{kk}|^{(n-1)/2}}\, d\mathcal{A}_{kk} \\ &= 2^{2h}\frac{\Gamma_2\left(\frac{n-1}{2}+h\right)}{\Gamma_2\left(\frac{n-1}{2}\right)}|\tfrac{1}{n}\Sigma_{kk}|^h\, 2^{2h} \\ &\quad \int_{\mathcal{A}_{kk}>0} \frac{1}{(tr(\mathcal{A}_{kk}))^{2h}} \frac{e^{-\frac{1}{2}tr\left(n\Sigma_{kk}^{-1}\mathcal{A}_{kk}\right)}|\mathcal{A}_{kk}|^{(n-4+2h)/2}}{2^{n-1+2h}\,\Gamma_2\left(\frac{n-1}{2}+h\right)|\frac{1}{n}\Sigma_{kk}|^{(n-1+2h)/2}}\, d\mathcal{A}_{kk} \\ &= 2^{4h}\frac{\Gamma_2\left(\frac{n-1}{2}+h\right)}{\Gamma_2\left(\frac{n-1}{2}\right)}|\tfrac{1}{n}\Sigma_{kk}|^h\, E\left((tr\ A_{kk})^{-2h}\right) \end{aligned}$$

for $A_{kk} \sim W_2\left(n-1+2h, \frac{1}{n}\Sigma_{kk}\right)$.

But, since under $H_{0b|a}$ in (5.341), Σ_{kk} is diagonal, with both diagonal elements equal to σ_k^2, the diagonal elements of A_{kk} are independent, with the i-th diagonal element of A_{kk}, $a_{kk(i)} \sim W_1\left(n-1+2h, \frac{1}{n}\sigma_k^2\right)$, so that we have $tr(A_{kk}) \sim W_1\left(2(n-1)+4h, \frac{1}{n}\sigma_k^2\right)$, and as such, from (5.493),

$$E\left((tr\ A_{kk})^{-2h}\right) = 2^{-2h}\left(\tfrac{1}{n}\sigma_k^2\right)^{-2h}\frac{\Gamma(n-1+2h-2h)}{\Gamma(n-1+2h)},$$

and thus, taking into account that under $H_{0b|a}$ in (5.341), $|\Sigma_{kk}| = \left(\sigma_k^2\right)^2$, we may write

$$E\left(\Lambda_{bk|a}^h\right) = 2^{4h}\frac{\Gamma_2\left(\frac{n-1}{2}+h\right)}{\Gamma_2\left(\frac{n-1}{2}\right)}\left(\left(\tfrac{1}{n}\sigma_k^2\right)^2\right)^h 2^{-2h}\left(\tfrac{1}{n}\sigma_k^2\right)^{-2h}\frac{\Gamma(n-1+2h-2h)}{\Gamma(n-1+2h)}$$
$$= 2^{2h}\frac{\Gamma_2\left(\frac{n-1}{2}+h\right)}{\Gamma_2\left(\frac{n-1}{2}\right)}\frac{\Gamma(n-1+2h-2h)}{\Gamma(n-1+2h)}.$$

But then, applying to the ratio of multivariate Gamma functions of order $p = 2$ the definition of the multivariate Gamma function in (5.492), and using for the ratio of Gamma functions the duplication formula for the Gamma function, which may be directly obtained from (3.5), for $n = 2$ in that expression, giving

$$\Gamma(2z) = \Gamma(z)\,\Gamma\left(z+\tfrac{1}{2}\right)(2\pi)^{-1/2}\,2^{2z-\frac{1}{2}}, \qquad (5.583)$$

this finally yields

$$E\left(\Lambda_{bk|a}^h\right) = \left\{\prod_{j=1}^{2}\frac{\Gamma\left(\frac{n-1}{2}+h-\frac{j-1}{2}\right)}{\Gamma\left(\frac{n-1}{2}-\frac{j-1}{2}\right)}\right\}2^{-2h}\frac{\Gamma\left(\frac{n-1}{2}\right)}{\Gamma\left(\frac{n-1}{2}+h\right)}\frac{\Gamma\left(\frac{n}{2}\right)}{\Gamma\left(\frac{n}{2}+r\right)}$$
$$= \frac{\Gamma\left(\frac{n-2}{2}+h\right)}{\Gamma\left(\frac{n-2}{2}\right)}\frac{\Gamma\left(\frac{n}{2}\right)}{\Gamma\left(\frac{n}{2}+h\right)}, \qquad (h > -n/2+1)$$

which shows that $\Lambda_{bk|a}$ has the same distribution as that of a $Beta\left(\frac{n-2}{2}, 1\right)$ r.v. because its moments match the moments of such a r.v. and its range being bounded, its moments determine its distribution.

Appendix 17: An Alternative Way to Obtain the l.r.t. Statistic in (5.418)

Since the elements of the matrix U in Sect. 5.2.1, with elements given by (5.357), are not function of the elements in Σ, and since for a circular matrix Σ_{cp}, of dimensions $p\times p$ the matrix $U\Sigma U'$ is a diagonal matrix, with

$$U\Sigma_{cp}U' = diag(\tau_1, \tau_2, \ldots, \tau_p)$$

where, for $j = 2, \ldots, p - m$, with $m = \lfloor p/2 \rfloor$,

$$\tau_j = \tau_{p-j+2}\,, \tag{5.584}$$

testing the null hypothesis of circularity in (5.417) is equivalent to testing the null hypothesis

$$H_0 : U\Sigma U' = diag(\tau_1, \tau_2, \ldots, \tau_p) \tag{5.585}$$

with τ_j $(j = 2, \ldots, p - m)$ satisfying the relation in (5.584).

Then the null hypothesis in (5.585) may be decomposed as

$$H_{0b|a} \circ H_{0a}$$

where

$$H_{0a} : U\Sigma U' = diag(\sigma_1^2, \ldots, \sigma_p^2)$$

is the null hypothesis of independence of the p variables in $U\underline{X}$, and

$$H_{0b|a} : \sigma_j^2 = \sigma_{p-j+2}^2\,, \quad j = 2, \ldots, p - m$$
$$\text{assuming } H_{0a}\,.$$

From Sect. 5.1.11, the $(2/n)$-th power of the l.r.t. statistic to test H_{0a} is

$$\Lambda_a = \frac{|V|}{\prod_{j=1}^{p} v_{jj}}$$

where, with A being the m.l.e. of Σ, we take $V = nUAU'$, with v_{jj} being its j-th diagonal element. Then, for example, from expression (2.13) in Marques et al. (2011), where the matrices denoted there by A_k are the matrices of sums of products and sums of squares of deviations from the sample means, that is, are equal to n times the m.l.e.'s, the $(2/n)$-th power of the l.r.t. statistic to test $H_{0b|a}$ is, for $V = nU'AU$,

$$\Lambda_{b|a} = \prod_{j=2}^{p-m} 2^2 \,\frac{v_{jj}\, v_{p-j+2,p-j+2}}{\left(v_{jj} + v_{p-j+2,p-j+2}\right)^2}\,.$$

As such, using Lemma 10.3.1 in Anderson (2003), we have that the $(2/n)$-th power of the l.r.t. statistic to test H_0 in (5.585) is,

$$\begin{aligned}\Lambda=\Lambda_a\,\Lambda_{b|a}&=2^{2(p-m-1)}\frac{|V|}{\prod_{j=1}^{p}v_{jj}}\prod_{j=2}^{p-m}\frac{v_{jj}\;v_{p-j+2,p-j+2}}{\left(v_{jj}+v_{p-j+2,p-j+2}\right)^2}\\&=2^{2(p-m-1)}\frac{|V|}{v_{11}\left(v_{m+1,m+1}\right)^{1-mod(p,2)}}\prod_{j=2}^{p-m}\frac{1}{\left(v_{jj}+v_{p-j+2,p-j+2}\right)^2}\\&=\frac{|V|}{\prod_{j=1}^{p}v_j^*}\end{aligned} \tag{5.586}$$

for

$$mod(p,2)=\begin{cases}0 & \text{if } p \text{ is even}\\ 1 & \text{if } p \text{ is odd}\end{cases}$$

and v_j^* defined as in (5.419).

It is then not too hard to obtain the h-th moment of Λ, under H_0 in (5.585), since taking into account that $V=nUAU'\sim W_p\left(n-1,U\Sigma U'\right)$, we may write, from the expression for the p.d.f. of the Wishart $W_p(n,\Sigma)$ distribution in (5.491), and taking into account that $(U\Sigma U')^{-1}=U'\Sigma^{-1}U$ and $|U\Sigma U'|=|\Sigma|$,

$$\begin{aligned}E\left(\Lambda^h\right)&=2^{2(p-m-1)h}\int_{\mathcal{V}>0}\frac{|\mathcal{V}|^h}{\prod_{j=1}^{p}\mathfrak{v}_j^h}\,\frac{e^{-\frac{1}{2}tr(U'\Sigma^{-1}U\mathcal{V})}|\mathcal{V}|^{(n-2-p)/2}}{2^{(n-1)p/2}\Gamma_p(\frac{n-1}{2})\,|\Sigma|^{(n-1)/2}}\,d\mathcal{V}\\&=2^{2(p-m-1)h}\frac{\Gamma_p(\frac{n-1}{2}+h)}{\Gamma_p(\frac{n-1}{2})}\,2^{hp}\,|\Sigma|^h\\&\quad\int_{\mathcal{V}>0}\frac{1}{\prod_{j=1}^{p}\mathfrak{v}_j^h}\underbrace{\frac{e^{-\frac{1}{2}tr(U'\Sigma^{-1}U\mathcal{V})}|\mathcal{V}|^{(n-2+2h-p)/2}}{2^{(n-1+2h)p/2}\,\Gamma_p(\frac{n-1}{2}+h)\,|\Sigma|^{(n-1+2h)/2}}}_{\text{pdf of } W_p(n-1+2h,\,U\Sigma U')}\,d\mathcal{V}\\&=2^{2(p-m-1)h}\frac{\Gamma_p(\frac{n-1}{2}+h)}{\Gamma_p(\frac{n-1}{2})}\,2^{hp}\,|\Sigma|^h\,E\left(\prod_{j=1}^{p}v_j^{-h}\right)\end{aligned}$$

for $V\sim W(n-1+2h,U\Sigma U')$ and where, given the independence, under H_{0a}, of the diagonal elements of V, and the definition of the v_j in (5.419), referring to the second expression in (5.586), we may write

$$E\left(\prod_{j=1}^{p}v_j^{-h}\right)=E\left(v_{11}^{-h}\right)\left(E\left(v_{m+1,m+1}^{-h}\right)\right)^{1-mod(p,2)}\prod_{j=2}^{p-m}E\left((v_{jj}^*)^{-2h}\right)$$

where

$$v_{jj} \sim W_1(n-1+2h, \tau_j) \quad \text{and} \quad v_{jj}^* = v_{jj}+v_{p-j+2,p-j+2} \sim W_1(2n-2+4h, \tau_j)\,,$$

so that, since v_{jj} and v_{jj}^* are scalars, we have $v_{jj} = |v_{jj}|$ and $v_{jj}^* = |v_{jj}^*|$, and thus, from (5.493),

$$E\left(v_{jj}^{-h}\right) = 2^{-h}\,\tau_j^{-h}\,\frac{\Gamma\left(\frac{n-1+2h}{2}-h\right)}{\Gamma\left(\frac{n-1+2h}{2}\right)}$$

and

$$E\left((v_{jj}^*)^{-2h}\right) = 2^{-2h}\,\tau_j^{-2h}\,\frac{\Gamma\left(\frac{2n-2+4h}{2}-2h\right)}{\Gamma\left(\frac{2n-2+4h}{2}\right)}\,.$$

Moreover, under H_0, given the relation in (5.584), we have

$$|\Sigma|^h = |U\Sigma U'|^h = \prod_{j=1}^{p}\tau_j^{\,h} = \tau_1^{\,h}\left(\tau_{m+1}^{\,h}\right)^{1-mod(p,2)}\prod_{j=2}^{p-m}\tau_j^{\,2h}$$

so that, using (5.492) and the duplication formula for the Gamma function in (5.583), we may finally write

$$\begin{aligned}
E\left(\Lambda^h\right) &= 2^{2(p-m-1)h}\left\{\prod_{j=1}^{p}\frac{\Gamma\left(\frac{n-j}{2}+h\right)}{\Gamma\left(\frac{n-j}{2}\right)}\right\}\frac{\Gamma\left(\frac{n-1}{2}\right)}{\Gamma\left(\frac{n-1}{2}+h\right)}\left(\frac{\Gamma\left(\frac{n-1}{2}\right)}{\Gamma\left(\frac{n-1}{2}+h\right)}\right)^{1-mod(p,2)}\\
&\qquad\times\left\{\prod_{j=2}^{p-m}\frac{\Gamma(n-1)}{\Gamma(n-1+2h)}\right\}\\
&= 2^{2(p-m-1)h}\left\{\prod_{j=2}^{p}\frac{\Gamma\left(\frac{n-j}{2}+h\right)}{\Gamma\left(\frac{n-j}{2}\right)}\right\}\left(\frac{\Gamma\left(\frac{n-1}{2}\right)}{\Gamma\left(\frac{n-1}{2}+h\right)}\right)^{1-p \bmod 2}\\
&\qquad\times\left\{\prod_{j=2}^{p-m}\frac{\Gamma\left(\frac{n-1}{2}\right)\Gamma\left(\frac{n}{2}\right)}{\Gamma\left(\frac{n-1}{2}+h\right)\Gamma\left(\frac{n}{2}+h\right)}2^{-2h}\right\}\\
&= \left\{\prod_{j=1}^{m}\frac{\Gamma\left(\frac{n-1-j}{2}+h\right)\,\Gamma\left(\frac{n-1}{2}\right)}{\Gamma\left(\frac{n-1-j}{2}\right)\,\Gamma\left(\frac{n-1}{2}+h\right)}\right\}\left\{\prod_{j=m+1}^{p-1}\frac{\Gamma\left(\frac{n-1-j}{2}+h\right)\,\Gamma\left(\frac{n}{2}\right)}{\Gamma\left(\frac{n-1-j}{2}\right)\Gamma\left(\frac{n}{2}+h\right)}\right\},
\end{aligned}$$

which for $h > -(n-p)/2$ is the expression for the h-th moment of the product of Beta r.v.'s in (5.420) and (5.421), and which, given that the range of Λ is bounded, determines the distribution of Λ as given by (5.420) and (5.421).

References

Abdi, H.: Bonferroni and Šidák corrections for multiple comparisons. In: Salkind, N. (ed.) Encyclopedia of Measurement and Statistics. Thousand Oaks, Sage, CA (2007)
Anderson, T.W.: An Introduction to Multivariate Statistical Analysis, 3rd edn. Wiley, New York (2003)
Anderson, T.W., Fang, K.-T.: Inference in multivariate elliptically contoured distributions based on maximum likelihood. In: Fang, K.-T., Anderson, T.W. (eds) Statistical Inference in Elliptically Contoured and Related Distributions, pp. 201–216. Allerton Press, Inc., New York (1990)
Anderson, T.W., Fang, K.-T., Hsu, H.: Maximum-likelihood estimates and likelihood-ratio criteria for multivariate elliptically contoured distributions. Can. J. Stat. **14**, 55–59 (1986)
Arnold, B.C., Coelho, C.A., Marques, F.J.: The distribution of the product of powers of independent uniform random variables – a simple but useful tool to address and better understand the structure of some distributions. J. Multivar. Anal. **113**, 19–36 (2013)
Bacon-Sone, J., Fung, W.K.: A new graphical method for detecting single and multiple outliers in univariate and multivariate data. J. R. Stat. Soc. Ser. C (Appl. Stat.) **36**, 153–162 (1987)
Bartlett, M.S.: Further aspects of the theory of multiple regression. Proc. Camb. Philos. Soc. **34**, 33–40 (1938)
Bartlett, M.S.: A note on tests of significance in multivariate analysis. Proc. Camb. Philos. Soc. **35**, 180–185 (1939)
Bartlett, M.S.: The goodness of fit of a single hypothetical discriminant function in the case of several groups. Ann. Eugen. **16**, 199–214 (1951)
Bock, R.D.: Multivariate analysis of variance of repeated measurements. In: Harris, C.W. (ed.) Problems in Measuring Change, pp. 85–103. University of Wisconsin Press, Madison, Wisconsin (1963)
Box, G.E.P.: Problems in the analysis of growth and wear curves. Biometrics **6**, 362–389 (1950)
Brillinger, D.R.: Time Series: Data Analysis and Theory. SIAM, Philadelphia (2001)
Caroni, C., Prescott, P.: Sequential application of Wilks's multivariate outlier test. J. R. Stat. Soc. Ser. C (Appl. Stat.) **41**, 355–364 (1992)
Carroll, J.D.: Generalization of canonical correlation analysis to three or more sets of variables. In: Proceedings 76th Annual Convention APA, pp. 227–228 (1978)
Chmielewski, M.A.: Invariant scale matrix hypothesis tests under elliptically symmetry. J. Multivar. Anal. **10**, 343–350 (1980)
Coelho, C.A.: The Generalized Canonical Analysis. Ph.D. Thesis, The University of Michigan, Ann Arbor, MI, USA (1992)
Coelho, C.A.: The generalized integer Gamma distribution – a basis for distributions in multivariate statistics. J. Multivar. Anal. **64**, 86–102 (1998)
Coelho, C.A.: Addendum to the paper "The generalized integer Gamma distribution – a basis for distributions in multivariate statistics". J. Multivar. Anal. **69**, 281–285 (1999)
Coelho, C.A.: The generalized near-integer Gamma distribution: a basis for 'near-exact' approximations to the distribution of statistics which are the product of an odd number of independent Beta random variables. J. Multivar. Anal. **89**, 191–218 (2004)
Coelho, C.A.: The likelihood ratio test for equality of mean vectors with compound symmetric covariance matrices. In: Gervasi, O., Murgante, B., Misra, S., Borruso, G., Torre, C.M., Rocha, A.M.A.C., Taniar, D., Apduhan, B.O., Stankova, E., Cuzzocrea, A. (eds.) Computational Science and Its Applications. Lecture Notes in Computer Science 10408, vol. V, pp. 20–32. Springer, New York (2017)
Coelho, C.A.: Likelihood ratio tests for equality of mean vectors with circular covariance matrices. In: Oliveira, T.A., Kitsos, C., Oliveira, A., Grilo, L.M. (eds.) Recent Studies on Risk Analysis and Statistical Modeling, pp. 255–269. Springer, New York (2018)

Coelho, C.A., Marques, F.J.: The advantage of decomposing elaborate hypotheses on covariance matrices into conditionally independent hypotheses in building near-exact distributions for the test statistics. Linear Algebra Appl. **430**, 2592–2606 (2009)

Coelho, C.A., Marques, F.J.: Near-exact distributions for the likelihood ratio test statistic to test equality of several variance-covariance matrices in elliptically contoured distributions. Comput. Stat. **27**, 627–659 (2012)

Coelho, C.A., Marques, F.J.: The multi-sample block-scalar sphericity test: exact and near-exact distributions for its likelihood ratio test statistic. Commun. Stat. Theory Methods **42**, 1153–1175 (2013)

Coelho, C.A., Marques, F.J.: A double decomposition of the test of independence of sets of variables that allows for a modeling view of this test. In: Mathematical Methods in Science and Mechanics. Mathematics and Computers in Science and Engineering Series. WSEAS Press (2014)

Coelho, C.A., Roy, A.: Testing the hypothesis of a block compound symmetric covariance matrix for elliptically contoured distributions. Test **26** 308–330 (2017)

Coelho, C.A., Arnold, B.C., Marques, F.J.: Near-exact distributions for certain likelihood ratio test statistics. J. Stat. Theory Pract. **4**, 711–725 (memorial issue in honor of H. C. Gupta, guest-edited by C. R. Rao) (2010)

Coelho, C.A., Marques, F.J., Oliveira, S.: The exact distribution of the likelihood ratio test statistic used to test simultaneously the equality of means and circularity of the covariance matrix. AIP Conf. Proc. **1558**, 789–792 (2013)

Coelho, C.A., Arnold, B.C., Marques, F.J.: The exact and near-exact distributions of the main likelihood ratio test statistics used in the complex multivariate normal setting. TEST **24**, 386–416 (+ 14 pp. of supplementary material – online resource) (2015)

Coelho, C.A., Marques, F.J., Oliveira, S.: Near-exact distributions for likelihood ratio statistics used in the simultaneous test of conditions on mean vectors and patterns of covariance matrices. Math. Probl. Eng. article ID 8975902, 25 pp. (2016)

Davis, A.W.: On the differential equation for Meijer's $G^{p,0}_{p,p}$ function, and further tables of Wilks's likelihood ratio criterion. Biometrika **66**, 519–531 (1979)

Davis, A.W., Field, J.B.F.: Tables of some multivariate test criteria. Tech. Rep. no. 32, Division of Mathematical Statistics, C.S.I.R.O., Canberra, Australia (1971)

Fang, K.T., Zhang, Y.-T.: Generalized Multivariate Analysis. Springer, New York (1990)

Fang, C., Krishnaiah, P.R., Nagarsenker, B.N.: Asymptotic distributions of the likelihood ratio test statistics for covariance structures of the complex multivariate normal distributions. J. Multivar. Anal. **12**, 597–611 (1982)

Fisher, R.A.: The use of multiple measurement in taxonomic problems. Ann. Eugen. **7**, 179–188 (1936)

Golub, G.H., van Loan, C.F.: Matrix Computations, 3rd edn. The Johns Hopkins University Press, Baltimore, MA (1996)

Goodman, N.R.: On the Joint Estimation of the Spectra, Cospectrum and Quadrature Spectrum of a Two-Dimensional Stationary Gaussian Process. Scientific Paper No. 10, Engineering Statistics Laboratory, New York University/Ph.D. Dissertation, Princeton University (1957)

Goodman, N.R.: Statistical Analysis Based on a Certain Multivariate Complex Gaussian Distribution (An Introduction). Ann. Math. Stat. **34**, 152–177 (1963a)

Goodman, N.R.: The distribution of the determinant of a complex wishart distributed matrix. Ann. Math. Stat. **34**, 178–180 (1963b)

Gupta, A.K.: Distribution of Wilks' likelihood-ratio criterion in the complex case. Ann. Inst. Stat. Math. **23**, 77–87 (1971)

Healy, M.J.R.: Experiments for comparing growth curves. Abstract No. 752. Biometrics **17**, 333 (1961)

Ito, K., Schull, W.J.: On the robustness of the T_0^2 test in multivariate analysis of variance when variance-covariance matrices are not equal. Biometrika **51**, 71–82 (1964)

James, A.T.: Distributions of matrix variates and latent roots derived from normal samples. Ann. Math. Stat. **35**, 475–501 (1964)

Jensen, S.T.: Covariance hypotheses which are linear in both the covariance and the inverse covariance. Ann. Stat. **16**, 302–322 (1988)

Jensen, D.R., Good, I.J.: Invariant distributions associated with matrix laws under structural symmetry. J. R. Stat. Soc. Ser. B **43**, 327–332 (1981)

Johnson, R., Wichern, D.: Applied Multivariate Statistical Analysis, 6th edn. Pearson New International Edition, Pearson Education Limited, Edinburgh Gate, Harlow, Essex (2014)

Kariya, T.: Robustness of multivariate tests. Ann. Stat. **9**, 1267–1275 (1981)

Kettenring, J.R.: Canonical analysis of several sets of variables. Biometrika **58**, 433–451 (1971)

Khatri, C.G.: On a MANOVA model applied to problems in growth curve. Inst. of Statitics, Univ. of North Carolina, Mimeograph Series No. 399 (1964)

Khatri, C.G.: Classical statistical analysis based on a certain multivariate complex gaussian distribution. Ann. Math. Stat. **36**, 98–114 (1965)

Khatri, C.G.: A note on a MANOVA model applied to problems in growth curve. Ann. Inst. Stat. Math **18**, 75–86 (1966)

Khatri, C.G.: Testing some covariance structures under a growth curve model. J. Multivar. Anal. **3**, 102–116 (1973)

Knapp, T.R.: Canonical correlation analysis: a general parametric significance-testing system. Psychol. Bull. **85**, 410–416 (1978)

Kramer, C.Y., Jensen, D.R.: Fundamentals of multivariate analysis, part IV – analysis of variance for balanced experiments. J. Qual. Technol. **2**, 32–40 (1970)

Kres, H.: Statistical Tables for Multivariate Analysis – A Handbook with References to Applications. Springer Series in Statistics. Springer, New York (1983)

Krishnaiah, P.R., Lee, J.C., Chang, T.C.: The distributions of the likelihood ratio statistics for tests of certain covariance structures of complex multivariate normal populations. Biometrika **63**, 543–549 (1976)

Kshirsagar, A.M.: Goodness of fit of a discriminant function from the vector space of dummy variables. J. R. Stat. Soc. Ser. B **33**, 111–116 (1971)

Kshirsagar, A.M.: Multivariate Analysis. Marcel Dekker, New York (1972)

Kshirsagar, A.M., Smith, W.B.: Growth Curves. Marcel Dekker, New York (1995)

Lawley, D.N.: Tests of significance in canonical analysis. Biometrika **46**, 59–66 (1959)

Lee, Y.S.: Some results on the distribution of Wilks's likelihood ratio criterion. Biometrika **59**, 649–664 (1972)

Lee, Y.S., Chang, T.C., Krishnaiah, P.R.: Approximations to the distributions of the likelihood ratio statistics for testing certain structures on the covariance matrices of real multivariate Normal populations. In: Krishnaiah, P.R. (ed.) Multivariate Analysis – IV, pp. 105–118. North-Holland Publishing Company, Amsterdam (1977)

Leech, F.B., Healy, M.J.R.: The analysis of experiments on growth rate. Biometrics **15**, 98–106 (1959)

Lv, X.-G., Huang, T.-Z.: A note on inversion of Toeplitz matrices. Appl. Math. Lett. **20**, 1189–1193 (2007)

Marques, F.J., Coelho, C.A.: Near-exact distributions for the sphericity likelihood ratio test statistic. J. Statist. Plann. Inference **138**, 726–741 (2008)

Marques, F.J., Coelho, C.A.: Obtaining the exact and near-exact distributions of the likelihood ratio statistic to test circular symmetry through the use of characteristic functions. Comput. Stat. **28**, 2091–2115 (2013)

Marques, F.J., Coelho, C.A., Arnold, B.C.: A general near-exact distribution theory for the most common likelihood ratio test statistics used in multivariate analysis. TEST **20**, 180–203 (2011)

Mathai, A.M.: A few remarks about some recent articles on the exact distributions of multivariate test criteria: I. Ann. Inst. Stati. Math. **25**, 557–566 (1973)

Mathai, A.M.: On distributions of test statistics in the complex Gaussian case. In: Developments in Statistics and Its Applications, King Saud University Riyadh, pp. 277–285 (1983)

Mathai, A.M., Katiyar, R.S.: Exact percentage points for testing independence. Biometrika **66**, 353–356 (1979)

Morrison, D.F.: Multivariate Statistical Methods, 4th edn. Duxbury Press, Duxbury (2005)

Muirhead, R.J.: Aspects of Multivariate Statistical Theory, 2nd edn. Wiley, New York (2005)

Olkin, I.: Testing and estimation for structures which are circularly symmetric in blocks. ETS Research Bulletin Series, RB-72-41. Wiley, New York (1972)

Olkin, I., Press, S.J.: Testing and estimation for a circular stationary model. Ann. Math. Stat. **40**, 1358–1373 (1969)

Olson, C.L.: Comparative robustness of six tests in multivariate analysis of variance. J. Amer. Stat. Assoc. **69**, 894–908 (1974)

Pearson, E.S., Hartley, H.O.: Biometrika Tables for Statisticians, vol. II. Cambridge University Press, Cambridge (1972)

Pillai, K.C.S., Gupta, A.K.: On the exact distribution of Wilks's criterion. Biometrika **56**, 109–118 (1969)

Potthoff, R.F., Roy, S.N.: A generalized multivariate analysis of variance model useful especially for growth curve problems. Biometrika **51**, 313–326 (1964)

Posten, H.O.: Analysis of variance and analysis of regression with more than one response. In: Proc. Symposium on Applications of Statistics and Computers to Fuel and Lubricant Problems, pp. 91–109 (1962)

Rao, C.R.: Some statistical methods for comparison of growth curves. Biometrics **14**, 1–17 (1958)

Rao, C.R.: Some problems involving linear hypothesis in multivariate analysis. Biometrika **46**, 49–58 (1959)

Rencher, A.C.: Multivariate Statistical Inference and Applications. Wiley, New York (1998)

Rencher, A.C.: Methods of Multivariate Analysis, 2nd edn. Wiley, New York (2002)

Rencher, A.C., Christensen, W.F.: Methods of Multivariate Analysis, 3rd edn. Wiley, New York (2012)

Schatzoff, M.: Exact distributions of Wilks' likelihood ratio criterion. Biometrika **53**, 347–358 (1966)

SenGupta, A.: Generalized Canonical Variables. In: Wiley StatsRef: Statistical Reference Online. Wiley, New York (2014). http://onlinelibrary.wiley.com/doi/10.1002/9781118445112.stat02670/abstract

Seber, G.A.F.: Multivariate Observations. Wiley, New York (1984)

Seber, G.A.F.: A Matrix Handbook for Statisticians. Wiley, Hoboken, NJ (2008)

Šidák, Z.K.: Rectangular confidence regions for the means of multivariate normal distributions. J. Am. Stat. Assoc. **62**, 626–633 (1967)

Szatrowski, T.D.: Explicit solutions, one iteration convergence and averaging in the multivariate Normal estimation problem for patterned means and covariances. Ann. Inst. Stat. Math. **30**, 81–88 (1978)

Szatrowski, T.D.: Necessary and sufficient conditions for explicit solutions in the multivariate Normal estimation problem for patterned means and covariances. Ann. Stat. **8**, 802–810 (1980)

Timm, N.H.: Applied Multivariate Analysis, Springer, New York (2002)

Travers, R.M.W.: The use of a discriminant function in the treatment of psychological group differences. Psychometrika **4**, 25–32 (1939)

Wald, A., Brookner, R.J.: On the distribution of Wilks' statistic for testing the independence of several groups of variates. Ann. Math. Stat. **12**, 137–152 (1941)

Wall, F.J.: The Generalized Variance Ratio or U-Statistic. The Dikewood Corporation, Albuquerque, New Mexico (1967)

Wilks, S.S.: Certain generalizations in the analysis of variance. Biometrika **24**, 471–494 (1932)

Wilks, S.S.: On the independence of k sets of normally distributed statistical variables. Econometrica **3**, 309–326 (1935)

Wilks, S.S.: Sample criteria for testing equality of means, equality of variances, and equality of covariances in a Normal multivariate distribution. Ann. Math. Stat. **17**, 257–281 (1946)

Wilks, S.S.: Multivariate statistical outliers. Sankhya Ser. A **25**, 407–426 (1963)

Wishart, J.: The generalised product moment distribution in samples from a normal multivariate population. Biometrika **20A**, 32–52 (1928)
Wishart, J.: Growth rate determinations in nutrition studies with the bacon pig, and their analysis. Biometrika **30**, 16–28 (1938)
Wooding, R.A.: The Multivariate Distribution of Complex Normal Variables. Biometrika **43**, 212–215 (1956)

Chapter 6
Mathematica®, MAXIMA, and R Packages to Implement the Likelihood Ratio Tests and Compute the Distributions in the Previous Chapter

Abstract In this chapter details are given on the packages programmed in Mathematica®, MAXIMA, and R for the implementation of all the tests addressed in Chap. 5. Functions and modules were developed for several tasks such as the computation of p-values and quantiles for each test and to obtain the computed value of the statistics and corresponding p-values from data stored in a data file. As such, also functions and modules used to read these data files, which may have several different internal structures, were programmed. Details are given on the use of all functions and modules, and comparisons are established among the three packages. The full content of all three packages is available on the book's supplementary material web site.

Keywords Modules · Packages · p-Value computation · p-Value computation from data sets · Quantile computation · Reading data files

6.1 Introduction

In order to allow the user to easily implement all tests addressed in Chap. 5, three packages were programmed, using three different softwares, each one with its own peculiarities. There is a package programmed in Mathematica® which is indeed the one that exhibits better, that is, lower computation times among all three packages. However, the authors also think it to be advantageous to have packages with similar functionalities, but programmed in freeware softwares. The freeware softwares chosen were MAXIMA and R, given their capability to use extended precision, that is, precision that goes beyond the common double precision, and given the fact that in addition they yield the correct results. There are other freeware softwares which although being able to handle some form of extended precision, end up giving wrong results, most often due to the fact that the implementation of the extended precision is done on top of the double precision engine.

C. A. Coelho, B. C. Arnold, *Finite Form Representations for Meijer G and Fox H Functions*, Lecture Notes in Statistics 223,
https://doi.org/10.1007/978-3-030-28790-0_6

All three packages will always give exactly the same results, as it is shown in Sect. 6.4.1, although in MAXIMA and R the very last decimal places may be spurious.

Technically, all package modules are actually functions in MAXIMA as well as in R, while in Mathematica® some of them are functions and the other are modules. Anyway, for the sake of simplicity in the exposition, we will refer to all of them hereon as "modules."

Although when doing some elementary computations in each of these three softwares one will have to be aware of the bells and whistles of the correct handling of values in order to obtain the desired precision, the way all package modules are implemented takes away from the user all such worries, unless some previous computation has to be done on the modules' arguments, as it is exemplified in Sect. 6.4.1, with the use of the modules to compute the GIG and EGIG p.d.f. and c.d.f.

In Fig. 6.1 we may see the problems we may encounter when trying to compute the exact value of 5/3.3333333333, with 34 exact decimal places, using the software Mathematica®. We may see how only the last three commands in Fig. 6.1 give the correct result, by giving all input in its rationalized form or by giving explicitly the desired precision with the attribute "`", as it is done in the command before the last one, or yet by using the function `Rat` (which was programmed exactly to help overcoming these problems and which is included in the package `EGIG` which is made available for the computations associated with the contents of the present book—see also Sect. 6.4.1). We may see that the Mathematica® native

```
SetPrecision[5 / 3.3333333333, 35]
1.5000000000150000012411055649863556

5 / SetPrecision[3.3333333333, 35]
1.5000000000149999346278740861695220

SetPrecision[5 / Rationalize[3.3333333333], 35]
1.5000000000150000012411055649863556

SetPrecision[5 / Rationalize[3.3333333333, 0], 35]
1.5000000000149999346289348929851322

SetPrecision[5 / (33 333 333 333 / 10^10), 35]
1.5000000000150000000015000000000015

SetPrecision[5 / 3.3333333333`35, 35]
1.5000000000150000000015000000000015

SetPrecision[5 / Rat[3.3333333333], 35]
1.5000000000150000000015000000000015
```

Fig. 6.1 Different commands and corresponding results obtained when trying to compute 5/3.3333333333 with 34 exact decimal places in Mathematica®

function `Rationalize` is not able to overcome the problem. Among the last three commands in Fig. 6.1 the last one is the one that is recommended to be used.

MAXIMA was chosen since, besides being a freeware software with all the basic tools needed, it also has now a friendly interface for Windows, but more importantly, it is able to handle exact rationals and it has hardwired in its base software the extended precision mechanism, through the use of the `bfloat` "function," or just by adding to the number we want to treat as an extended precision number, a scientific notation exponent which uses a `b` instead of the commonly used `e`. For example, if we set the system variable `fpprec` to 35, either by writing an input command like

```
fpprec:35
```

where the colon is used in MAXIMA to assign values to variables, or by using the menu *Numeric/Set Precision* feature and inserting the value 35, then the inputs in Fig. 6.2 will give the corresponding outputs.

We may see that the first command in Fig. 6.2 only gives the double precision result. The second and third commands in this figure are not able to give the correct result because we are trying to implement the extended precision on a value which is originally in double or floating point precision and we may see that the native function `rat` is also not able to produce the correct result, while the function `rat2` which was programmed to be included in the MAXIMA `EGIG` package is able

Fig. 6.2 Different commands and corresponding results obtained when trying to compute 5/3.3333333333 with 34 exact decimal places in MAXIMA

```
5/3.3333333333;
1.500000000015

bfloat(5/3.3333333333);
1.500000000015000001241105564986355 6b0

5/bfloat(3.3333333333);
1.5000000000149999346278740861695 22b0

bfloat(5/rat(3.3333333333));
1.50000000001499993462893489298513 22b0

bfloat(5/rat2(3.3333333333));
1.500000000015000000000150000000000 15b0

bfloat(5/(33333333333/10^10));
1.500000000015000000000150000000000 15b0

5/3.3333333333b0;
1.500000000015000000000150000000000 15b0
```

to solve the problem. Anyway, the recommended syntax is the one used in the last command, by concatenating "b0" to the value that is given in floating point precision. We may also see how it is enough that just one of the input values is in extended precision for the result to be also in extended precision.

The software R was chosen given its great popularity and its ability to handle extended precision, once the package Rmpfr (which stands for 'multiple precision floating-point reliable' in R) is loaded.

In R, when not using the Rmpfr package, we may ask for a number of digits to be displayed which may go up to 22, although only the first 16 are correct, given that double precision is used. This feature is implemented with the command

```
options(digits=22)
```

and it will only affect the display of common double precision values and not the display of the extended precision values, whose number of digits displayed will only be a function of the precision used, as it is illustrated in Fig. 6.3. In this figure we may analyze the commands used when trying to compute 5/3.3333333333 in R, with 34 exact decimal places.

We may see how the first three commands that use the function mpfr in Fig. 6.3 only give a double precision value. This is so because we are indeed requiring the mpfr function to act on double precision values, being three typical cases

Fig. 6.3 Different commands and corresponding results obtained when trying to compute 5/3.3333333333 with 34 exact decimal places in R

```
> options(digits=22)
> 5/3.3333333333
[1] 1.500000000015
> 5/(33333333333/10^10)
[1] 1.500000000015
> mpfr(5/3.3333333333,113)
1 'mpfr' number of precision  113   bits
[1] 1.500000000015000012411055649863556
> 5/mpfr(3.3333333333,113)
1 'mpfr' number of precision  113   bits
[1] 1.500000000014999934627874086169522206
> 5/mpfr(33333333333/10^10,113)
1 'mpfr' number of precision  113   bits
[1] 1.500000000014999934627874086169522206
> 5/(mpfr(33333333333,113)/mpfr(10^10,113))
1 'mpfr' number of precision  113   bits
[1] 1.500000000015000000001500000000016
> 5/mpfr("3.3333333333",113)
1 'mpfr' number of precision  113   bits
[1] 1.500000000015000000001500000000016
> 5/mpfr("3.3333333333",120)
1 'mpfr' number of precision  120   bits
[1] 1.5000000000150000000001500000000015007
> mpfr(5/"3.3333333333",120)
Error in 5/"3.3333333333" : non-numeric
argument to binary operator
> mpfr("5/3.3333333333",120)
Error in mpfr.default("5/3.3333333333", 120) :
  str2mpfr1_list(x, *): x[1] cannot be made
  into MPFR
```

of attempts at implementing extended precision upon a value that is in double precision. Even the third command that uses function `mpfr` does not work correctly because R is not able to handle exact rationals, always converting them first to floating point or double precision values.

The first argument to the function `mpfr` is the value we want to handle in extended precision and the second argument is an integer specifying the number of bits of precision. In Fig. 6.3 we first used 113 bits of precision, which gives us an extended precision value with 34 decimal places, since

$$113 \log_{10} 2 \approx 34$$

which is indeed the number of decimal places in our result. In order to avoid the need for this conversions between number of bits and number of decimal digits of precision, all R functions programmed for the package `EGIG` which use precision as an argument will use this argument as the number of decimal digits and not as the number of bits. The conversion is done internally in the code of those functions.

But the fourth and fifth commands that use function `mpfr` are then able to give more or less the result we are looking for, only with the last decimal place wrong. The solution is to use a couple more bits of precision, leaving the last decimal place still not right, but giving us 34 decimal places that are correct. Wrapping the floating point value we want to use with extended precision in double quotes is indeed the smartest way to handle the problem. However, one cannot wrap in double quotes any mathematical operations, but only floating point values, with any number of decimal places. There is no need to do this to integer values since the `mpfr` function handles integers and long integers in its exact form.

We should note that while functions in R and MAXIMA use common curved brackets around their arguments, Mathematica® uses square brackets. Also, while a simple "Enter" (or "Return") will execute the commands in R, in Mathematica® and MAXIMA they are executed with "Shift-Return" or alternatively by hitting the "Enter" touch in the numeric keypad, although this may be changed through the options menu in MAXIMA, to make commands executable just by hitting "Return" (or "Enter").

6.2 Loading the Packages and Getting Help

From this section on we will address the three softwares used in their alphabetical order, that is, Mathematica®, then MAXIMA followed by R.

In all three softwares the package that handles all the tests addressed in the previous chapter is called `EGIG`. In Mathematica® it is loaded either with the command

```
<<EGIG`
```

or alternatively by using the command

```
Needs["EGIG`"]
```

which will load the package as soon as a function or module from this package is used. The package file named `EGIG.m` has to be located in the sub-folder `Packages` of the Mathematica® installation folder.

In MAXIMA, loading the package is done with a call to the function `load`, with a command like

```
load(EGIG)
```

and in R, once the package file is placed in the sub-directory `library` of the R installation folder, the package is loaded through the R menu `Packages/Load Package`, selecting then the name of the package from the list that is provided.

In Mathematica® one may obtain help on a given function or module by typing its name preceded by a question mark, which may or may not have a blank space separating it from the function or module name. For example if one wants to get help on the module `ReadFileR`, which will be addressed in more detail in the next section, one should type the command in Fig. 6.4 to obtain the help information displayed as output in this figure.

In MAXIMA one may get help on a function, or on a set of functions with a common text in their names, and also on their arguments, by using `Ctrl-K` or `Shift-Ctrl-K`. For example if one knows that the function that one wants to use has `Read` in its name, then one may type `Read` and then `Ctrl-K` and will get a list of all the functions whose name has `Read` in it. Then one may choose one of the functions from that list. Let us suppose it was the function `ReadFileR`. Then

```
?ReadFileR
```

ReadFileR[file,index1,index2] reads the contents of the data file 'file' (in case its contents are real, not complex, data values), where 'file' has to be preceded by the path, in case the file is not located in the current working directory. This is the only mandatory argument of this module. The data file has to have a structure with rows corresponding to observations and columns to variables and it has to consist of either one only data matrix which may be preceded by any number of blank or text lines and which may be followed by any number of blank or text lines, or otherwise by several submatrices, all with the same number of data columns, but possibly with different numbers of rows, each one corresponding to a different sample and separated from each other by one or more blank or text lines.

– If index1 is given the value 1, it will allow the user to select variables and/or samples and

– If index2 is given the value 1 it will allow the user to select observations for each sample.

Fig. 6.4 Example of usage of the help feature for module `ReadFileR` in Mathematica®

```
exampleEGIG(ReadFileR);
 "Usage: ReadFileR(file,index1,index2)"
 "       file -> name of the data file, given as a string and preceded
                by the path in case the file is in a different
                directory from the current directory (only
                mandatory argument)"
 "       index1 -> if given the value 1, it will allow the user to
            select variables and/or samples (optional argument)"
 "       index2 -> if given the value 1 it will allow the user to
            select observations for each sample (optional argument)."
 "This function reads the contents of the data file 'file' (in case its
 contents are real, not complex, data values), and where 'file' has
 to be preceded by the path, in case the file is not located in the
 current working directory. This is the only mandatory argument of
 this module. The data file has to have a structure with rows
 corresponding to observations and columns to variables and it has
 to consist of either one only data matrix which may be preceded by
 any number of blank or text lines and which may be followed by any
 number of blank or text lines, or otherwise by several submatrices,
 all with the same number of data columns, but possibly with different
 numbers of rows, each one corresponding to a different sample
 and separated from each other by one or more blank or text lines."
```

Fig. 6.5 Example of usage of the help feature for function `ReadFileR` in MAXIMA

if one wants to know the arguments for this function and their order, one may type `Shift-Ctrl-K` and will get

```
ReadFileR(<file>,<[index]>)
```

with `<file>` highlighted, showing that it is the part to be replaced with the mandatory argument name, which has to be the name of the data file to be read. The argument `[index]`, enclosed in brackets, indicates that it is an optional argument. Actually it stands for the set of all optional arguments of the function, which in this case are the two optional arguments described in detail in Fig. 6.5. One might indeed have used `Shift-Ctrl-K` right after typing `Read`, obtaining a list of all the available functions whose name has `Read` in it, but in this list now the functions come already with the indication of their arguments.

Further, a particular help file was prepared along with the MAXIMA package which uses a version of the MAXIMA function `example` called `exampleEGIG`, which, if called using as argument the name of a function in the package, will display the necessary information on the function and, if appropriate, at least one example of usage of the same function. In Figs. 6.5 and 6.6 we may view the output obtained when calling this function with arguments respectively the package functions `ReadFileR` and `PvalEqMeanVecR`, the first one used to read data files with real values, occasionally split into different samples, and the second one used to compute p-values for the statistic in Sect. 5.1.1.

```
exampleEGIG(PvalEqMeanVecR);
"This function computes the p-value for a computed value of the statistic in
                                                             Section 5.1.1"
"Usage: PvalEqMeanVecR(n,p,q,x,pd,prec)"
"        n -> sample size; p -> number of variables; q -> number of samples;
         x -> computed value of the statistic;
         pd -> precision digits (optional argument with a default value of 10):
               number of digits used in printing the p-value;
         prec -> number of precision digits used in computations
                (by default the module automatically sets this value to 1500);"
"Example 1: (where the command is followed by a ';')"
PvalEqMeanVecR(15,4,5,0.123)
 0.2495967175
2.4959671751283258998989563141[1444 digits]0554334664878997965211331182b-1
"Example 2: (where the command is followed by a '$')"
PvalEqMeanVecR(15,4,5,0.123)
 0.2495967175
"Example 3: (where the command is followed by a '$')"
PvalEqMeanVecR(15,4,5,0.123,20)
 0.24959671751283258999
```

Fig. 6.6 Example of usage of the help feature for function `PvalEqMeanVecR` in MAXIMA

In R, for a quick help, a function with the name `functions` is built into the package, which if used without any argument, that is, just as

`functions()`

will display a list with the names of all the functions available in the package, together with their arguments. This function may also be used with a string argument, and in that case it will display a list with the names of all the functions in the package with that string in their names, together with their arguments. Thus, for example, in order to display the function `ReadFileR` one only has to use for example a command like

`functions("Read")`

obtaining as output a list of all the functions in the package that have `Read` in their names, together with their arguments.

There is also the possibility of obtaining the common R help through the use of the function `help`, using as argument the function onto which one wants to obtain the information. This will give a more extensive description of the function capabilities, output, and functioning. One may yet use the function `apropos` using as argument a string to obtain a list of all functions with that string in their names.

6.3 Modules to Read Data Files with One or More Samples

Each one of the three softwares, Mathematica®, MAXIMA, and R, have their own functions to read data files. However, these functions are generally only adequate to read simple data files structured as a single table or with all data values spread on a single or several rows. Since we are interested in being able to read data files with a structure similar to one of the files in Fig. 6.7, we decided to program in these three softwares functions that are able to read such data files, as well as other data files which will give rise to a set of sub-samples organized into matrices.

The possible existence of blank lines or lines with text in between the data rows in files with a similar structure to those in Fig. 6.7, to indicate the variables names or the sample being displayed, poses some not so easy to overcome problems in reading such data files. Moreover, we want those functions to be able to understand the multi-sample structure of such data files. Although in all formulations in Chap. 5 we have always used data matrices which are structured with their rows corresponding to the variables and their columns corresponding to the observation units, all data file reading modules assume data files whose columns correspond to the variables in the study and rows that correspond to the observation units. However, some of the modules or functions that compute the l.r. statistics and the corresponding p-values from data files may use data files in the transposed position. See the Sect. 6.4.

ex_1.dat

```
 Names of variables
X1 X2 X3 X4
 1st sample
23.4  56.7   21.2   36.5
34.5  23.4  28.9   25.4
21.2  33.3  56.7    25.6
 2nd sample
 X1      X2      X3      X4
56.8   56.4   25.6   43.3

23.4 56.8  23.4   52.2
33.5  45.8   23.6     42.2
33.7  38.9   56.8     53.2
 4th sample
43.2   45.6   34.5   32.2
56.4   24.7   78.3   34.3
74.5   83.5   84.3   52.3
24.5   25.7   37.4   52.3
35.7   74.5   84.6   83.3
fdfdfdfd
dgdg                        .
25  26  27  28
hjhjhjhjhjhj
```

ex_2.dat

```
 X1    X2    X3     X4

23.4  56.7  21.2  36.5
34.5  23.4  28.9  25.4
21.2  33.3  56.7  25.6
56.8  56.4  25.6  43.3
23.4  56.8  23.4  52.2
33.5  45.8  23.6  42.2
33.7  38.9  56.8  53.2
43.2  45.6  34.5  32.2
56.4  24.7  78.3  34.3
74.5  83.5  84.3  52.3
24.5  25.7  37.4  52.3
35.7  74.5  84.6  83.3
```

ex_3.dat

```
16.4  49.7  14.2  29.5  5
27.5  16.4  21.9  18.4  3
14.2  26.3  49.7  18.6  5
56.8  56.4  25.6  39.3  3
23.4  56.8  23.4  52.2  2
33.5  45.8  23.6  42.2  1
33.7  38.9  56.8  53.2  2
47.2  49.6  38.5  36.2  2
60.4  28.7  82.3  38.3  1
78.5  87.5  88.3  56.3  1
28.5  29.7  41.4  56.3  5
39.7  78.5  88.6  87.3  3
```

Fig. 6.7 Data files used to exemplify the use of the functions and modules used to read data files

Similar to what happens with all other modules, the modules used to read data files were set with identical names in all three softwares.

6.3.1 Modules `ReadFileR` and `ReadFileC`

Modules `ReadFileR` and `ReadFileC` are the end-user modules to read multi-sample data files where all samples have the same number of variables and they are the reading-file modules used in all `PvalData` modules that use multi-sample data files where all samples are supposed to have been taken on the same variables.

Module `ReadFileR` is able to read data files with real values, such as the ones in Fig. 6.7. These files have to have their rows corresponding to the observations and the columns corresponding to the measured or observed variables, although if the file is structured as a single table it may then have rows corresponding to the variables measured and the columns corresponding to the observations taken. In case there is more than one sample and one wants to individualize these samples, then each sample should start with at least one blank line or a line of text separating it from the previous one, with the first sample being allowed to start at the very beginning of the file, and with all samples having an equal number of measured or observed variables.

Module `ReadFileR` is intentionally programmed to read only files with real values, while the companion `ReadFileC` module is able to read both complex and real data files. Both have a similar functioning, with the same optional and mandatory arguments. They have a single mandatory argument which is the name of the data file to be read, and two optional arguments, the first of which, if given a value of 1, allows the user to select variables and/or samples from the file, while the second one, if given a value of 1, will allow the user to select observations from each sample.

Whatever the structure of the data file is, as long as it conforms with one of the structures in Fig. 6.7, modules `ReadFileR` and `ReadFileC` will always produce as output an array with two components, the first of which is itself an array, whose elements are the individual samples in the file, the second component being a list with the sample sizes.

If the file to be read has the samples already individualized, as it is the case of file `ex_1.dat` in Fig. 6.7, then if module `ReadFileR` is used without any of its optional arguments, it will just go ahead and read the file, without posing any questions. This is illustrated in Fig. 6.8, where the Mathematica® version of module `ReadFileR` is used to read file `ex_1.dat`, storing the result in a variable called `mat`, for later use.

The use of a semicolon after the first command in Fig. 6.8 precludes the display of the unformatted contents of the variable `mat`. This second command in Fig. 6.8 displays then the contents of the first component of the variable `mat`, displaying the individual samples in matrix form, while the last command displays the contents of the second component of the variable `mat`, which is the list with the sample sizes.

```
mat = ReadFileR[path <> "ex_1.dat"];

 There are 5 samples, with sizes {3, 1, 3, 5, 1} and 4 variables

Table[MatrixForm[mat[[1]][[j]]], {j, 1, 5}]

 (23.4 56.7 21.2 36.5)
{(34.5 23.4 28.9 25.4), ( 56.8 56.4 25.6 43.3 ),
 (21.2 33.3 56.7 25.6)

                        (43.2 45.6 34.5 32.2)
 (23.4 56.8 23.4 52.2)  (56.4 24.7 78.3 34.3)
 (33.5 45.8 23.6 42.2), (74.5 83.5 84.3 52.3), ( 25 26 27 28 )}
 (33.7 38.9 56.8 53.2)  (24.5 25.7 37.4 52.3)
                        (35.7 74.5 84.6 83.3)

mat[[2]]

{3, 1, 3, 5, 1}
```

Fig. 6.8 Reading file `ex_1.dat` with the Mathematica® version of module `ReadFileR`, without using any options and storing the result in a variable for later use

In Fig. 6.9 is illustrated the use of the optional arguments of module `ReadFileR` when the MAXIMA version of this module is used to read the multi-sample data file `ex_1.dat` in Fig. 6.7, using the option that allows for variable and sample selection and also the option that allows for the selection of observations in each sample. We may note that MAXIMA, recognizing that the contents of the first component of the output produced by module `ReadFileR` is a list of matrices, automatically displays these in matrix format, producing a very nice to read output.

In Fig. 6.10 we have the output and the interaction produced when the R version of `ReadFileR` is used to read file `ex_1.dat` with the same options that were used in Fig. 6.9 with thc MAXIMA version.

When the file to be read is similar to files `ex_2.dat` or `ex_3.dat` in Fig. 6.7, which are organized in a single table, then module `ReadFileR` asks the user whether there is a column with the sample assignments, as it happens with the fifth column of file `ex_3.dat` in Fig. 6.7. In case the answer is negative, the module asks the user to provide a list with the sample sizes. The sum of the values given in this list is checked against the total number of observations in the file and in case the two values do not match the user is asked to provide a new list with the sample sizes. This is illustrated in Fig. 6.11 where `ReadFileR` is used to read file `ex_2.dat`, without using any options.

In Fig. 6.12 is shown the interaction with the user when reading file `ex_2.dat` with `ReadFileR` in Mathematica®, while using the first optional argument with a value of 1, in order to allow for the selection of variables and samples and in Fig. 6.13 is shown the interaction with the user and the output obtained when reading file `ex_3.dat` with `ReadFileR` in MAXIMA.

```
ReadFileR(sconcat(path,"ex_1.dat"));

There are 5 samples, with sizes: [3,1,3,5,1] and 4 variables
```

$$
[[\begin{bmatrix} 23.4 & 56.7 & 21.2 & 36.5 \\ 34.5 & 23.4 & 28.9 & 25.4 \\ 21.2 & 33.3 & 56.7 & 25.6 \end{bmatrix}, \begin{bmatrix} 56.8 & 56.4 & 25.6 & 43.3 \end{bmatrix}, \begin{bmatrix} 23.4 & 56.8 & 23.4 & 52.2 \\ 33.5 & 45.8 & 23.6 & 42.2 \\ 33.7 & 38.9 & 56.8 & 53.2 \end{bmatrix}, \begin{bmatrix} 43.2 & 45.6 & 34.5 & 32.2 \\ 56.4 & 24.7 & 78.3 & 34.3 \\ 74.5 & 83.5 & 84.3 & 52.3 \\ 24.5 & 25.7 & 37.4 & 52.3 \\ 35.7 & 74.5 & 84.6 & 83.3 \end{bmatrix},
$$

$$
\begin{bmatrix} 25 & 26 & 27 & 28 \end{bmatrix}], [3,1,3,5,1]]
$$

```
ReadFileR(sconcat(path,"ex_1.dat"),1,1);
Do you want to select samples ? (1=Yes) 1;
Please insert a list, inside square brackets,
 with the numbers of the samples you want to keep, separated by commas: [1,3,4];
Do you want to select variables ? (1=Yes) 1;
Please insert a list, inside square brackets,
 with the numbers of the variables you want to keep, separated by commas: [1,2,4];

There are 3 samples, with sizes: [3,3,5] and 3 variables
Original samples [1,3,4]
Original variables [1,2,4]


Please give a double list with the elements to be kept in each sample:
                                                [[1,2],[1,3],[1,3,5]];
There are now 3 samples, with sizes: [2,2,3] and 3 variables
```

$$
[[\begin{bmatrix} 23.4 & 56.7 & 36.5 \\ 34.5 & 23.4 & 25.4 \end{bmatrix}, \begin{bmatrix} 23.4 & 56.8 & 52.2 \\ 33.7 & 38.9 & 53.2 \end{bmatrix}, \begin{bmatrix} 43.2 & 45.6 & 32.2 \\ 74.5 & 83.5 & 52.3 \\ 35.7 & 74.5 & 83.3 \end{bmatrix}], [2,2,3]]
$$

Fig. 6.9 Reading the data file ex_1.dat in Fig. 6.7 with ReadFileR in MAXIMA using both optional arguments in order to be able to select variables, samples, and observations in each sample

We will hereon illustrate the functioning of the file reading modules with the MAXIMA versions of the modules because this is the software that produces the output and user-interaction in a more convenient, easy to read, and concise form.

As already stated, module ReadFileC has a usage similar to that of module ReadFileR. However, in each one of the three softwares, although fairly similar, the definition of complex values has its own peculiarities. This means that a data file with complex values written to be read with one of the three softwares will not be adequate to be read by any of the other two softwares.

As such one has to know precisely how complex values are defined in each of the three softwares. In any one of them complex values to be read from a file are to be written as a real, floating-point or rational, plus or minus another real, floating-point or rational times the square-root of minus one. It is the representation of this square root of minus one that differs from one software to the other. In Mathematica® it is represented by a capital I, in MAXIMA by *%i, and in R by i. Data files with complex data values appropriately expressed in each of the three softwares are

```
> ReadFileR(paste(path,"ex_1.dat",sep=""),1,1)
Do you want to select samples ? (1=Yes) 1: 1
Please insert a list with the numbers of the samples
                you want to keep, separated by spaces: 1: 1 3 4
Do you want to select variables ? (1=Yes) 1: 1
Please insert a list with the numbers of the variables
                you want to keep, separated by spaces: 1: 1 2 4

 There are  3  samples, with sizes:  3 3 5
 and 3 variables

 Original samples  1 3 4
 Original variables  1 2 4

Please insert in each line a list with the observations to keep
 in each sample, separated by spaces:
1: 1 2
1: 1 3
1: 1 3 5

 There are now  3  samples, with sizes:  2 2 3
 and 3 variables

[[1]]
[[1]][[1]]
  [,1] [,2] [,3]
1 23.4 56.7 36.5
2 34.5 23.4 25.4

[[1]][[2]]
  [,1] [,2] [,3]
5 23.4 56.8 52.2
7 33.7 38.9 53.2

[[1]][[3]]
   [,1] [,2] [,3]
8  43.2 45.6 32.2
10 74.5 83.5 52.3
12 35.7 74.5 83.3


[[2]]
[1] 2 2 3
```

Fig. 6.10 Reading the data file `ex_1.dat` in Fig. 6.7 with `ReadFileR` in R using both optional arguments in order to be able to select variables, samples, and observations in each sample

displayed in Fig. 6.14, named as `ex1_comp_`, followed by the first letters of the name of the corresponding software.

In Fig. 6.15 we may see how one can read the file `ex1_comp_Max.dat` with `ReadFileC` in MAXIMA and can view the output produced.

```
ReadFileR(sconcat(path,"ex_2.dat"));
Is there a column in the data file with the sample assignments ? (1-Yes) 0;
Please give the sample sizes for the q samples as a list with [n1,n2,...,nq]: [3,5,5];
The sum of the sample sizes you provided and the total sample size in the data file
                                                                        do not match
Please give the sample sizes for the q samples as a list with [n1,n2,...,nq]: [3,4,5];

There are 3 samples, with sizes: [3,4,5] and 4 variables

                                           [43.2 45.6 34.5 32.2]
                   [56.8 56.4 25.6 43.3]   [56.4 24.7 78.3 34.3]
 [[[23.4 56.7 21.2 36.5]  [23.4 56.8 23.4 52.2]
   [34.5 23.4 28.9 25.4], [33.5 45.8 23.6 42.2], [74.5 83.5 84.3 52.3]], [3,4,5]]
   [21.2 33.3 56.7 25.6]  [33.7 38.9 56.8 53.2]   [24.5 25.7 37.4 52.3]
                                           [35.7 74.5 84.6 83.3]
```

Fig. 6.11 Reading the data file `ex_2.dat` in Fig. 6.7 with `ReadFileR` in MAXIMA, without using any optional arguments

6.3.2 *Modules* **ReadFiledifpR** *and* **ReadFiledifpC**

Modules `ReadFiledifpR` and `ReadFiledifpC` are the modules to be used to read files with multiple samples with possibly different numbers of columns. These files have to be organized like file `ex_1.dat` in Fig. 6.7, or rather, as file `ex1_n.dat` in Fig. 6.16.

As happens with modules `ReadFileR` and `ReadFileC`, while the first of these modules only reads files with real values, the second is able to read files with complex or real values. The arguments for these modules are exactly the same as those for the `ReadFileR` and `ReadFileC` modules, with the optional arguments having a similar functioning. The only exception is that if the first optional argument is given the value 1, and if we choose to select variables, then now the user will have to provide a double list, specifying which variables are to be kept for each of the samples, since these samples may now correspond to different variables.

Figure 6.17 illustrates the functioning and use of the MAXIMA version of module `ReadFiledifpR` in reading file `ex1_n.dat` in Fig. 6.16, using both optional arguments, the one to select samples and/or variables and the one to select observations from each sample.

6.3.3 *Modules* **ReadFile1sR** *and* **ReadFile1sC**

Modules `ReadFile1sR` and `ReadFile1sC` are the modules used to read files which correspond to a single sample or which are to be converted to a single sample. Once again module `ReadFile1sR` is used for data files with real values, while for data files with complex values module `ReadFile1sC` is the one that has to be used, although it may also handle real data files.

```
ReadFileR[path <> "ex_2.dat", 1];
```

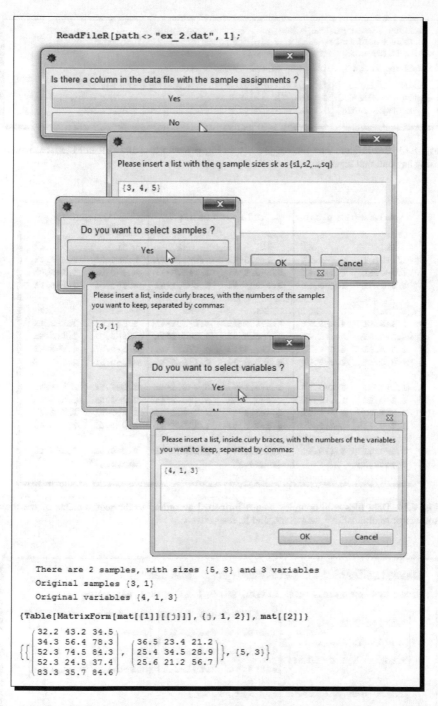

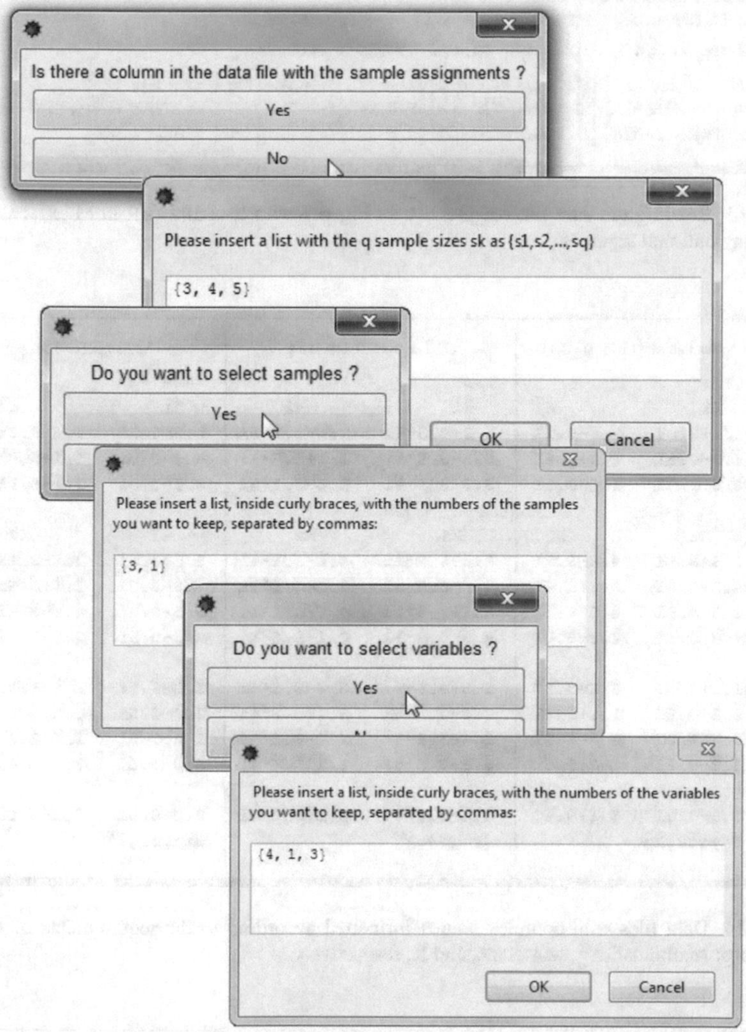


```
There are 2 samples, with sizes {5, 3} and 3 variables
Original samples {3, 1}
Original variables {4, 1, 3}

{Table[MatrixForm[mat[[1]][[j]]], {j, 1, 2}], mat[[2]]}

    32.2 43.2 34.5
    34.3 56.4 78.3     36.5 23.4 21.2
{{  52.3 74.5 84.3  ,  25.4 34.5 28.9  }, {5, 3}}
    52.3 24.5 37.4     25.6 21.2 56.7
    83.3 35.7 84.6
```

Fig. 6.12 Reading the data file `ex_2.dat` in Fig. 6.7 with `ReadFileR` with the software Mathematica®

```
ReadFileR(sconcat(path,"ex_3.dat"));
Is there a column in the data file with the sample assignments ? (1-Yes) 1;
Which is the column in the data file with the sample assignments ? 5;
There are 4 samples, with sizes: [3,3,3,3] and 4 variables
```

$$[[\begin{bmatrix}33.5 & 45.8 & 23.6 & 42.2\\ 60.4 & 28.7 & 82.3 & 38.3\\ 78.5 & 87.5 & 88.3 & 56.3\end{bmatrix},\begin{bmatrix}23.4 & 56.8 & 23.4 & 52.2\\ 33.7 & 38.9 & 56.8 & 53.2\\ 47.2 & 49.6 & 38.5 & 36.2\end{bmatrix},\begin{bmatrix}27.5 & 16.4 & 21.9 & 18.4\\ 56.8 & 56.4 & 25.6 & 39.3\\ 39.7 & 78.5 & 88.6 & 87.3\end{bmatrix},\begin{bmatrix}16.4 & 49.7 & 14.2 & 29.5\\ 14.2 & 26.3 & 49.7 & 18.6\\ 28.5 & 29.7 & 41.4 & 56.3\end{bmatrix}],[3,3,3,3]]$$

Fig. 6.13 Reading the data file `ex_3.dat` in Fig. 6.7 with `ReadFileR` in MAXIMA, without using any optional arguments

ex1_comp_Math.dat		ex1_comp_Max.dat		ex1_comp_R.dat	
vars		vars		vars	
X1	X2	X1	X2	X1	X2
1.2+0.5I	2.6-3.1I	1.2+0.5*%i	2.6-3.1*%i	1.2+0.5i	2.6-3.1i
5.3-0.5I	1.1+5.6I	5.3-0.5*%i	1.1+5.6*%i	5.3-0.5i	1.1+5.6i
3.5+3.5I	1.6-3.1I	3.5+3.5*%i	1.6-3.1*%i	3.5+3.5i	1.6-3.1i
X1	X2	X1	X2	X1	X2
2.5+3.5I	4.6-3.1I	2.5+3.5*%i	4.6-3.1*%i	2.5+3.5i	4.6-3.1i
4.3-0.5I	2.1+5.6I	4.3-0.5*%i	2.1+5.6*%i	4.3-0.5i	2.1+5.6i
3.5+3.5I	4.6-3.1I	3.5+3.5*%i	4.6-3.1*%i	3.5+3.5i	4.6-3.1i
4.3-0.5I	2.1+5.6I	4.3-0.5*%i	2.1+5.6*%i	4.3-0.5i	2.1+5.6i
1.5+3.5I	3.6-3.1I	1.5+3.5*%i	3.6-3.1*%i	1.5+3.5i	3.6-3.1i
2.3-0.5I	1.1+5.6I	2.3-0.5*%i	1.1+5.6*%i	2.3-0.5i	1.1+5.6i
1.5+3.5I	2.6-3.1I	1.5+3.5*%i	2.6-3.1*%i	1.5+3.5i	2.6-3.1i
5.3-0.5I	3.1+5.6I	5.3-0.5*%i	3.1+5.6*%i	5.3-0.5i	3.1+5.6i
2.3-0.5I	3.1+5.6I	2.3-0.5*%i	3.1+5.6*%i	2.3-0.5i	3.1+5.6i
dsjgdsgkj		dsjgdsgkj		dsjgdsgkj	

Fig. 6.14 Data files with complex values formatted according to the requirements of the three softwares: Mathematica®, MAXIMA, and R, respectively

```
ReadFileC(sconcat(path,"ex1_comp_Max.dat"));
There are 4 samples, with sizes: [3,4,4,1] and 2 variables
```

$$[[\begin{bmatrix}0.5\%i+1.2 & 2.6-3.1\%i\\ 5.3-0.5\%i & 5.6\%i+1.1\\ 3.5\%i+3.5 & 1.6-3.1\%i\end{bmatrix},\begin{bmatrix}3.5\%i+2.5 & 4.6-3.1\%i\\ 4.3-0.5\%i & 5.6\%i+2.1\\ 3.5\%i+3.5 & 4.6-3.1\%i\\ 4.3-0.5\%i & 5.6\%i+2.1\end{bmatrix},\begin{bmatrix}3.5\%i+1.5 & 3.6-3.1\%i\\ 2.3-0.5\%i & 5.6\%i+1.1\\ 3.5\%i+1.5 & 2.6-3.1\%i\\ 5.3-0.5\%i & 5.6\%i+3.1\end{bmatrix},$$
$$\begin{bmatrix}2.3-0.5\%i & 5.6\%i+3.1\end{bmatrix}],[3,4,4,1]]$$

Fig. 6.15 Reading file `ex1_comp_Max.dat` in Fig. 6.14 with `ReadFileC` in MAXIMA

```
 Names of variables
X1 X2 X3 X4 X5
 1st sample
23.4  56.7   21.2   36.5  24.3
34.5  23.4  28.9   25.4   5.46
21.2  33.3  56.7    25.6  7.23
 2nd sample
 X1      X2      X3      X4
56.8    56.4    25.6    43.3

23.4 56.8  23.4   52.2
33.5  45.8   23.6    42.2
33.7  38.9   56.8    53.2
 4th sample
43.2   45.6   34.5
56.4   24.7   78.3
74.5   83.5   84.3
24.5   25.7   37.4
35.7   74.5   84.6
fdfdfdfd
dgdg
25  26  27  28
hjhjhjhjhjhj
```

Fig. 6.16 Data file `ex1_n.dat` displaying samples with different numbers of variables

```
(%i410) ReadFiledifpR(sconcat(path,"ex1_n.dat"),1,1);
     Do you want to select samples ? (1=Yes) 1;
     Please insert a list, inside square brackets,
      with the numbers of the samples you want to keep, separated by commas:  [1,3,4];
     Do you want to select variables ? (1=Yes) 1;
     Please insert a double list, inside square brackets,
      with the numbers of the variables you want to keep for each sample,
       separated by commas:  [[1,2,3],[1,4],[1,2,3]];

     There are 3 samples, with sizes:  [3,3,5] and [3,2,3] variables
     Original samples [1,3,4]
     Original variables  [[1,2,3],[1,4],[1,2,3]]

  Please give a double list with the elements to be kept in each sample: [[1,2],[2,3],[1,3,5]];

     There are now 3 samples, with sizes:  [2,2,3] and [3,2,3] variables

                                                [43.2 45.6 34.5]
          [23.4 56.7 21.2]  [33.5 42.2]         [              ]
(%o410) [[[              ], [         ],        [74.5 83.5 84.3]], [2,2,3]]
          [34.5 23.4 28.9]  [33.7 53.2]         [              ]
                                                [35.7 74.5 84.6]
```

Fig. 6.17 Reading the data file `ex1_n.dat` in Fig. 6.16 with `ReadFiledifpR` in MAXIMA

If one of these modules is used on a file with a single table structure, such as that of the files `ex_2.dat` or `ex_3.dat` in Fig. 6.7, without any optional arguments, it produces a table corresponding to a single sample with variables corresponding to all the columns in the file and observations corresponding to all rows in the file. A first and a second optional arguments may be used, with a value of 1, to allow selection of columns and rows from the original data file, as illustrated in Fig. 6.18.

```
ReadFile1sR(sconcat(path,"ex_3.dat"),1,1);
Do you want to select variables ? (1=Yes) 1;
Please insert a list, inside square brackets,
 with the numbers of the variables you want to keep, separated by commas: [2,3,4];

There is 1 sample with size 12 , and 3 variables
Original variables [2,3,4]

Please give a list, inside square brackets, with the observations
 to be kept in the sample: [1,3,5,7,9];
 ⎡49.7 14.2 29.5⎤
 ⎢              ⎥
 ⎢26.3 49.7 18.6⎥
 ⎢              ⎥
[⎢56.8 23.4 52.2⎥,5]
 ⎢              ⎥
 ⎢38.9 56.8 53.2⎥
 ⎢              ⎥
 ⎣28.7 82.3 38.3⎦
```

Fig. 6.18 Reading the data file `ex_3.dat` in Fig. 6.7 with `ReadFile1sR` in MAXIMA

```
ReadFile1sR(sconcat(path,"ex1_n.dat"),1,1);
 The file you are reading seems to have more than one sample;
Do you want to (1) join all them into a single sample, or
               (2) select a few samples or just one of them ? 2;
Insert the number of the sample you want to keep, or a list,
 inside square brackets, with the numbers of the samples you want to keep,
   separated by commas: [2,3,5];
Please insert a list, inside square brackets,
 with the numbers of the variables you want to keep, separated by commas: [1,2,3];

There is 1 sample with size [5] , and 3 variables

Please give a list, inside square brackets, with the observations
 to be kept in the sample: [1,3,5];
 ⎡56.8 56.4 25.6⎤
[⎢33.5 45.8 23.6⎥,3]
 ⎣ 25   26   27 ⎦
```

Fig. 6.19 Reading the data file `ex1_n.dat` in Fig. 6.16 with `ReadFile1sR` in MAXIMA

If a file with a clear multi-sample structure as file `ex_1.dat` in Fig. 6.7 is used, then a message is issued warning that a multi-sample file is being used and a first question is placed to the user asking if he/she wants to select just one or a few samples to be used as the new sample, and if the user choice is this second one, a second question is placed, asking for the order number or numbers of the sample or samples to be selected. Even files with different number of variables per sample, as file `ex1_n.dat` in Fig. 6.16 may be used, as long as the samples we are joining together have the same number of variables, as is illustrated in Fig. 6.19.

6.4 Computational End-User Functions and Modules

The Mathematica® package and the packages programmed in MAXIMA and R are programmed using two different approaches, but which indeed closely reflect the approach followed in Chap. 5 for all the tests addressed there.

All three packages are programmed taking as a basis the replacement of the Meijer *G* functions with the corresponding EGIG formulation, but while the Mathematica® package is programmed based on the formulation that comes out of the theorems and corollaries in Chaps. 3 and 4, the packages programmed in MAXIMA and R were programmed directly based on the GIG/EGIG p.d.f. and c.d.f. representations for the distribution of each test statistic given in each subsection in Chap. 5. This gives indeed a further means of checking the results stated in those subsections since a given module will always give the same result no matter which package is used, possibly only with a difference in the very last decimal places, when extended precision is called for, usually meaning that a very small number of the last decimal places obtained with either MAXIMA or R are spurious.

6.4.1 Modules to Compute the GIG and EGIG p.d.f. and c.d.f.

Although not really directly necessary to the end-user who is only interested in computing p-values or quantiles for the tests addressed in Chap. 5, modules to compute the p.d.f. and c.d.f. of the GIG and EGIG distributions are also provided. These modules are respectively called `GIGpdf`, `GIGcdf`, `EGIGpdf`, and `EGIGcdf`, and are actually the basis for all computations of quantiles and p-values in the three packages.

Each of these modules has three mandatory arguments which are: (1) the set of integer shape parameters, (2) the set of rate parameters, and (3) the running value for the random variable, where the p.d.f. or the c.d.f. are to be computed. They also have a fourth argument which is optional and to which an integer value may be given, setting the number of precision digits with which the result is to be displayed. This argument has a default value of 50.

In Fig. 6.20 we may examine the way we may use the function or module `GIGcdf` to obtain values for the c.d.f. of the GIG distribution. We may see that the functions or modules in all three softwares give the same values, at least up to the last three or four decimal places. While all digits given by Mathematica® are correct, the last 3 to 4 digits given by MAXIMA and R are usually spurious, but it also happens that for the example addressed in Fig. 6.20 Mathematica® gives 4 digits less than the number of digits we ask for, but assures that the digits outputted are indeed correct.

Since Mathematica® and MAXIMA are able to handle exact rationals, we might think that it would be better to give the values for the last two mandatory arguments of these modules in the rationalized form, but this is not necessary since these

Mathematica®

```
GIGcdf[{2, 5, 7}, {2.3, 5.6, 3.4}, 3.5]
GIGcdf[{2, 5, 7}, {2.3, 5.6, 3.4}, 3.5, 50]
GIGcdf[{2, 5, 7}, {2.3, 5.6, 3.4}, 3.5, 55]

0.41675408382364462925824464785707908352179l0902

0.416754083823644629258244647857079083521791090 2

0.41675408382364462925824464785707908352179109022218 3
```

MAXIMA

```
GIGcdf([2,5,7],[2.3,5.6,3.4],3.5);
GIGcdf([2,5,7],[2.3,5.6,3.4],3.5,50);
GIGcdf([2,5,7],[2.3,5.6,3.4],3.5,55);
4.16754083823644629258244647857079083521791090223 71b-1
4.16754083823644629258244647857079083521791090223 71b-1
4.16754083823644629258244647857079083521791090222182 7712b-1
```

R

```
> GIGcdf(c(2,5,7),c(2.3,5.6,3.4),3.5)
1 'mpfr' number of precision  166   bits
[1] 0.416754083823644629258244647857079083521791090226432
> GIGcdf(c(2,5,7),c(2.3,5.6,3.4),3.5,50)
1 'mpfr' number of precision  166   bits
[1] 0.416754083823644629258244647857079083521791090226432
> GIGcdf(c(2,5,7),c(2.3,5.6,3.4),3.5,55)
1 'mpfr' number of precision  182   bits
[1] 0.41675408382364462925824464785707908352179109022218272662
```

Fig. 6.20 Computing values for the GIG c.d.f. with the function or module `GIGcdf` in Mathematica®, MAXIMA, and R

modules are programmed in such a way that we will obtain the same output values either by giving the last two mandatory arguments as rationals or as real floating point values. We may see from the results in Fig. 6.21 that giving all arguments rationalized gives exactly the same results as those in Fig. 6.20 in all three softwares, with the only exception of the last three decimal places in MAXIMA, which anyway are spurious.

There is however some care to be taken when providing arguments to these modules with a number of digits that exceeds 17 in MAXIMA or 15 in R. In order to be handled with the adequate precision, these arguments have to be given in MAXIMA immediately followed by `b0` and in R wrapped in double or single quotes. We may take a look at Fig. 6.22 to see that while the first commands in MAXIMA and R do not give the correct result, which is the one obtained with the Mathematica®

Mathematica®

```
GIGcdf[{2, 5, 7}, {23/10, 56/10, 34/10}, 35/10]

0.41675408382364462925824464785707908352179109 02
```

MAXIMA

```
GIGcdf([2,5,7],[23/10,56/10,34/10],35/10);
4.167540838236446292582446478570790835217910902 2448b-1
```

R

```
> GIGcdf(c(2,5,7),c(23/10,56/10,34/10),35/10)
1 'mpfr' number of precision  166   bits
[1] 0.41675408382364462925824464785707908352179109 0226432
```

Fig. 6.21 Computing values for the GIG c.d.f. with `GIGcdf` in Mathematica®, MAXIMA, and R using rational arguments (results to be compared with the ones in Fig. 6.20)

Mathematica®

```
GIGcdf[{2, 5, 7}, {23/10, 56/10, 34/10}, 1.23456789123456789]

0.00036457962850943810330935649344190482446248
```

MAXIMA

```
GIGcdf([2,5,7],[23/10,56/10,34/10],1.23456789123456789);
3.6457962850943784803777534924623378398300022705588b-4

GIGcdf([2,5,7],[23/10,56/10,34/10],1.23456789123456789b0);
3.6457962850943810330935649344190482446248244896525b-4
```

R

```
> GIGcdf(c(2,5,7),c(23/10,56/10,34/10),1.23456789123456789)
1 'mpfr' number of precision  166   bits
[1] 0.00036457962850944408800975887407289505917254370153 2781
> GIGcdf(c(2,5,7),c(23/10,56/10,34/10),"1.23456789123456789")
1 'mpfr' number of precision  166   bits
[1] 0.00036457962850943810330935649344190482446248329768 8319
```

Fig. 6.22 Computing values for the GIG c.d.f. with `GIGcdf` in Mathematica®, MAXIMA, and R using arguments with a large number of digits

version, the second commands in each of these two softwares give the same result as the command in Mathematica®.

However, if for some reason one wants to use one of these modules with the third argument given as the result of some function, instead of just a real value, some extra care has to be taken, in order to obtain the desired precision. In order to illustrate this, let us suppose that we would like to use the function or module `EGIGcdf` to obtain the values of the GIG c.d.f. in Figs. 6.20 and 6.21. In Fig. 6.23 it is illustrated how this has to be accomplished. We may note how the first command in each one of the three softwares does not yield the desired precision, since the values obtained do not correctly match the ones in Figs. 6.20 and 6.21 which have the correct precision. But, the second command in each of the three softwares is able to yield the correct value. We should note how in Mathematica® we have to use the argument of the exponential function in a rationalized form. One alternative would be to use the function `Rat`, with `Rat[3.5]` as the argument of the exponential

Mathematica®

```
1 - EGIGcdf[{2, 5, 7}, {2.3, 5.6, 3.4}, Exp[-3.5]]

0.4167540838236446388018130376674092812138959163

1 - EGIGcdf[{2, 5, 7}, {2.3, 5.6, 3.4}, Exp[-35 / 10]]

0.4167540838236446292582446478570790835217910902
```

MAXIMA

```
roundn(1-EGIGcdf([2,5,7],[2.3,5.6,3.4],exp(-3.5)),50);
4.16754083823644538097353595622385475629120019769 81b-1

roundn(1-EGIGcdf([2,5,7],[2.3,5.6,3.4],exp(-3.5b0)),50);
4.1675408382364462925824464785707908352179109022371b-1
```

R

```
> 1-EGIGcdf(c(2,5,7),c(2.3,5.6,3.4),exp(-3.5))
1 'mpfr' number of precision  166   bits
[1] 0.416754083823644638801813037667409281213895916276285
> 1-EGIGcdf(c(2,5,7),c(2.3,5.6,3.4),exp(mpfr("-3.5",50*log(10,2))))
1 'mpfr' number of precision  166   bits
[1] 0.416754083823644629258244647857079083521791090226432
> zr<-mpfr("-3.5",50*log(10,2))
> 1-EGIGcdf(c(2,5,7),c(2.3,5.6,3.4),exp(zr))
1 'mpfr' number of precision  166   bits
[1] 0.416754083823644629258244647857079083521791090226432
```

Fig. 6.23 Using the function or module `EGIGcdf` to compute values of the c.d.f. of the EGIG distribution and to obtain values of the c.d.f. of the GIG distribution

function, which would indeed be the preferable way to handle the problem. In MAXIMA we have to give the argument of the exponential function followed by `b0` in order to make it a `bigfloat` or extended precision value, and in R we have to make the argument of the exponential also an extended precision value, now by using appropriately the `mpfr` function, as referred in Sect. 6.1, or alternatively by using the set of two commands that follow the second R command. In each of the three softwares actually all comes to the fact that we need to give the exponential function a value which in Mathematica® is a rational or in the other two softwares is an extended precision value, with the required precision, since otherwise the value that comes out of the computation of the exponential function will be just a double precision floating-point value, which does not have the necessary precision for further calculations.

We may also notice the use of the function `roundn` in MAXIMA. This function is programmed to take as first argument the value to be rounded, preferably as an extended precision value, and as second argument the number of decimal places to which we want the first argument to be rounded to, giving as a result a "bigfloat" value. Its use is necessary since in MAXIMA, due to implementation details too long to be explained here, if the call to any of the modules `GIGpdf`, `GIGcdf`, `EGIGpdf`, or `EGIGcdf` is included in any further computation we will get a result with at least the triple of the digits specified by their optional argument or by its default value of 50 is assured to be correct. Also, we may notice that in R, when setting the precision of the `mpfr` extended precision value, the second argument used is the required precision of the value in bits. As such, we set this argument to 50 times $\log_2(10)$ in order to obtain 50 digits of precision. Of course it would work well if we had used any larger value, which means that we could have just used 50 as the second argument of `mpfr`.

6.4.2 Modules to Assist the Implementation of the Tests in Chap. 5

As already mentioned at the beginning of Chap. 5, for each test addressed in that chapter, five end-user functions or modules were built: `PDF<name>`, `CDF<name>`, `Pval<name>`, `Quant<name>`, and `PvalData<name>`, where `<name>` is one of the acronyms which, enclosed in square brackets, immediately follow each subsection title in Chap. 5. To make it simpler, these names are also listed in Table 6.1, along with the corresponding Chap. 5 subsection numbers.

In MAXIMA and R there are also functions named `Pval<name>c`. These functions compute the p-values for computed values of the statistics, without doing any pretty printing of the p-value, that is, the value is outputted just as it comes out of the computations, while the functions `Pval<name>` produce a formatted output of the p-value. Actually in these two softwares functions `Pval<name>c` are the computational modules used by the functions `Pval<name>`.

Table 6.1 Particular arguments of functions or modules for subsections in Chap. 5

Chap. 5 Subsec.	<name>	Arguments	Chap. 5 Subsec.	<name>	Arguments
5.1.1	EqMeanVecR	n, p, q	5.1.22	EqMeanVecCirc	n, p, q
5.1.2	NullMeanVecR	n, p, q	5.1.23	NullMeanVecCirc	n, p, q
5.1.3	ProfParR	n, p, q	5.1.24	EqMeanVecCS	n, p, q
5.1.4	MatEVR	n, p, q, p^*, q^*	5.1.25	NullMeanVecCS	n, p, q
5.1.5	EqMeanVecC	n, p, q	5.1.26	EqMeanVecSph	n, p, q
5.1.6	NullMeanVecC	n, p, q	5.1.27	NullMeanVecSph	n, p, q
5.1.7	ProfParC	n, p, q	5.1.28	ProfParCirc	n, p, q
5.1.8	MatEVC	n, p, q, p^*, q^*	5.1.29	ProfParCS	n, p, q
5.1.9	Ind2R	$n, p1, p2$	5.1.30	ProfParSph	n, p, q
5.1.10	Ind2C	$n, p1, p2$	5.2.1	CircOddp	n, p
5.1.11	IndR	n, pk	5.2.2	CircMeanOddp	n, p
5.1.12	IndC	n, pk	5.2.3	CircS	nk, pk
5.1.13	OutR	n, p, k	5.2.4	CircMeanSOddp	nk, pk
5.1.14	OutC	n, p, k	5.3.1	CircMeanEvenp	n, p
5.1.15	SymEqR	n, p	5.3.2	CircMeanSEvenp	nk, pk
5.1.16	CompSymEqR	n, p	5.3.3	CircMeans	nk, pk
5.1.17	SymSphEq	n	5.3.4	IndCircMeans	n, pk
5.1.18	CompSymSphEq	n	5.3.5	IndCircMeans1Odd	n, pk
5.1.19	SymEqC	n, p	5.3.6	IndCircSph	n, p
5.1.20	CompSymEqC	n, p	5.3.7	IndCircSphEqMean	n, p
5.1.21	BSSph	n, m			

Functions `PDF<name>c`, `CDF<name>c`, `Pval<name>c`, `Pval<name>`, and `Quant<name>` use as first mandatory arguments the arguments listed in Table 6.1, in the order they are listed, according to the test they refer to, and as such, according to the `<name>` they bear. In Table 6.1 these arguments are listed along with the subsections in Chap. 5 and the `<name>` they refer to. Besides these arguments these modules also have a further mandatory argument which for modules `PDF<name>c`, `CDF<name>c`, `Pval<name>c`, and `Pval<name>` is the computed value of the statistic and for module `Quant<name>` is the α-value for which the quantile is to be computed. In Table 6.2 is explained the meaning of the arguments in Table 6.1.

In Mathematica®, modules `Pval<name>` have a fourth and a fifth arguments, which are optional. These are the number of digits used to print the p-value and the number of precision digits used in the computation of the p-value. These optional arguments have default values of 10 and 50, respectively. If the second of these arguments is not given, or in case it is given its default value of 50, then if the module finds 50 to be too much of a small number of precision digits for the computations, the module will look for what it finds to be an adequate number of precision digits for the computation. In case this argument is given a value different from 50, the module will use this as the number of precision digits for the computations.

Table 6.2 Meaning of the arguments in Table 6.1

Argument	Meaning
n	Overall sample size
p	Number of variables
q	Number of samples
p^*	Number of variables after transformation
q^*	Number of samples after transformation
$p1$	Number of variables of first subset
$p2$	Number of variables of second subset
pk	List or vector with the numbers of variables of each subset
k	Number of outliers to be tested for
m	Number of subsets of variables
nk	List or vector with the sample sizes

Fig. 6.24 Examples of Mathematica® commands that use the module `PvalEqMeanVecR` to compute a p-value for the test in Sect. 5.1.1

```
PvalEqMeanVecR[50, 14, 7, .06056]
PvalEqMeanVecR[50, 14, 7, .06056, 50, 20]
PvalEqMeanVecR[50, 14, 7, .06056, , 30]
0.05000717984667696176l

0.05000717984667696176l

0.0500071798466769617609773438937
```

In Fig. 6.24 we may see some examples of commands that use the Mathematica® module `PvalEqMeanVecR` to compute the p-value for a computed value of 0.06056 for the l.r.t. statistic for the test of equality of mean vectors treated in Sect. 5.1.1 of Chap. 5, in a case where the overall sample size is equal to 50, the number of variables 14, and the number of populations involved is 7.

The first of these commands uses the default values for the numbers of precision digits for the computation and for the number of digits used to print the p-value. The second command does exactly the same, although now by stating these values explicitly, while the third command asks for a printed value with 30 digits.

In MAXIMA and R the functions `Pval<name>` have just one optional argument, with default value of 20, which is the number of decimal places to be used in printing the p-value, with the number of digits used in the computations being at least double of this value, in such a way that it is assured for most common cases that this precision is enough for the required number of digits in the output. In Fig. 6.25 there is a set of commands that in MAXIMA use the function `PvalEqMeanVecR` to compute a p-value for the test in Sect. 5.1.1 for the same values of n, p, and q that are used in Fig. 6.24. Output values in Fig. 6.25 are to be compared with those in Fig. 6.24. We may see that the values exactly match, except for the fact that in MAXIMA and R the optional argument gives the number of decimal places used to

```
PvalEqMeanVecR(50,14,7,.06056)$
PvalEqMeanVecR(50,14,7,.06056,20)$
PvalEqMeanVecR(50,14,7,.06056,30)$
  0.0500071798466769617 6
  0.0500071798466769617 6
  0.05000717984667696176097 7343894
```

Fig. 6.25 Examples of MAXIMA commands that use module `PvalEqMeanVecR` to compute a p-value for the test in Sect. 5.1.1

```
PvalEqMeanVecR(50,14,7,.06056);
  0.0500071798466769617 6
5.0007179846676961760977342742623008865589420765291b-2

pv:PvalEqMeanVecR(50,14,7,.06056)$
  0.0500071798466769617 6

pv;
5.0007179846676961760977342742623008865589420765291b-2
```

Fig. 6.26 Examples of alternative MAXIMA commands that use the module `PvalEqMeanVecR` to compute a p-value for the test in Sect. 5.1.1 and store the computed non-formatted value for later use

```
> PvalEqMeanVecR(50,14,7,.06056)
0.0500071798466769617 61
> PvalEqMeanVecR(50,14,7,.06056,20)
0.0500071798466769617 61
> PvalEqMeanVecR(50,14,7,.06056,30)
0.05000717984667696176097 73438937
```

Fig. 6.27 Examples of R commands that use the module `PvalEqMeanVecR` to compute a p-value for the test in Sect. 5.1.1

print the p-value, while in Mathematica® it gives, more precisely, the number of digits used in printing the value.

In MAXIMA all functions `Pval<name>`, besides printing the p-value with an easily readable format, also output the computed value itself, so that following each command in Fig. 6.25 with a dollar sign prevents this computed value from being outputted. We may see in Fig. 6.26 how not placing a dollar sign right after the command that uses the MAXIMA function `PvalEqMeanVecR` will give as output the formatted value of the p-value together with the computed value, this one printed as a `bigfloat` or extended precision value, and how one can store this non-formatted value in a variable for later use.

In R everything works in a similar way, except that in this software functions `Pval<name>` use an invisible output attribute, apparently only printing the p-value in a similar manner as the MAXIMA function does, but in reality also yielding an invisible computed non-formatted value which, as it happens in MAXIMA, may be stored in a variable for later use. In Fig. 6.27 we can see that the p-values obtained with the R function exactly match the values obtained with Mathematica® and also those obtained with MAXIMA.

In Fig. 6.28 we can see the existence and use of the invisible outputted non-formatted value of the `Pval<name>` R functions and how it can be stored in a variable named `pv` for later use. Only about half of the digits of this non-formatted

```
> pv<-PvalEqMeanVecR(50,14,7,.06056)
0.0500071798466769 61761
> pv
1 'mpfr' number of precision  265   bits
[1] 0.05000717984667696176097734389367430073235 87
```

Fig. 6.28 R commands that use the module `PvalEqMeanVecR` to compute a p-value for the test in Sect. 5.1.1 and store the computed non-formatted value for later use

```
pv = PvalEqMeanVecR[50, 14, 7, .06056];
pv

0.0500071798466769 61761
```

Fig. 6.29 Mathematica® commands that use the module `PvalEqMeanVecR` to compute a p-value for the test in Sect. 5.1.1 and store the computed non-formatted value for later use

Mathematica®

```
Timing[PvalEqMeanVecR[50, 14, 7, .06056]]

{0.015600, 0.0500071798466769 61761}
```

Maxima

```
PvalEqMeanVecR(50,14,7,.06056)$
  0.050007179846676961 76
Evaluation took 0.0470 seconds (0.0460 elapsed)
```

R

```
> system.time(PvalEqMeanVecR(50,14,7,.06056))
0.0500071798466769 61761
   user  system elapsed
   2.01    0.00    2.03
```

Fig. 6.30 Computing times for module `PvalEqMeanVecR` to compute the p-value for $n = 50$, $p = 14$, and $q = 7$ for the test in Sect. 5.1.1

value are visible in Fig. 6.28, since this value is an extended precision value with $265 \log_{10}(2) \approx 80$ digits.

Of course in Mathematica® one can also store the output value in a variable for later use, as illustrated in Fig. 6.29, but this value will only have the number of digits defined by the value of the first optional argument of module `Pval<name>` or its default value.

Among all three softwares, Mathematica® is the one that delivers smaller computing times (Fig. 6.30).

Modules `Quant<name>` have a set of 3 or 4 mandatory arguments which are, in this order, the arguments listed in Table 6.1, followed by the α-value for the quantile, which has, of course, to be a value between 0 and 1. Then, they have a set of 4 optional arguments, to be given in this order:

(1) the *epsilon* value, an upper bound for the difference between two consecutive iteration values for the quantile, with a default value of 10^{-11} (this value is never allowed to be larger than 10^{-3});
(2) the number of precision digits used in the computations, with a default value that equals twice the nearest integer to the negative base 10 logarithm of *epsilon* in (1), with a minimum of 50,
(3) an initial value, with a default value of 1/2,
(4) the number of decimal places with which the outputted value is to be printed with, which is never smaller than 3 and which defaults to the integer part of the negative base 10 logarithm of *epsilon* in (1); in Mathematica® and R this is, more precisely, the number of significant digits, rather than the number of decimal places.

These optional arguments have to be given in this order, which was so chosen since it seemed to be the order corresponding to their expected frequency of use, that is, with the optional arguments listed first being those that are thought to be potentially the most used ones.

In case one wants to use an optional argument which is listed above after other optional arguments, then all previous optional arguments have to be either given or signaled by leaving in Mathematica® and R an empty slot, or in MAXIMA by giving it the value [ ], but with no need to list any optional arguments that in the list above appear in a subsequent position. That is, if for example in Mathematica® one wants to compute the 0.05 quantile for the test of equality of mean vectors in Sect. 5.1.1, for a case where the overall sample size is 50, and there are 14 variables and 7 samples, using all default values for the optional arguments, then one only needs to type

```
QuantEqMeanVecR[50,14,7,0.05]
```
(6.1)

while if one wants to compute the same quantile but using an *epsilon* of 10^{-20}, with all other optional arguments with their default values, one should type

```
QuantEqMeanVecR[50,14,7,0.05,10^-20]
```
(6.2)

with no need to indicate or signal any other optional arguments.

However, if one would like to compute the same quantile with the default values for *epsilon* and for the number of precision digits for the computations but using a starting value of 0.06, one would have to use a command like

```
QuantEqMeanVecR[50,14,7,0.05, , ,0.06]
```

in Mathematica®, or

```
QuantEqMeanVecR(50,14,7,0.05, , ,0.06)
```

in R, or

```
QuantEqMeanVecR(50,14,7,0.05,[],[],0.06)
```

in MAXIMA.

Although the default values used for these four optional arguments seem to be adequate for the general use we want to make of the Quant<name> modules, there may be several different reasons that may lead us to use for these arguments values which are different from their default values.

While the use of smaller values of *epsilon* will lead to more precise values for the quantiles, larger values will speed up the determination of the approximate value of the quantile, especially in cases where the sample size is very large, or the number of variables or samples involved is large. The number of precision digits used in the computations may need to be changed only in cases where the computations are so complicated that more precision is needed, which would be cases with very large sample sizes and with large numbers of variables and samples involved, or otherwise cases where the overall sample size hardly exceeds the sum of the number of variables and of samples, and these two numbers are quite large. In any case this will be necessary only in very demanding situations, since the modules or functions Quant<name> internally adjust the default precision as a function of more demanding circumstances. This is the reason why in MAXIMA the functions Quant<name> print, together with the value of the quantile, also the value used for the system variable `fpprec`. In case more precision is necessary, then the second optional argument should be made larger than the `fpprec` value that was outputted. In R if the invisible output value is inspected, after being stored in some variable, it is possible to see how many bits of precision were used in its computation. Then a value larger than this value times $\log_2(10)$ should be used as the number of digits required for the computations in case more precision is required. For the initial value, a value different from the default value may be given in order to help speed up the computation process and in cases where, given the very large values of the overall sample size, and of the numbers of variables and samples, some precision issues may arise. In any case, not only does this require some knowledge of what will be a good starting value, in order to not make the computations even harder, but also, as we will see with the examples given shortly, there are usually no precision issues even for rather large values of the sample sizes, number of variables and number of samples involved. Finally, the number of decimal places used to print the quantiles, which in Mathematica® will be, more precisely, the number of significant digits printed, will by default match the negative logarithm base 10 of *epsilon*, which is the minimum number of decimal places that are exactly determined. For example for an *epsilon* value of 10^{-16} we will have at least 16 decimal places of the quantile exactly determined. This value may be changed, usually to a slightly larger value, since indeed the number of correct decimal places in the value obtained for the quantile commonly exceeds the value that appears as the negative exponent of the base 10 of the scientific notation of *epsilon*.

Mathematica®

```
QuantEqMeanVecR[50, 14, 7, .05]
QuantEqMeanVecR[50, 14, 7, .05, 10^-16]
QuantEqMeanVecR[50, 14, 7, .05, 10^-30]

0.060558301728

0.06055830172764450

0.06055830172764449917887461487 79
```

MAXIMA

```
QuantEqMeanVecR(50,14,7,.05)$
QuantEqMeanVecR(50,14,7,.05,10^-16)$
QuantEqMeanVecR(50,14,7,.05,10^-30)$
 0.0605583017
 0.0605583017276445
 0.060558301727644499178874614878
```

R

```
> QuantEqMeanVecR(50,14,7,.05)
0.060558301728
> QuantEqMeanVecR(50,14,7,.05,10^-16)
0.06055830172764450
> QuantEqMeanVecR(50,14,7,.05,10^-30)
0.0605583017276444991788746148779
```

Fig. 6.31 Quantiles obtained with Mathematica®, MAXIMA, and R for the test of equality of mean vectors in Sect. 5.1.1 using module or function QuantEqMeanVecR, with different *epsilon* values

From the examples in Fig. 6.31 we may see how the three softwares give once again exactly the same results and how these results match what we would expect, since from the results of the computation of the p-values done with the function or module PvalEqMeanVecR, the value of 0.06056 had already given a p-value quite close to 0.05. The only difference being that with MAXIMA we obtain one decimal place less than with Mathematica® or R, since with this software the negative base ten logarithm of *epsilon* gives the number of decimal places with which the result is printed by default, instead of the number of significant digits after the decimal point, which is what happens with Mathematica® and R.

The modules Quant<name> programmed in Mathematica® use a first step with an improved bisection type algorithm and then a final step with a Newton-Raphson type algorithm to obtain the quantiles, while the functions Quant<name> programmed in MAXIMA and R use an intermediate step with an improved secant type method. Due to this fact, although the values obtained for the quantiles, when using the default value for the fourth optional argument, will always match, when we use for this argument a value larger than its default value, or when we look at the invisible outputted value in R, which, as it happened with the Pval<name> functions, may be looked at after printing the value of the variable where these may be stored, or when looking at the outputted value in MAXIMA when the Quant<name> commands are not followed by a dollar sign, the last digits will most often not match. This fact is due to slightly different implementations of these functions and also to the different implementations of the extended precision numbers in these softwares. Anyway, the required precision directly related to the value of *epsilon* used is always assured. Indeed even when using the default value for *epsilon*, most often the value obtained for the quantile has many more exact decimal places than the ones that are printed by default. This may be seen for example by

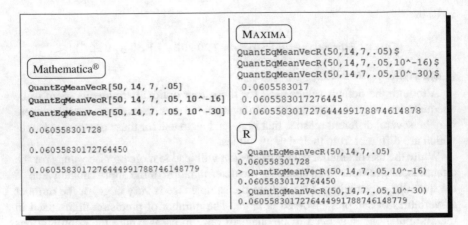

Fig. 6.32 Quantiles obtained with Mathematica® for the test of equality of mean vectors in Sect. 5.1.1 using module `QuantEqMeanVecR`, with different *epsilon* values and different values, larger than the default value, for the fourth optional argument

```
QuantEqMeanVecR[50, 14, 7, .05, 10^-16]
QuantEqMeanVecR[50, 14, 7, .05, , , , 16]
QuantEqMeanVecR[50, 14, 7, .05, 10^-7, , , 16]

0.06055830172764450
0.06055830172764450
0.06055830172764485

QuantEqMeanVecR[50, 14, 7, .05, 10^-25]
QuantEqMeanVecR[50, 14, 7, .05, 10^-40]
QuantEqMeanVecR[50, 14, 7, .05, 10^-50]

0.0605558301727644499178887461
0.060555830172764449917887461487790066190407
0.0605558301727644499178874614877900661904071582605644

QuantEqMeanVecR[50, 14, 7, .05, 10^-16, , , 25]
QuantEqMeanVecR[50, 14, 7, .05, 10^-16, , , 40]
QuantEqMeanVecR[50, 14, 7, .05, 10^-16, 70, , 40]
QuantEqMeanVecR[50, 14, 7, .05, 10^-16, 200, , 50]

0.0605558301727644499178887461
0.060555830172764449917887461487790066190407
0.060555830172764449917887461487790066190407
0.0605558301727644499178874614877900661904071582605644
```

Fig. 6.33 Quantiles obtained with MAXIMA for the test of equality of mean vectors in Sect. 5.1.1 using module `QuantEqMeanVecR`, with different *epsilon* values and different values, larger than the default value, for the fourth optional argument

```
QuantEqMeanVecR(50,14,7,.05,10^-16)$
QuantEqMeanVecR(50,14,7,.05,[],[],[],16)$
QuantEqMeanVecR(50,14,7,.05,10^-7,[],[],16)$
 0.0605583017276445 (fpprec: 50)
 0.0605583017299454 (fpprec: 50)
 0.0605583017276445 (fpprec: 50)

QuantEqMeanVecR(50,14,7,.05,10^-25)$
QuantEqMeanVecR(50,14,7,.05,10^-40)$
QuantEqMeanVecR(50,14,7,.05,10^-50)$
 0.060558301727644491788746 (fpprec: 50)
 0.06055830172764449917887461487790066190 41 (fpprec: 80)
 0.0605583017276444991788746148779006619040715826056 4
                                                   (fpprec: 100)
QuantEqMeanVecR(50,14,7,.05,10^-16,[],[],25)$
QuantEqMeanVecR(50,14,7,.05,10^-16,[],[],40)$
QuantEqMeanVecR(50,14,7,.05,10^-16,200,[],50)$
 0.060558301727644491788746 (fpprec: 50)
 0.0605583017276444991788746148779006619041 (fpprec: 80)
 0.06055830172764449917887461487790066190407215993032
                                                   (fpprec: 200)
```

using a set of commands as the ones in Figs. 6.32, 6.33, and 6.34 where the main aim in using a value for the fourth optional argument larger than its default value is to illustrate the precision with which the numerical approximations for the exact quantiles are computed.

For each of the three softwares, the first set of commands illustrates that even using the default value of 10^{-6} for the *epsilon* parameter, which, as already remarked, is an upper bound for the difference between two successive iterated values for the quantile, we may obtain in Mathematica® the correct value for the first 16 decimal places of the quantile, while in MAXIMA and R we will need to use a value of 10^{-7} in order to obtain this value. The second set of commands obtains and displays the quantiles for *epsilon* values of 10^{-25}, 10^{-40}, and 10^{-50}. We confirm once again that all three softwares give exactly the same results, apart from

```
> QuantEqMeanVecR(50,14,7,.05,10^-16)
0.06055830172764450
> QuantEqMeanVecR(50,14,7,.05,,,,16)
0.06055830173024717
> QuantEqMeanVecR(50,14,7,.05,10^-7,,,16)
0.06055830172764450
>
> QuantEqMeanVecR(50,14,7,.05,10^-25)
0.0605583017276444991788746l
> QuantEqMeanVecR(50,14,7,.05,10^-40)
0.06055830172764449917887461487790066190407
> QuantEqMeanVecR(50,14,7,.05,10^-50)
0.060558301727644499178874614877900661904071582605644
>
> QuantEqMeanVecR(50,14,7,.05,10^-16,,,25)
0.0605583017276444991788746l
> QuantEqMeanVecR(50,14,7,.05,10^-16,,,40)
0.06055830172764449917887461487790066190407
> QuantEqMeanVecR(50,14,7,.05,10^-16,200,,50)
0.060558301727644499178874614877900661904072525514963
```

Fig. 6.34 Quantiles obtained with R for the test of equality of mean vectors in Sect. 5.1.1 using module `QuantEqMeanVecR`, with different *epsilon* values and different values, larger than the default value, for the fourth optional argument

the fact that Mathematica® and R give one digit more than MAXIMA. The third set of commands then shows that using an *epsilon* value of 10^{-16} actually gives us 40 correct significant digits in the quantile, although in Mathematica® we have to use a value of 70 for the second optional parameter, which sets the number of precision digits to be used in the computations, in order to obtain 40 correct significant digits. But the last command in this set of commands shows that indeed using an *epsilon* value of 10^{-16} we will not be able to obtain more than 40 correct significant digits in the quantile, no matter how many digits of precision we use in the computations. This has to be accomplished by giving a smaller value for *epsilon*, with an *epsilon* value of 10^{-50} ensuring that this desideratum is accomplished. In Mathematica® even the use of an *epsilon* value of 10^{-100} will give us the quantile value in about 3/4 of a second in an Intel® Core™ Duo P8700 CPU running at 2.53 GHz with the Windows 7 Home Premium operating system.

Although this procedure, of using a larger value than its default value for the fourth optional parameter, which specifies the number of decimal places of the quantile to be printed, may enable us to obtain, say, more correct decimal places than those we asked for, and as such although this may enable us to obtain this number of decimal places in a slightly lesser computation time, this procedure is commonly not recommended, since anyway we would need to compute the quantile with a larger precision to be able to know how many of those decimal places are really correct, so that the use of a non-default value for this fourth argument, larger than its default value has to be done with some care and parsimony.

In Figs. 6.35 and 6.36 we may see how we can check how precise are the quantiles we obtain with the `Quant<name>` modules or functions. In these two figures the "reconstitution" of the p-value is done for the 0.05 quantile obtained for an *epsilon* value of 10^{-50}.

From Figs. 6.35 and 6.36 we may see that while in Mathematica® by storing in a variable the quantile value obtained by using the `QuantEqMeanVecR` module with no further optional arguments we just only get the very same value that we get printed out from the execution of the module, while in MAXIMA and R by doing the

```
qq = QuantEqMeanVecR[50, 14, 7, .05, 10^-50];
qq

0.06055830172764449917887461487790066190407158260564 4

PvalEqMeanVecR[50, 14, 7, qq]
PvalEqMeanVecR[50, 14, 7, qq, , 80]
5/100 - PvalEqMeanVecR[50, 14, 7, qq, , 80]

0.05000000000000000000000

0.05000000000000000000000000000000000000000000000000018702072892747137603568376589 3

-1.870207289274713760356837658 93 × 10^-52

qq2 = QuantEqMeanVecR[50, 14, 7, .05, 10^-50, , , 100];
qq2

0.060558301727644499178874614877900661904071582605643955761322733807235664165026604\
 6326182807290245697 6

PvalEqMeanVecR[50, 14, 7, qq2, , 100]
5/100 - PvalEqMeanVecR[50, 14, 7, qq2, , 100]

0.0499999999999999999999999999999999999999999999999999999999999999999999999999826660\
 55725344066714442445

1.7333944274655933285557555 × 10^-76
```

Fig. 6.35 Quantiles obtained with Mathematica® for the test of equality of mean vectors in Sect. 5.1.1 using module QuantEqMeanVecR, with an *epsilon* value of 10^{-50} and different values, larger than the default value, for the fourth optional argument

same we get a value with many more decimal places than those that were printed out when the QuantEqMeanVecR function is executed.

Differences in computing times among the three softwares are even more evident when using the Quant<name> modules or functions, since these modules or functions usually use several calls to the p.d.f. and c.d.f. computing modules or functions (Fig. 6.37).

In Mathematica®, modules Quant<name> have a further fifth optional argument which if given the value 1 will make the module use a slightly different algorithm which deals with precision issues in a slightly different way. The use of this optional argument with a value of 1 will in most cases not change in any way the results obtained but in most common cases it will usually lead to higher computation times. However, in situations which may be very exigent from the point of view of the number of precision digits necessary for the computations, the use of this optional argument with a value of 1 may indeed lead to smaller computation times. See Fig. 6.38 where the 0.05 quantile for the l.r.t. statistic used in Sect. 5.1.1 for $n = 50$ and $p = 14$, $q = 7$, is computed with an *epsilon* value of 10^{-50}, thus requiring at least 50 exact decimal places in the quantile. As also shown in the same Fig. 6.38, if we would use instead an *epsilon* value of 10^{-200}, thus requiring at least 200 exact decimal places in the quantile, both alternatives will keep giving the same result, but with the one that uses this fifth optional argument with a value of 1 displaying now a clearly shorter computation time.

MAXIMA

```
qq:QuantEqMeanVecR(50,14,7,.05,10^-50);
 0.06055830172764449917887461487790066190407158260564
6.055830172764449917887461487790066190407158260564395576132273380723566416506316767045369307787441529b-2
qq;
6.055830172764449917887461487790066190407158260564395576132273380723566416506316767045369307787441529b-2
PvalEqMeanVecR(50,14,7,qq)$
PvalEqMeanVecR(50,14,7,qq,80)$
5/100-PvalEqMeanVecRc(50,14,7,qq,80);
  0.0500000000
  0.04999999999999999999999999999999999999999999999999999999999999999999999999998123
1.8767777883370616050482623821[1445 digits]006216907147691912918405870-77
```

R

```
> qq<-QuantEqMeanVecR(50,14,7,.05,10^-50)
0.060558301727644499178874614877900661904071582605644
> qq
1 'mpfr' number of precision  1103   bits
[1] 0.06055830172764449917887461487790066190407158260564395576132273380723566416506760708031776224915124329906665770164З
> PvalEqMeanVecR(50,14,7,qq)
0.05000000000
> PvalEqMeanVecR(50,14,7,qq,110)
0.050000000000000000000000000000000000000000000000000000000000000000000000000000000000000000000000000000001428734239
> pv2<-PvalEqMeanVecR(50,14,7,qq,110)
0.050000000000000000000000000000000000000000000000000000000000000000000000000000000000000000000000000000001428734239
> pv2
1 'mpfr' number of precision  1103   bits
[1] 0.050000000000000000000000000000000000000000000000000000000000000000000000000000000000000000000000000000001428734239102
> print(5/100-pv2,digits=20)
1 'mpfr' number of precision  1103   bits
[1] 2.7755575615628913511e-18
> print(mpfr(".05",1103)-pv2,digits=20)
1 'mpfr' number of precision  1103   bits
[1] -1.4287342391028437278e-102
```

Fig. 6.36 Quantiles obtained with Mathematica® for the test of equality of mean vectors in Sect. 5.1.1 using module `QuantEqMeanVecR`, with different *epsilon* values and different values, larger than the default value, for the fourth optional argument

```
Timing[QuantEqMeanVecR[50, 14, 7, .05, 10^-50]]

{0.577204, 0.0605583017276444991788746148779006619040715826056 44}

Timing[QuantEqMeanVecR[50, 14, 7, .05, 10^-50, , , , 1]]

{0.327602, 0.060558301727644499178874614877900661904071582605644}

Timing[QuantEqMeanVecR[50, 14, 7, .05, 10^-200]]

{5.132433,
 0.0605583017276444991788746148779006619040715826056439557613227\
  338072356641650676070803177622491512432987286988386588268 5079\
  027329784412598816179592607774478018193314538352520829083 1088\
  345783167567184843}

Timing[QuantEqMeanVecR[50, 14, 7, .05, 10^-200, , , , 1]]

{0.967206,
 0.0605583017276444991788746148779006619040715826056439557613227\
  3380723566416506760708031776224915124329872869883865882685079\
  0273297844125988161795926077744780181933145383525208290831088\
  345783167567184843}
```

Fig. 6.37 Computation times for quantiles obtained for the test statistic in Sect. 5.1.1 of Chap. 5 obtained with the Mathematica® module `QuantEqMeanVecR`, without and with the fifth optional argument

Modules or functions `PvalData<name>`, as their names indicate, are used to compute p-values for a computed value of the l.r.t. statistic obtained from data values stored in a data file. These modules or functions will use one of the data file reading modules or functions to read the data file, according to the nature of the data required to implement the corresponding test.

See Table 6.3 for a list of the file reading modules or functions used in the `PvalData<name>` module for each subsection in Chap. 5.

The only interaction that these modules have with the user pertains to the data file reading modules that they use. Besides printing the computed value of the l.r.t. statistic and the corresponding p-value, these modules also print some summary information about the size of samples and number of variables in the analysis.

These modules use as only mandatory argument the name of the data file and as optional arguments, in this order: (1) an index which if given the value 1 allows for the selection of variables or observations, (2) an index which if given the value 1 allows for the selection of the elements to be taken in each sample, (3) the number of precision digits to be used in computing the value of the l.r.t. statistic (with a default value of 50), and (4) the number of digits used to print the p-value (with a default value of 10).

a)
```
PvalDataEqMeanVecR[path <> "ex_2.dat", , , , 35]
 There are 3 samples, with sizes {3, 4, 5}
                                        and 4 variables
Computed value of Λ: 0.2099507759
p-value: 0.17881013976425131534279445544155654
```

b)
```
PvalDataEqMeanVecR[path <> "ex_2.dat", , , , , cov → 1, norm → 1]
 There are 3 samples, with sizes {3, 4, 5} and 4 variables
Sample 1 is too small to carry out a normality test
Sample 2 is too small to carry out a normality test
```

		Statistic	P-Value
Sample 3	Anderson-Darling	3.14059	0.977271
	Kolmogorov-Smirnov	3.13401	0.976567
	Kuiper	1.83129	0.389023
	Mardia Combined	15.1707	0.182143
	Mardia Kurtosis	−2.22051	0.182927
	Mardia Skewness	5.12	0.241667
	Pearson χ^2	2.49228	0.79576
	Shapiro-Wilk	0.552182	0.000126531
	Watson U^2	2.45792	0.778769

```
There are 2 samples which are too small
   to carry out a test of equality of covariance matrices
Computed value of Λ: 0.2099507759
p-value: 0.1788101398
```

Fig. 6.38 Examples of use of the Mathematica® module `PvalDataEqMeanVecR` with the file `ex_2.dat` in Fig. 6.7: (**a**) using the default value for the first three optional arguments and a value of 35 for the fourth optional argument; (**b**) using the default values for the optional arguments and requiring the test of equality of covariance matrices and the multivariate normality tests, in a situation where some samples have a size that does not exceed the number of variables involved

In Mathematica®, these modules have a further optional argument called `norm`, which has to be called by its name. This argument can only be used when running version 8 or later of Mathematica® and if given the value 1 (one), this will entail the implementation of the whole set of multivariate normality tests available in Mathematica®, on each of the samples being used. We should note here that for samples of complex r.v.'s the multivariate normality tests are carried out separately for the real and imaginary parts, without checking on the form of the covariance matrix, that is, without checking if the covariance matrix has the structure in (5.516).

Also only in Mathematica®, the `PvalData<name>` modules that refer to the l.r. tests in Sects. 5.1.1, 5.1.3, 5.1.5, 5.1.7, and 5.1.22–5.1.30 have a fifth optional argument, which has also to be called by its name `cov` and which if given the

Table 6.3 Data file reading modules or functions used in the PvalData<name> modules or functions with the indication of the Chap. 5 subsection test where they are used—see Table 6.1 for the corresponding <name> linkage

File reading module or function	Subsection in Chap. 5 of the corresponding test
ReadFileR	5.1.1, 5.1.2, 5.1.3, 5.1.22, 5.1.23, 5.1.24, 5.1.25, 5.1.26, 5.1.27, 5.1.28
	5.1.29, 5.1.30
ReadFileC	5.1.5, 5.1.6, 5.1.7
ReadFile1sR	5.1.4, 5.1.9, 5.1.11, 5.1.13, 5.1.15, 5.1.16, 5.1.17, 5.1.18, 5.1.21, 5.2.1,
	5.2.2, 5.3.1, 5.3.4, 5.3.5 , 5.3.6, 5.3.7
ReadFile1sC	5.1.8, 5.1.10, 5.1.12, 5.1.14, 5.1.19, 5.1.20
ReadFiledifpR	5.2.3, 5.2.4, 5.3.2, 5.3.3

value of 1 (one) will implement the l.r. test for equality of covariance matrices by using a very accurate near-exact approximation for the distribution of the l.r.t. statistic, which gives a very good approximation even for very small sample sizes. The near-exact distribution used with the modules PvalDataEqMeanVecR, PvalDataNullMeanVecR, and PvalDataProfParR is the one detailed in Coelho et al. (2010) that matches four exact moments, while the near-exact distribution used with the modules PvalDataEqMeanVecC, PvalDataNullMean VecC, and PvalDataProfParC is the one detailed in Coelho et al. (2015, Secs. 4,5) that also matches four exact moments. Although this option is also available for the tests in Sects. 5.1.22–5.1.30 in Chap. 5, the test of equality of covariance matrices being carried out does not take into account the structure of the matrices being tested because such a test is not currently available in the literature for covariance matrices with circular, compound symmetric, or spherical structures.

If used together, the relative order in which the two optional arguments norm and cov are given is irrelevant, since they are referred to by their names, although they have to appear after the other two optional arguments, even if these are not used, which means that, if for example, none of the first two non-named optional arguments is used, then norm and/or cov have to appear after two empty slots separated by commas. See Fig. 6.37 for details.

It should be noted that the multivariate normality tests will be carried out only for samples which have a size that is larger than the number of variables involved, while the test of equality of covariance matrices will only be carried out if all samples have a size that is larger than the number of variables involved. In cases in which not all samples have a larger dimension than the number of variables in the analysis and the multivariate normality tests are required a warning is issued for those samples for which the normality tests are not possible to be carried out, and in case the test of equality of covariance matrices is requested in such a situation, also a warning is issued, indicating how many samples have a size smaller than the number of variables (see Fig. 6.37 for details).

It may seem that for the Mathematica® modules it might have been easier for the user to have all four optional arguments called for by their names, instead of the choice that was made. However, the implementation was chosen to be done in this way so that the modules or functions would have similar syntax in all three softwares used.

In order for it to be possible to carry out the test of equality of covariance matrices, all samples have to have a dimension that is larger than the number of variables in the analysis. In Fig. 5.13 in Chap. 5 we have the command used to carry out such a test on the well-known *Iris* data set, together with the multivariate normality tests. By looking at the results obtained one may be surprised by the extremely low p-value obtained for the test of equality of covariance matrices, which shows that this hypothesis should be rejected.

References

Coelho, C.A., Marques, F.J.: Near-exact distributions for the likelihood ratio test statistic to test equality of several variance-covariance matrices in elliptically contoured distributions. Comput. Stat. **27**, 627–659 (2012)

Coelho, C.A., Arnold, B.C., Marques, F.J.: Near-exact distributions for certain likelihood ratio test statistics. J. Stat. Theory Pract. **4**, 711–725 (memorial issue in honor of H. C. Gupta, guest-edited by C. R. Rao) (2010)

Coelho, C.A., Arnold, B.C., Marques, F.J.: The exact and near-exact distributions of the main likelihood ratio test statistics used in the complex multivariate normal setting. TEST **24**, 386–416 (2015) (+14 pp. of supplementary material – online resource)

Marques, F.J., Coelho, C.A., Arnold, B.C.: A general near-exact distribution theory for the most common likelihood ratio test statistics used in multivariate analysis. TEST **20**, 180–203 (2011)

Chapter 7
Approximate Finite Forms for the Cases Not Covered by the Finite Representation Approach

Abstract In this chapter the authors set the guidelines to approach cases not covered by the finite form representations studied in the book, give new Mellin inversion formulas for both the p.d.f. and the c.d.f., and develop sharp upper bounds on the difference between the exact and approximate representations for the Meijer G functions as well as for the differences between the exact and approximate p.d.f.'s and c.d.f.'s of the product of independent Beta r.v.'s.

Keywords Alternative approaches · Near-exact distributions · Other cases · Simulation

In some of the examples exhibited in Chap. 5 there were certain restrictions on the parameters (e.g., p even or q odd) which must be satisfied if the finite representation approach is to be used. It is then reasonable to ask what if these restrictions are not satisfied?

The answer is that if the restrictions are not applicable, although the special function approach is theoretically applicable in all cases, impractical computation times will generally be encountered, except in very simple cases. Unless sample sizes and values of other parameters as the number of variables or the number of populations involved are really small, the computation of exact values of percentiles, p-values, and distribution functions will be very hard and time consuming.

There are two ways out of this dilemma. One of them, which is elementary in nature and computer intensive but which is always available, is to simulate a given number of products of independent Beta r.v.'s which correspond to the null distribution of the statistic and then evaluate the proportion of these simulated values that are less than the value of the observed statistic and take this as a simulated p-value. However, in order to obtain an accurate enough p-value one may need in some more complicated cases a very large number of simulated values, in the order of a few million.

Anyway, the values obtained with the simulation approach will never be as accurate as those that may be obtained by using a sharp near-exact approximation

C. A. Coelho, B. C. Arnold, *Finite Form Representations for Meijer G and Fox H Functions*, Lecture Notes in Statistics 223,
https://doi.org/10.1007/978-3-030-28790-0_7

approach where the "major" part of the integrand function of the Meijer G function is left unchanged while a "smaller" part is asymptotically approximated. We will shortly explain precisely what we mean here by "major" and "smaller."

Such approximations may be obtained by using the fact that any Beta distributed r.v.

$$Y \sim Beta(a, b)$$

may always be written as the product of two independent Beta distributed r.v.'s, say Y_1 and Y_2 where for $b^* < b$,

$$Y_1 \sim Beta(a, b^*) \quad \text{and} \quad Y_2 \sim Beta(a + b^*, b - b^*)\,.$$

This fact may be easily checked by writing the h-th moment of Y for $h > -a$, which fully identifies the distribution of Y, as

$$\begin{aligned} E(Y^h) &= \frac{\Gamma(a+b)}{\Gamma(a)} \frac{\Gamma(a+h)}{\Gamma(a+b+h)} \\ &= \frac{\Gamma(a+b)}{\Gamma(a+b^*)} \frac{\Gamma(a+b^*+h)}{\Gamma(a+b+h)} \frac{\Gamma(a+b^*)}{\Gamma(a)} \frac{\Gamma(a+h)}{\Gamma(a+b^*+h)} \\ &= E(Y_2^h)\, E(Y_1^h) \end{aligned}$$

and it may be used to obtain sharp and manageable approximations for Meijer $G^{p,0}_{p,p}(\,\cdot\,|\,\cdot\,)$ and $G^{p,1}_{p+1,p+1}(\,\cdot\,|\,\cdot\,)$ functions for which the parameters do not satisfy the conditions in Theorems 3.1–3.3.

Let us consider a product of Beta r.v.'s similar to those in Theorems 3.1–3.3 but where the second parameters do not satisfy any of the relations in the statement of those theorems. That is, let us consider the r.v.

$$Z = \prod_{\nu=1}^{m^*} \prod_{\ell=1}^{n_\nu} \prod_{j=1}^{k_{\nu\ell}} Y_{\nu\ell j}\,, \tag{7.1}$$

where

$$Y_{\nu\ell j} \sim Beta(a_{\nu\ell j}, b_{\nu\ell j})$$

are independent r.v.'s, with $a_{\nu\ell j}, b_{\nu\ell j} \in \mathbb{R}^+$ but not necessarily satisfying any of the conditions in Theorems 3.1 to 3.3.

Let us then define

$$b^*_{\nu\ell j} = \lfloor b_{\nu\ell j} \rfloor$$

and consider

$$Z = Z_1 Z_2\,,$$

where Z_1 and Z_2 are two independent r.v.'s, with

$$Z_1 = \prod_{\nu=1}^{m^*}\prod_{\ell=1}^{n_\nu}\prod_{j=1}^{k_{\nu\ell}} Y^*_{\nu\ell j} \quad \text{and} \quad Z_2 = \prod_{\nu=1}^{m^*}\prod_{\ell=1}^{n_\nu}\prod_{j=1}^{k_{\nu\ell}} Y^{**}_{\nu\ell j}$$

where, for $\nu = 1, \dots, m^*$; $\ell = 1, \dots, n_\nu$ and $j = 1, \dots, k_{\nu\ell}$,

$$Y^*_{\nu\ell j} \sim Beta(a_{\nu\ell j}, b^*_{\nu\ell j}) \quad \text{and} \quad Y^{**}_{\nu\ell j} \sim Beta(a_{\nu\ell j} + b^*_{\nu\ell j}, b_{\nu\ell j} - b^*_{\nu\ell j})$$

form two sets of mutually independent r.v.'s.

Then let us take $W = -\log Z$, $W_1 = -\log Z_1$ and $W_2 = -\log Z_2$. Given the independence of Z_1 and Z_2, we will have the c.f. (characteristic function) of W given by

$$\Phi_W(t) = E(e^{\mathrm{i}tW}) = E\left(e^{\mathrm{i}tW_1 + \mathrm{i}tW_2}\right) = E\left(e^{\mathrm{i}tW_1}\right) E\left(e^{\mathrm{i}tW_2}\right) = \Phi_{W_1}(t)\,\Phi_{W_2}(t)$$

where, using (3.6),

$$\begin{aligned}
\Phi_{W_1}(t) &= E\left(e^{\mathrm{i}tW_1}\right) = E\left(Z_1^{-\mathrm{i}t}\right) = \prod_{\nu=1}^{m^*}\prod_{\ell=1}^{n_\nu}\prod_{j=1}^{k_{\nu\ell}} E\left(\left(Y^*_{\nu\ell j}\right)^{-\mathrm{i}t}\right) \\
&= \prod_{\nu=1}^{m^*}\prod_{\ell=1}^{n_\nu}\prod_{j=1}^{k_{\nu\ell}} \frac{\Gamma(a_{\nu\ell j} + b^*_{\nu\ell j})}{\Gamma(a_{\nu\ell j})} \frac{\Gamma(a_{\nu\ell j} - \mathrm{i}t)}{\Gamma(a_{\nu\ell j} + b^*_{\nu\ell j} - \mathrm{i}t)} \qquad (7.2) \\
&= \prod_{\nu=1}^{m^*}\prod_{\ell=1}^{n_\nu}\prod_{j=1}^{k_{\nu\ell}} \prod_{\eta=0}^{b^*_{\nu\ell j}-1} (a_{\nu\ell j} + \eta)\,(a_{\nu\ell j} + \eta - \mathrm{i}t)^{-1},
\end{aligned}$$

which is the c.f. of a GIG distribution with depth at most $g = \sum_{\nu=1}^{m^*}\sum_{\ell=1}^{n_\nu}\sum_{j=1}^{k_{\nu\ell}} b^*_{\nu\ell j}$ with rate parameters all equal to 1, in the case in which all parameters $a_{\nu\ell j} + \eta$ are different, or the c.f. of a GIG distribution of depth $g + r - \sum_{q=1}^{r} p_q$ in case there are r groups, each one of them with p_q ($q = 1, \dots, r$) equal rate parameters $a_{\nu\ell j} + \eta$, and shape parameters equal to the multiplicities of the corresponding rate parameters.

Then we would also have

$$\begin{aligned}
\Phi_{W_2}(t) &= E\left(e^{\mathrm{i}tW_2}\right) = E\left(Z_2^{-\mathrm{i}t}\right) = \prod_{\nu=1}^{m^*}\prod_{\ell=1}^{n_\nu}\prod_{j=1}^{k_{\nu\ell}} E\left(\left(Y^{**}_{\nu\ell j}\right)^{-\mathrm{i}t}\right) \\
&= \prod_{\nu=1}^{m^*}\prod_{\ell=1}^{n_\nu}\prod_{j=1}^{k_{\nu\ell}} \frac{\Gamma(a_{\nu\ell j} + b_{\nu\ell j})}{\Gamma(a_{\nu\ell j} + b^*_{\nu\ell j})} \frac{\Gamma(a_{\nu\ell j} + b^*_{\nu\ell j} - \mathrm{i}t)}{\Gamma(a_{\nu\ell j} + b_{\nu\ell j} - \mathrm{i}t)}
\end{aligned}$$

which is the c.f. of the sum of $\sum_{\nu=1}^{m^*}\sum_{\ell=1}^{n_\nu} k_{\nu\ell}$ independent $Logbeta(a_{\nu\ell j}+b^*_{\nu\ell j}, b_{\nu\ell j}-b^*_{\nu\ell j})$ r.v.'s, that is, r.v.'s whose exponential has a $Beta(a_{\nu\ell j}+b^*_{\nu\ell j}, b_{\nu\ell j}-b^*_{\nu\ell j})$ distribution.

Since for the distribution of W_1 we have a nice finite form representation, while this is not the case for W_2, we will leave $\Phi_{W_1}(t)$ untouched and we will asymptotically approximate $\Phi_{W_2}(t)$. We will say that $\Phi_{W_1}(t)$ is the major part of $\Phi_W(t)$ and $\Phi_{W_2}(t)$ the smaller part, since, for $b_{\nu\ell j} > 1$, we will have

$$\int_{-\infty}^{+\infty} \left|\Phi_W(t) - \Phi_{W_1}(t)\right| dt \ll \int_{-\infty}^{+\infty} \left|\Phi_W(t) - \Phi_{W_2}(t)\right| dt . \tag{7.3}$$

In order to obtain a sharp but simple approximation for $\Phi_{W_2}(t)$ we will first make use of a result which comes out from expressions (11), (12), and (14) in Sects. 5 and 6 of Tricomi and Erdélyi (1951). This will enable us to approximate the c.f. of a $Logbeta(a, b)$ r.v., for increasing a, by the c.f. of an infinite mixture of $\Gamma(b+k, a)$ $(k = 0, 1, \dots)$ distributions, by writing

$$\frac{\Gamma(a-\mathrm{i}t)}{\Gamma(a+b-\mathrm{i}t)} \simeq \sum_{k=0}^{\infty} p_k(b)\,(a-\mathrm{i}t)^{-b-k}$$

with $p_0(b) = 1$ and

$$p_k(b) = \frac{1}{k}\sum_{m=0}^{k-1}\left(\frac{\Gamma(1-b-m)}{\Gamma(-b-k)(k-m+1)!} + (-1)^{k+m}\, b^{k-m+1}\right) p_m(b)\,, \qquad k \in \mathbb{N}\,.$$

Then, using a somewhat heuristic approach, we may, for

$$r = \sum_{\nu=1}^{m^*}\sum_{\ell=1}^{n_\nu}\sum_{j=1}^{k_{\nu\ell}} b_{\nu\ell j} - b^*_{\nu\ell j}\,, \tag{7.4}$$

approximate $\Phi_{W_2}(t)$ by

$$\widetilde{\Phi}(t) = \sum_{k=0}^{r^*} \pi_k (\lambda^*)^{r+k} (\lambda^* - \mathrm{i}t)^{-(r+k)}$$

which is the c.f. of a finite mixture of $r^* + 1$ Gamma distributions with shape parameters equal to $r + k$, for $k = 0, \dots, r^*$, and a common rate parameter λ^* which is the rate parameter in

$$\Phi^*(t) = \theta(\lambda^*)^{r_1}(\lambda^* - \mathrm{i}t)^{-r_1} + (1-\theta)(\lambda^*)^{r_2}(\lambda^* - \mathrm{i}t)^{-r_2}\,.$$

The value for λ^* is obtained by solving the system of equations

$$\frac{\delta h}{\delta t^h}\Phi^*(t)\bigg|_{t=0} = \frac{\delta h}{\delta t^h}\Phi_{W_2}(t)\bigg|_{t=0}, \qquad h=1,\ldots,4,$$

for λ^*, r_1, r_2, and θ, keeping then only the value for λ^*. This system of equations usually admits many solutions, from which we retain the one that presents real values for all four parameters and that furthermore presents for r_1 the solution which has the closest value to r in (7.4). Then the weights π_k, for $k=0,\ldots,r^*-1$, are obtained from the numerical solution of the system of r^* equations

$$\frac{\delta h}{\delta t^h}\widetilde{\Phi}_2(t)\bigg|_{t=0} = \frac{\delta h}{\delta t^h}\Phi_{W_2}(t)\bigg|_{t=0}, \qquad h=1,\ldots,r^*,$$

and taking then $\pi_{r^*} = 1 - \sum_{k=0}^{r^*-1}\pi_k$.

By proceeding in this way we will use

$$\begin{aligned}
\widetilde{\Phi}_W(t) &= \Phi_{W_1}(t)\,\widetilde{\Phi}_2(t)\\
&= \left\{\prod_{\nu=1}^{m^*}\prod_{\ell=1}^{n_\nu}\prod_{j=1}^{k_{\nu\ell}}\prod_{\eta=1}^{b^*_{\nu\ell j}-1}(a_{\nu\ell j}+\eta)(a_{\nu\ell j}+\eta-\mathrm{i}t)^{-1}\right\}\sum_{k=0}^{r^*}\pi_k(\lambda^*)^{r+k}(\lambda^*-\mathrm{i}t)^{-(r+k)}\\
&= \sum_{k=0}^{r^*}\pi_k(\lambda^*)^{r+k}(\lambda^*-\mathrm{i}t)^{-(r+k)}\prod_{\nu=1}^{m^*}\prod_{\ell=1}^{n_\nu}\prod_{j=1}^{k_{\nu\ell}}\prod_{\eta=1}^{b^*_{\nu\ell j}-1}(a_{\nu\ell j}+\eta)(a_{\nu\ell j}+\eta-\mathrm{i}t)^{-1}
\end{aligned} \tag{7.5}$$

as a near-exact c.f. for W. $\widetilde{\Phi}_W(t)$ is the c.f. of a mixture of r^*+1 Generalized Near-Integer Gamma (GNIG) distribution of depth equal to 1 plus that of the GIG distribution of W_1, with the same rate and shape parameters, to which is added the rate parameter λ^*, and its associated shape parameter $r+k$ ($k=0,\ldots,r^*$).

As such the corresponding near-exact distribution of Z, using the same contraction set notation used in Sect. 2.1 of Chap. 2, will have p.d.f. and c.d.f. given by

$$f_Z^*(z) = \sum_{k=0}^{r^*}\pi_k f^{GNIG}\left(-\log z\,\middle|\,\overset{\approx}{\left\{\{1\}_{\substack{\nu=1:m^*\\ \ell=1:n_\nu\\ j=1:k_{\nu\ell}\\ \eta=1:b^*_{\nu\ell j}-1}}\right\}}, r;\ \overset{\sim}{\left\{\{a_{\nu\ell j}+\eta\}_{\substack{\nu=1:m^*\\ \ell=1:n_\nu\\ j=1:k_{\nu\ell}\\ \eta=1:b^*_{\nu\ell j}-1}}\right\}}, \lambda^*; g^*\right)\frac{1}{z}$$

and

$$F_Z^*(z) = 1-\sum_{k=0}^{r^*}\pi_k F^{GNIG}\left(-\log z\,\middle|\,\overset{\approx}{\left\{\{1\}_{\substack{\nu=1:m^*\\ \ell=1:n_\nu\\ j=1:k_{\nu\ell}\\ \eta=1:b^*_{\nu\ell j}-1}}\right\}}, r;\ \overset{\sim}{\left\{\{a_{\nu\ell j}+\eta\}_{\substack{\nu=1:m^*\\ \ell=1:n_\nu\\ j=1:k_{\nu\ell}\\ \eta=1:b^*_{\nu\ell j}-1}}\right\}}, \lambda^*; g^*\right)$$

where $f^{GNIG}(\cdot)$ and $F^{GNIG}(\cdot)$ denote respectively the p.d.f. and the c.d.f. of the GNIG distribution, according to the notation in appendix, and where, as in Sect. 2.1 in Chap. 2,

$$\widetilde{\left\{ \{a_{\nu\ell j}+\eta\}_{\substack{\nu=1:m^* \\ \ell=1:n_\nu \\ j=1:k_{\nu\ell} \\ \eta=1:b^*_{\nu\ell j}-1}} \right\}}$$

denotes the set of different $a_{\nu\ell j}+\eta$ values, and

$$\approx\!\!\left\{ \{1\}_{\substack{\nu=1:m^* \\ \ell=1:n_\nu \\ j=1:k_{\nu\ell} \\ \eta=1:b^*_{\nu\ell j}-1}} \right\}$$

denotes the set of the corresponding frequencies, while the exact p.d.f. and c.d.f. of Z in terms of the Meijer G function are, from expressions (2.4) and (2.6) in Sect. 2.1, respectively given by

$$f_Z(z) = \left\{ \prod_{\nu=1}^{m^*} \prod_{\ell=1}^{n_\nu} \prod_{j=1}^{k_{\nu\ell}} \frac{\Gamma(a_{\nu\ell j}+b_{\nu\ell j})}{\Gamma(a_{\nu\ell j})} \right\} G^{p^*,0}_{p^*,p^*} \left(\begin{array}{c} \{a_{\nu\ell j}+b_{\nu\ell j}-1\}_{\substack{\nu=1:m^* \\ \ell=1:n_\nu \\ j=1:k_{\nu\ell}}} \\ \{a_{\nu\ell j}-1\}_{\substack{\nu=1:m^* \\ \ell=1:n_\nu \\ j=1:k_{\nu\ell}}} \end{array} \middle|\, z \right) \tag{7.6}$$

and

$$F_Z(z) = \left\{ \prod_{\nu=1}^{m^*} \prod_{\ell=1}^{n_\nu} \prod_{j=1}^{k_{\nu\ell}} \frac{\Gamma(a_{\nu\ell j}+b_{\nu\ell j})}{\Gamma(a_{\nu\ell j})} \right\} G^{p^*,1}_{p^*+1,p^*+1} \left(\begin{array}{c} \left\{ 1, \{a_{\nu\ell j}+b_{\nu\ell j}\}_{\substack{\nu=1:m^* \\ \ell=1:n_\nu \\ j=1:k_{\nu\ell}}} \right\} \\ \left\{ \{a_{\nu\ell j}\}_{\substack{\nu=1:m^* \\ \ell=1:n_\nu \\ j=1:k_{\nu\ell}}}, 0 \right\} \end{array} \middle|\, z \right). \tag{7.7}$$

As such we will use $f_Z^*(z)$ as an approximation for $f_Z(z)$ and $F_Z^*(z)$ as an approximation for $F_Z(z)$, or, equivalently, we will use

$$G^{p^*,0}_{p^*,p^*}\left(\begin{array}{c|c} \left\{a_{\nu\ell j}+b_{\nu\ell j}-1\right\}_{\substack{\nu=1:m^*\\ \ell=1:n_\nu\\ j=1:k_{\nu\ell}}} & \\ & z \\ \left\{a_{\nu\ell j}-1\right\}_{\substack{\nu=1:m^*\\ \ell=1:n_\nu\\ j=1:k_{\nu\ell}}} & \end{array}\right)$$

$$\simeq \widetilde{G}^{p^*,0}_{p^*,p^*}\left(\begin{array}{c|c} : & z \end{array}\right)=\left\{\prod_{\nu=1}^{m^*}\prod_{\ell=1}^{n_\nu}\prod_{j=1}^{k_{\nu\ell}}\frac{\Gamma(a_{\nu\ell j})}{\Gamma(a_{\nu\ell j}+b_{\nu\ell j})}\right\} f_Z^*(z) \tag{7.8}$$

and

$$G^{p^*,1}_{p^*+1,p^*+1}\left(\begin{array}{c|c} \left\{1,\left\{a_{\nu\ell j}+b_{\nu\ell j}\right\}_{\substack{\nu=1:m^*\\ \ell=1:n_\nu\\ j=1:k_{\nu\ell}}}\right\} & \\ & z \\ \left\{\left\{a_{\nu\ell j}\right\}_{\substack{\nu=1:m^*\\ \ell=1:n_\nu\\ j=1:k_{\nu\ell}}},0\right\} & \end{array}\right)$$

$$\simeq \widetilde{G}^{p^*,1}_{p^*+1,p^*+1}\left(\begin{array}{c|c} : & z \end{array}\right)=\left\{\prod_{\nu=1}^{m^*}\prod_{\ell=1}^{n_\nu}\prod_{j=1}^{k_{\nu\ell}}\frac{\Gamma(a_{\nu\ell j})}{\Gamma(a_{\nu\ell j}+b_{\nu\ell j})}\right\} F_Z^*(z)\,, \tag{7.9}$$

where $\widetilde{G}^{p^*,0}_{p^*,p^*}\left(\begin{array}{c|c} : & z \end{array}\right)$ and $\widetilde{G}^{p^*,1}_{p^*+1,p^*+1}\left(\begin{array}{c|c} : & z \end{array}\right)$ are not legitimate Meijer G functions but rather just shorthand notations which will be used in the next subsection.

7.1 Upper Bounds on the Error of the Approximations for the Meijer *G* Functions

Upper bounds on the errors we incur in when using the approximations in (7.8) and (7.9), that is, upper bounds on the difference between the Meijer G functions $G^{p^*,0}_{p^*,p^*}(\,:\,|\,z)$ or $G^{p^*,1}_{p^*+1,p^*+1}(\,:\,|\,z)$ and the approximations provided by $\widetilde{G}^{p^*,0}_{p^*,p^*}(\,:\,|\,z)$ and $\widetilde{G}^{p^*,1}_{p^*+1,p^*+1}(\,:\,|\,z)$, may be obtained from the corresponding inverse Mellin transforms.

Actually it is possible to obtain an inverse Mellin transform for the p.d.f. of a nonnegative r.v. in a much more manageable form than the usual inverse Mellin transform in (2.3) in Chap. 2. Indeed, it is also possible to obtain an inverse Mellin transform for the c.d.f. of such an r.v. in a very manageable form.

We may note that taking $W = -\log Z$, actually as already noted above, we have, for $t \in \mathbb{R}$ and $s = -\mathrm{i}t\ (\in \mathbb{C})$,

$$\Phi_W(t) = E\left(e^{\mathrm{i}tW}\right) = E\left(Z^{-\mathrm{i}t}\right) = M_Z(-\mathrm{i}t+1) = M_Z(s+1)$$

where

$$M_Z(s) = E\left(Z^{s-1}\right), \quad s \in \mathbb{C}$$

denotes the Mellin transform of the nonnegative r.v. $Z = e^{-W}$.

As such, from the usual inversion formula for the c.f., relative to the p.d.f., we may obtain the p.d.f. of W as

$$f_W(w) = \frac{1}{2\pi}\int_{-\infty}^{+\infty} e^{-\mathrm{i}tw}\,\Phi_W(t)\,dt$$

and, if we now take $Z = e^{-W}$, the p.d.f. of Z as

$$f_Z(z) = \frac{1}{2\pi}\int_{-\infty}^{+\infty} z^{\mathrm{i}t}\,M_Z(-\mathrm{i}t)\,dt\,, \quad (t \in \mathbb{R}) \tag{7.10}$$

where $Im\left(z^{\mathrm{i}t}M_Z(-\mathrm{i}t)\right)$ is an odd function in t and $Re\left(z^{\mathrm{i}t}M_Z(-\mathrm{i}t)\right)$ is an even function, so that we may indeed write (7.10) as

$$f_Z(z) = \frac{1}{\pi}\int_0^{+\infty} Re\left(z^{\mathrm{i}t}M_Z(-\mathrm{i}t)\right)dt\,, \tag{7.11}$$

which gives a much more manageable inverse Mellin transform for the p.d.f. than the usual inverse Mellin transform in expression (2.3) in Chap. 2.

For Z in (7.1) we have

$$M_Z(\mathrm{i}t) = \Phi_W(t-1) = \prod_{\nu=1}^{m^*}\prod_{\ell=1}^{n_\nu}\prod_{j=1}^{k_{\nu\ell}} \frac{\Gamma(a_{\nu\ell j}+b_{\nu\ell j})}{\Gamma(a_{\nu\ell j})}\,\frac{\Gamma(a_{\nu\ell j}-\mathrm{i}(t-1))}{\Gamma(a_{\nu\ell j}+b_{\nu\ell j}-\mathrm{i}(t-1))}\,,$$

but we have a similar inversion formula to that in (7.11) for $f_Z^*(z)$, the approximate or near-exact p.d.f. of Z, with

$$f_Z^*(z) = \frac{1}{\pi}\int_0^{+\infty} Re\left(z^{\mathrm{i}t}\,M_Z^*(-\mathrm{i}t)\right)dt \tag{7.12}$$

where

$$M_Z^*(\mathrm{i}t) = \widetilde{\Phi}_W(t-1)$$

for $\widetilde{\Phi}_W(\cdot)$ given by (7.5).

Then, if one wants to obtain an upper bound for $|f_Z(z) - f_Z^*(z)|$ for a given value of z ($\in]0,1[$), one may obtain it from (7.11) and (7.12) as

$$\begin{aligned} |f_Z(z) - f_Z^*(z)| &= \frac{1}{\pi}\left|\int_0^{+\infty} \left(Re\left(z^{\mathrm{i}t} M_Z(-\mathrm{i}t)\right) - Re\left(z^{\mathrm{i}t} M_Z^*(-\mathrm{i}t)\right)\right) dt\right| \\ &\leq \frac{1}{\pi}\int_0^{+\infty} \left|\left(Re\left(z^{\mathrm{i}t} M_Z(-\mathrm{i}t)\right) - Re\left(z^{\mathrm{i}t} M_Z^*(-\mathrm{i}t)\right)\right)\right| dt = \Delta_1^*, \end{aligned} \tag{7.13}$$

but if one wants an upper bound on $|f_Z(z) - f_Z^*(z)|$ which is valid for any $z \in]0,1[$, then one will have to use a Mellin inversion formula of the type displayed in (7.10), since then one will be able to write

$$\begin{aligned} |f_Z(z) - f_Z^*(z)| &= \frac{1}{2\pi}\left|\int_{-\infty}^{+\infty} z^{\mathrm{i}t}\left(M_Z(-\mathrm{i}t) - M_Z^*(-\mathrm{i}t)\right) dt\right| \\ &\leq \frac{1}{2\pi}\int_{-\infty}^{+\infty} \left|z^{\mathrm{i}t}\left(M_Z(-\mathrm{i}t) - M_Z^*(-\mathrm{i}t)\right)\right| dt \\ &= \frac{1}{2\pi}\int_{-\infty}^{+\infty} \left|M_Z(-\mathrm{i}t) - M_Z^*(-\mathrm{i}t)\right| dt \\ &= \frac{1}{\pi}\int_0^{+\infty} \left|M_Z(-\mathrm{i}t) - M_Z^*(-\mathrm{i}t)\right| dt \end{aligned} \tag{7.14}$$

which will be a very sharp upper bound on $|f_Z(z) - f_Z^*(z)|$.

Given that the right-hand side of (7.14) is no longer a function of z, one may write

$$\max_{0<z<1} |f_Z(z) - f_Z^*(z)| \leq \Delta_1 \tag{7.15}$$

for

$$\Delta_1 = \frac{1}{\pi}\int_0^{+\infty} \left|M_Z(-\mathrm{i}t) - M_Z^*(-\mathrm{i}t)\right| dt\,. \tag{7.16}$$

Then, from (7.6), (7.8), and (7.13) we may write, for a given $z \in]0,1[$,

$$\left| G_{p^*,p^*}^{p^*,0}\left(\begin{array}{c} \cdot \\ \cdot \end{array} \middle| z \right) - \widetilde{G}_{p^*,p^*}^{p^*,0}\left(\begin{array}{c} \cdot \\ \cdot \end{array} \middle| z \right) \right| \leq \left\{ \prod_{\nu=1}^{m^*} \prod_{\ell=1}^{n_\nu} \prod_{j=1}^{k_{\nu\ell}} \frac{\Gamma(a_{\nu\ell j})}{\Gamma(a_{\nu\ell j} + b_{\nu\ell j})} \right\} \Delta_1^*$$

and

$$\max_{0<z<1}\left|G^{p^*,0}_{p^*,p^*}\left(\begin{array}{c}\cdot\\ \cdot\end{array}\bigg|\, z\right)-\widetilde{G}^{p^*,0}_{p^*,p^*}\left(\begin{array}{c}\cdot\\ \cdot\end{array}\bigg|\, z\right)\right| \leq \left\{\prod_{\nu=1}^{m^*}\prod_{\ell=1}^{n_\nu}\prod_{j=1}^{k_{\nu\ell}}\frac{\Gamma(a_{\nu\ell j})}{\Gamma(a_{\nu\ell j}+b_{\nu\ell j})}\right\}\Delta_1,$$

for Δ_1^* in (7.13) and Δ_1 in (7.16).

It is also possible to obtain a manageable inverse Mellin transform for the c.d.f. based on the Gil-Pelaez (1951) inversion formula for the c.d.f.. This inversion formula may be written as

$$F_W(w)=\frac{1}{2}-\frac{1}{2\pi}\int_{-\infty}^{+\infty}\frac{e^{-\mathrm{i}tw}\,\Phi_W(t)}{\mathrm{i}t}\,dt$$

which then yields for $Z=e^{-W}$

$$F_Z(z)=\frac{1}{2}+\frac{1}{2\pi}\int_{-\infty}^{+\infty}\frac{z^{\mathrm{i}t}\,M_Z(-\mathrm{i}t+1)}{\mathrm{i}t}\,dt \tag{7.17}$$

where we have $Im\left(\frac{z^{\mathrm{i}t}\,M_Z(-\mathrm{i}t+1)}{\mathrm{i}t}\right)$ as an odd function and $Re\left(\frac{z^{\mathrm{i}t}\,M_Z(-\mathrm{i}t+1)}{\mathrm{i}t}\right)$ as an even function, so that we may write

$$F_Z(z)=\frac{1}{2}+\frac{1}{\pi}\int_0^{+\infty}Re\left(\frac{z^{\mathrm{i}t}\,M_Z(-\mathrm{i}t+1)}{\mathrm{i}t}\right)dt\,. \tag{7.18}$$

An upper bound on $|F_Z(z)-F_Z^*(z)|$ for a given $z\in]0,1[$ may then be obtained from the inversion formula in (7.18) as

$$\begin{aligned}\left|F_Z(z)-F_Z^*(z)\right| &= \frac{1}{\pi}\left|\int_0^{+\infty}\left(Re\left(\frac{z^{\mathrm{i}t}\,M_Z(-\mathrm{i}t+1)}{\mathrm{i}t}\right)-Re\left(\frac{z^{\mathrm{i}t}\,M_Z^*(-\mathrm{i}t+1)}{\mathrm{i}t}\right)\right)dt\right|\\ &\leq \frac{1}{\pi}\int_0^{+\infty}\left|Re\left(\frac{z^{\mathrm{i}t}\,M_Z(-\mathrm{i}t+1)}{\mathrm{i}t}\right)-Re\left(\frac{z^{\mathrm{i}t}\,M_Z^*(-\mathrm{i}t+1)}{\mathrm{i}t}\right)\right|dt=\Delta_2^*\end{aligned} \tag{7.19}$$

while for any $z\in]0,1[$ we have, from (7.17),

$$\begin{aligned}\left|F_Z(z)-F_Z^*(z)\right| &= \frac{1}{2\pi}\left|\int_{-\infty}^{+\infty}z^{\mathrm{i}t}\,\frac{M_Z(-\mathrm{i}t+1)-M_Z^*(-\mathrm{i}t+1)}{\mathrm{i}t}\,dt\right|\\ &\leq \frac{1}{2\pi}\int_{-\infty}^{+\infty}\left|\frac{z^{\mathrm{i}t}}{\mathrm{i}t}\left(M_Z(-\mathrm{i}t+1)-M_Z^*(-\mathrm{i}t+1)\right)\right|dt\\ &= \frac{1}{2\pi}\int_{-\infty}^{+\infty}\left|\frac{M_Z(-\mathrm{i}t+1)-M_Z^*(-\mathrm{i}t+1)}{\mathrm{i}t}\right|dt\\ &= \frac{1}{\pi}\int_0^{+\infty}\left|\frac{M_Z(-\mathrm{i}t+1)-M_Z^*(-\mathrm{i}t+1)}{\mathrm{i}t}\right|dt\,.\end{aligned} \tag{7.20}$$

Since once again the right-hand side of (7.20) is not a function of z, for

$$\Delta_2 = \frac{1}{\pi}\int_0^{+\infty}\left|\frac{M_Z(-\mathrm{i}t+1)-M_Z^*(-\mathrm{i}t+1)}{\mathrm{i}t}\right| dt$$

we may write

$$\max_{0<z<1}\left|F_Z(z)-F_Z^*(z)\right| \le \Delta_2$$

and

$$\max_{0<z<1}\left|G^{p^*,1}_{p^*+1,p^*+1}\left(\begin{array}{c}:\\ .\end{array}\middle|\, z\right)-\widetilde{G}^{p^*,1}_{p^*+1,p^*+1}\left(\begin{array}{c}:\\ .\end{array}\middle|\, z\right)\right| \le \left\{\prod_{\nu=1}^{m^*}\prod_{\ell=1}^{n_\nu}\prod_{j=1}^{k_{\nu\ell}}\frac{\Gamma(a_{\nu\ell j})}{\Gamma(a_{\nu\ell j}+b_{\nu\ell j})}\right\}\Delta_2\,,$$

or, for a given $z\in]0,1[$, write

$$\left|G^{p^*,1}_{p^*+1,p^*+1}\left(\begin{array}{c}:\\ .\end{array}\middle|\, z\right)-\widetilde{G}^{p^*,1}_{p^*+1,p^*+1}\left(\begin{array}{c}:\\ .\end{array}\middle|\, z\right)\right| \le \left\{\prod_{\nu=1}^{m^*}\prod_{\ell=1}^{n_\nu}\prod_{j=1}^{k_{\nu\ell}}\frac{\Gamma(a_{\nu\ell j})}{\Gamma(a_{\nu\ell j}+b_{\nu\ell j})}\right\}\Delta_2^*\,.$$

This approximation approach is somewhat equivalent to the use of near-exact distributions. These are asymptotic distributions built using a different philosophy in approximating the exact c.f. of the statistic. The technique combines an adequately developed decomposition of the c.f., most often a factorization, with the action of keeping the "major" part of this c.f. unchanged, and replacing the remaining "smaller" part by an adequate asymptotic approximation, where "major" part and "smaller" part of $\Phi_W(t)$, the c.f. of the statistic under study, are to be taken respectively as $\Phi_{W_1}(t)$ and $\Phi_{W_2}(t)$ in (7.3). All this is done in order to obtain a manageable and very well-fitting approximating distribution, which may be used to compute near-exact quantiles or p-values. When appropriately developed, these distributions perform very well even for very small samples and besides being asymptotic in terms of sample size they are also asymptotic for increasing numbers of variables used and, in the multi-sample cases, also for increasing number of populations involved, being particularly useful in situations where it is not possible to obtain the exact distribution in a manageable form.

Near-exact distributions have already been developed for many of the l.r.t. statistics discussed in Chap. 5 for the situations for which finite form representations are not available. For the statistic in Sect. 5.1.1, for cases where p is odd and q is even, near-exact distributions are available in Coelho et al. (2010) and Marques et al. (2011). Near-exact distributions for the statistic in Sect. 5.1.9, for both p_1 and p_2 odd or the statistic in Sect. 5.1.11 when two or more sets of variables have an odd number of variables, have been discussed in Coelho (2004), Alberto and Coelho (2007), Grilo and Coelho (2007, 2010a,b, 2012), Coelho et al. (2010), and Marques et al. (2011).

Near-exact distributions have also been already developed for a range of other l.r.t. statistics, including some that may be used in tests of structured covariance matrices and others for use when dealing with tests on mean vectors with structured covariance matrices, some of which yield near-exact distributions for l.r.t. statistics in many other subsections in Chap. 5. Coelho and Roy (2017) address the l.r.t. for block compound symmetry and as such the near-exact distributions in that paper may be used for the statistic associated with the test in Sect. 5.1.15 for the case of odd p or for the case of more than two sets of variables. Coelho and Marques (2013a) handle the l.r.t. for block-scalar sphericity and the near-exact distributions in that reference may be used for the statistic associated with a similar test to the one in Sect. 5.1.21 where blocks of different dimensions or blocks of dimensions different from 2 are considered. Near-exact distributions developed in Coelho (2018) may be used for the l.r.t. statistics in Sects. 5.1.22 and 5.1.23, respectively for the cases of even q and odd q, while the ones in Coelho (2017) may be used for the l.r.t. statistic in Sect. 5.1.24 for the case of even q. Marques and Coelho (2013a) provide near-exact distributions for the l.r.t. statistic in Sect. 5.2.1 for even p, while the results in Marques and Coelho (2008, 2013b), Coelho et al. (2010, Sects. 3, 5), and Coelho et al. (2016, Sects. 2, 6) may be used to build near-exact distributions for the test statistics in Sects. 5.3.6 and 5.3.7, based on the developments in Coelho and Marques (2009). Some further near-exact distributions for l.r.t. statistics which may be used for tests of covariance structures both under the real and the complex multivariate setting may be found in (Coelho and Marques, 2010, 2012, 2013b; Marques and Coelho, 2010, 2011, 2012a,b,c, 2013c, 2015; Coelho et al., 2015, 2016; Correia et al., 2018). Near-exact distributions for a number of other statistics may be found in (Coelho, 2006; Coelho and Mexia, 2010; Coelho and Marques, 2011; Grilo and Coelho, 2013; Coelho and Arnold, 2014; Marques and Coelho, 2016; Marques et al., 2015, 2017).

If in (7.1) all or some of the Beta r.v.'s $Y_{\nu j\ell}$ are raised to some (positive) power, then one would have to resort to the use of Fox's H function, but the whole approximation process would remain essentially the same, with the only small difference that for those r.v.'s which would appear raised to a given power, the terms it in the c.f. of $W = -\log Z$ would appear multiplied by that same power, both in $\Phi_{W_1}(t)$ and in $\Phi_{W_2}(t)$, it being then necessary to divide each of the two factors in the last expression of (7.2) (considering the second one without its power -1), by that power to make $\Phi_{W_1}(t)$ to correspond to the c.f. of a GIG distribution.

Appendix: Expressions for the Probability Density and Cumulative Distribution Functions of the GNIG Distribution

Let W be a r.v. with a GIG distribution of depth p, with rate parameters $\lambda_1, \ldots, \lambda_p$ and shape parameters $r_1, \ldots, r_p \in \mathbb{N}$ and let $W^* \sim \Gamma(r, \lambda^*)$, with $r \in \mathbb{R}^+ \backslash \mathbb{N}$. Let further W and W^* be two independent r.v.'s. Then the r.v.

$$Y = W + W^*$$

has a Generalized Near-Integer Gamma (GNIG) distribution (Coelho, 2004) of depth $p+1$, with rate parameters $\lambda_1, \ldots, \lambda_p$ and λ^* and corresponding shape parameters $r_1, \ldots, r_p$ and r, with p.d.f.

$$\begin{aligned} f_Y(y) &= f^{GNIG}\Big(y \,|\, \{r_j\}_{j=1:p}, r; \{\lambda_j\}_{j=1:p}, \lambda^*; p+1\Big) \\ &= K(\lambda^*)^r \sum_{j=1}^{p} e^{-\lambda_j y} \sum_{k=1}^{r_j} \left\{ c_{j,k} \frac{\Gamma(k)}{\Gamma(k+r)} y^{k+r-1} \, {}_1F_1\big(r, k+r, -(\lambda^* - \lambda_j)y\big) \right\} \end{aligned}$$

and c.d.f.

$$\begin{aligned} F_Y(y) &= F^{GNIG}\Big(y \,|\, \{r_j\}_{j=1:p}, r; \{\lambda_j\}_{j=1:p}, \lambda^*; p+1\Big) \\ &= \frac{(\lambda^*)^r y^r}{\Gamma(r+1)} \, {}_1F_1\big(r, r+1, -\lambda^* y\big) \\ &\quad -K(\lambda^*)^r \sum_{j=1}^{p} e^{-\lambda_j y} \sum_{k=1}^{r_j} c^*_{j,k} \sum_{i=0}^{k-1} \left\{ \frac{y^{r+i} \lambda_j^i}{\Gamma(r+1+i)} \, {}_1F_1\big(r, r+1+i, -(\lambda^* - \lambda_j)y\big) \right\} \end{aligned}$$

for $z > 0$, where

$$K = \prod_{j=1}^{p} \lambda_j^{r_j}, \qquad c^*_{j,k} = \frac{c_{j,k}}{\lambda_j^k} \Gamma(k)$$

and

$${}_1F_1(a, b, y) = \sum_{i=0}^{\infty} \frac{\Gamma(a+i)}{\Gamma(b+i)} \frac{z^i}{i!}$$

is the Kummer confluent hypergeometric function, which is a function that is quickly and correctly computed even in extended precision by software such as Mathematica® and MAXIMA.

References

Alberto, R.P., Coelho, C.A.: Study of the quality of several asymptotic and near-exact approximations based on moments for the distribution of the Wilks Lambda statistic. J. Statist. Plann. Inference **137**, 1612–1626 (2007)

Coelho, C.A.: The Generalized Near-Integer Gamma distribution: a basis for 'near-exact' approximations to the distribution of statistics which are the product of an odd number of independent Beta random variables. J. Multivar. Anal. **89**, 191–218 (2004)

Coelho, C.A.: The exact and near-exact distributions of the product of independent Beta random variables whose second parameter is rational. J. Combin. Inform. System Sci. **31**, 21–44 (2006)

Coelho, C.A.: The likelihood ratio test for equality of mean vectors with compound symmetric covariance matrices. In: Gervasi, O., Murgante, B., Misra, S., Borruso, G., Torre, C.M., Rocha, A.M.A.C., Taniar, D., Apduhan, B.O., Stankova, E., Cuzzocrea, A. (eds.) Computational Science and Its Applications. Lecture Notes in Computer Science 10408, vol. V, pp. 20–32. Springer, New York (2017)

Coelho, C.A.: Likelihood ratio tests for equality of mean vectors with circular covariance matrices. In: Oliveira, T.A., Kitsos, C., Oliveira, A., Grilo, L.M. (eds.) Recent Studies on Risk Analysis and Statistical Modeling, pp. 255–269. Springer, New York (2018)

Coelho, C.A., Arnold, B.C.: On the exact and near-exact distributions of the product of generalized Gamma random variables and the generalized variance. Commun. Statist. Theory Methods **43**, 2007–2033 (2014)

Coelho, C.A., Marques, F.J.: The advantage of decomposing elaborate hypotheses on covariance matrices into conditionally independent hypotheses in building near-exact distributions for the test statistics. Linear Algebra Appl. **430**, 2592–2606 (2009)

Coelho, C.A., Marques, F.J.: Near-exact distributions for the independence and sphericity likelihood ratio test statistics. J. Multivar. Anal. **101**, 583–593 (2010)

Coelho, C.A., Marques, F.J.: On the exact, asymptotic and near-exact distributions for the likelihood ratio statistics to test equality of several Exponential distributions. AIP Conf. Proc. **1389**, 1471–1474 (2011)

Coelho, C.A., Marques, F.J.: Near-exact distributions for the likelihood ratio test statistic to test equality of several variance-covariance matrices in elliptically contoured distributions. Comput. Statist. **27**, 627–659 (2012)

Coelho, C.A., Marques, F.J.: The multi-sample block-scalar sphericity test: exact and near-exact distributions for its likelihood ratio test statistic. Commun. Statist. Theory Methods **42**, 1153–1175 (2013a)

Coelho, C.A., Marques, F.J.: Near-exact distributions for the block equicorrelation and equivariance likelihood ratio test statistic. AIP Conf. Proc. **1557**, 429–433 (2013b)

Coelho, C.A., Mexia, J.T.: Product and Ratio of Generalized Gamma-Ratio Random Variables: Exact and Near-exact Distributions - Applications. Lambert Academic Publishing AG & Co. KG, Saarbrücken, Germany (2010). ISBN: 978-3-8383-5846-8

Coelho, C.A., Roy, A.: Testing the hypothesis of a block compound symmetric covariance matrix for elliptically contoured distributions. TEST **26**, 308–330 (2017)

Coelho, C.A., Arnold, B.C., Marques, F.J.: Near-exact distributions for certain likelihood ratio test statistics. J. Stat. Theory Pract. **4**, 711–725 (2010) (invited paper for the special memorial issue in honor of H. C. Gupta, guest-edited by C. R. Rao)

Coelho, C.A., Arnold, B.C., Marques, F.J.: The exact and near-exact distributions of the main likelihood ratio test statistics used in the complex multivariate normal setting. TEST **24**, 386–416 (2015)

Coelho, C.A., Marques, F.J., Oliveira, S.: Near-Exact Distributions for Likelihood Ratio Statistics Used in the Simultaneous Test of Conditions on Mean Vectors and Patterns of Covariance Matrices. Math. Problems in Engineering, Article ID 8975902 (2016). https://doi.org/10.1155/2016/8975902

Correia, B.R., Coelho, C.A., Marques, F.J.: Likelihood ratio test for the hyper-block matrix sphericity covariance structure – characterization of the exact distribution and development of near-exact distributions for the test statistic. REVSTAT **16**, 365–403 (2018)

Gil-Pelaez, J.: Note on the inversion Theorem. Biometrika **38**, 481–482 (1951)

Grilo, L.M., Coelho, C.A.: Development and study of two near-exact approximations to the distribution of the product of an odd number of independent Beta random variables. J. Statist. Plann. Inference **137**, 1560–1575 (2007)

Grilo, L.M., Coelho, C.A.: The exact and near-exact distributions for the Wilks Lambda statistic used in the test of independence of two sets of variables. Amer. J. Math. Management Sci. **30**, 111–145 (2010a)

Grilo, L.M., Coelho, C.A.: Near-exact distributions for the generalized Wilks Lambda statistic. Discuss. Math. Probab. Stat. **30**, 53–86 (2010b)

Grilo, L.M., Coelho, C.A.: A family of near-exact distributions based on truncations of the exact distribution for the generalized Wilks Lambda statistic. Commun. Statist. Theory Methods **41**, 2321–2341 (2012)

Grilo, L.M., Coelho, C.A.: Near-exact distributions for the likelihood ratio statistic used to test the reality of a covariance matrix. AIP Conf. Proc. **1558**, 793–796 (2013)

Marques, F.J., Coelho, C.A.: Near-exact distributions for the sphericity likelihood ratio test statistic. J. Statist. Plann. Inference **138**, 726–741 (2008)

Marques, F.J., Coelho, C.A.: The exact and near-exact distributions of the likelihood ratio statistic for the block sphericity test. AIP Conf. Proc. **1281**, 1237–1240 (2010)

Marques, F.J., Coelho, C.A.: The multi-sample block-matrix sphericity test. AIP Conf. Proc. **1389**, 1479–1482 (2011)

Marques, F.J., Coelho, C.A.: The block sphericity test – exact and near-exact distributions for the likelihood ratio statistic. Math. Methods Appl. Sci. **35**, 373–383 (2012a)

Marques, F.J., Coelho, C.A.: The multi-sample independence test. AIP Conf. Proc. **1479**, 1129–1132 (2012b)

Marques, F.J., Coelho, C.A.: Near-exact distributions for the likelihood ratio test statistic of the multi-sample block-matrix sphericity test. Appl. Math. Comput. **219**, 2861–2874 (2012c)

Marques, F.J., Coelho, C.A.: Obtaining the exact and near-exact distributions of the likelihood ratio statistic to test circular symmetry through the use of characteristic functions. Comput. Statist. **28**, 2091–2115 (2013a)

Marques, F.J., Coelho, C.A.: The multisample block-diagonal equicorrelation and equivariance test. AIP Conf. Proc. **1558**, 797–800 (2013b)

Marques, F.J., Coelho, C.A.: The multi-sample block-scalar sphericity test under the complex multivariate normal case. AIP Conf. Proc. **1557**, 420–423 (2013c)

Marques, F.J., Coelho, C.A.: Near-exact distributions for the likelihood ratio test statistic for testing multisample independence — the real and complex cases. J. Statist. Theory Practice **9**, 37–58 (2015)

Marques, F.J., Coelho, C.A.: Near-exact Distributions for Positive Linear Combinations of Independent Non-central Gamma Random Variables. AIP Conf. Proc. **1738**, 190005–1–190005-4 (2016). https://doi.org/10.1063/1.4951972

Marques, F.J., Coelho, C.A., Arnold, B.C.: A general near-exact distribution theory for the most common likelihood ratio test statistics used in Multivariate Analysis. TEST **20**, 180–203 (2011)

Marques, F.J., Coelho, C.A., Carvalho, M.: On the distribution of linear combinations of independent Gumbel random variables. Stat. Comput. **25**, 683–701 (2015)

Marques, F.J., Coelho, C.A., Rodrigues, P.C.: Testing the equality of several linear regression models. Comput. Stat. **32**, 1453–1480 (2017)

Tricomi F.G., Erdélyi A.: The asymptotic expansion of a ratio of gamma functions. Pac. J. Math. **1**, 133–142 (1951)

Index

Note: Page numbers in *italics* denote Appendix section

C. A. Coelho, B. C. Arnold, *Finite Form Representations for Meijer G and Fox H Functions*, Lecture Notes in Statistics 223,
https://doi.org/10.1007/978-3-030-28790-0